·········· **Vickers Mobile Hydraulics Manual**

ISBN 0-9634162-5-1
Copywrite © 1998 by Vickers, Incorporated
Training Center
2730 Research Drive
Rochester Hills, Michigan 48309-3570

No part of this book may be reproduced in any form or by any means without the prior written consent of the Publisher

·········· Preface

The first Vickers Mobile Hydraulics Manual was published by the Vickers Training Center in 1967. Its acceptance by the mobile community was exceptional. Its last revision, prior to the current re-invention you now hold, was in 1979. It was last reprinted in 1993.

Requests for a hydraulics manual for the mobile industry come into the Vickers Training Center weekly. The need is evident for a manual that will provide both basic and mature knowledge of hydraulics as it is applied to the unique demands of mobile machinery. The 1967 manual would no longer suffice because of the rapid advancements of technology in the intervening years.

The audience for a manual of this type is a broad cross-section of all levels of mobile product and application experience. Some people may be just starting out in the field; others may have a good knowledge of hydraulics, but want more information on its application to mobile equipment; and a few may be accomplished in the design, application, production, maintenance and repair of mobile machinery. This manual was written with the intent to address this entire assortment of knowledge and experience.

The first three chapters encompass an introduction to the nature of hydraulics and the basic principles of hydraulics and electricity. Chapters four through 17 address all of the elements found in hydraulic systems. These chapters start with where the work is done (the actuators), and work back through typical circuits to explore the functions and interaction of valves, pumps, conductors and fittings, reservoirs, fluids and the care of fluids. Finally, it all comes together with a chapter on circuitry. Here you can explore some rather clever and unique approaches to solving hydraulic application problems.

It is hoped that this manual will be the first of many iterations - a document that lives and grows, with the input of its readers - keeping pace with the rapidly advancing technologies of power and motion control.

Acknowledgements

As with any massive undertaking, this Mobile Hydraulics Manual took several months of "thinking" time - called planning in business circles, and then some organizing time. Outlines were written, schedules were developed, budgets were set.... all of those neat project things that need to be done to create a smooth-running, coordinated effort. Things did run pretty smoothly. They just took a heckuva long time. Several times longer than originally expected.

The actual writing started in 1996, with a lot of help from some very talented people. I can't say enough about the contributions of some very important people in the world of mobile hydraulics, they are:

Peter Everts, a Certified Fluid Power Specialist, Certified Mobile Hydraulic Mechanic, and currently a hydraulics instructor with Certified Fluid Consultants, wrote the material for chapters 16 and 17. Pete is an accomplished writer, as you will see. His years with Vickers, including the Training Center, show in his presentation of two complex subjects; Fluids and Fluid Conditioning. Pete also proofread all of the chapters, an invaluable contribution.

Professor John Cundiff of the College of Agriculture and Life Sciences, Biological Systems Engineering Department at Virginia Polytechnic Institute, Blacksburgh, VA, did an exceptionally thorough job on chapter 13, Hydrostatic Transmissions. John has one of the more complete hydraulics courses of any college or university - probably worldwide, and has a very unique "four-square" hydrostatic transmission test facility to demonstrate the application and operation of HSTs.

John Schueller, Ph.D., P.E., Associate Professor of Mechanical Engineering at the University of Florida, Gainesville, demonstrated his acumen on chapter 11, Fixed Displacement Pumps. It is an extensive treatment of the subject, what you would expect from a man who teaches the subject in a very advanced technical facility. When I last met with John, he and his students were developing an electronically controlled hydraulic machine that navigates using the global positioning satellites.

Joseph Vacca, a Vickers retiree after at least 100 years of service and now residing in East St. Louis, IL, shows his depth of knowledge on Directional Valves (chapter 6), Cartridge Valves (chapter 8) and Auxiliary Valves (chapter 9). After many administrative years in both mobile and industrial hydraulics, he capped his career with a few years as operations director at Vickers Carol Stream plant, where the screw-in cartridge valves are designed and produced. His knowledge of the intricacies and use of valves becomes obvious.

Many other people were involved, in one way or another, who wanted to contribute to a hydraulic manual specifically dedicated to the mobile industry. Al Rylandsholm of Artex Industries, Ltd., Port Moody, BC, Canada, created many of the illustrations used; Rich Dettloff, a former employee of the Vickers Training Center, who is now with San Joaquin Delta Community College in Stockton, CA, did some of the illustrations; Jody Wren and Jerry Zmyslowski, both of Vickers application engineering group, provided many of the circuit drawings in chapter 18; Mary Samulski, manager of the Vickers publications group, personally dedicated herself to formatting all of this material into a readable book form; Connie Crawford, also of the Vickers publications group, contributed her extensive talents to the formatting process; all of the staff and instructors of the Training Center gave support whenever called upon; and by all means, Frank Garner, global education manager for Vickers, made sure that the funds and the incentive were there to be dedicated to a Mobile Hydraulics Manual that would serve the industry well. He consistently provided the encouragement that made it possible.

Frederick C. Wood
Training Manager
Vickers Training Center

Table of Contents

Chapter 1
Introduction to Mobile Fluid Power

Pascal's Law	1
Fluid Power Advantages	5
Pump Inlet Conditions	6
Fluid Pumps	12
Pump Inlet Characteristics	15
Force, Work, Energy And Power	17

Chapter 2
Basic Hydraulic Principles

Hydraulic Fluids	27
Hydraulic Terminology	29
Series And Parallel Circuitry	38
Open Circuit / Closed Circuit Systems	43

Chapter 3
Basic Electrical Principles

Electrohydraulics	45
Hydraulic / Electric Analogies	46
Electromagnetics	54
Rectifiers	67
Pulse Width Modulation	74
Conductors	75
Closed Loop Vs. Open Loop Circuitry	79

Chapter 4
Linear Actuators

Construction	83
Types Of Cylinders	89
Cylinder Ratings	94
Metering Cylinder Flow	98
Regeneration	100

Chapter 5
Rotary Actuators

Hydraulic Motor Characteristics	108
High Speed Hydraulic Motors	110
High Torque, Low Speed Motors	116
Variable Displacement Motors	121
Operating Parameters	124

Chapter 6
Directional Control Valves

Check Valves	132
Mobile Spool Type Directional Control Valves	138
Valve Spools	143
Basic Features	148
Specialty Features	154
Remote Control Of Hydraulic Valve Functions	157
Non-spool Directional Control Valves	160
Electrohydraulic Directional Control Valves	164

Table Of Contents

Chapter 7 **Electronic** **Controls**	Basic Control Concepts	167
	Electronic Control Variations	172
	Control Concepts	176
Chapter 8 **Cartridge Valves**	Cartridge Valve Concept	181
	Screw-in Cartridge Valves	185
	Functional Characteristics	189
	Slip-in Cartridge Valves	203
	Valve Configurations Available	205
	Slip-in Cartridge Valve Application Configurations	208
Chapter 9 **Auxiliary Valves**	Pressure Controls	215
	Load Control Valves	230
	Flow Control	234
Chapter 10 **Accumulators**	Accumulator Types	242
	Accumulator Applications	246
Chapter 11 **Fixed** **Displacement** **Pumps**	Types Of Pumps	251
	Vane Pump Design	255
	Gear Pumps	260
	Piston Pumps	262
	Operation Of Fixed Displacement Pumps	265
Chapter 12 **Variable** **Displacement** **Pumps**	Variable Displacement Pumps	276
	Inline Piston Pump Controls	281
Chapter 13 **Hydrostatic** **Transmissions**	Advantages Of Hydrostatic Transmissions	296
	Classification Of Hydrostatic Transmissions	307
	Closed Circuit Hydrostatic Transmissions	309
	Review Of Pump And Motor Operating Characteristics	314
	Operation Of Closed Circuit, Closed Loop Hydrostatic Transmissions	323
	Basic Concepts In Traction	325
Chapter 14 **Fluid Conductors** **and Connectors**	Selection Criteria	330
	Hose	333
	Steel Tubing	334
	Mating Connectors	336
	Quick-disconnect Couplings	337
Chapter 15 **Reservoirs**	Basic Functions Of A Reservoir	339
	Reservoir Location	341
	Reservoir Components	346
	Pressurized Reservoirs	349

Table Of Contents

Chapter 16 **Hydraulic Fluids**	Purposes Of Hydraulic Fluids	352
	Fluid Properties	356
	Additives	361
	Fluid Types	364
Chapter 17 **Fluid Conditioning**	Defining Contamination	369
	Sources Of Contamination	375
	Results Of Contamination	382
	Principles Of Systemic Contamination Control	386
Chapter 18 **Circuits**	System Flow Control	401
	Alternate Pilot Pressure Source	403
	Prime Mover Protection	404
	Proportional Cooling Fan Drive	406
	Float Position Using CMX Valve	408
	Two Pump Coordination System	410
	Fixed Pump Load Sensing System	412
	Meter-out Using Counterbalance Valves	414
	Hydrostatic Propel Traction Control	416
	Steering Mode Selection	418
	Water Pump Flow Control	420
	Intermittent High Pressure Operation	422
	Closed Loop Velocity Control	424
	Regeneration Selector Circuit	426
	Propel Creep Control	428
Appendix	Definition of Technical Terms	431
	Abbreviations	443
	Metric Conversion Factors	445
	Pressure Conversion Factors	453
	Volume Conversion Factors	453
	Atmospheric Properties	453
	Oil Viscosity Recommendations	454
	Oil Flow Capacity of Tubing	454
	ISO/ANSI Basic Symbols for Fluid Power Equipment and Systems	456
	Comparative Viscosity Classification	458
	Flow Capacity of Piping	459
Index		461

CHAPTER 1 .. Introduction to Mobile Fluid Power

In its most basic definition, hydraulics is the use of liquids to perform a task. By this definition, the use of hydraulics to simplify work, multiply human efforts and transmit power goes back as far as recorded history. There is evidence that man used simple forms of hydraulics to build pyramids, as well as to transport people and material. In more recent history, humans used liquids to power grist mills and grinding stones, and then to power machinery in factories.

Nature has used water power to carve out its splendor over many centuries, and is still doing it. Every flowing river, every rainfall, every movement of the tides, every eruption of a geyser, changes the appearance and structure of the earth's surface. This action is nature's hydraulic system.

The human circulatory system (as well as that of most of nature's creatures) is a hydraulic system. More than that, it is a very sophisticated closed-circuit hydraulic system.

The point is, hydraulics was not an invention; it was a discovery. Man has developed many devices that put liquids to modern use, and these are inventions. However, they incorporate a natural phenomenon that has existed since liquids.

Pascal's Law

Blaise Pascal (1623-1662), a French philosopher and mathematician, is credited with applying some basic mathematical principles to hydraulics that are still in use today. These principles were expressed in a paper issued in 1647 titled "Recit de la Grand Experience de L'Equilibre de Liqueurs". Proof of the difficulty Pascal had in gaining acceptance of his work is that the paper was not published until 1663, a year after his death.

Furthermore, it was almost 150 years later before Pascal's principles were put to practical industrial use. The first significant application of hydraulic principles was done by Joseph Brahma (1748-1814). Brahma obtained a patent in 1795 for a hydraulic press, based very closely on the sketches and drawings of Pascal done more than a century-and-a-half earlier.

The culmination of Pascal's efforts is the primary law of hydraulics, known as Pascal's Law, and is stated:

> *Pressure applied to a confined liquid is transmitted undiminished in all directions, and acts with equal force on all equal areas, and at right angles to those areas.*

This principle, also referred to as the laws of confined fluids, is best demonstrated by considering the result of driving a stopper into a full glass bottle (Figure 1.1).

Figure 1.1 Applying pressure to a liquid

Because liquid is essentially incompressible, and forces are transmitted undiminished throughout the liquid and act equally on equal areas of the bottle, and the area of the body of the bottle is much greater than the neck, the body will break with a relatively light force on the stopper. Figure 1.2 illustrates this phenomenon.

Figure 1.2 Container bursting due to pressure

Figure 1.3 illustrates the relationship of areas that causes a greater force on the body of the bottle than is applied to the neck. In this illustration, the neck of the bottle has a cross sectional area of one square inch (which can be written as 1 in^2). When the pressure created by this force is transmitted throughout the fluid, it influences all adjacent areas with equal magnitude. It stands to reason that a larger area (a greater number of square inches) will be subjected to a higher combined force.

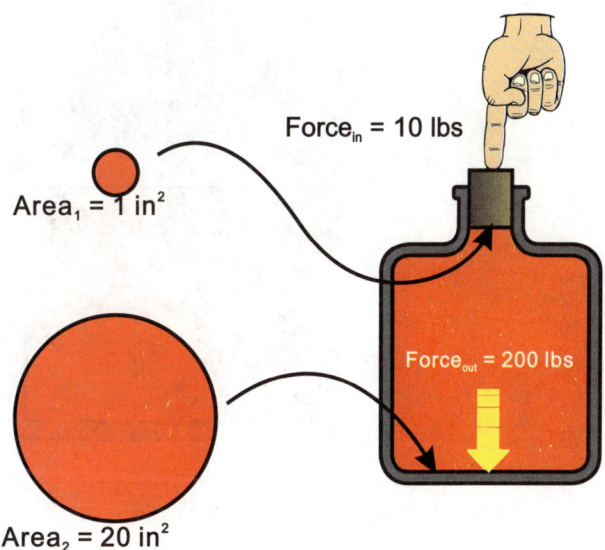

Figure 1.3 Pressure, area, force relationship

The bottom of the bottle in Figure 1.3 has a total area of 20 in^2 as shown, and the force applied by the liquid is 10 pounds per in^2. Therefore, the combined force over the entire bottom area is the sum of 10 pounds acting on each of the 20 square inch areas. Because there are 20 areas, and 10 pounds on each, the combined force at the bottom of the bottle is 200 pounds.

This relationship is represented by the following formula:

Formula 1-1

Force = Pressure x Area

This formula states that the force applied to an area under pressure is the product of the pressure times the area. In customary U.S. units, force is usually defined in pounds, pressure in pounds-per-square-inch (lbs/in^2, or psi), and area in square inches (in^2).

In the metric system, force is usually defined in Newtons (N), pressure in Bar or Kilopascals (Kp), and area in square centimeters (cm^2). The relationship between force, pressure and area is the same as in the above formula.

The same relationship is used to determine the pressure in a fluid resulting from a force applied to it. Figure 1.4 shows a weight being supported by fluid over a 10 in^2 area. By rearranging the above formula, the fluid pressure of 20 psi can be determined by:

Pressure = Force ÷ Area

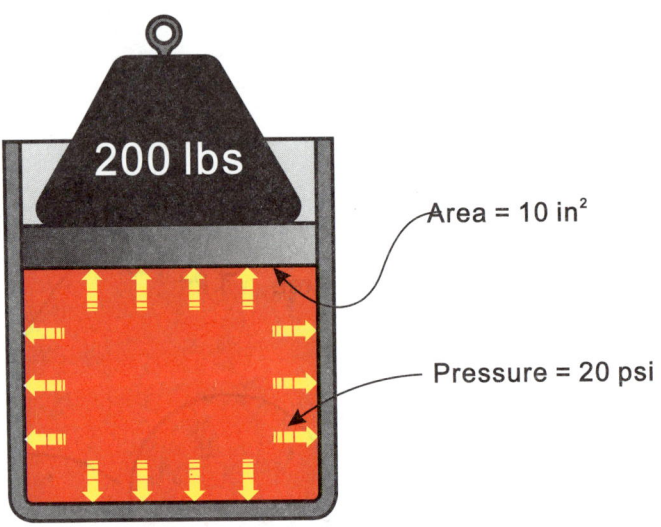

Figure 1.4 Pressure created by weight

Pascal demonstrated the practical use of his laws with illustrations such as that shown in Figure 1.5. This diagram shows how, by applying the same principle described above, a small input force applied against a small area can result in a large force by enlarging the output area.

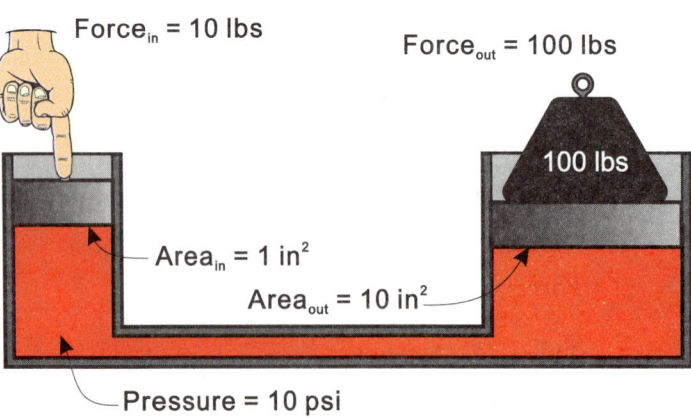

Figure 1.5 Transmitting force by fluid

This pressure, applied to the larger output area, will produce a larger force as determined by the formula on the previous page. Thus, a method of multiplying force, much the same as with a pry-bar or lever, is accomplished using fluid as the medium.

FLUID POWER ADVANTAGES

Multiplying forces is only one advantage of using fluid to transmit power. As the diagram in Figure 1.5 shows, the forces do not have to be transmitted in a straight line (linearly). Force can be transmitted around corners or in any other non-linear fashion while being amplified. Fluid power is truly a flexible power transmission concept.

Actually, fluid power is the transmission of power from an essentially stationary, rotary source (an electric motor or an internal combustion engine) to a remotely positioned rotary (circular) or linear (straight line) force amplifying device called an actuator. Fluid power can also be looked upon as part of the transformation process of converting a benign form of potential energy (electricity or fuel) to an active mechanical form (linear or rotary force and power).

Once the basic energy is converted to fluid power, other advantages exist:

1. Forces can be easily altered by changing their direction or reversing them.

2. Protective devices can be added that will allow the load operating equipment to stall, but prevent the prime mover (motor or engine) from being overloaded and the equipment components from being excessively stressed.

3. The speed of different components on a machine, such as the boom and winch of a crane, can be controlled independently of each other, as well as independently of the prime mover speed.

A complete hydraulic system consists of a reservoir of fluid, a hydraulic pump driven by an internal combustion (IC) engine or an electric motor, a system of valves to control and direct the output flow of the pump, and actuators that apply the forces to conduct the work being performed. Figure 1.6 is a simplified illustration of these major components.

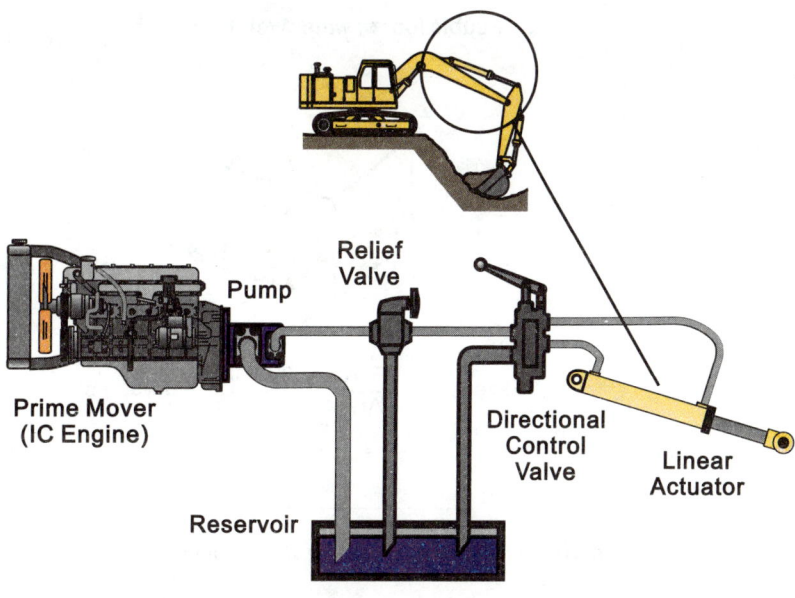

Figure 1.6 Simplified Hydraulic Circuit

Chapter 1 Introduction to Mobile Fluid Power

PUMP INLET CONDITIONS

The system fluid is forced out of the reservoir into the inlet side of a pump by the sum of several pressures that act on the fluid (Figure 1.7). The first pressure is the one caused by the weight of the fluid; the second is caused by the weight of the atmosphere; a third may be present if a pressurized reservoir is employed.

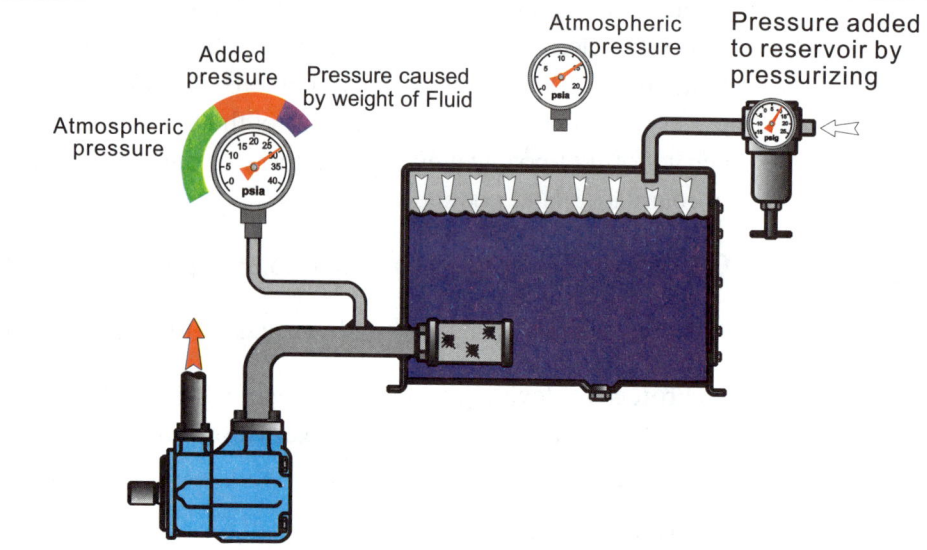

Figure 1.7 Pressure at reservoir outlet

Fluid Weight

A cubic foot of water weighs about 62.4 pounds. This weight acts downward due to the force of gravity, and causes a pressure at the bottom of the fluid. Figure 1.8 shows how this weight is distributed across the entire bottom of the water volume. In this example, the entire weight is supported by an area measuring one foot by one foot, or one ft^2.

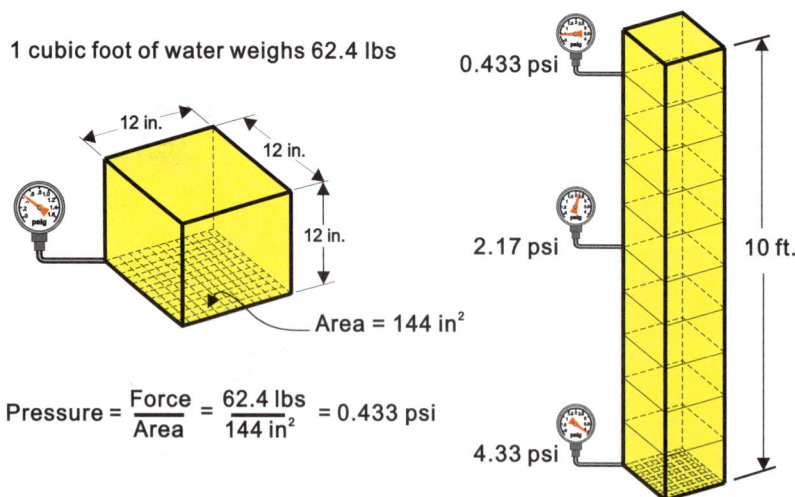

Figure 1.8 Pressure caused by weight of water

There are 144 square inches in one square foot (12 inches x 12 inches). Therefore, the weight of water acting on one square inch at the bottom of the cube is 62.4 lbs/144in^2. This equals 0.433 pounds per in^2. In terms of pressure, this is 0.433 pounds per square inch (psi) for a column of water one foot tall.

A two foot tall column of water would weigh twice as much spread over the same area, or 0.866 psi.

This is the same pressure we feel on our eardrums when we swim under water, and our experience tells us that the pressure increases with depth. The pressure can be expressed as follows:

Formula 1-2

Pressure (psi) = Water depth (ft) x 0.433 psi per foot of depth

Other fluids behave the same as water, the difference being relative to the difference in weight of the fluids. The difference is usually defined by the Specific Gravity of the fluid (SG), which is the ratio of the fluid's weight to the weight of water.

Formula 1-3

SG_x = Weight of fluid x ÷ Weight of water

A typical specific gravity for oil used in hydraulic systems is 0.92, meaning the weight of the oil is 92% of the weight of water. The relationship of the first formula then becomes:

Formula 1-4

Pressure (psi) = $Fluid_x$ Depth (ft) x 0.433 psi/ft water x SG_x

Pure water weighs 62.4 pounds per cubic foot at 4°C, the temperature at which it is most dense. The weight will be slightly less at higher temperatures, but the difference is generally ignored in hydraulic calculations.

Typical hydraulic oil in a reservoir creates a pressure of 0.4 psi per foot of height, as illustrated in Figure 1.9. This pressure at the bottom of a reservoir helps to push the fluid out of the reservoir and into the inlet of a hydraulic pump if the pump inlet is below the fluid level.

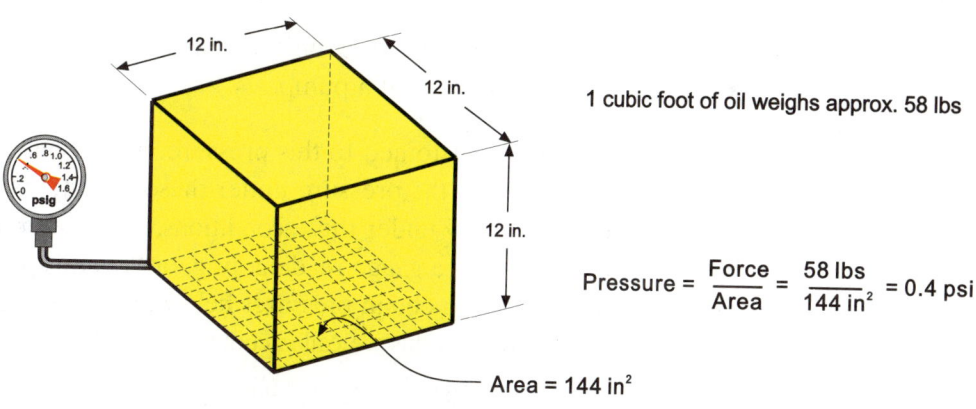

Figure 1.9 Pressure caused by the weight of oil

Chapter 1 Introduction to Mobile Fluid Power

Atmospheric Pressure

We generally do not think of air as having weight. Any reasonable quantity of it is so light that we usually ignore the weight. And, because we exist in the atmosphere continuously, we never think of the fact that the air above us has significant weight.

A column of air measuring one inch by one inch across (1 square inch of area), and extending from the earth's surface at sea level to the extreme of the atmosphere, would actually have a significant weight. This weight, on an average day, is 14.7 pounds, as illustrated in Figure 1.10. Therefore, the pressure that continuously exists at sea level due to the weight of the air above us is 14.7 psi. This Figure is referred to as a standard atmosphere, or the atmospheric pressure on a typical day at sea level.

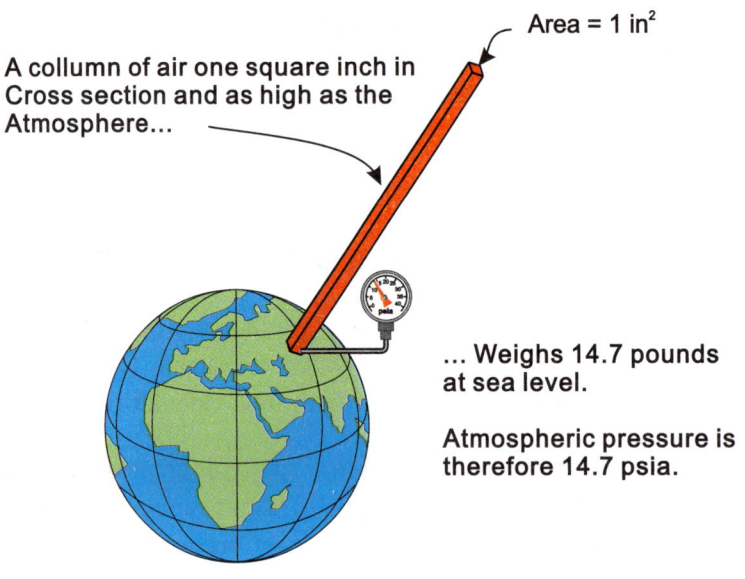

Figure 1.10 Weight of air causes atmospheric pressure

This pressure, acting on the reservoir fluid, also helps to push fluid out of the reservoir and into the inlet of a pump.

Because we are so accustomed to this pressure, and because it exists all the time all around us, we consider the pressure under these conditions to be "zero". Pressure gauges also read "zero" under these conditions, so we refer to the standard atmospheric pressure as a gauge reading; psig, or "pounds-per-square-inch-gauge." It is, of course, possible to obtain pressures below this level by removing some of the atmospheric pressure, and this we call vacuum.

By removing all of the atmospheric pressure, we would arrive at a "new" zero, and this we call "absolute zero". Absolute zero is 14.7 psi below gauge zero, and is considered a perfect vacuum (see Figure 1.11). There is no pressure below absolute zero.

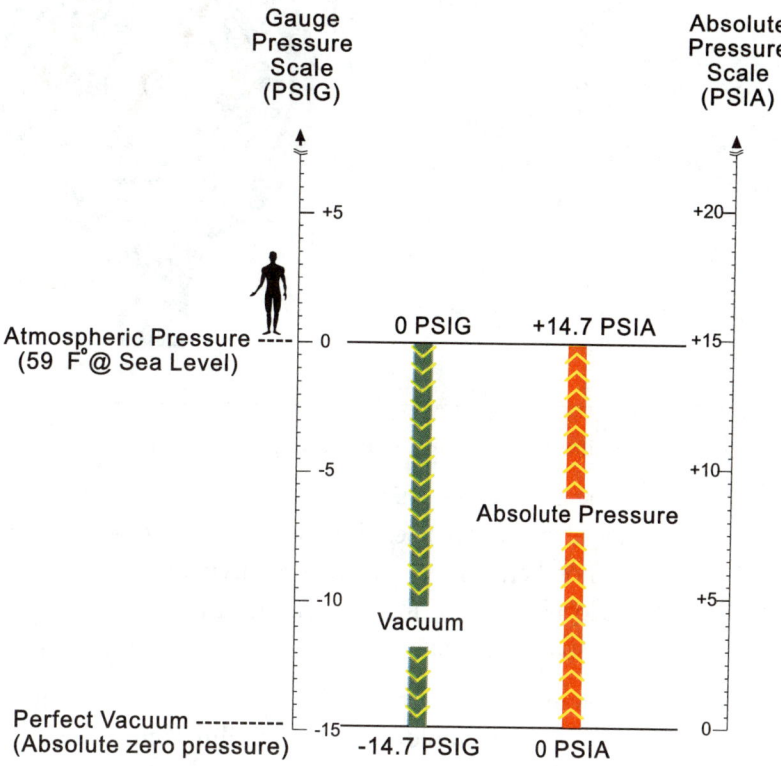

Figure 1.11 Gauge and absolute pressure

To differentiate between the two pressures, we always label absolute pressure "psia", standing for "pounds-per-square-inch-absolute". This means that the zero for this pressure is absolute zero, and all positive pressure readings start from this level. If the pressure starts at atmospheric pressure as the "zero", then it is designated gauge pressure, or psig.

NOTE

If there is no designation, and the pressure is labeled as psi, then it is assumed that the pressure is psig.

Figure 1.12 shows the relationship between gauge pressures, absolute pressures and vacuum all illustrated on one gauge dial.

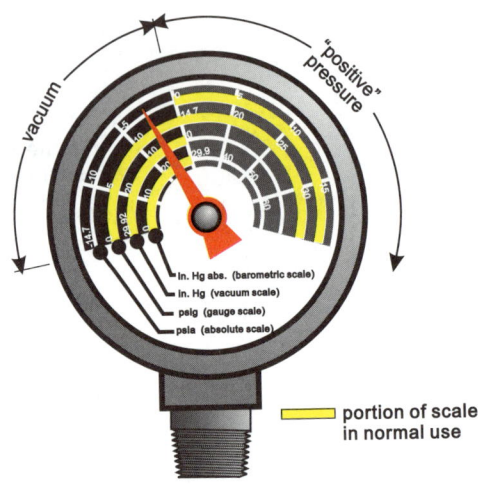

Figure 1.12 Absolute, gauge and vacuum scales

Barometric Pressure

One can see now that as we move above sea level, such as up a mountain, the column of air above us becomes shorter, and thus the weight of the air above us becomes less. The atmospheric pressure is then reduced, and the air is not compressed as much. We recognize this as "thin" air at higher altitudes, and we feel a shortness of breath; the reason being that we get less air into our lungs each time we inhale.

It is important to recognize this phenomenon; at higher altitudes, the atmospheric pressure available to help push fluid out of the bottom of a hydraulic reservoir and into the inlet of a pump is less than at lower altitudes.

Atmospheric pressure is measured by use of a barometer, and this is illustrated in Figure 1.13. A tube full of mercury is inverted in a pool of mercury as shown. The mercury will fall out of the tube until it reaches a specific height. The space above the mercury in the tube will become a perfect vacuum of 0 psia. The height of the mercury in the tube will correspond to atmospheric pressure, because it is atmospheric pressure that is preventing the mercury from falling the rest of the way out of the tube.

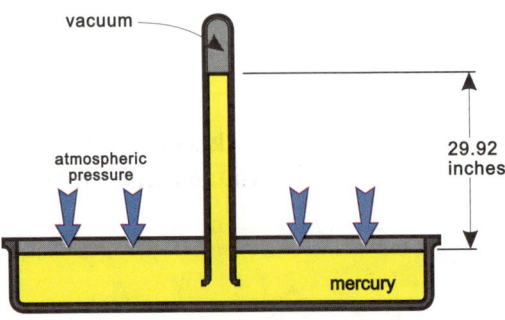

Figure 1.13 Barometer principle

At standard atmospheric pressure of 14.7 psi, mercury will fall in the tube until it reaches a height of 29.92 inches above the pool. As the atmospheric pressure changes (due to climate or altitude change), the height of the mercury will change accordingly. The relationship between the height of the mercury column and gauge or absolute pressure is:

Formula 1-5

1 inch of Mercury (1" Hg) = 0.491 psi

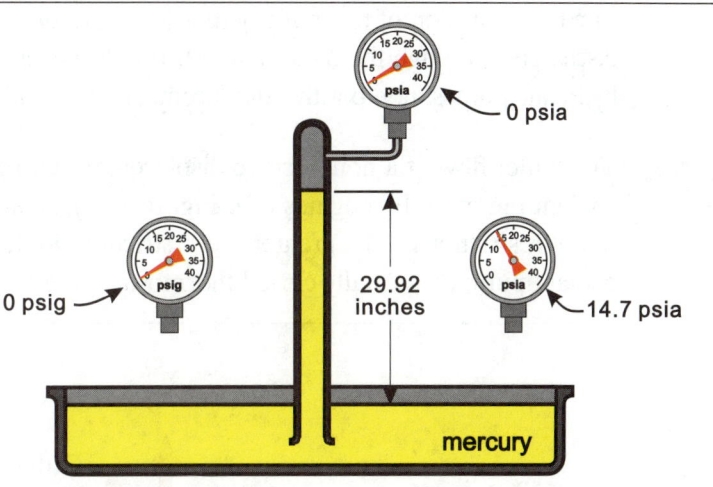

Figure 1.14 Pressure and vacuum relationship

The relationship is further demonstrated by the illustration in Figures 1.14 and 1.15.

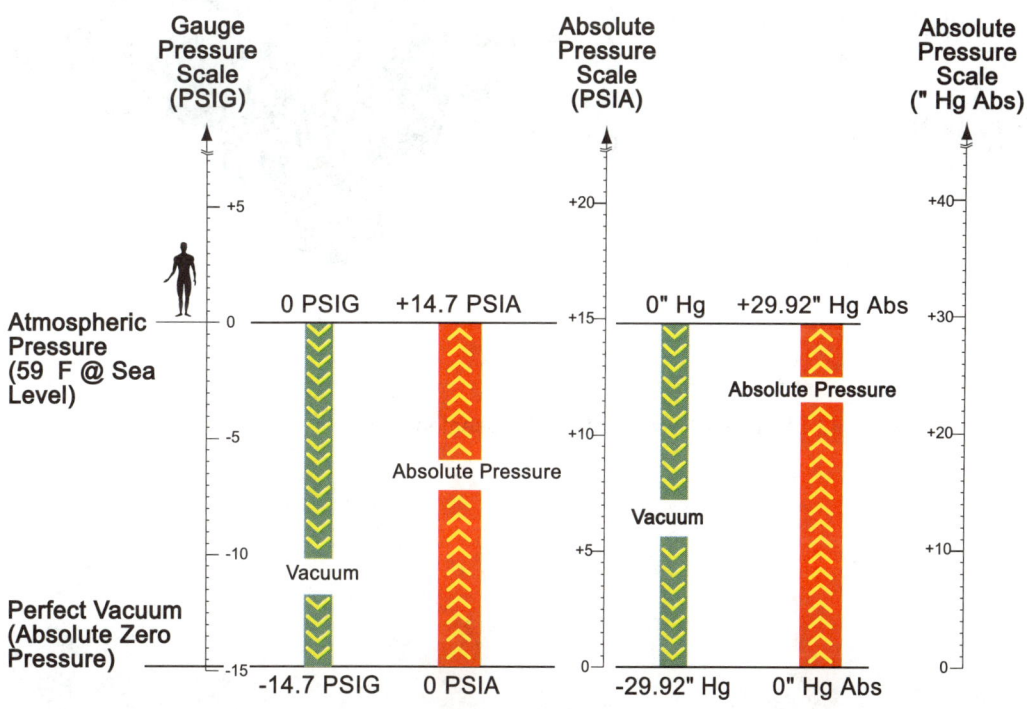

Figure 1.15 Gauge and absolute pressure comparison

Chapter 1 Introduction to Mobile Fluid Power

Pressurized Reservoirs

In addition to the pressures created by atmosphere and the weight of the reservoir fluid, a third pressure may be superimposed into the reservoir to assist pushing fluid into a pump inlet. There are several ways that this can be accomplished which will be covered in Chapter 14. It is important to note that if the sum of the fluid weight and atmospheric pressure are not sufficient to fill the pump inlet, a third source of pressure can sometimes be added.

FLUID PUMPS

The basic function of hydraulic pumps is to take fluid that is provided at the inlet, and discharge it through the outlet into a hydraulic system. There are two broad categories of hydraulic pumps; non-positive displacement and positive displacement.

Non-Positive Displacement Pumps

The outlet flow of a non-positive displacement pump, best characterized by the coolant pump on IC engines (see Figure 1.16), is entirely dependent on the inlet and outlet restrictions. The greater the restriction on the outlet side (in the case of the coolant pump, a partially closed thermostat), the less flow the pump will discharge.

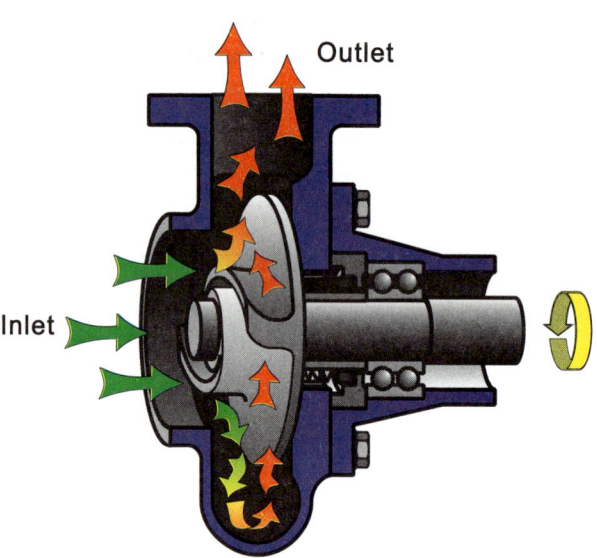

Figure 1.16 Non-positive displacement pump

The output flow can even be stopped if the restriction (thermostat) becomes closed, or the coolant passages become clogged, as in Figure 1.17.

For hydraulic systems, this would not be a very reliable source of flow. Non-positive displacement pumps are seldom used in hydraulic systems because they cannot be relied upon to provide fluid under all conditions.

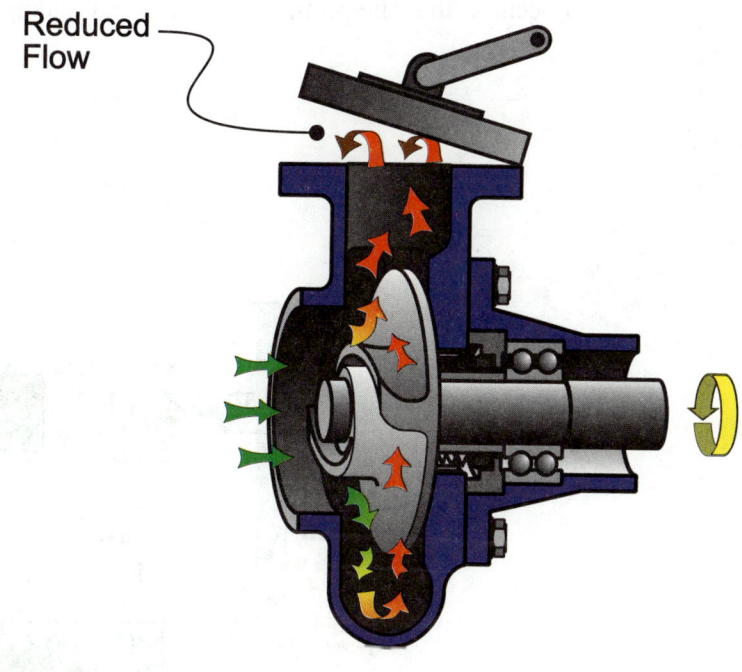

Figure 1.17 Restricted flow from non-positive displacement pump

Chapter 1 Introduction to Mobile Fluid Power

Positive Displacement Pump

A positive displacement pump will discharge a specified amount of fluid during each revolution or stroke (such as the hand pump in Figure 1.18), almost regardless of the restriction on the outlet side. Because of this characteristic, positive displacement pumps are nearly always the pump of choice in hydraulic systems.

Positive displacement hydraulic pumps are designated by their volume of displacement, such as gallons per minute or cubic inches per revolution. This designation is usually a theoretical displacement, and does not allow for any losses that may occur within the pump due to internal leakage.

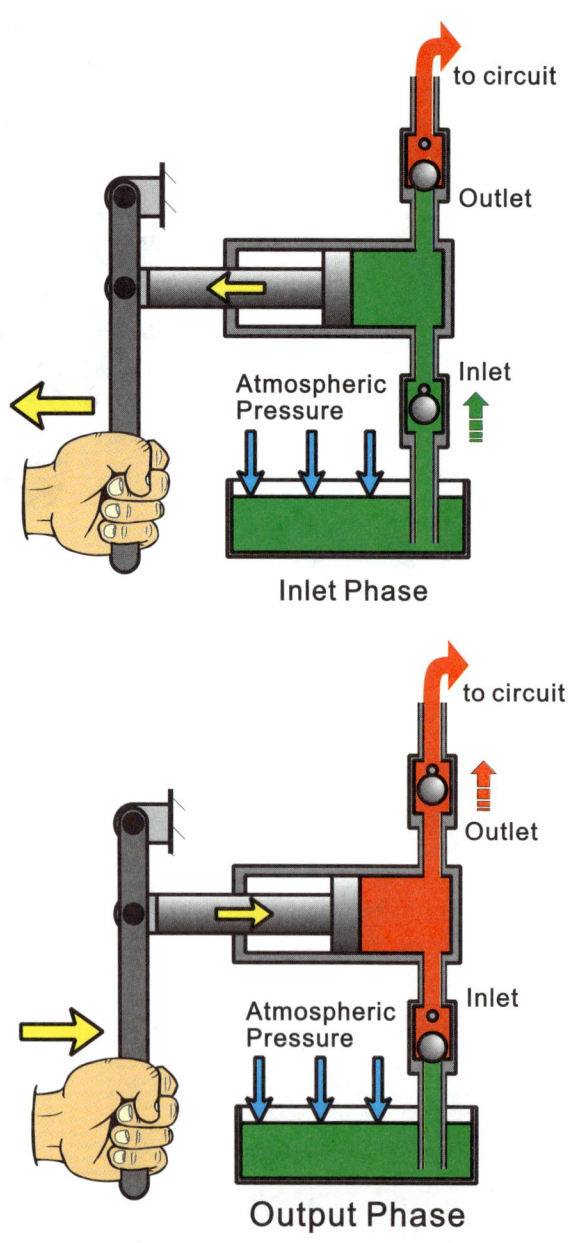

Figure 1.18 Positive displacement pump

PUMP INLET CHARACTERISTICS

With each element of fluid that is discharged from a hydraulic pump, an equal amount must be available at the inlet side to replace it. The availability of the fluid at the inlet is entirely dependent upon the reservoir pressures that force the fluid into the pump.

The larger the pump, and the faster the pump runs, the more fluid is needed to replace the amount that is discharged. This will depend upon there being adequate pressure in the reservoir to force fluid into the pump. Without sufficient pressure, "starvation" of the pump will occur, and this will cause severe damage to the pump components, and ultimately pump failure.

There are many factors that can hinder the flow of fluid between the reservoir and the pump:

- A fluid line that is too small for the volume of fluid going through it.
- A clogged outlet on the reservoir.
- A pump that is located too far away from the reservoir, or too far above it.
- A fluid that is too viscous to flow easily.

When one or more of these conditions exist to the point that "starvation" of the pump begins to occur, they must be corrected immediately.

Aeration and Cavitation

Inadequate inlet conditions to a hydraulic pump will result in either of two phenomenon that are very destructive; aeration and cavitation. They are very similar in their results, that is, eroded internal components of the pump, but they are caused by two different conditions.

Aeration is caused by the introduction of air into the fluid and into the inlet of the pump. The air can come from a leak in one of the inlet lines or fittings, as in Figure 1.19; it can come from air being introduced into the fluid in the reservoir or in the hydraulic system and not given time to escape; or it can exist in the fluid in a dissolved, or "entrained", form, being held in place by the molecular structure of the fluid. Regardless of the source of air, it must be minimized to prevent damage to the pump.

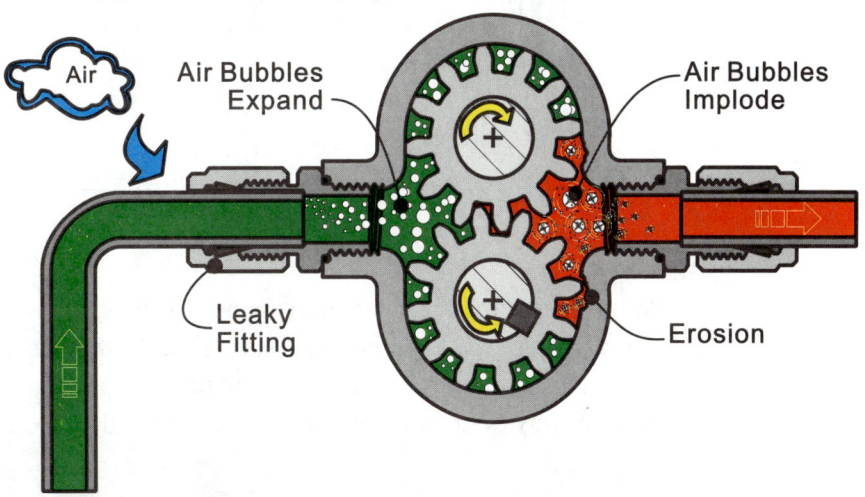

Figure 1.19 Aeration of hydraulic pump

Fluid entering the inlet side of a pump will be at a low pressure (vacuum), and will exit the pump at a higher pressure. The low pressure at the inlet will cause any air bubbles in the fluid to expand; the lower the pressure, the greater the expansion. As the fluid travels through the pump to the outlet side, pressure on the fluid will increase and cause the air bubbles to collapse. This collapsing occurs very rapidly, called implosion, and releases a high level of energy. This concentration of energy will erode metal components inside the pump, and cause total pump failure in a very short time. This is illustrated in Figure 1.19.

Cavitation, on the other hand, will occur with or without the presence of air. This phenomenon occurs when the pressure (vacuum) at the pump inlet falls below the vapor pressure of the fluid. The fluid will essentially "boil", forming bubbles of fluid vapor, as illustrated in Figure 1.20. These vapor bubbles will act much the same as the air bubbles during aeration, expanding and then contracting with a rather violent force, imploding against the internal metal components of the pump. The resulting erosion is severe, and will cause total pump failure very quickly.

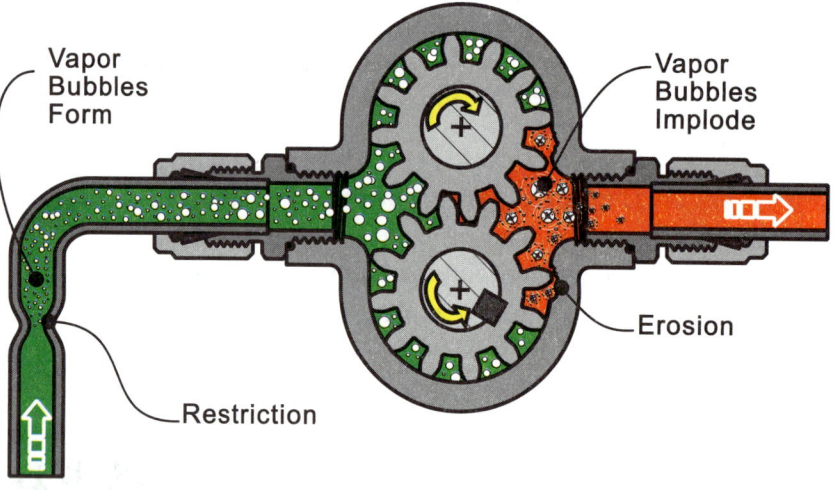

Figure 1.20 Cavitation of hydraulic pump

Both aeration and cavitation are very noisy events, sounding much as though a handful of marbles had been tossed into the pump inlet. Both phenomenon are aggravated by increased pressure at the outlet, and by reduced pressure (increased vacuum) at the inlet. Both are very destructive, and must be eliminated as soon as they are detected.

Systems

Once the fluid is discharged from the pump, it is available to be directed to an actuator, under the intended conditions and at the intended rate, where it will apply the necessary forces to conduct the work required. There are a large number of hydraulic components used to accomplish this activity, all of which will be covered in later chapters in this manual. They can be broken down into four basic categories:

Pressure controls - components that regulate the maximum pressure that can occur in a hydraulic system, or a part of the system. Reducing, counterbalance, circuit relief, etc.

Flow controls - components that regulate the maximum amount of flow that will be allowed in a system, or in a part of the system.

Directional controls - components that allow a path for the fluid to be selected and directs the fluid down that path.

Actuators - components that convert the fluid energy into linear (straight line) or rotary (rotating) forces.

Frequently, single components will consist of two or more of the above components. For example, a single directional control valve frequently includes a pressure control, and may occasionally include a flow control as well.

FORCE, WORK, ENERGY AND POWER

We would have little use for hydraulic systems if they did not have practical benefits for us. In fact, whether in its simplest form or in a complicated, sophisticated form, we utilize hydraulic systems to apply forces, to do work, and to conserve energy through the judicious use of power.

Each of these terms; force, work, energy and power, has a specific meaning in the process of converting fuel into the performance of tasks.

Force

A force is a push against something, as depicted in Figure 1.21. It is usually measured in pounds (customary U.S.) or Newtons (Metric), but it can also be measured in ounces, tons, grams, kilograms, or any other expression that is commonly used to designate weight. A force can be linear (a push in a straight line), or it can be rotational (a circular force, or twist). It can also be stationary or moving.

Figure 1.21 Force

An example of a linear force is that caused by the weight of an object that is hanging from a boom, as shown in Figure 1.22. The weight acts straight down, due to gravity, with a force equal to its weight. It is applying a linear force to the cable that is supporting it.

Figure 1.22 Weight causing a force

An example of a rotational force is that applied to a pipe by a wrench while turning, as shown in Figure 1.23.

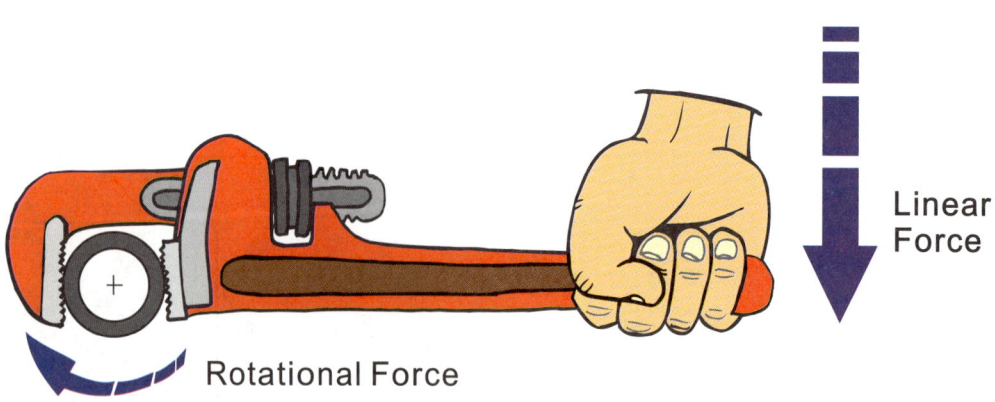

Figure 1.23 Linear and rotational forces

Work

Work is accomplished when something is moved. A force can be applied to something, even at a very high level, but if the object that the force is being applied to does not move, no work is done.

If the object does move, we can determine the amount of work that has been done by determining the force applied, and the distance the object moved, as shown in Figure 1.24. The work involved is determined by the formula:

Formula 1-6
Work = Distance x Force

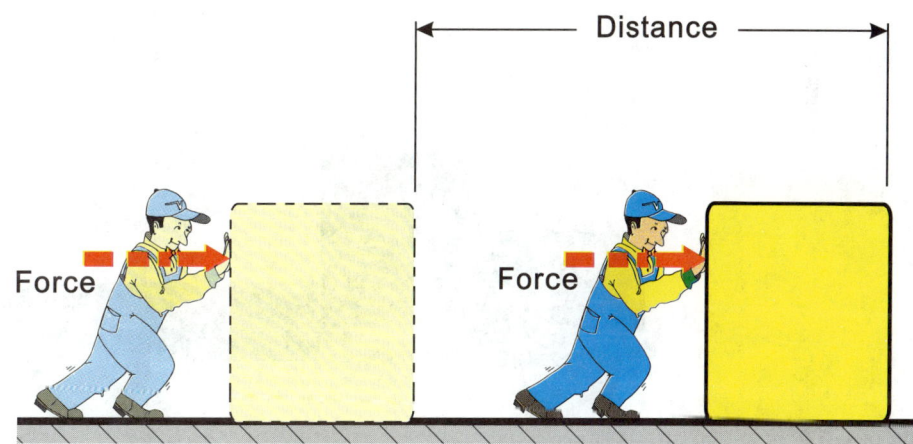

Figure 1.24 Work equals force x distance moved

In the customary U.S. system, we usually measure the force in pounds and the distance in feet. Work is then expressed in "foot-pounds". It is not uncommon to express work in terms of "inch-pounds" or "inch-ounces", although "foot-pounds" is the most commonly used and preferred term.

In the metric system, force is generally expressed in Kilograms or Newtons, and the distance in meters. Work is then expressed in "Kilogram-meters" or "Newton-meters".

Chapter 1 Introduction to Mobile Fluid Power

Energy

Energy is the potential to do work. It is like work, and is measured and described in the same terms. The difference is that energy is a form of work that is dormant until it is released.

Referring again to a weight hanging on a boom, a specific amount of work was expended to raise the weight off the ground and place it there (see Figure 1.25). That work, using the opposite formula, is the force applied (which in this case is equal to the weight) times the distance over which the force was applied (which in this case is the height the weight was lifted). This weight now has the capacity to do an equal amount of work by moving from its height back down to the ground. As long as the weight is hanging on the boom, it is doing no work, but it has the capacity to do work, and this is called energy.

Figure 1.25 Raised weight resulting in potential energy

Recognizing that the energy "stored" in the weight is equal to the work expended to pick the weight up, gives rise to the law of physics called "Conservation of Energy":

Energy can neither be created nor destroyed, although it can be changed from one form to another.

Examples of energy changing form are all around us; electric energy changing to mechanical energy through an electric motor; fuel changing to heat energy when it is ignited, which can change further to mechanical energy when it drives an internal combustion engine; and hydraulic energy changing to heat energy when it flows across a relief valve.

The heat energy in this last example is frequently referred to as "waste", which it may well be if there is no convenient way to use this heat when it is generated. However, it is an example of energy changing form, during a process.

Power

A large object can be moved a fixed distance either very slowly or very rapidly, as shown in Figure 1.26. The result is the same amount of work being done in both cases, because time is not a factor in the concept of work. The same force, over a like distance, requires the same amount of work whether it is done over a few seconds or over a day. Obviously, there is a difference between the two situations, because one is more productive than the other.

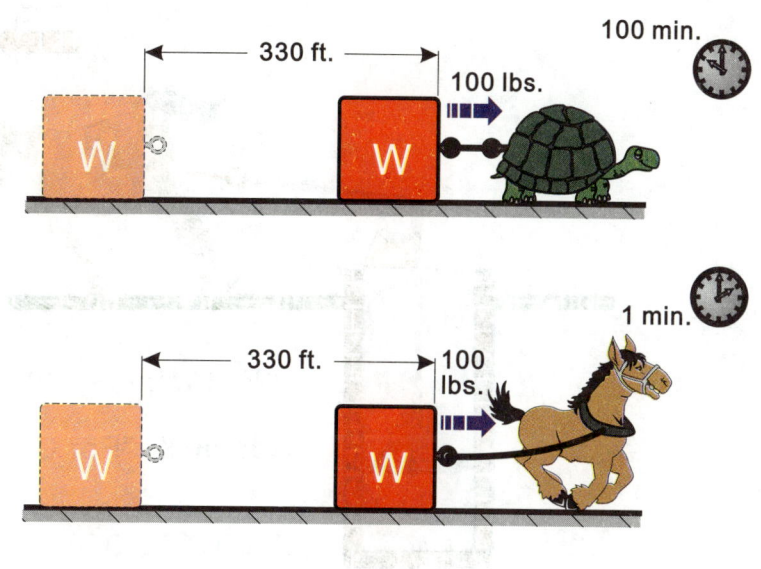

Figure 1.26 Equal work over different time spans

The difference is the time involved. Including the time element in our calculation provides a measure of power. The same amount of work done over a shorter period of time requires more power.

James Watt (1736-1819), A Scottish inventor and the inventor of the steam engine, understood the relevance of power when he devised a method of comparing his new invention to the work facilitator of the day, the horse. With a pulley arrangement and a series of weights hanging down a well as shown in Figure 1.27, he attempted to learn what a typical horse could pull consistently at a reasonable speed. He concluded that an average horse could raise 100 pounds out of the well at a rate of $2^1/_2$ miles per hour, or 220 feet per minute.

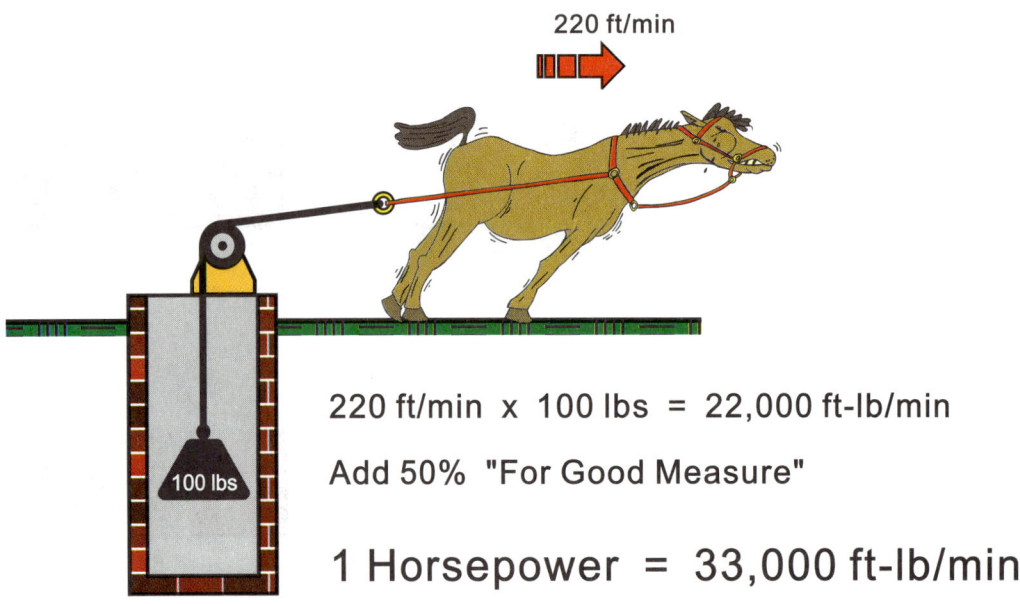

Figure 1.27 James Watt's determination of horsepower

The product of speed (or velocity) and weight gave a figure of 22,000 foot-pounds per minute as a reasonable amount of work that a horse can do. To eliminate any possible accusations of favoritism toward his new invention, he increased this number by 50% to provide a work-rate of 33,000 foot-pounds per minute, and this he called a horsepower. As unscientific as this experiment may have been, the universal standard for power today is still:

Formula 1-7

1 Horsepower = 33,000 foot pounds per minute

Put another way, moving an object with a force of 33,000 pounds for a distance of one foot over a time span of one minute requires one horsepower.

Other relationships of force, distance and time are shown in Figure 1.28. The relationship to a rotational force is shown in Figure 1.29.

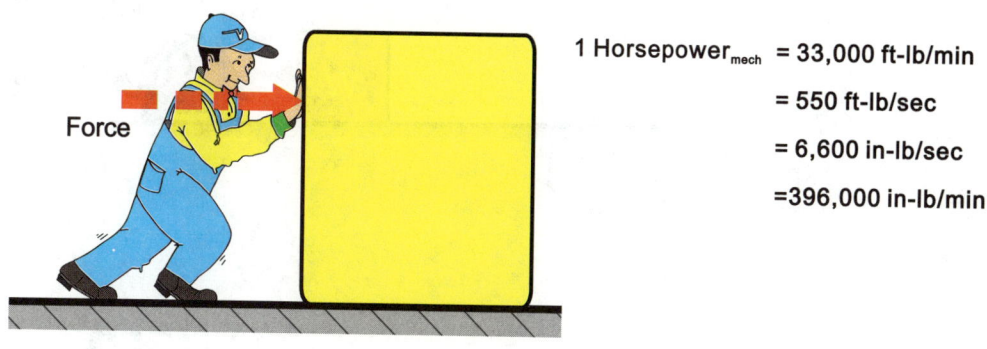

Figure 1.28 Horsepower equivalents

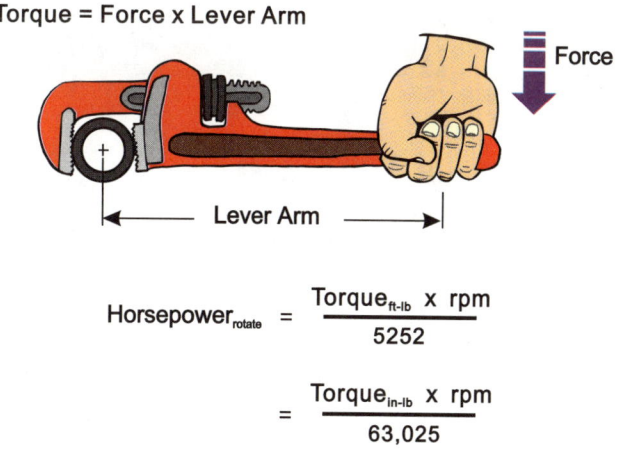

$$\text{Horsepower}_{rotate} = \frac{\text{Torque}_{ft\text{-}lb} \times \text{rpm}}{5252}$$

$$= \frac{\text{Torque}_{in\text{-}lb} \times \text{rpm}}{63,025}$$

Figure 1.29 Torsional relationship to horsepower

And now we can define the difference between the two cases of equal work, over different periods of time. Figure 1.30 illustrates the calculations.

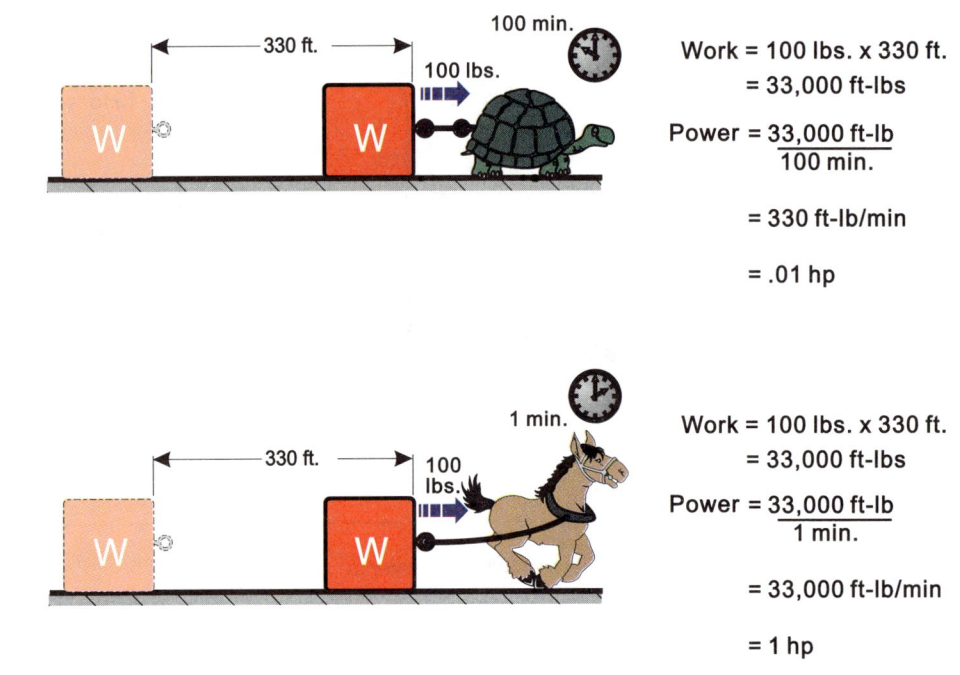

Figure 1.30 Power is the rate of doing work

When a time factor is included in all other forms of energy, they can all be related to horsepower. Both heat and electricity have their horsepower equivalents as shown in Figure 1.31, and hydraulic horsepower is defined in Figure 1.32. These relationships are also described in Figure 1.33, which relates heat, electricity, hydraulics and rotational forces to the basic standard of horsepower.

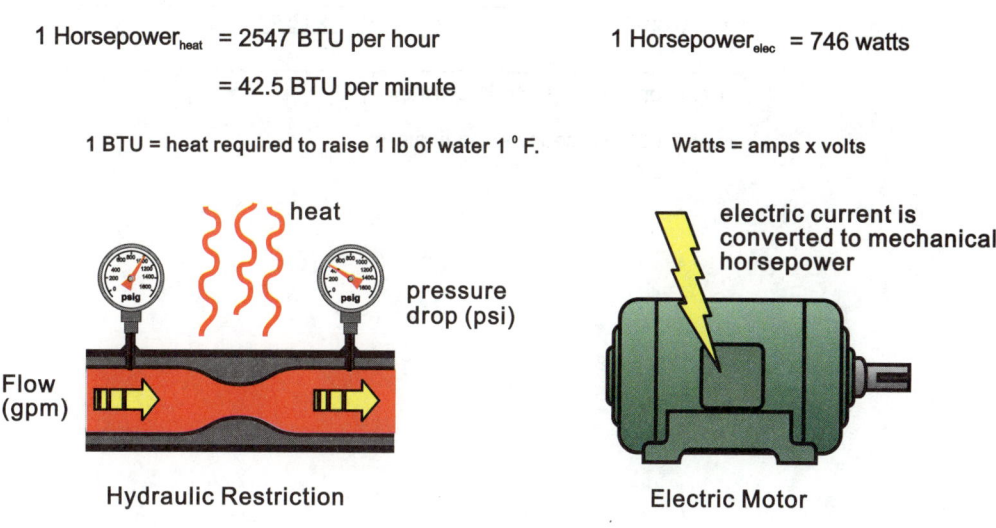

Figure 1.31 Horsepower relationship to heat and electricity

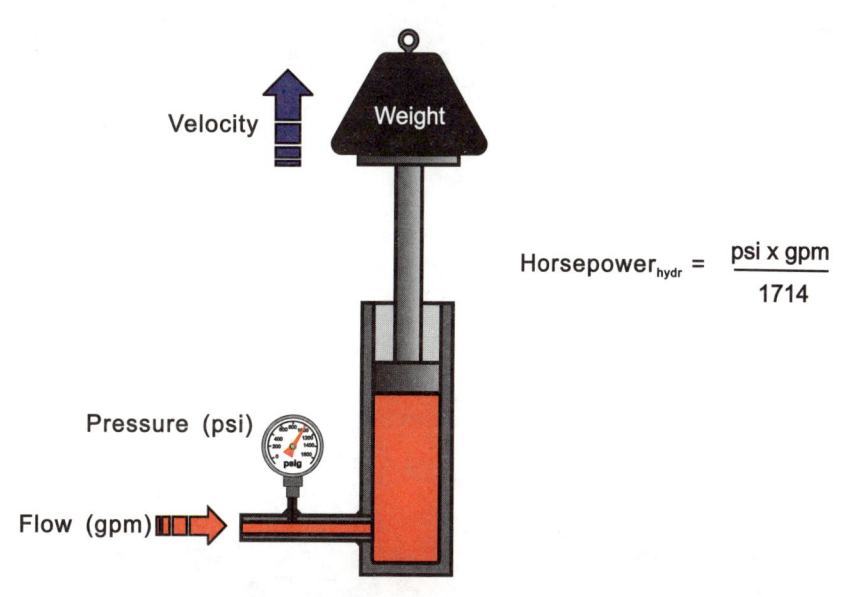

Figure 1.32 Hydraulic horsepower

Chapter 1 Introduction to Mobile Fluid Power

	Power Relationships	
	Customary U.S. Units (Horsepower)	**SI Units (Kilowatt)**
Mechanical	33,000 ft.-lbs. per min. 550 ft.-lbs. per second	1000 Newton-Meters per sec
Electrical	$\dfrac{\text{Amps x Volts}}{746}$	$\dfrac{\text{Amps x Volts}}{1000}$
Heat	42.44 BTU/minute	56.84 BTU/minute
Hydraulic	$\dfrac{\text{PSI x GPM}}{1714}$	$\dfrac{\text{BAR x LPM}}{600}$
Rotational	$\dfrac{\text{lb.-ft. x RPM}}{5252}$	$\dfrac{\text{Newton meters x RPM}}{9550}$

1 Horsepower = 0.746 Kilowatts

Figure 1.33 Horsepower equivalents

CHAPTER 2 .. Basic Hydraulic Principles

As with any technology, the basic language and principles of hydraulics, fluid power, must be understood before advancing to subjects of increasing depth of operation, requirements, component function, design, and troubleshooting and repair. This understanding is also critical to proper safety practices.

This chapter covers the fundamental principles of hydraulic operation and introduces subjects that are referred to in more depth in subsequent chapters. It is absolutely essential that hydraulics users be familiar with these principles in order to properly take advantage of the inherent benefits of hydraulic and electrohydraulic power and motion control.

The efficiency of repair and maintenance personnel is increased exponentially with fundamental knowledge of the fluid power principles that govern the design, operation, useful life, safety and maintenance of fluid power systems. Without this knowledge, operation and repair will take longer, cost more and could be dangerous to machine operators and machine repair personnel.

The depth of information offered by the remainder of this text is enhanced by the understanding of the "how, what and why" presented in this chapter.

HYDRAULIC FLUIDS

Purpose of Hydraulic Fluids

In Chapter 1, we learned that the basic definition of hydraulics is the use of liquids to perform a task. By applying pressure to a liquid, we can amplify forces and transmit forces and power over long distances, over circuitous routes and through unpleasant environments to perform useful work while decreasing man's effort and increasing his productivity.

Many types of liquids are used in hydraulic systems for many reasons, depending on the task and the working environment. Chapter 16 will describe some of the specific characteristics of various fluids used in hydraulic systems, but all perform some basic functions:

> **First**, the fluid is used to transmit forces and power through conduits to an actuator where work can be done.
>
> **Second,** the fluid is a lubricating medium for the hydraulic components used in the circuit.
>
> **Third**, the fluid is a cooling medium, carrying heat away from the "hot spots" in the hydraulic circuit or components and discharging it elsewhere.
>
> And fourth, the fluid seals clearances between the moving parts of components to increase efficiencies and reduce the heat created by excess leakage.

Chapter 2 Basic Hydraulic Principles

Hydraulic Fluid Compressibility

Hydraulic fluids are generally considered incompressible. In fact, they do compress slightly when under pressure, but compressibility is small enough that it is reasonable to ignore under most circumstances. Hydraulic fluids will compress approximately 0.4% of its volume for every 1000 psi pressure.

Specific Gravity (SG)

One of the important criteria of a hydraulic fluid is its weight. As pointed out in Chapter 1, the heavier the fluid, the greater the pressure will be at the bottom of a reservoir. Coincidentally, a heavier fluid requires a greater pressure to push it into the inlet of a pump. The weight of fluids is generally defined by its relationship with the weight of water. This relationship is called "Specific Gravity", and can be expressed as follows:

Formula 2-1

$$\text{Specific gravity of fluid X} = \frac{\text{Weight of fluid x}}{\text{Weight of water}}$$

A typical hydraulic oil weighs between 55 and 58 pounds per cubic foot. Water weighs about 62.4 pounds per cubic foot. The range of specific gravity for typical hydraulic oils is then:

$$SG_1 = \frac{55}{62.4} = 0.88 \qquad SG_2 = \frac{58}{62.4} = 0.93$$

The above values result in the following statement:

The specific gravity of typical hydraulic fluids is between 0.88 and 0.93.

Specific gravity is sometimes stated as a percent figure. This may be a more meaningful way to express these numbers, because we are essentially saying that the weight of typical hydraulic fluids is between 88% and 93% of the weight of water.

Specific Heat (SH)

The specific heat of a substance is the amount of heat required to raise a set amount of the substance a fixed temperature. In customary U.S. units, it is the number of BTU's (British Thermal Units) required to raise the temperature of one pound of the substance one degree Fahrenheit. In metric units, it is the number of calories required to raise the temperature of one gram of the substance one degree Centigrade. Regardless of which system of units is used, the numeric value of the Specific Heat is the same.

For pure water, the value of specific heat is 1, meaning:

One BTU will raise the temperature of one pound of water one degree F.

or

One calorie will raise the temperature of one gram of water one degree C.

Because the specific heat of water is 1, the amount of heat required to raise the temperature of other materials can be related to water in the same way that specific gravity can relate its weight. For example, a typical specific heat of oil is 0.51. This means that it only takes 51% as much heat to raise the temperature of a quantity of oil as it does an equal amount of water.

Chapter 2 Basic Hydraulic Principles

Furthermore, the law of conservation of energy, which states that energy can be neither created nor destroyed, means that the same amount of heat will be given off a substance when it is cooled, as was required to heat it. Therefore, if one pound of water is cooled one degree F, then it will give off one BTU. That is to say, one BTU of heat must be absorbed by something else, such as the surrounding air.

Because the specific heat of oil is approximately half the specific heat of water, then one pound of water will increase temperature one degree F by cooling almost two pounds of oil by one degree F.

HYDRAULIC TERMINOLOGY

Area (A)

Area is a designated surface measurement, and is two dimensional. The area of a surface that measures one inch by one inch is one square inch, or 1 in^2.

It is easy to determine the area of a square or rectangle, in that it is merely the product of two sides. Irregular areas become more complex, but fortunately it is not necessary to determine those areas under normal circumstances.

A common area to be determined in hydraulics is a circle, such as the base of a cylinder or the inside of a hose. Circular areas are made easy for us by the use of the mathematical term "Pi", which is designated by the symbol π and has a value of 3.1416[1]:

Formula 2-2

$$\begin{aligned}\text{Area}_{circle} &= \pi \times r^2 \\ &= \pi \times \frac{D^2}{4} \\ &= D^2 \times 0.7854\end{aligned}$$

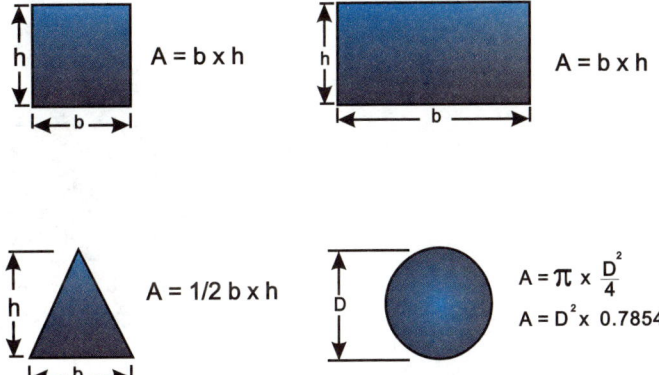

Figure 2.1 Areas of common shapes

Chapter 2 Basic Hydraulic Principles

The Figure 2.1 provides formulas for determining areas of the most common surface shapes:

(1) Pi is a Greek letter denoted by the symbol π, and represents a mathematical relationship between the radius of a circle and the circumference. It is usually defined in two, three or four decimal places; the 8-place value is 3.14159265.

Pressure (psi)

Pressure is a unit force; that is, it is the force applied per unit of area. In the customary U.S. system, it is commonly measured in pounds-per-square-inch (lbs/in^2, or psi), although pounds-per-square-foot (lbs/ft^2) may also be used. In the metric system, bar is the more common pressure designation, equal to approximately 14.5 lbs/in^2, but is not necessarily a universal designation. The international standard for pressure in metric units is the kilopascal, which is equivalent to 1000 Newtons-per-square-meter (N/m^2). 100 kilopascals is equal to one bar, or 14.5 lbs/in^2.

Note that one bar (14.5 psi) is very close to one standard atmosphere (14.7 psi) of pressure, but is not equal. Actually, the origination of bar (a contraction of barye) as a unit of pressure had nothing to do with atmospheric (or barometric) pressure. It is approximately equal to one kilogram per square centimeter.

In hydraulics, pressure is the unit force applied to or by a fluid. It is exerted in all directions in the fluid, and impinges at right angles (90 degrees) on any area it comes in contact with. An object immersed in a fluid will be subjected to the force on all surfaces that the fluid touches, as will all parts of the fluid container (see Figure 2.2).

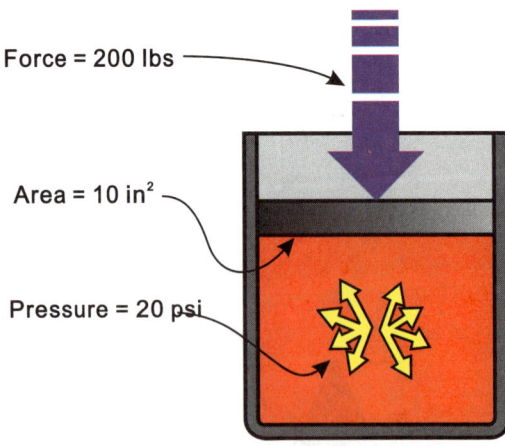

Figure 2.2 Pressure is exerted in all directions by a fluid

Pressure can be calculated by dividing the force applied by the area that the force is applied against (Figure 2.3):

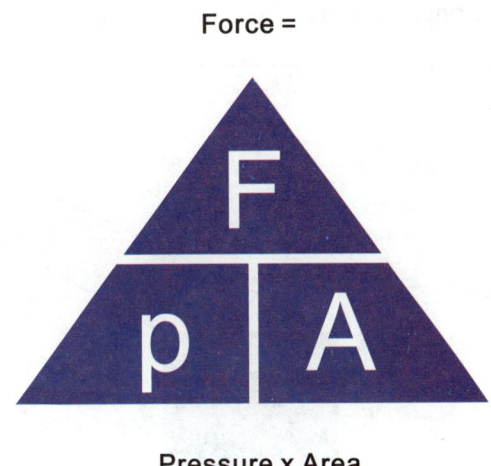

Figure 2.3 Force / pressure / area relationship

Formula 2-3

Pressure = Force / Area

If the force is in pounds, and the area against which the force is being applied is measured in square inches, the resulting pressure is in lbs/in^2, and designated PSI.

Volume (V)

Volume is three dimensional, and refers to the total quantity of a substance or object. In hydraulics, volume is generally expressed in cubic inches (in^3) in customary U.S. units or in cubic centimeters (cm^3) in metric units. Cubic feet, cubic yards or cubic meters are not uncommon terms.

Probably the most common U.S. unit of volume when referring to fluids is "gallon", which consists of 231 cubic inches (231 in^3). The equivalent metric unit would be "liter", which consists of 1000 cubic centimeters (1000 cm^3).

Several common volume relationships are shown in the following table:

Volume		Gallon	in^3	liter	cm^3
1 Gallon	=	1	231	3.785	3785.0
1 in^3	=	0.0043	1	0.0164	16.387
1 Liter	=	0.264	61.02	1	1000
1 cm^3	=	0.0003	0.061	.001	1

Table 2-1. Volume relationships

The determination of the volume of a cube or a rectangular body is the base dimension times the height dimension times the length (b x h x L, Figure 2.4). Keeping in mind that the area of one end is the base dimension times the height dimension (b x h), then the volume of the rectangular figure becomes simply the area of one end times the length (Figure 2.4).

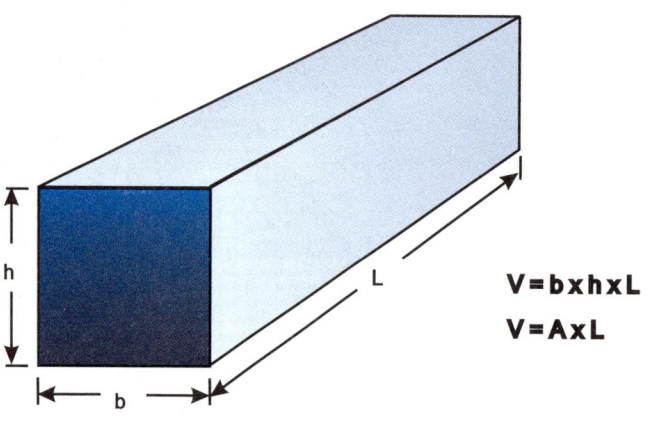

Figure 2.4 Volume of a rectangular shape

This is particularly convenient when calculating the volume of a cylindrical object, such as a hydraulic cylinder. Knowing the bore of the cylinder, one can easily calculate the cross sectional area from formula 2-3 (shown in Figure 2.1). The bore area times the effective length of the cylinder (in this case, stroke) gives the volume of fluid that the cylinder can contain when full (see Figure 2.5).

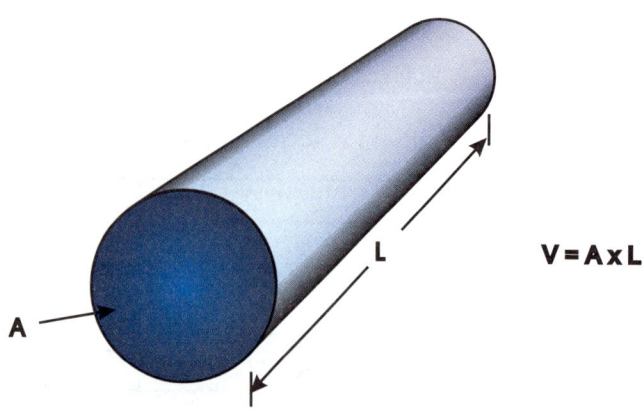

Figure 2.5 Volume of a cylindrical shape

Chapter 2 Basic Hydraulic Principles

Velocity (V)   Velocity is another word for speed; it is the speed at which something is moving in a defined direction. In hydraulics, we generally speak of the velocity of fluid in a hose, tube or through a component such as a valve.

In customary U.S. units, velocity is usually defined in feet-per-second (ft/sec). It is frequently defined in feet-per-minute (ft/min), inches-per-minute (in/min) or inches-per-second (in/sec) as well. In metric units, velocity may be defined in meters-per-second, meters per minute or centimeters-per-second (m/sec, m/min, cm/sec).

Flow (Q)   Flow is simply the movement of a quantity of fluid during a period of time. Fluids are confined in hydraulics, such as in hoses, tubes, reservoirs and components, so flow is the movement of a fluid through these confining elements.

In customary U.S. units, flow is designated by the letter "Q", and is usually expressed in gallons-per-minute, or GPM. It is also common, however, to express flow in cubic-inches-per-minute (in^3/min) or per-second (in^3/sec). In metric units, flow is expressed in liters-per-minute, or LPM, but may also be expressed in cubic-centimeters-per-minute (cm^3/min) or per-second (cm^3/sec).

Flow is basically the velocity of a quantity of fluid past a given point. To visualize this, consider a cross-sectional area of fluid inside a tube, as pictured in Figure 2.6. If this cross-sectional "slice" of fluid moved at the rate of one foot in one second, then it would "push" one foot of fluid ahead of it every second. The volume of that fluid is the cross sectional area times the length (per formula 2--4). The time, in this case, is one second. This gives rise to the basic formula for flow in hydraulics:

Formula 2–4

Flow = Area x Velocity, or $Q = A \: x \: v$

In using the above formula, one must be careful to use the correct dimensions (or units) so that they equal on both sides of the equation. For example, if the area is in square inches (in^2), then the velocity must be in inches-per-second (in/sec) or inches-per-minute (in/min). The flow will then be cubic-inches-per-second (in^3/sec) or cubic-inches-per-minute (in^3/min).

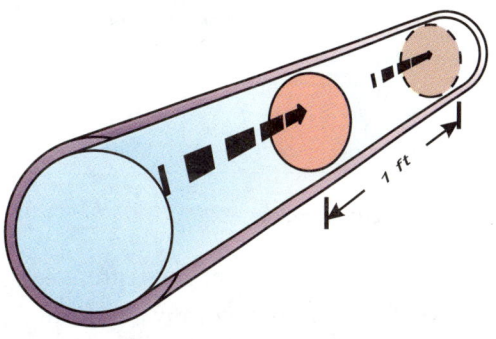

Figure 2.6 Fluid flow is the velocity of a cross sectional area (v x A)

Chapter 2 Basic Hydraulic Principles

The following table will help to convert dimensions from one format to another:

This Unit	Multiplied By	Divided By	Will Equal This Unit
gallons	231		in^3
in^3		231	gallons
Liters	1000		cm^3
cm^3		1000	Liters
gpm	231		in^3/min
lpm	1000		cm^3/min
in^3	16.387		cm^3
cm^3		16.387	in^3
in^2	6.452		cm^2
cm^2		6.452	in^2
in^3/sec	60		in^3/min
in^3/min		60	in^3/sec
seconds		60	minutes
minutes	60		seconds
ft^3	1728		in^3

Table 2.2 Conversions of common measures

Laminar Flow, Turbulent Flow

We would like to think of flow in a hydraulic system as a smooth transition of fluid from one point to another; all particles of the fluid would be moving parallel to all other particles, and there would be no turmoil within the fluid. This we would call laminar flow (Figure 2.7), and it is very desirable.

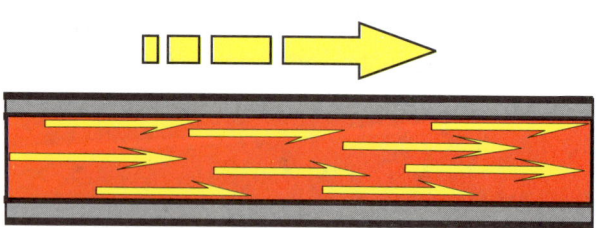

Figure 2.7 Laminar flow

In fact, hydraulic system flow often experiences more turmoil than we would like. Although the fluid generally moves in the direction we would like, it also travels through small conduits, across sharp-edged restrictions, through small orifices, around sharp bends, in fact, through all the places that have a tendency to cause anything but a nice, smooth transition.

Particles of the fluid are traveling helter-skelter among each other (see Figure 2.8), causing friction and inefficient movement. This type of flow, called turbulent flow, is undesirable and wasteful. Unfortunately, the economic and practical aspects of mobile fluid power result in most flow being in the turbulent variety.

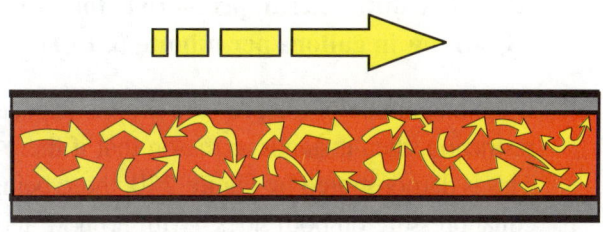

Figure 2.8 Turbulent flow

Pressure Drop

When fluid flows across an orifice, as in Figure 2.9, it loses some of its energy. This is reflected in a lower pressure at the downstream side of the orifice, as illustrated by the two gauges. The difference between the upstream and downstream pressure is called a pressure drop; it is the drop in pressure caused by the flow and the restriction (orifice).

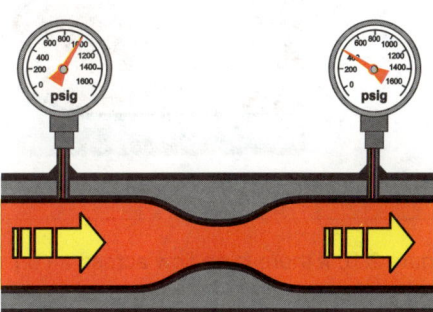

Figure 2.9 Flow past an orifice creates a pressure drop

The magnitude of the pressure drop will vary, depending upon:

- The rate of flow passing across the orifice
- The size of the orifice
- The ease with which the fluid will flow (viscosity).

The downstream flow must be the same as the upstream flow in Figure 2.9, because there is nowhere for the fluid to escape. However, if the pressure in the fluid is lower, then the energy in the fluid is less. Because, as we learned earlier, energy can not be destroyed, the difference in energy must be given off in the form of heat.

The amount of heat is predictable by the following formula:

Formula 2-5

BTU / hr = ΔP x Q x 1.485

Where

ΔP = The difference in pressure before and after the orifice
Q= Flow in gallons per minute (GPM)

If the magnitude of the pressure drop is dependent on the amount of flow passing the restriction, then it stands to reason that if there is no flow, there will be no pressure drop. This is demonstrated by Figure 2.10; there being no flow across the orifice will result in equal pressure on both sides. With no flow and no pressure drop, there will be no heat rejected due to a drop in energy, as demonstrated by formula 2-5.

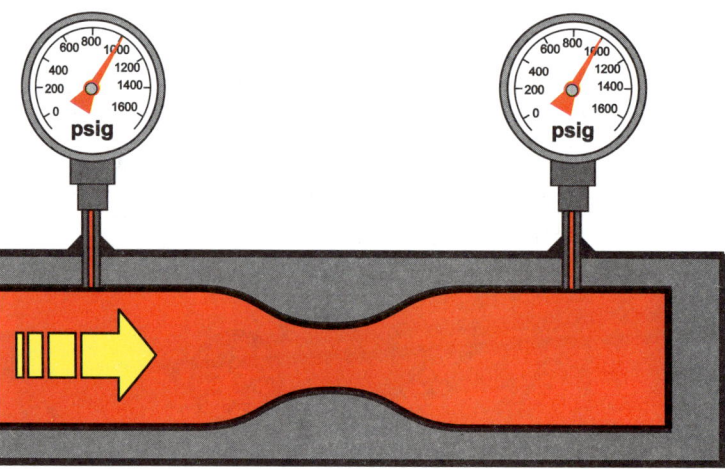

Figure 2.10 If there is no flow across an orifice, there is no pressure drop

This direct relationship between flow and pressure drop is an important consideration in hydraulics; if there is no flow between point A and point B, there will be no pressure drop. Conversely, if there is no difference in pressure between points A and B, there is no fluid flow between these two points.

Energy

Energy was defined in Chapter 1 as the capacity to do work, and that it cannot be created or destroyed. It can, as shown in the previous paragraphs, be converted from one form to another, or it can be converted into work. Under all circumstances, the amount of energy is always the same, and the amount of work performed by it is equal, provided we account for any "losses" that take place in the process (such as the "loss" of heat given off by the process).

Daniel Bernoulli (1700-1782), a Dutch-born Swiss mathematical physicist, defined two types of energy that occur in a hydraulic system; potential energy and kinetic energy.

Potential energy is the energy resulting from position or configuration. Kinetic energy is the energy resulting from motion.

A fluid in a hydraulic system can contain either or both types of energy. Fluid that has been raised to an elevated position, or is pressurized, contains potential energy. Fluid that is moving contains kinetic energy. Either or both can be released in the form of work, in which case the energy in the fluid will decrease in exact proportion to the amount of work done.

Bernoulli further proved that the sum of the kinetic and potential energy in a moving, incompressible fluid will always remain the same. That is to say, in a fluid moving in a steady state, if the potential energy increases or decreases, the kinetic energy will increase or decrease, such that the sum of the two will always remain the same. This deduction is known as Bernoulli's Law, and was first published in 1738.

Bernoulli's Law can be demonstrated in a flow path as shown in Figure 2.11. As the cross section of the tube changes, the velocity of the fluid changes in inverse proportion (see formula 2-4). As the velocity of the fluid is reduced in the larger center area of the tube, the kinetic energy decreases. The potential energy, in the form of pressure within the fluid, will increase. As the fluid continues to flow to the smaller area of the tube, the fluid velocity (kinetic energy) increases to its original level, and the pressure (potential energy) decreases.

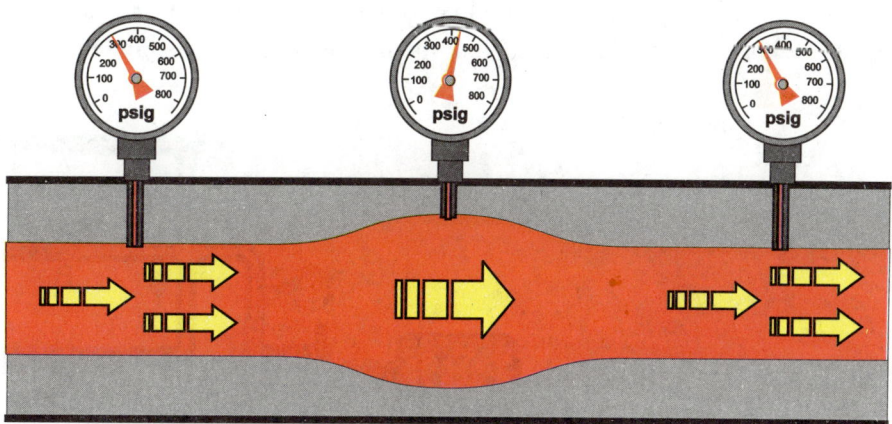

Figure 2.11 A demonstration of Bernoulli's principal

Bernoulli's principle must be taken into account in the design of hydraulic systems and components, because of the imbalance of forces that can result from sudden changes in fluid velocity.

Chapter 2 Basic Hydraulic Principles

An example is the flow of oil across a narrow metering area of a control valve as shown in Figure 2.12. Forces can develop that make the valve behave erratically because of the sudden changes in velocity and pressure during the transition through the metering area.

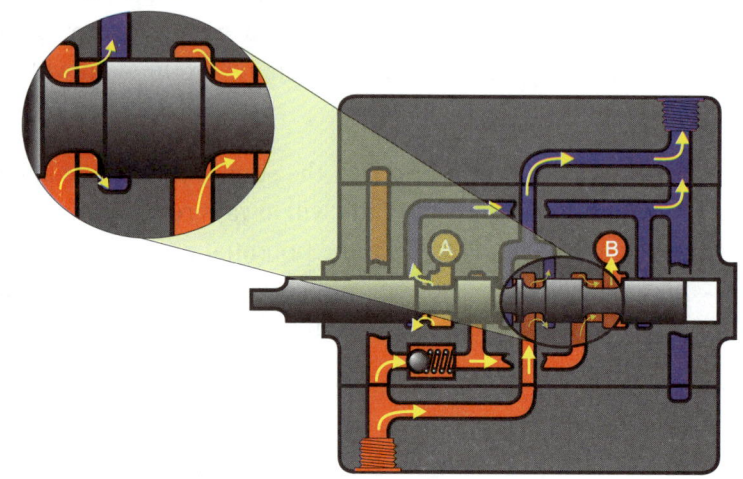

Figure 2.12 Fluid velocity changes can cause an imbalance of forces

Series and Parallel Circuitry

The pump flow in Figure 2.13 is equally available to three branches of the circuit, branches A, B and C. Branch A is unrestricted, and the flow is free to proceed directly to the reservoir at a very low pressure drop. Branch B has a restriction (orifice) in it, but the fluid is otherwise free to flow to the reservoir. Branch C has greater restriction than B (smaller orifice), but the fluid is otherwise also free to flow to the reservoir.

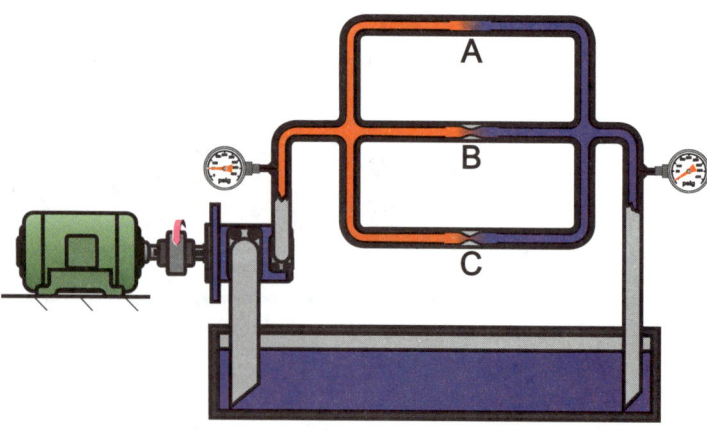

Figure 2.13 Parallel circuitry, where flow takes the path of least resistance

In this type of simplified circuit, it is easy to see that the majority of the pump flow will take path A to the reservoir, merely because it is the easier flow path to take. Some flow will move across B, and even less will move across C because of the restrictions. The pressure drop across the circuit is the pressure required to move the fluid through path A, the easiest course.

This simplified example illustrates a basic concept of hydraulics:

In a parallel circuit, flow will take the path of least resistance, and the circuit pressure is the pressure required to force the fluid through this path.

If the flow in path A were to be blocked, then pressure in the circuit would rise, because a greater pressure drop would be required to force the fluid across the restriction in branch B. Again, the majority of the flow would take the easiest path, in this case branch B, because the restriction size is larger than in C. Furthermore, the pressure in the circuit would be that required to force the fluid across branch B.

An alternative circuit is shown in Figure 2.14. Here, all fluid must first travel through each of the restrictions before flowing to the reservoir. This is a simplified representation of a series circuit. Flow does not have a choice, but must travel through each restriction and overcome the pressure drop of each one independently.

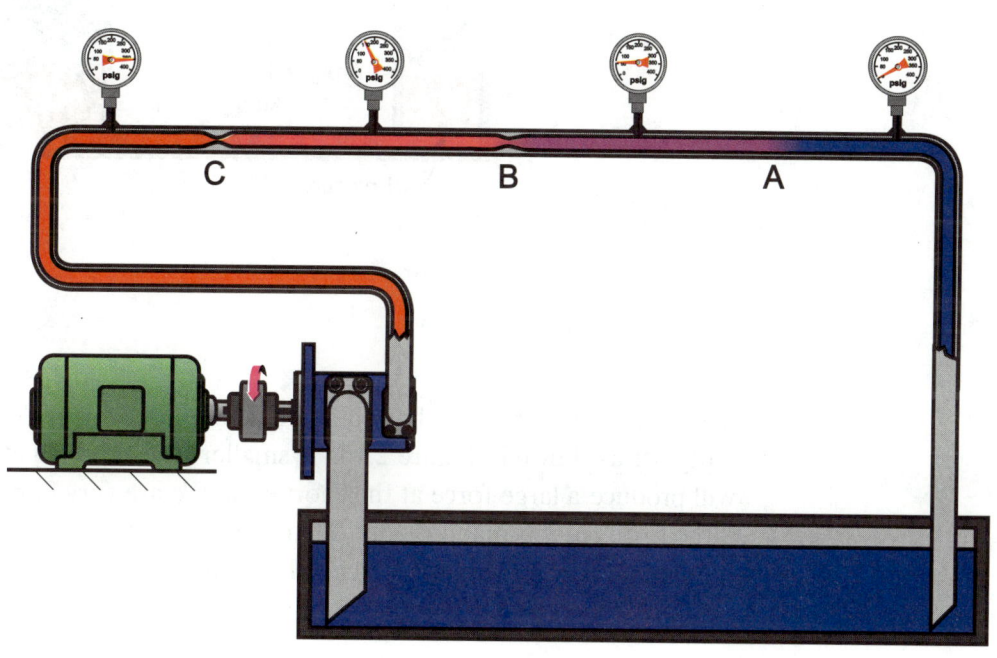

Figure 2.14 Series circuitry, where all pressure drops are additive

In this example, the pressure drop across area A, where there is no restriction, is about 50 psi. This means that pressure at the outlet of orifice B must be great enough to overcome the pressure drop of A.

The pressure drop required to force the fluid across orifice B is about 100 psi. Therefore, to have a 100 psi pressure differential between inlet and outlet of restriction B requires an input pressure of 150 psi, because the outlet pressure is 50 psi. The same concept is applied to restriction C, which requires a 150 psi pressure drop to force flow through it. This results in an inlet pressure at orifice C of 300 psi.

Chapter 2 Basic Hydraulic Principles

This example illustrates another basic concept of hydraulics:

In a series circuit, all flow must pass through all branches of the circuit, and the pressure drops across each element of the circuit are additive.

Hydraulic circuits are either parallel, series or some combination of parallel and series.

Hydraulic Leverage

It was illustrated in Chapter 1 that a significant advantage of hydraulics is the ability to multiply a force. A small force applied against a small area can result in a large force when the pressure is applied against a large area. This is demonstrated in Figure 2.15.

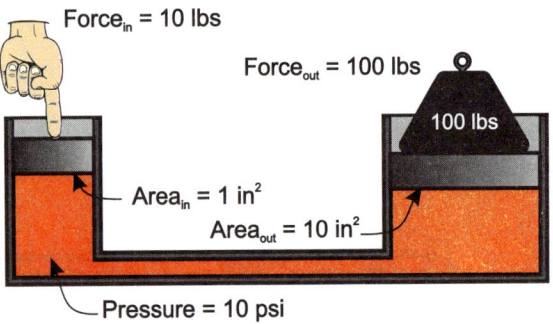

Figure 2.15 Hydraulic leverage

The effect is the same as that obtained through mechanical advantage, such as with a pry bar as shown in Figure 2.16. A smaller force applied at the long end of the bar will produce a large force at the short end. We are very familiar with producing large forces in this manner, and we are very accustomed to adjusting the lengths of the bar in order to obtain greater or lesser forces.

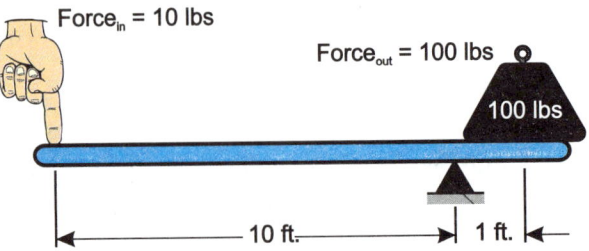

Figure 2.16 Mechanical leverage

What we can readily see in the pry bar, though, is that this force multiplication does not come free. What we gain in "leverage", or force, we lose in distance traveled. We know from experience that it is necessary to move the long end of the bar through a full arc, in order to obtain a short lift at the short end of the bar as demonstrated in figure 2.17. This "gain" and "loss", or advantage and disadvantage, is very predictable by the following relationship:

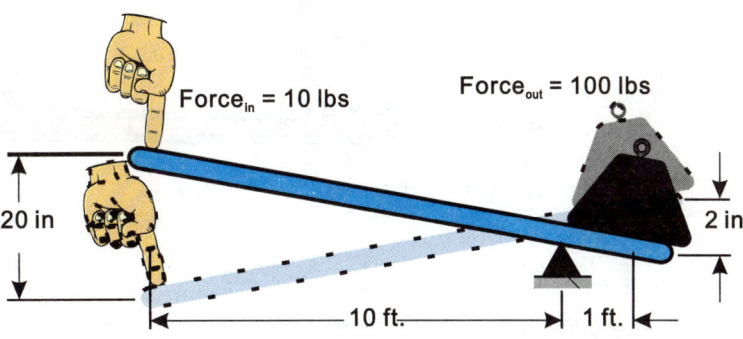

Figure 2.17 What is gained in force is lost in distance travelled

$$Force_{out} = Force_{in} \times \frac{Long\ Lever\ Arm}{Short\ Lever\ Arm}$$

and:

$$Distance_{out} = Distance_{in} \times \frac{Short\ Lever\ Arm}{Long\ Lever\ Arm}$$

Furthermore, knowing that the work performed is equal to the force times the distance, we find that the work at both ends of the lever is the same:

$Force_{in} \times Distance_{in} = Force_{out} \times Distance_{out}$

Chapter 2 Basic Hydraulic Principles

The hydraulic relationship is analogous to the mechanical relationship, except that the area ratio becomes the "lever arm". In the example of Figure 2.18, we can see that the above relationship of force and distance input equals the force and distance output, and the "gain" and "loss" is determined by the area ratio:

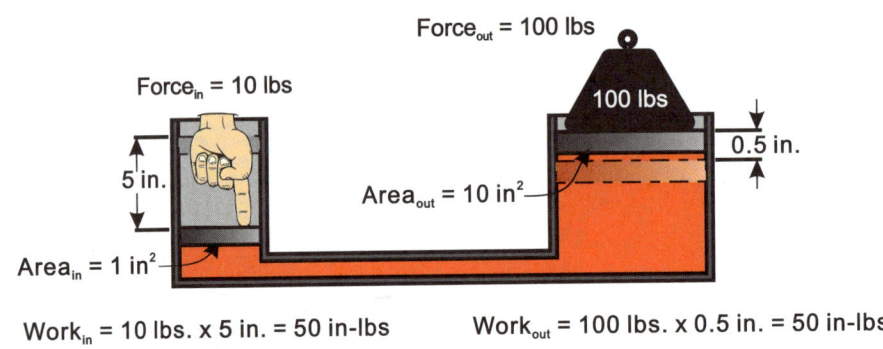

Figure 2.18 Hydraulic leverage gains in force but loses in distance travelled

$$\text{Force}_{out} = \text{Force}_{in} \times \frac{\text{Area}_{out}}{\text{Area}_{in}}$$

and:

$$\text{Distance}_{out} = \text{Distance}_{in} \times \frac{\text{Area}_{in}}{\text{Area}_{out}}$$

Because the length of time for the movement is the same on both ends of either the mechanical or hydraulic lever, we find that the power input equals the power output.

Efficiency

The above examples illustrate that the work input to a system, and the power input to a system, are equal to the work and power output of the system. This will only be true if there are no "losses" to the system between the input and the output. The practical matter is that there are always some "losses" that occur between the input and output, such that the work out and the power out will always be less than the work and power input. The difference between these two is usually expressed in terms of efficiency, which is a measure of the output relative to the input.

Efficiency is expressed as a percentage (%), and is determined by relating the input and output as follows:

$$\text{Efficiency} = \frac{\text{Output}}{\text{Input}} \times 100$$

Various forms of efficiency will be utilized throughout this manual, but in all cases, it will relate the system output to input as shown above.

Chapter 2 Basic Hydraulic Principles

Open Circuit / Closed Circuit Systems

The most common type of hydraulic system in use today is the open circuit system, shown in a simplified illustration in Figure 2.19. As shown, the hydraulic fluid is provided to a pump from a reservoir, and from the pump to an actuator. The appropriate valving to condition and direct the flow is not illustrated in this example.

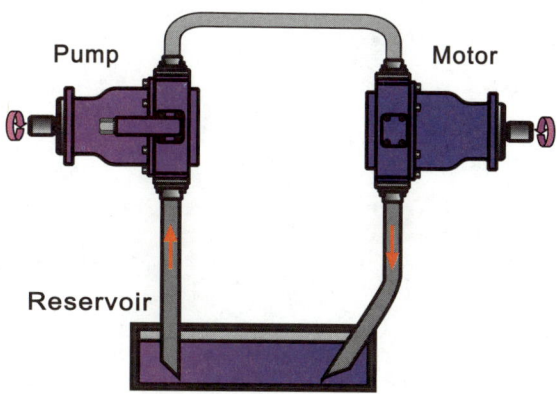

Figure 2.19 Open circuit concept

The actuator utilizes the fluid to perform a task, and then returns the fluid to the reservoir. Because most or all the fluid returns to this common source at the end of its work cycle, the system is designated an open circuit.

An alternative, and very common, type of system is the closed circuit as illustrated in Figure 2.20. In this system, also designated a hydrostatic transmission circuit, the fluid is returned from the actuator to the pump, where it is delivered back to the actuator again. Because the fluid does not pass through the "open" reservoir during each work cycle, the system is referred to as closed.

Again, the valving necessary for a closed circuit system to operate properly is not included in Figure 2.20 for illustration purposes. In practice, a small reservoir is needed, as well as relief and replenishing valves, to complete the circuit. These elements and a complete discussion of closed circuit systems will be provided in chapter 13.

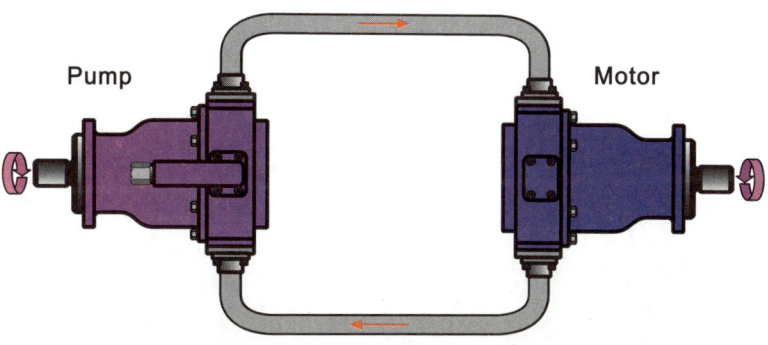

Figure 2.20 Closed circuit concept

Chapter 2 Basic Hydraulic Principles

CHAPTER 3 .. Basic Electrical Principles

The need for hydraulics users, designers and maintenance personnel to understand the basic concepts of electricity and electrical control of fluid power has increased incredibly over the last two decades. Increasing sophistication and capabilities in electric components and the combination of electric control and fluid power products have required OEMs, users and those responsible for maintenance of fluid power machines to learn the basic elements of all technologies involved.

This chapter deals with the basic concepts of electricity and function of electricity as it relates to power transmission and signal manipulation particularly with use in hydraulics. Basic electrical components and their theory of operation, as well as application, are covered to familiarize readers with their requirements, attributes and typical applications.

From Ohm's Law to solenoids, torque motors, and various sensing devices to amplifiers and open vs. closed loop control are discussed in easy to understand and relate terms and examples. Analogies between hydraulic and electric principles and components are drawn to illustrate the commonality as well as the critical differences between these two technologies. As with the previous chapter, basic hydraulic principles, much of the material covered in this chapter is referenced in subsequent chapters and dealt with in more depth.

Electrohydraulics

The use of hydraulics on modern mobile machinery more and more includes electrical activation and control. There are significant advantages that make the electrical transmission of power and signals attractive in the often confined spaces of today's mobile equipment:

- The conduit for electrical signals and commands can be relatively small compared to hydraulic hoses, tubes and fittings, taking up less space and being easier to route.
- The transmission of electrical signals is very fast, and is less subject to environmental temperature variations that can affect the speed of hydraulic signal transmission.
- Although hydraulic fluids are nearly incompressible, the small degree of compression can affect the speed and accuracy of feedback signals and the frequency of response of highly sophisticated control circuits. This is of no concern with electricity.
- Miniaturization of electrical components allows very complex processing of signals to be performed in a very small space.
- Some command and control signals can be transmitted through the air, eliminating the need for conduit.

Chapter 3 Basic Electrical Principles

Other considerations in the management of power on mobile equipment makes the use of hydraulics more advantageous:

- There is a limit to the amount of electrical power that can be economically transmitted throughout machinery due to the size of the conduit and components required at higher power levels.
- The ability to electrically insulate a control point from the main machine structure becomes very expensive and limited in power.
- Electrical signals can sometimes be affected by airborne electrical transmissions.
- The amount of force and power available through electrical actuators is limited.

It is a natural conclusion to combine the attributes of electricity and hydraulics in order to have the signal management capability of electricity and the force and power transmission capability of hydraulics. The Vickers term for this combination is Electrohydraulics.

Because of the greater use of electrohydraulics on mobile machinery, a manual addressing mobile hydraulics must include basic electricity and electrical control components.

Hydraulic / Electric Analogies

It is often helpful for the individual who understands hydraulic principles, or for the individual who understands electrical principles, to appreciate analogies between the two technologies. Figure 3.1 shows a simplified hydraulic circuit and its electrical counterpart. The function of each of the paired components is very similar.

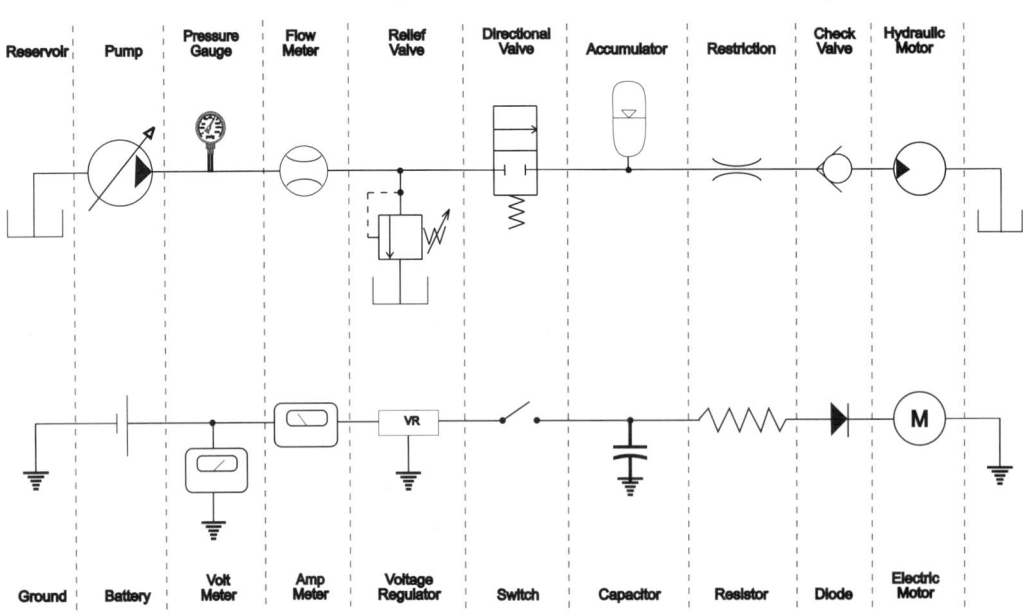

Figure 3.1 Hydraulic / Electric analogy

The important concepts illustrated are that flow, pressure and restrictions in a hydraulic circuit are analogous to current, voltage and resistance in an electrical circuit.

Pressure is the "driver" of fluid in a hydraulic circuit and causes flow to occur. Voltage is the "driver" in an electrical circuit, and causes current to flow.

Flow is the movement of fluid in a hydraulic circuit; current is the movement of electrons in an electrical circuit. Flow is measured in volume per unit of time (such as gallons-per-minute), and current is measured in electrons per unit of time. 6.23×10^{18} electrons flowing past a point equals one coulomb, and one coulomb-per-second is equal to one ampere, the measure of current flow.

Restrictions in a hydraulic circuit are a resistance to the flow of fluid. The greater the restriction (smaller orifice), the greater the pressure drop across it. A resistor in an electrical circuit resists the current flow. The higher the resistance, the greater the voltage drop across it.

BASIC ELECTRICAL CONCEPTS

Law of Electric Charges

All materials are made up of atoms. Atoms consist of electrons, protons and neutrons. Electrons are electrically charged particles within the atom. In certain kinds of material, some of the electrons are free to move around, and travel from atom to atom. These are called "free" electrons, and the material is called a conductor. Metals fall into this category.

In other materials, the electrons are not free, and remain attached to the atom to which they belong. These materials are called insulators, or dielectrics. Rubber is an example of this kind of material.

When a conductor is in a neutral state, the electrons are free to move around at random. When an electrically charged object is brought close to, or in contact with, the conductor, the electrons move in accordance with a defined rule developed by the French physicist Charles A. de Coulomb (1736-1806) in 1785:

Like charges repel each other and unlike charges attract each other.

Coulomb first discovered that there were two types of electric charges, which he called "positive" and "negative". Experimentally, he found that when two spheres of like charge were brought together, they would repel each other (Figure 3.2). Further, he found that two spheres of opposite charge would attract each other. Through much experimentation, he established a force relationship between the intensity of the charge on the spheres and the repelling or attracting forces.

Figure 3.2 Like charges repel, unlike charges attract

Chapter 3 Basic Electrical Principles

Figure 3.3 shows the lines of electric intensity between spheres charged equally positive, equally negative and equally opposite. The direction of the lines is also illustrated, showing that electric intensity flows from the positive charge to the negative charge. The amount of the charge is measured in coulombs, which is given as 6.23×10^{18} electrons.

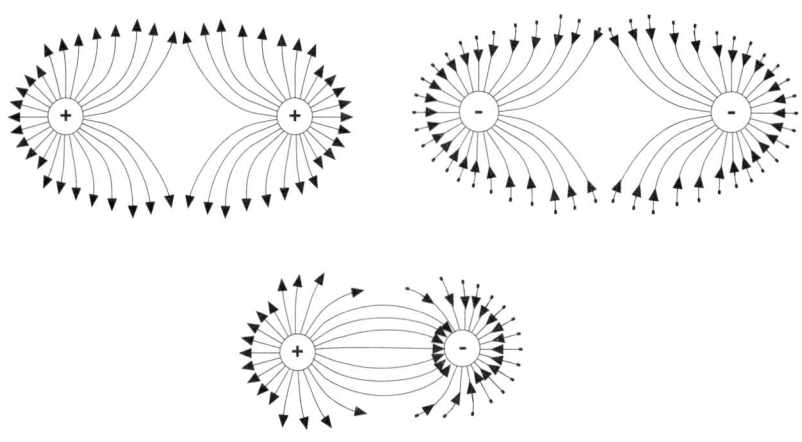

Figure 3.3 Lines of electric intensity

Current

The direction and quantity of electric charges can be controlled through a conductor, much the same as the direction and quantity of hydraulic fluid is controlled through a hose or tube. The resulting flow of charges is electric current, and is measured by the number of electrons that pass a given point during a given period of time.

The standard unit of measurement for current is the ampere, named after the French physicist Andre' Ampe're (1775-1836). It represents the flow of one coulomb of electrons past a given point in one second.

The instrument used to measure current is called an ammeter.

Electric Potential (EMF)

In order for electrons to be moved between two points, along a wire for example, a certain amount of work must be done on them. This would be equivalent to applying a force on a fluid to move it from one point to another through a tube. The force applied to the fluid results in pressure, and it must be higher at one point in the tube than at another in order for the fluid to move. The force on the fluid times the distance the fluid moves results in work done.

A force is applied to electrons in a conductor as well, and the force must be higher on one end of the conductor than on the other. The electrons will always move from a point of high force, called potential, to a point of lower force. The force applied, over a distance, is the work performed on the electrons.

Because the study of electricity was mostly done in the metric system of measurements, the units of force and distance are the Newton and the meter. One Newton of force over a distance of one meter results in one Newton-meter of work, which is also called a joule. One joule applied to one coulomb is called a volt, after Count Alessandro Volta (1745-1847), inventor of the voltaic cell (battery).

Thus, a voltage difference between two points along a conductor, is referred to as the potential difference, or the electromotive force (EMF). The potential difference between the two points is measured in volts, and is commonly called voltage drop or voltage.

The instrument used to measure voltage is a voltmeter.

Resistance

The relationship between voltage and current was developed by Georg Simon Ohm (1787-1854) in 1826. Ohm determined that if voltage in a circuit was continuously maintained by a source of EMF, then the current would be continuous. This resulted in Ohm's Law, which states:

For a given conductor at a given temperature, the current is directly proportional to the difference of potential between the ends of the conductor.

Resistance is the opposition to current. Just as an orifice resists the flow of fluid in a hydraulic system, resistance resists the flow of electrons in an electrical circuit. It is measured in ohms, and is designated by the letter R. The symbol for ohms is the Greek letter Omega (Ω).

One ohm is defined as the resistance that exists if it requires a potential of one volt to produce a current of one amp in a circuit. Thus, the relationship developed by Ohm to express his Ohm's Law is:

Formula 3-1

$$R = \frac{V}{I}$$

Perhaps a more useful way of expressing this relationship is:

Formula 3-2

$$V = I \times R$$

This expression states that it will take a voltage value of V to force a current I through a resistance with a value of R. This is merely a different way of expressing Ohm's Law.

All circuits have some resistance, even if it is only the resistance of the wire used as the conductor. A light bulb, for example, is a very fine piece of wire, called a filament, that has a high resistance. The high resistance causes the filament to heat up and glow, providing the light.

Most circuits are more complex, having many types and values of resistances throughout. The circuit characteristics of voltage, current and amperage can still be determined by the formula 3-2, however, because all the resistances can be mathematically combined into one value for calculation purposes.

Resistances aligned in *series* require that the current must pass through all of them in sequence, much as hydraulic fluid must pass through all orifices aligned in series (Chapter 2, Figure 2.14). In the hydraulic circuit, we learned that the pressure required to force the fluid through all of the orifices was the sum of the pressure required for each one individually. The electric requirement is analogous; the upstream voltage must be sufficient to push current through each resistor individually, and the total voltage is the sum of the individual voltage drops (Figure 3.4). Therefore, the resistance in formula 3-2 would be the sum of all the resistance in series:

Resistance In Series

Formula 3-3

$$R_t = R_1 + R_2 + R_3 + ... R_n$$

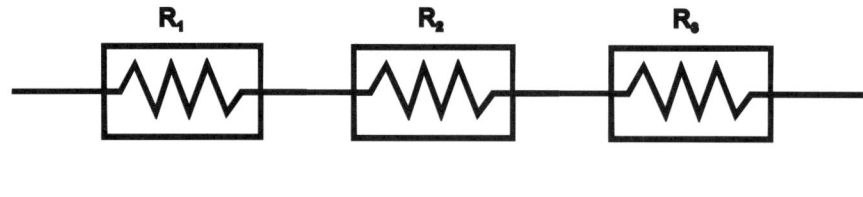

Figure 3.4 Series resistance

Resistances that are in a *parallel* configuration, however, present a different situation, but still analogous to the parallel hydraulic circuit (Chapter 2, Figure 2.13). Current, just like fluid flow, will seek the path of least resistance through the circuit. More current will travel through the low resistance branch, and the least current will travel through the highest resistance branch (Figure 3.5).

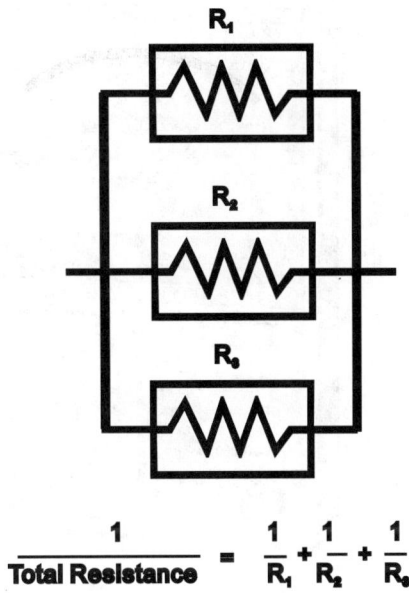

$$\frac{1}{\text{Total Resistance}} = \frac{1}{R_1} + \frac{1}{R_2} + \frac{1}{R_3}$$

Figure 3.5 Parallel resistance

A parallel group of resistances will have a single resistance equivalent for purposes of determining circuit characteristics, and this can be determined by the following formula:

Resistance In Parallel

Formula 3-4

$$\frac{1}{R_t} = \frac{1}{R_1} + \frac{1}{R_2} + \frac{1}{R_3} + \ldots \frac{1}{R_n}$$

By combining the application of formulas 3-3 and 3-4 throughout a circuit, the total resistance can be determined. The remaining circuit characteristics of voltage and current can then be determined by formula 3-2.

The instrument used to measure resistance is called an ohmmeter.

Chapter 3 Basic Electrical Principles

Figure 3.6 shows combination instruments, most common in electrical work, that measure each of the three basic characteristics of circuits; voltage, current and resistance. It is called a Volt-Ohm-Milliamp meter (VOM), or a Multi-meter, and is available with either an analog or digital readout.

 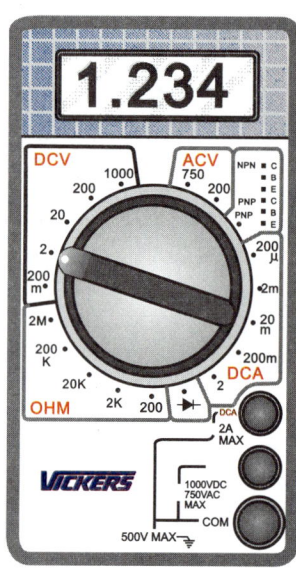

Figure 3.6 Analog and digital multi-meters

Power

Power was described in Chapter 1 as the time-rate of doing work. The unit of work in the metric system of measurements (the system in which most electrical relationships are defined) is the joule, equal to a force of one Newton over a distance of one meter. Performing one joule of work in one second is defined as a Watt (W). Since a Watt is the time-rate of doing work, it is a measure of power.

As shown earlier in this chapter, the unit of work in electricity is also a joule, and one joule applied to one coulomb is the definition of one volt. It was also shown that one coulomb passing a given point in one second, is defined as one ampere. Because these terms contain the elements of work and time in electricity, the following formula results:

Power$_{Watts}$ = Volts x Amperes

Formula 3-5

W = V x I

Because power in electricity is directly related to power in mechanics or hydraulics through the force-distance-time relationship, there is a direct correlation between them:

Formula 3-6

1 Horsepower = 746 Watts

We find from this that the watt is a rather small unit to express power, particularly in mobile hydraulic circuits associated with multiple-horsepower engines. Therefore, the kilowatt (KW) is generally used to express electrical power, which is equal to 1000 watts. Formula 3-6 then becomes:

Formula 3-7

1 HP = 0.746 KW

In the metric system, power is expressed in watts or kilowatts. In customary U.S. units of measurement, power is expressed in Horsepower. Formulas 3-6 and 3-7 relate the two systems.

Direct and Alternating Current

Throughout this chapter, the type of current that has been discussed flows in one direction through a conductor. This is called *Direct current* (DC), and results from a constant EMF source such as a battery. Direct current is the type most commonly used on mobile machinery, because the source is usually a battery.

The most common type of current in the U.S. is *alternating current* (AC), one that alternates its direction at a fixed frequency. Alternating current is popular for two reasons:

- A generator naturally produces alternating current as the armature constantly passes the coils; each pass first produces increasing positive current, and then decreasing positive current to increasing negative current, and back again. Without some form of commutator or diode arrangement, alternating current would be the output of all generators.

- Transformers can be used with alternating current to increase or decrease the magnitude of voltage or current without the use of moving parts. There is no DC counterpart to the transformer.

Voltage will parallel the alternating current in a circuit, just as it will parallel a direct current. The wave form and frequency will be the same.

Current and voltage is produced in a sinusoidal wave pattern (also called a "sine wave") as shown in Figure 3.7. This figure also illustrates the positive and negative nature of the pattern, and compares it to a direct current pattern as would be generated by a battery.

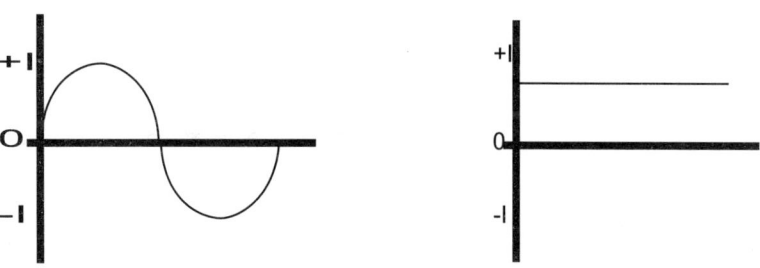

Figure 3.7 AC and DC current

Chapter 3 Basic Electrical Principles

Electromagnetics In mobile machinery, electricity is used to cause lights to light up, motors to turn or solenoids to operate. As mentioned earlier, electricity will create light by heating up a high resistance filament until it glows. Motors and solenoids use a different principle, called electromagnetics.

DC Solenoids Current flowing through a conductor creates a magnetic field around the conductor. This field is usually of little significance to us because it is very weak and not easily detected. The field is made up of magnetic lines, called flux lines, that circle the conductor, as shown in Figure 3.8. The direction of these flux lines is based on the direction of the current in the conductor.

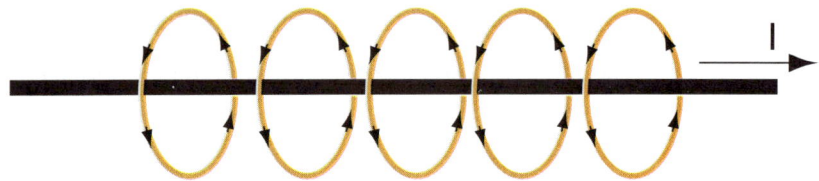

Figure 3.8 Flux lines around a conductor

The direction the flux lines move is based on the "Right Hand Rule", meaning if the thumb of the right hand is facing the direction of the current flow in a conductor, the fingers will be facing the direction the flux lines are moving around the conductor (see Figure 3.9).

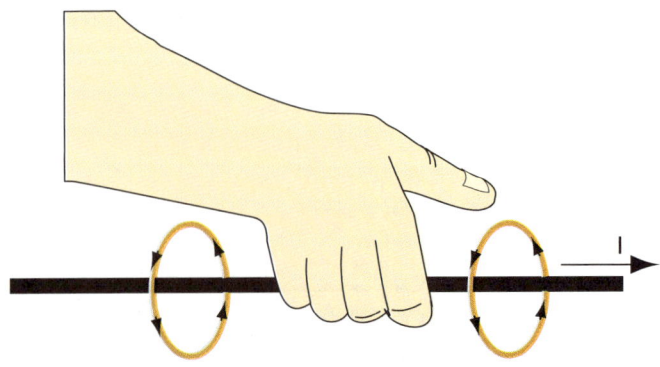

Figure 3.9 Right hand rule

Although the effect of flux lines near a single conducting wire is individually weak, it can be enhanced by coiling the wire. If a conducting wire is wound into a circle, the magnetic property becomes greater, as illustrated in Figure 3.10. By circling the wire several times into a coil, each circle is additive to the next so that the resulting magnetic force can be substantial.

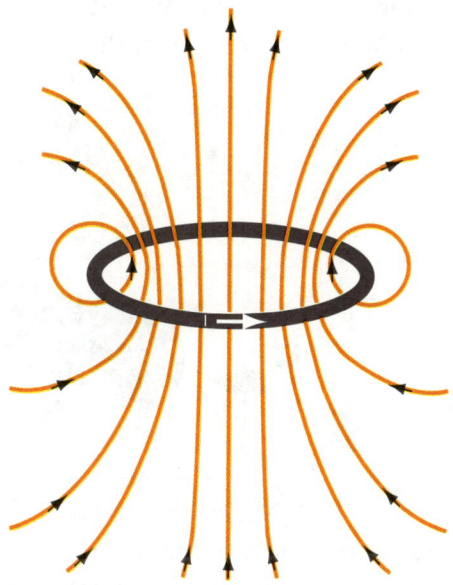

Figure 3.10 Flux lines magnify around a wire coil

Furthermore, if the wire is wound around a soft iron core, such as a tube, the magnetic properties are enhanced. The core will become magnetized, and the two ends of the core are differentiated by labeling them North and South, as shown in Figure 3.11. This labeling reflects the nature of magnets; left free to orient themselves, they will turn so that one end (the North end) will face geographically North.

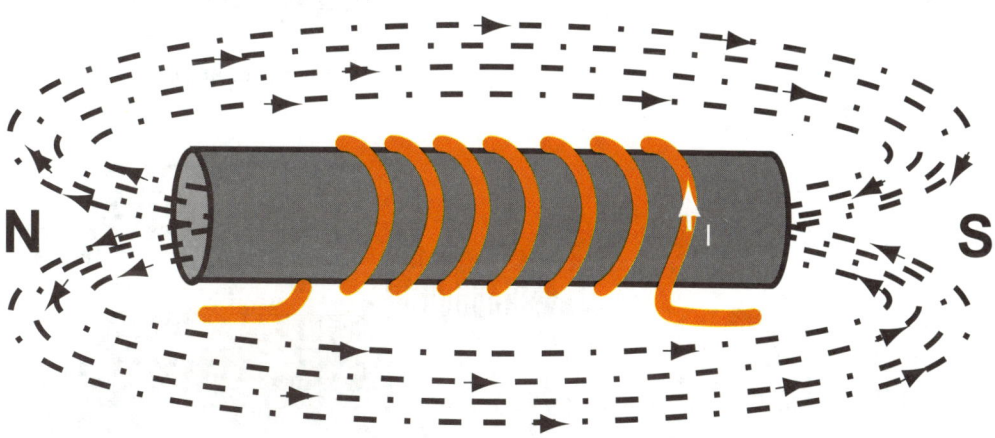

Figure 3.11 Iron core becomes magnetized

The magnetic properties are further enhanced if the coil is encased in iron. Thus, the magnetic force that results from a coil of wire with a current flowing through it is based on the magnitude of the current, the number of coils and the presence of an iron core or encasement. The direction of the force, that is, the North and South orientation of the force, depends upon the direction of the current in the coil.

One may note at this point that the "Right Hand Rule" now transposes to a coil; if the fingers of the right hand face the direction of current flow in a coil, the thumb will point in the direction of the magnetic force, North to South (Figure 3.12).

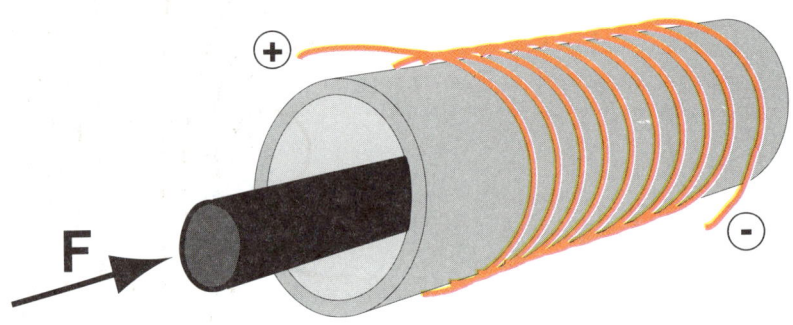

Figure 3.12 Magnetic force created by a coil

An iron rod placed in the center of the coil will become magnetized, and will center itself longitudinally when current is applied to the coil. This centering force is the essence of a solenoid, and results in the transformation of electrical energy into a linear mechanical force, as shown in Figure 3.13.

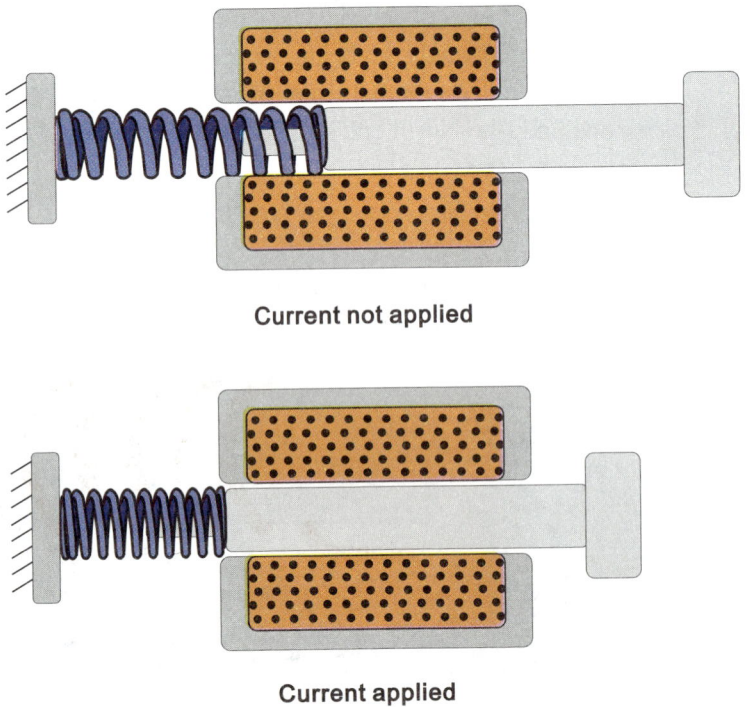

Figure 3.13 Centering force of a coil

It should be noted at this point that it does not matter which direction the current is flowing in the coil to impart a centering force to the core. By reversing the current flow, the magnetic properties of the core also reverse, and the core is still pulled to the center of the coil.

Proportional Solenoids

Proportional solenoids operate in much the same way as the on-off solenoids of the previous section, except that the input current is varied. The force on the core, and therefore the core movement, is proportional to this current.

As with standard solenoids, proportional solenoids will only move in one direction. If two direction motion is necessary, then two solenoids are required.

Force Motor

A force motor provides a proportional, bi-directional, linear motion in response to an electric signal. This motion is achieved with a single coil; two coils are required for bi-directional motion with solenoids or proportional solenoids. Force motors are frequently used to operate pilot operated control valves.

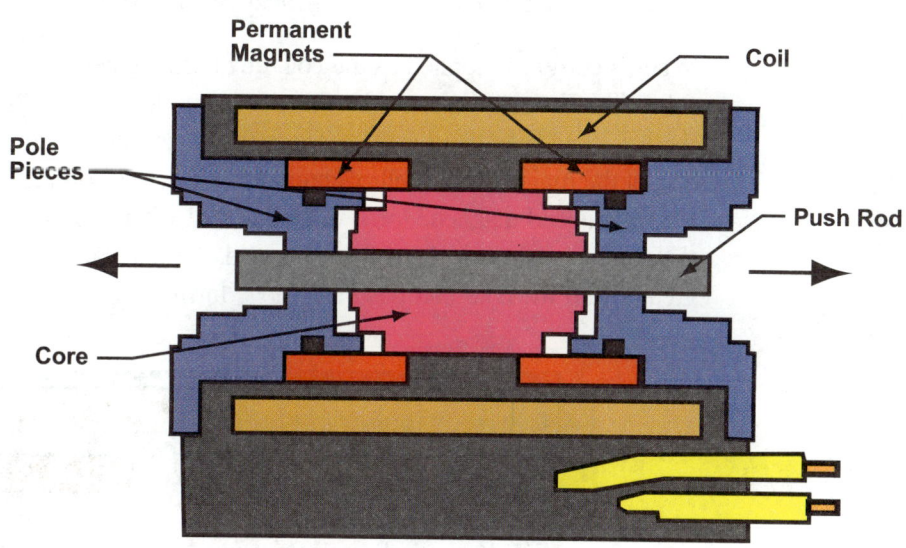

Figure 3.14 Force motor components

The components of a force motor are shown in Figure 3.14. The core, which is connected to a push-rod, is held in the center, or neutral, position by two permanent magnets.

The flux lines from the magnets travel through the pole pieces, magnetizing them equally, which provides the centering force. Figure 3.15 illustrates the flux line paths.

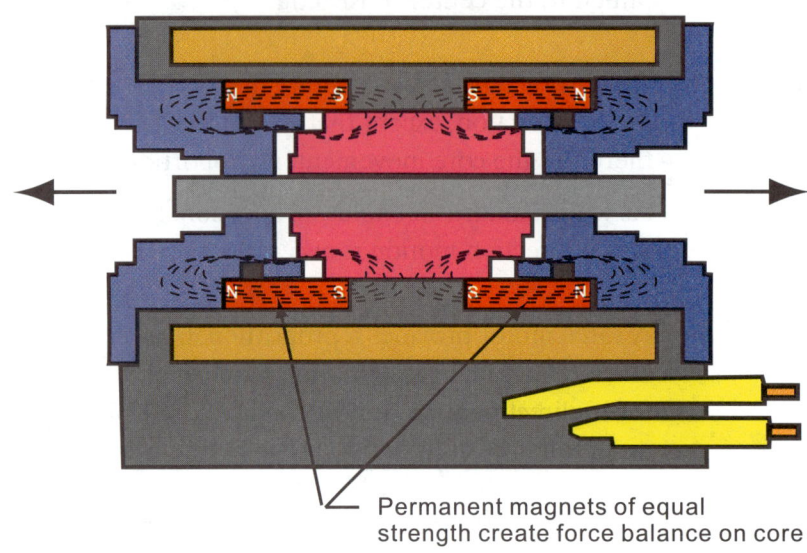

Figure 3.15 Equal flux lines create a neutral centering force

When a current is applied to the coil, flux lines are generated in the same way as with a solenoid coil. The direction the flux lines take depends on the direction of the current in the coil (the "right-hand rule"). These flux lines will be additive to the flux lines on one end of the magnets, and will counteract the flux lines on the opposite end (Figure 3.16). An unbalanced magnetic force will result that will move the core and the push-rod. As the current applied to the coil is increased, the unbalanced magnetic force becomes greater, and the core and push-rod move further.

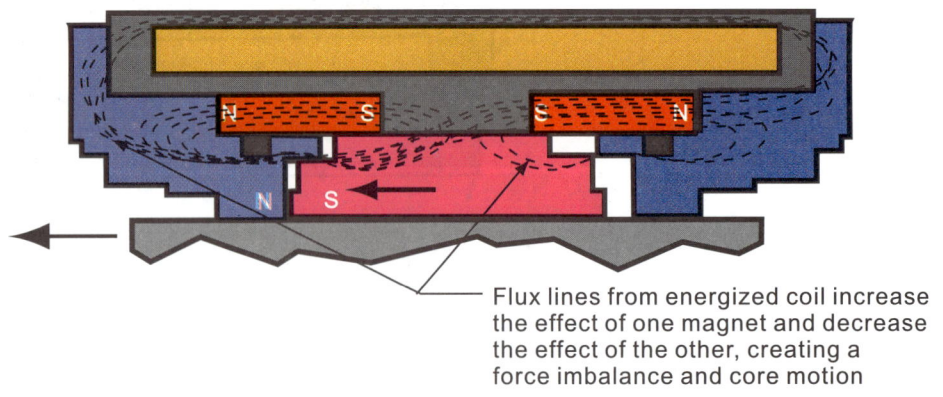

Figure 3.16 Energized coil creates an unbalances force

Reversing the direction of the current in the coil reverses the direction of the flux lines. The core and push-rod will then move an equal amount in the opposite direction. Proper shaping of the pole pieces results in a linear, proportional movement of the core and push-rod with respect to the direction and magnitude of current applied to the coil.

Torque Motor

A torque motor provides a proportional, bi-directional, rotational motion in response to an electric signal. A torque motor differs from a force motor in that two coils are used, and the output motion is rotational rather than linear. Torque motors are typically used to operate servo valves.

The components of a torque motor are shown in Figure 3.17. The permanent magnets are of equal strength, such that the core is held in its neutral, center position.

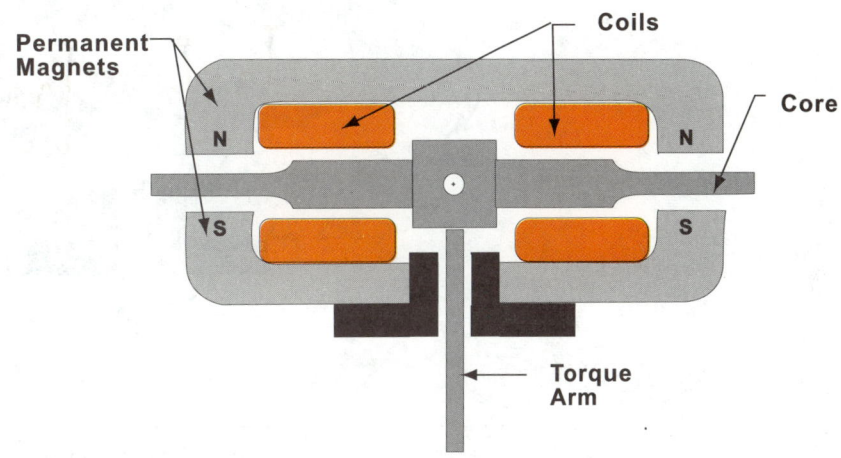

Figure 3.17 Torque motor components

When current is applied to the coils, the core is magnetized, one end becoming a North pole and the other end a South pole, depending upon the direction of current flow in the coils (the "right-hand rule"). As one end becomes a North pole, it is repelled by the North pole magnet, and is attracted by the South pole magnet. The direction of attraction is opposite on the South pole end of the core. Figure 3.18 illustrates the rotational motion of the core as current is applied to the coils.

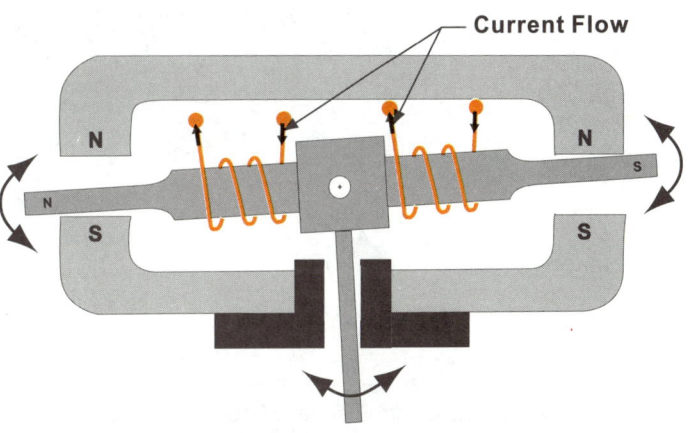

Figure 3.18 Coil current creates a rotational core movement

A small resistance against motion of the core will result in a proportional movement relative to the coil current. By reversing the direction of current flow in the coils (Figure 3.19), core motion will also reverse.

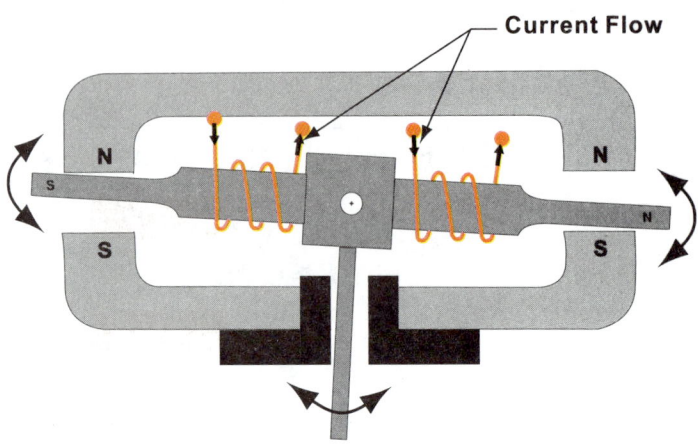

Figure 3.19 Reversing coil current reverses the core motion

DC Motor

A current passing through a conductor that is positioned at right angles to a magnetic field, will impart a force onto the conductor as shown in Figure 3.20. This force acts at right angles to both the conductor and the field. The force will act in the opposite direction when the current in the conductor is reversed.

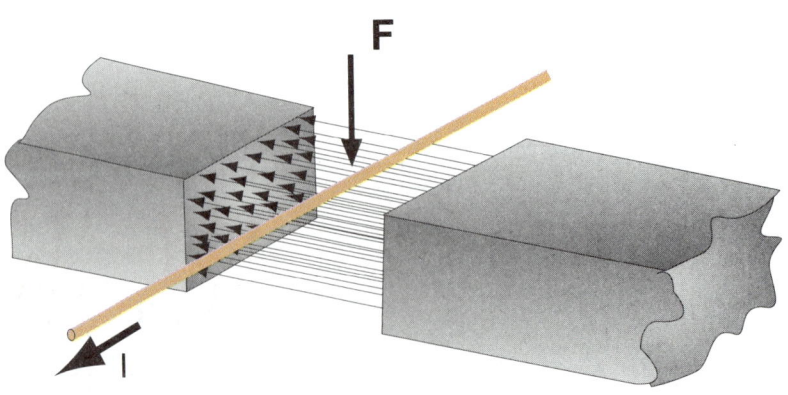

Figure 3.20 Electromagnetic force

Therefore, if the conductor is a loop, and passes through the magnetic field in both directions, as shown in Figure 3.21, two forces in opposite directions are created. These two forces will result in a circular force, or torque, as shown.

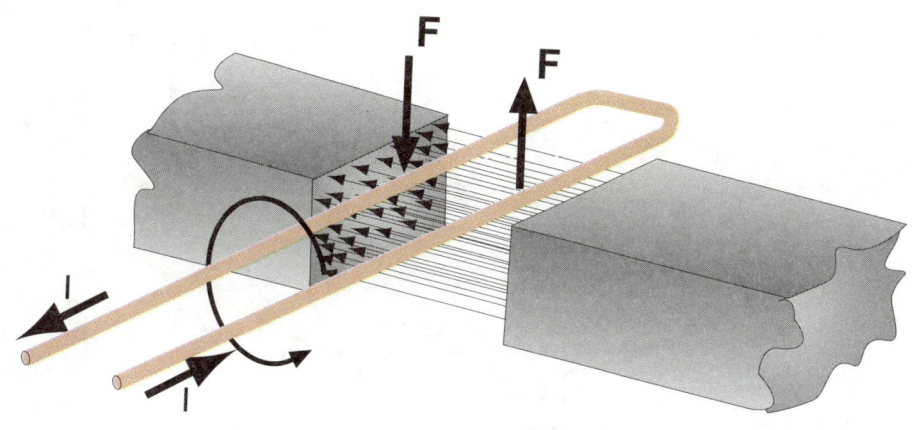

Figure 3.21 Rotational electromagnetic force

As the conductor loop revolves, the magnitude of the force that causes a rotating action reduces. The extreme is shown in Figure 3.22, where the conductor has rotated 90°, and the forces are no longer acting to cause a circular motion. By adding additional conductors, so that at least one is set is horizontal, or near horizontal, at all times, the point of rotation where no circular force is applied can be eliminated.

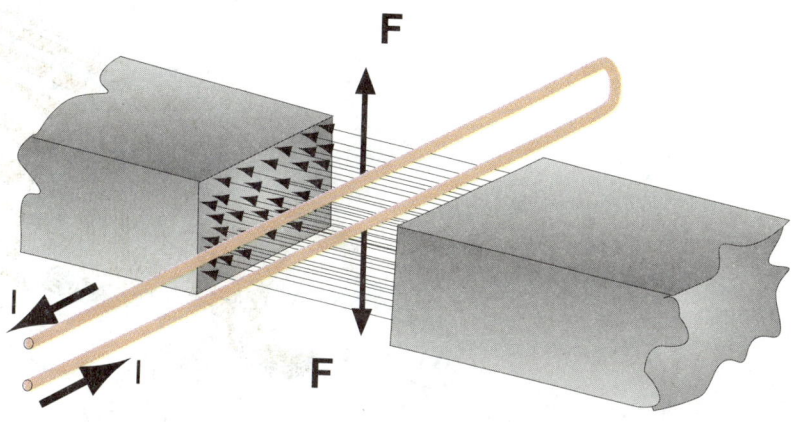

Figure 3.22 Rotational force decreases to zero in 90° increments

One can also notice that the direction of the current in the conductor(s) must be reversed as they rotate over-center, in order for the rotational force to continue acting in the same direction. This is done through a commutator arrangement such as illustrated in Figure 3.23.

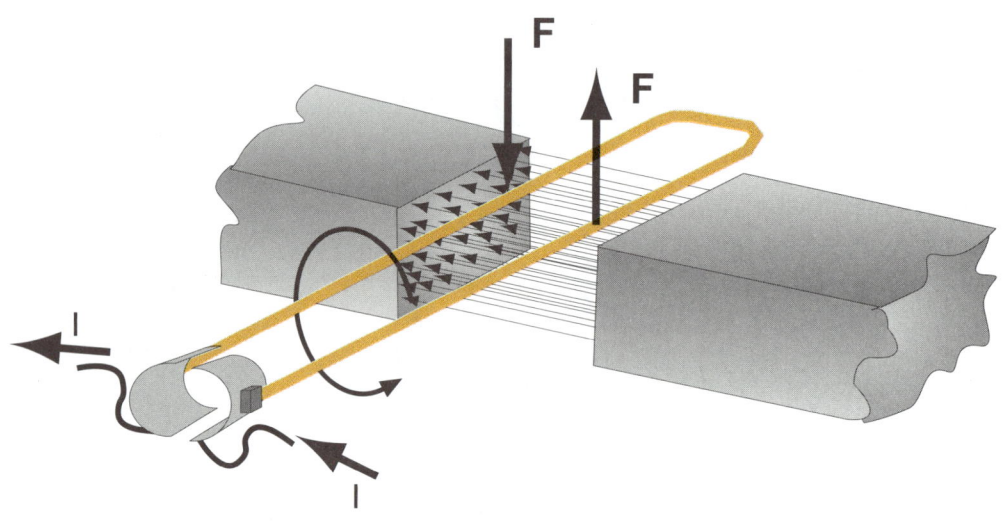

Figure 3.23 A commutator allows current to flow continuously in the same direction

A cross section of a DC motor is shown in Figure 3.24, where the more sophisticated design of the rotating conductors (armature) and commutator are illustrated, as well as the field windings that create the magnetic field through which the armature operates.

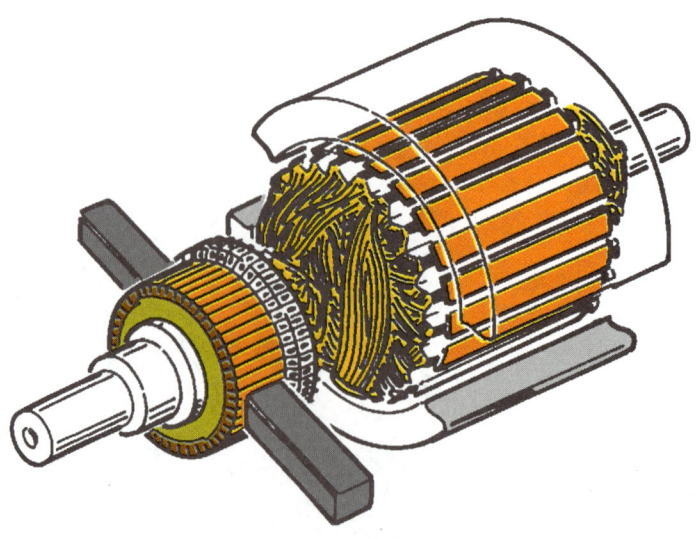

Figure 3.24 Conventional DC motor

Chapter 3 Basic Electrical Principles

DC Generators

A corollary to the basic principle of DC motor operation, Figure 3.20, is that passing a non-current carrying conductor through a magnetic field will generate a current in the conductor. The magnitude of this current will depend on the magnitude of the magnetic field and the speed and force applied to the movement of the conductor through the field. The forces illustrated in Figure 3.20 will be reversed in the current generating mode, as the forces are being applied *to* the conductor rather than *by* the conductor.

It takes little imagination at this point to apply the same concepts of a DC motor, Figures 3.21 through 3.23, to a DC generator. The conductors (armature) are now being rotated through the magnetic field, and the current that is being generated is taken off the commutator.

AC Generators

As each individual segment of the conductor in Figure 3.21 is rotated through the magnetic field, the current produced will be maximum at the point where the conductor is moving at 90° to the field, and minimum when the conductor is moving parallel to the field. Furthermore, as the conductor rotates 180°, the direction of the current reverses because the conductor is passing the field in the opposite direction. The commutator of Figure 3.23 is used to maintain singular current direction.

Without the commutator, the current flow would continuously cycle from maximum in one direction to maximum in the other. It would create a sinusoidal wave form, or sine wave, as shown in Figure 3.25. This "alternating current" is the basic type of electric power produced by a generator, and must be transformed by a commutator or a rectifier if direct current is the preferred form.

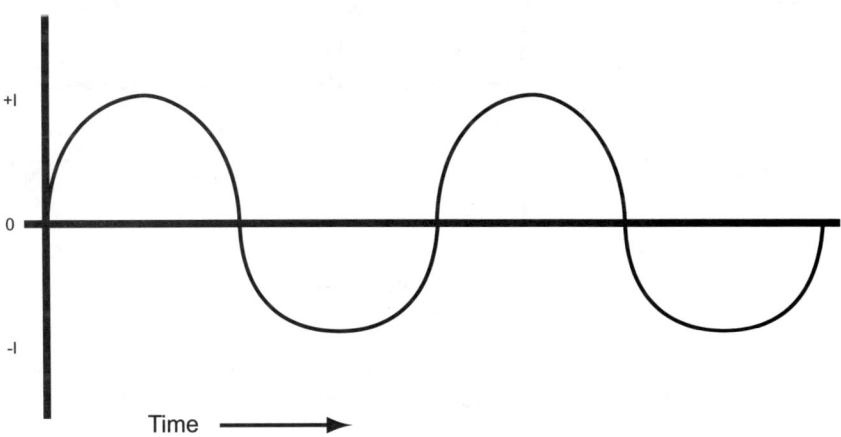

Figure 3.25 AC sine wave

Each complete wave form is called a cycle, and the length of the cycle, usually in seconds, is called a period. The number of times the cycle occurs during a segment of time, usually in a second, is called frequency. Typically, generators operate at 3000 or 3600 rpm, which results in an output frequency of 50 or 60 cycles per second, or cps.

Chapter 3 Basic Electrical Principles

Alternating current is the standard form of stationary electric power in the U.S. and in many other countries. Direct current is the standard form of electric power on mobile equipment, due to the need to maintain a battery as well as a battery being the primary source of electrical power.

AC Solenoids

As shown with a DC solenoid, the force that acts to center a core is the same regardless of which direction the current is flowing. Therefore, an alternating current can be applied to a solenoid with no apparent effect on the centering forces. Other effects occur, however, that must be addressed.

One effect of alternating current is caused by the wave form shown in Figure 3.25. Because the centering force does not depend upon the direction of the current, but on the magnitude, the resulting force on the solenoid core will vary as shown in Figure 3.26. This fluctuation will occur at twice the frequency of the alternating current, which is usually 50 or 60 cps. This high rate of change in the force is too rapid to allow the core to move much, even with the offsetting spring of Figure 3.13. Nevertheless, even a small amount of core motion at these high frequencies will cause a "hum" in the solenoid. This AC hum will not only be annoying, but also will cause a small amount of power loss to occur.

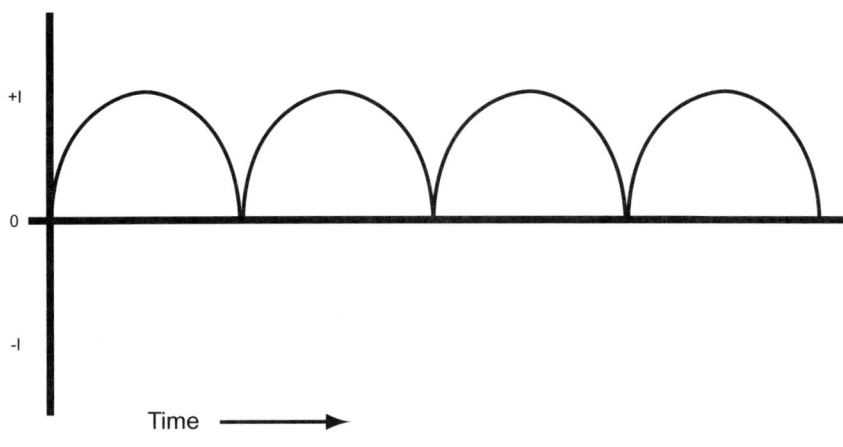

Figure 3.26 Centering force due to alternating current

The solution for this alternating current hum is to add a copper "shading ring" to the core or frame of the solenoid. This ring will magnetize and demagnetize the same as the core, but somewhat out of phase with it.

The magnetic field of the shading ring is sufficient to hold the core in place during the low part of the current cycle, as shown in Figure 3.27, effectively eliminating the core motion and the AC hum.

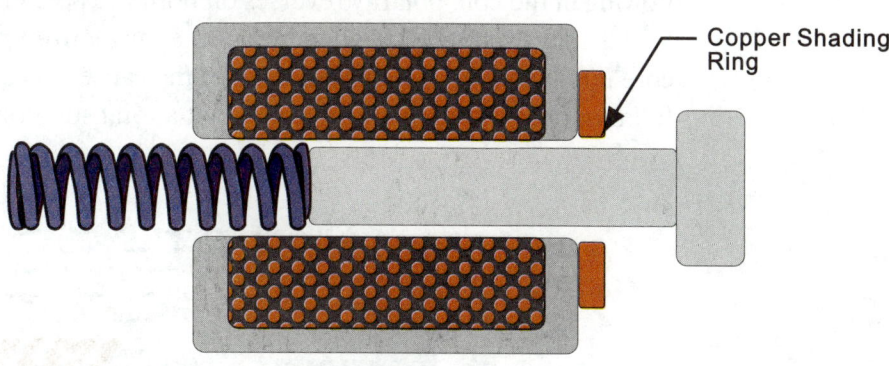

Figure 3.27 AC solenoid with shading ring

A second effect of alternating current on solenoids is the risk of overheating the coil if the core does not center properly. This is caused by the "inrush" current.

The impedance of a coil, which consists of a combination of the resistance of the coil and the inductive reactance (the resistance of the coil to a change in current flow), is minimum when the core is completely out of the coil. This causes a maximum current flow, the "inrush current", when the solenoid is first energized. The inductive reactance, and therefore the impedance, is at its maximum when the core becomes centered, causing a dramatic reduction in current flow. This lower current level is called the "holding current". Figure 3.28 illustrates the change in current flow with core position.

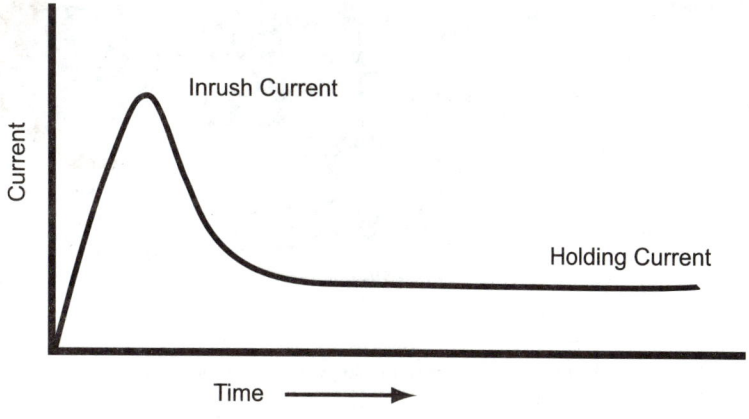

Figure 3.28 "Inrush" and "Holding" current for solenoids

Overheating can occur if a mechanical malfunction, either within or outside the solenoid, prevents the core from centering properly. The high inrush current is then maintained continuously on the coil, and "burn-out" can occur.

A partial solution to the mechanical malfunction cause of overheating is accomplished by adding a "pole piece" to the solenoid as shown in Figure 3.29. This pole piece becomes magnetized the same as the core, creating opposite polarities that are adjacent to each other. This occurs regardless of which way the current is flowing in the coil; polarity reverses on both the pole piece and the core as Figure 3.30 demonstrates. The added pull of the magnetized pole piece helps to pull the core into position, overcoming some of the causes of core "drag" that may prevent full motion. Although effective, it is not a total solution.

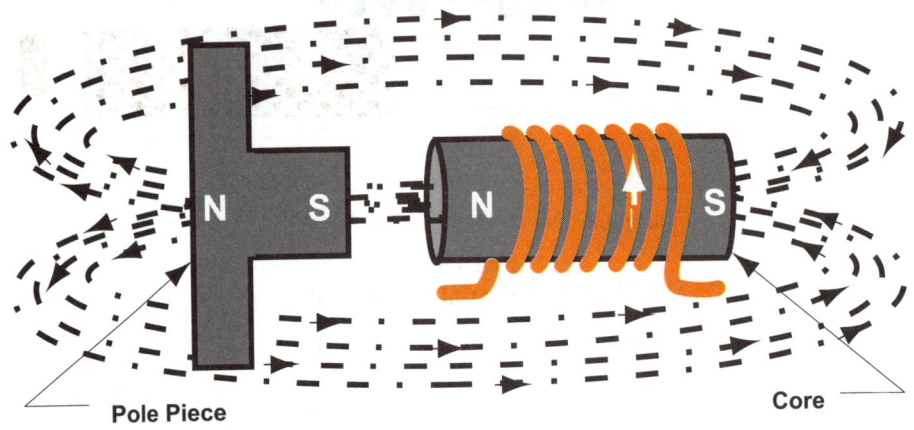

Figure 3.29 Solenoid with pole piece

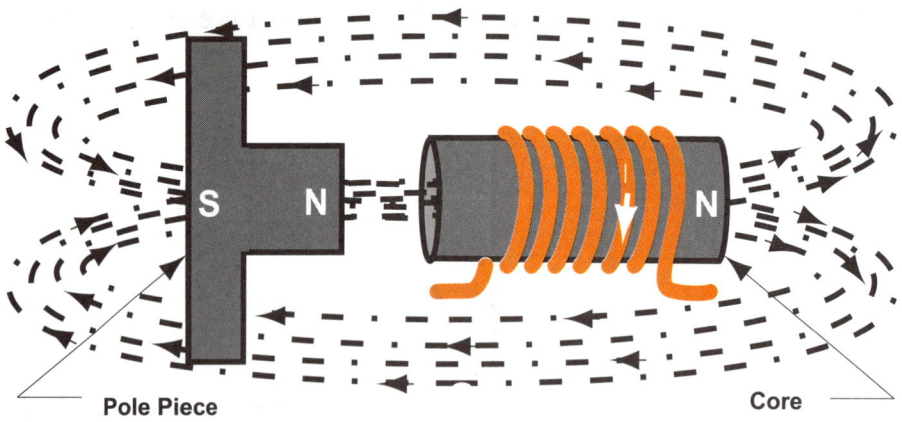

Figure 3.30 Reversing current reverses polarity, maintaining "pull"

The solution to this potential overheating is to rectify the alternating current, meaning to convert it into direct current, at the electrical input to the solenoid. Because direct current does not cause an inductive reactance in the coil, the inrush current does not exist, and burn-out is less likely to occur.

Note that by rectifying the current, the problem of solenoid hum is also eliminated. Today, nearly all AC solenoids used in hydraulic valves include rectifiers.

Rectifiers

A rectifier is a device that is used to convert an alternating current into a direct current. The primary component of a rectifier is a diode, the equivalent of a hydraulic check valve. A diode will allow current flow in one direction, but will block current flow in the opposite direction.

AC current flow has a wave form as shown in Figure 3.31. By connecting a diode into an electric circuit as shown in Figure 3.32, the current that passes to the load takes the form of the curve in Figure 3.33. The load receives a current in only one direction, the current in the opposite direction is blocked. Because the load is receiving only one-half of the current wave, this type of rectifier is referred to as a "half-wave rectifier".

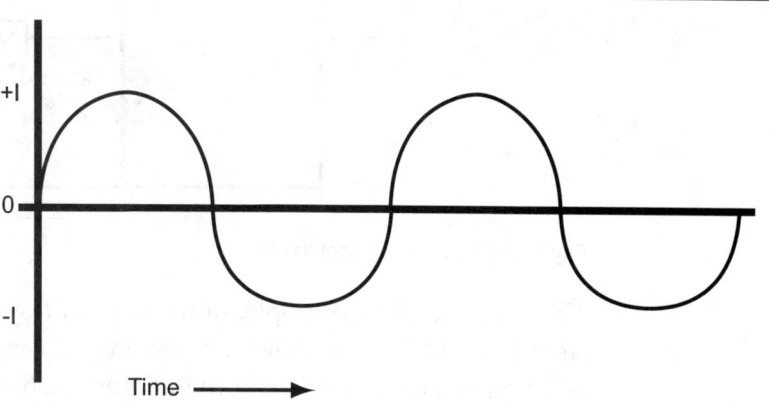

Figure 3.31 AC current wave form

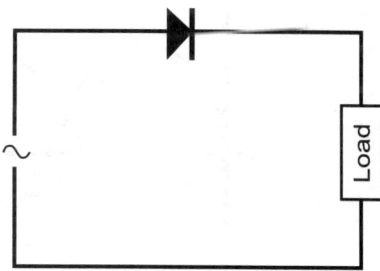

Figure 3.32 Half-wave rectifier

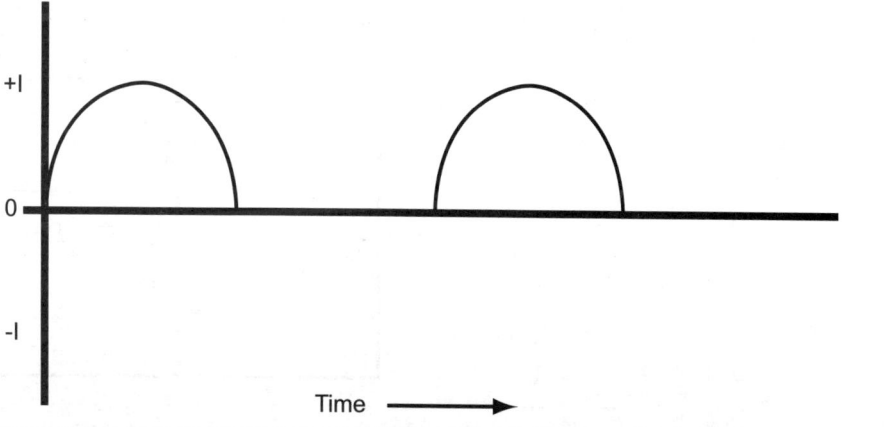

Figure 3.33 Half-wave rectification

An alternative circuit is shown in Figure 3.34. In this circuit, the load receives the current first from the positive wave through diodes 1 and 2, and then the current from the negative wave through diodes 3 and 4. Arranging the diodes in this manner results in the current passing through the load in the same direction, regardless of the direction of the alternating current. The resulting wave form is shown in Figure 3.35. This type of rectifier is called a "full-wave rectifier".

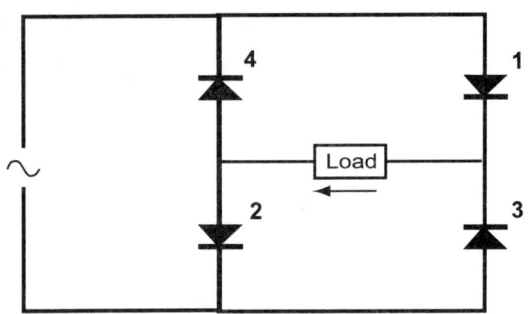

Figure 3.34 Full-wave rectifier

The wide variation, or ripple, of the current flow in Figure 3.35 is undesirable for smooth control of hydraulic components. It can be "smoothed" substantially by the addition of a capacitor as shown in Figure 3.36.

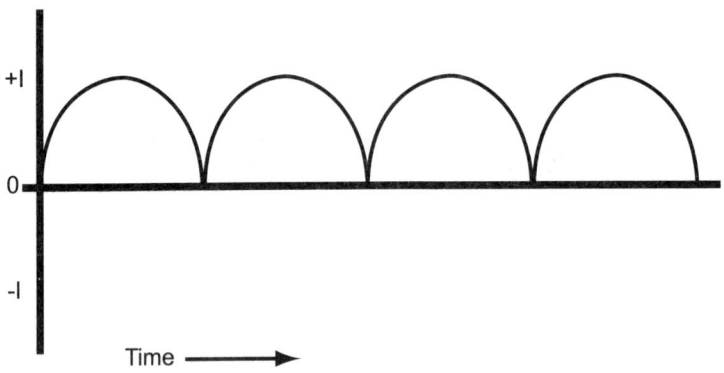

Figure 3.35 Full-wave rectification

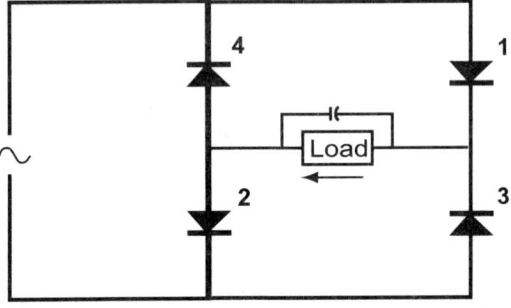

Figure 3.36 Full-wave rectifier with smoothing capacitor

This capacitor charges during the high current portion of the wave, and discharges during the low current portion, creating a smoothing action as shown in Figure 3.37. This slight ripple can be smoothed further by the addition of a voltage regulator, such that the circuit output will be a flat, steady direct current.

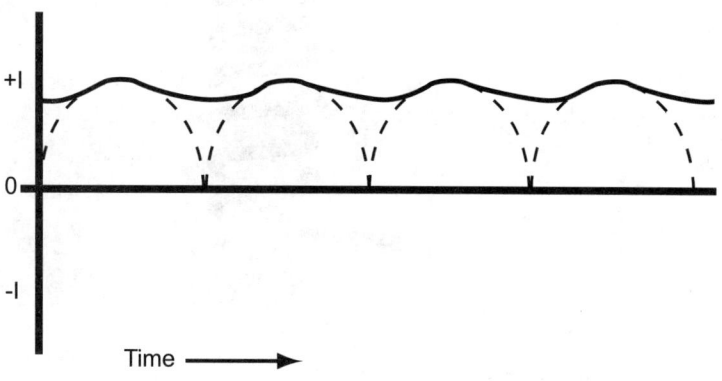

Figure 3.37 Rectified AC current with smoothing

Complete circuit rectification is shown diagrammatically in Figure 3.38.

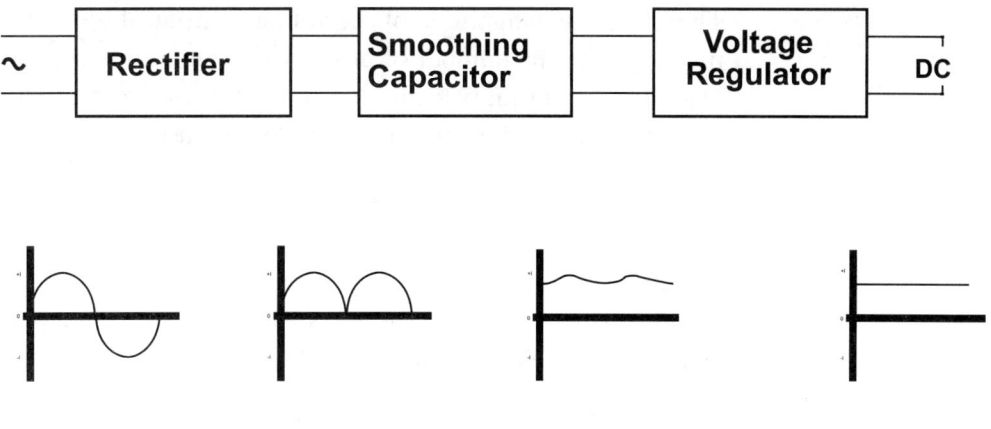

Figure 3.38 Fully rectified AC current

Chapter 3 Basic Electrical Principles

Transformers

A transformer takes the basic form of the diagram in Figure 3.39. It consists of a primary winding, a secondary winding and a core. Each winding contains a specific number of "turns", or wraps, of conductor around the core.

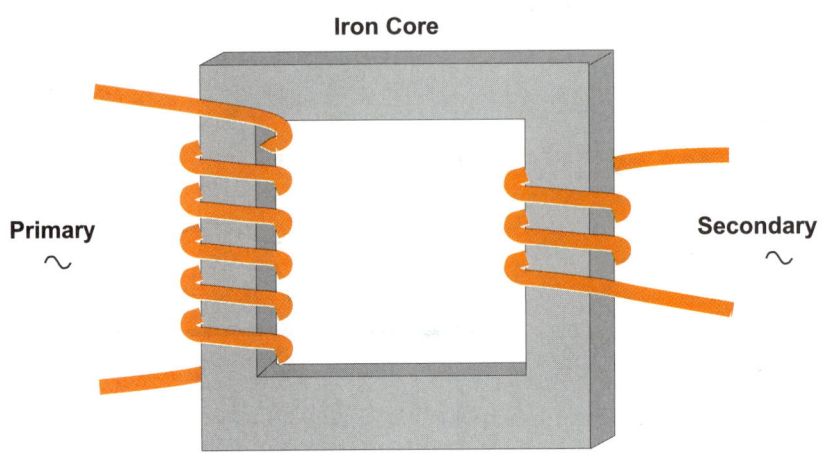

Figure 3.39 Basic transformer

Alternating current is applied to the primary winding, which will create a magnetic field in the core. This field will alternate with the incoming current, and the strength of the field will depend upon the number of turns in the primary winding. Further discussion of this concept was covered earlier in this chapter under solenoids.

The field generated by the primary windings will pass the secondary windings, and a voltage will be induced in the secondary windings. The strength of the voltage in each turn will be the same as in the primary winding, and the total voltage will be the sum of the voltage induced in each turn.

Therefore, the magnitude of the voltage output of the transformer will be in direct proportion to the number of primary and number of secondary turns. If there are twice as many primary turns as secondary turns (referred to as a turn ratio of 2), the output voltage will be one-half the input voltage.

Formula 3-8

$$V_2 = V_1 \times \frac{N_2}{N_1}$$

A transformer can be a step-up transformer, used to increase the voltage, or a step-down transformer, used to reduce the voltage.

Based on the law of conservation of energy (Chapter 1), the power into a transformer cannot be any greater or less than the power out, including any losses. Because electrical power is a product of the voltage and amperage (formula 3-5), then the current to a load on the output of a transformer will be the inverse of the turns ratio times the input current. It will also be the inverse of the voltage ratio:

Formula 3-9

$$V_1 I_1 = V_2 I_2 \qquad \frac{V_2}{V_1} = \frac{I_1}{I_2}$$

Amplifiers

The input signal to control solenoid valves on mobile equipment may come from a number of sources:

▶ Potentiometer - a variable resistance that is used to convert a physical position, either linear or rotary, into a corresponding electric signal.

▶ Pressure transducer - a device that provides an electric signal in direct proportion to a pressure imposed upon it.

▶ Temperature sensor - A device that provides an electrical output signal directly proportional to a temperature to which it is exposed.

▶ Tach-generator - A small, low voltage, DC generator that provides a voltage output in direct proportion to its running speed (RPM).

▶ Electric remote control - A manually operated mechanism for transmitting electric signals that directly relate to the position of lever(s).

▶ Microprocessor based control - A broad family of devices that provide electric output signals generated by the electronic processing of electric input signals. Devices such as digital controllers, programmable logic controllers (PLCs) and personal computers (PCs) are included in this category.

All of these devices provide a low power signal. The current, voltage or both must be amplified to provide sufficient power to control on-off or proportional solenoids, even small ones.

The key component of amplifiers is the transistor, a sophisticated electronic switching device. It is diagrammatically shown in Figure 3.40.

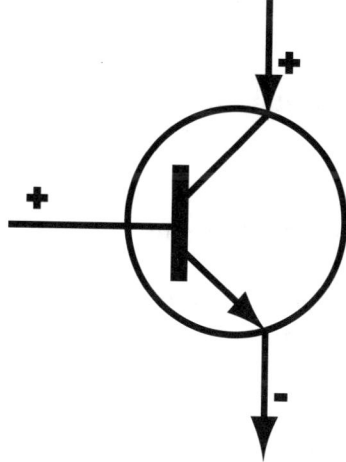

Figure 3.40 Transistor diagram

The basic concept of an amplifier is that power from a power supply will be directed to the output terminal in proportion to small voltages applied to the input terminal. Figure 3.41 shows this diagrammatically.

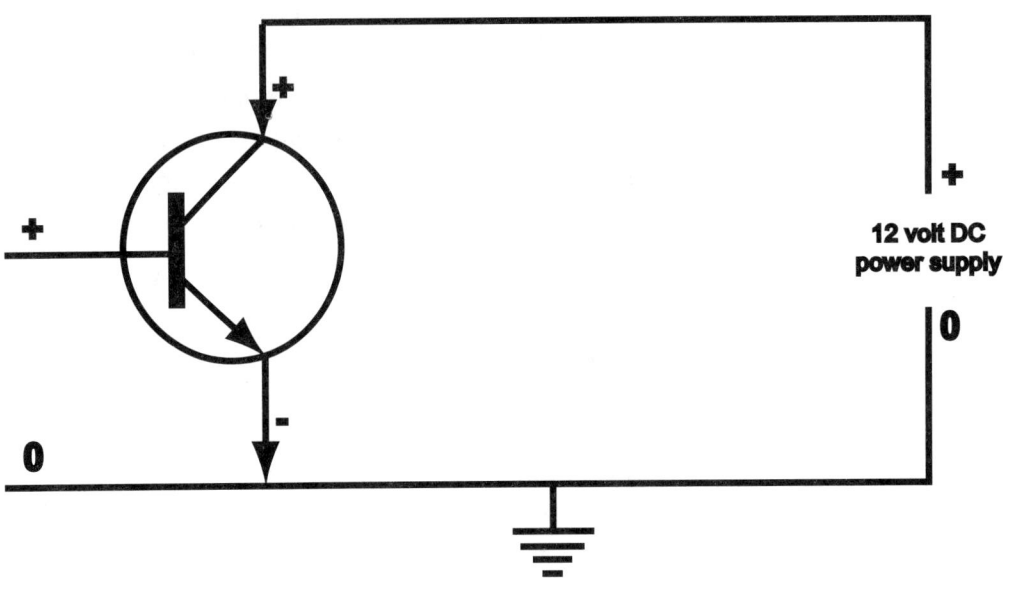

Figure 3.41 Amplifier circuit

The gain of the amplifier, that is, the ratio of the output to the input, is usually very high. The output is also normally inverted polarity, such that a small positive input signal will result in a large negative output signal. An amplifier output can then be defined as:

Formula 3-10

Voltage Output = Voltage Input x (- -A)

Where

A is the amplifier gain.

The amplifier input is the differential between two input terminals. Therefore, one of the terminals is usually connected to ground, or "zero".

An operational amplifier is one with the input terminals as shown in Figure 3.42. The input terminals are labeled "+" and "-"; these designations do not specify the input polarity, but have to do with the output polarity with respect to the input. The "+" terminal is also called the non-inverting terminal, and the "-" terminal is called the inverting terminal. Signals connected to the "+" terminal will provide like polarity at the output, and signals connected to the "-" terminal will provide opposite polarity at the output. If the same signal, either positive or negative, is applied to both terminals, the output will be zero.

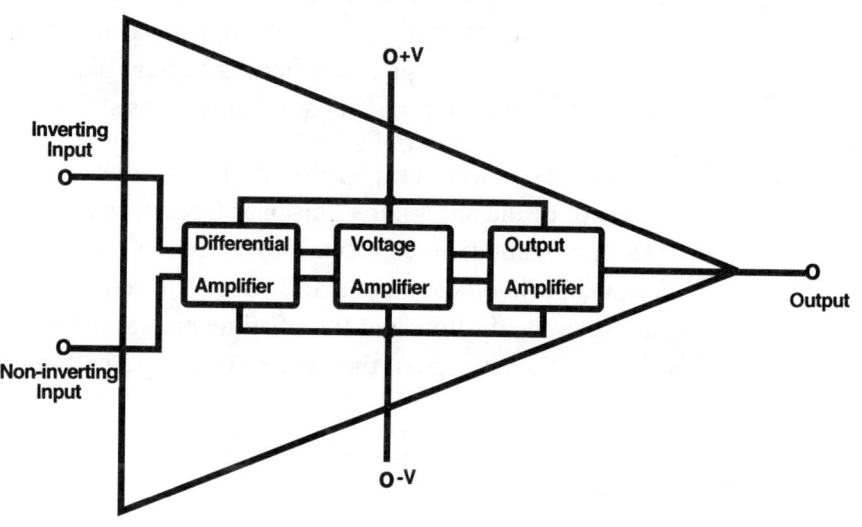

Figure 3.42 Basic operational amplifier

Operational amplifiers are also used for providing "feedback" signals to the input, a method of comparing the output result to the input, and using the "bias" signal to correct the input accordingly. A feedback arrangement is shown in Figure 3.43, and the resistors, R_f and R_i, can be adjusted to alter the gain of the amplifier.

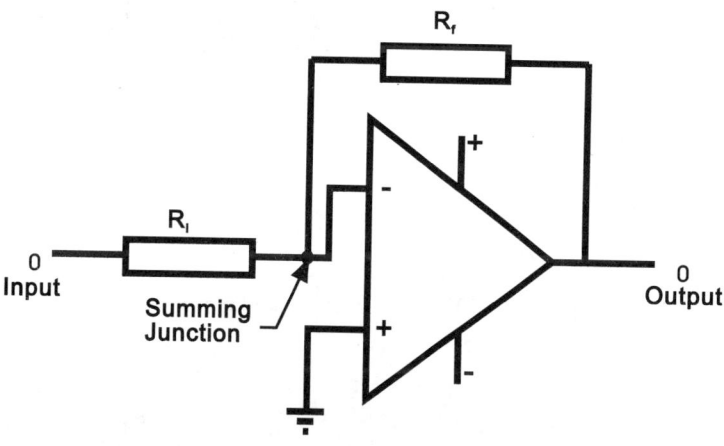

Figure 3.43 Operational amplifier with feedback

Pulse Width Modulation

Proportional solenoids are controlled by a variable voltage, typically supplied by the output transistor of an amplifier. The amplifier must act as a variable resistance, increasing and decreasing in value so that the power supply voltage will increase and decrease accordingly. This voltage drop is necessary in order to reduce the power to the solenoid, but the excess power will be dissipated as heat. The heat is not only a waste of energy, but it can be damaging to electronic components.

A technique called Pulse Width Modulation (PWM) will substantially reduce this heat build-up. PWM does not vary the power supply voltage passing through the amplifier. Instead, it works on the concept of switching the voltage either fully on or fully off. When the voltage is fully on, there is no voltage drop from the power supply across the transistor so there is no heat build-up. When the voltage is fully off, there is no power passing across the transistor, so there is no heat involved.

PWM relies on a high rate pulse frequency constantly applied to the output transistor of the amplifier. This base frequency is generally set about 1000 or more cycles per second, much too fast for a solenoid to react. Within this frequency, the power supply voltage is constantly being turned on and off. Figure 3.44 illustrates that when the on-time and the off-time are equal for each cycle, the resulting voltage is the average of the two, equal to one-half the power supply voltage.

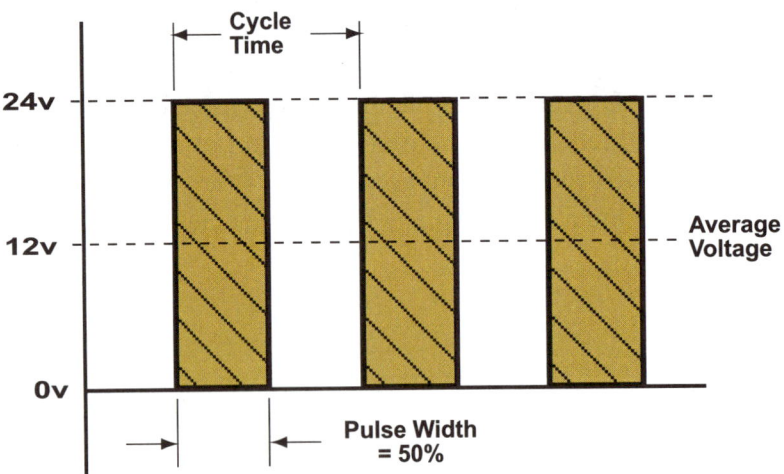

Figure 3.44 50% pulse width modulation

The average voltage level can be controlled by regulating the on time of each pulse in a series. Figure 3.45 illustrates several examples. By regulating the on-vs-off relationship, the average voltage passing to a solenoid is regulated, and very little power is wasted.

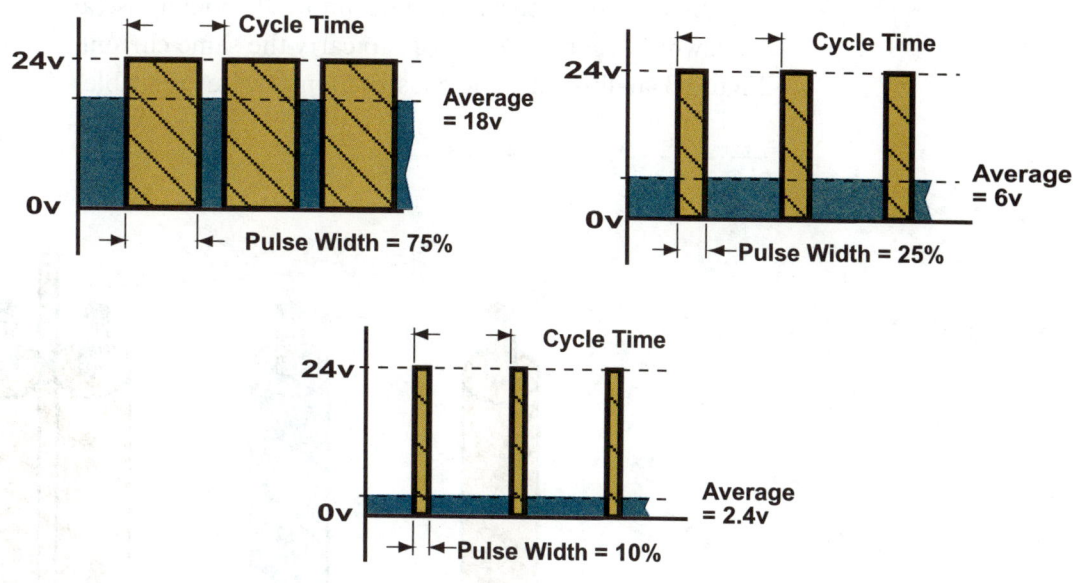

Figure 3.45 Varying pulse width varies the average voltage

Response time of a valve solenoid is very quick using PWM because the initial pulse, regardless of the intended voltage regulation level, will be maximum voltage (it is not the maximum voltage level that is being controlled, but the length of time the voltage level is "on"). The response time can be improved even further, for example, by setting the peak voltage at twice the solenoid rated voltage, and setting a 50% on-time for maximum average voltage value.

Pulse Width Modulation has rapidly gained popularity in the control of proportional and servo valves because of the reduced heat and power loss in the control circuit (resulting in a lower input power requirement), the faster response that can be expected and the ability to interface with any proportional/servo valve without modification.

Conductors

Electrical conductors are the path through which electrical charges are transferred from one point in a circuit to another. The material for conductors is selected for the ease with which current flows through it. Insulators and insulation, however, is selected for its characteristic of not passing current easily. Insulating material such as rubber, plastic or nylon is used to cover conductors to eliminate shocks and short circuits. The amount of current that can pass through an insulator is so small, it is considered non-existent.

Wires and cables are the most common types of conductors. Printed circuitry is also very common, but only in low power electronic applications.

Chapter 3 Basic Electrical Principles

Wires and Cables

Wires and cables come in a large variety of sizes and construction, each designed for specific application parameters and capacities. In general terms, a wire is a single conductor, and a cable is a grouping of two or more wires bound together.

Wires are of two constructions, solid and stranded. Solid wire is a single wire conductor; stranded wire is a group of small wires twisted together to form a single conductor. Stranded wire is much more flexible to use and install. It takes a somewhat larger stranded wire to carry the same current as a solid wire conductor. Figure 3.46 illustrates the construction of wire and cable.

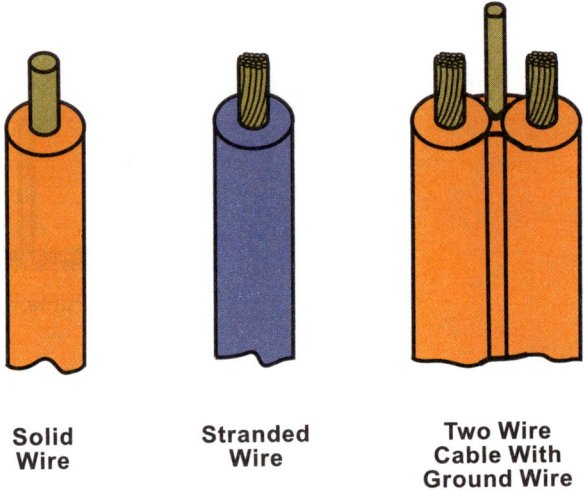

Solid Wire   Stranded Wire   Two Wire Cable With Ground Wire

Figure 3.46 Wire and cable types

Wire Gauge

Wires are designated by their "gauge" numbers, which is a linear numerical designation that relates inversely to the wire cross-sectional area. Wire diameter is measured in "mils" (one mil equals 0.001 inches), and the cross-sectional area is designated in "circular mils". A circular mil is the square of the diameter. For example, if a wire is 25 mils in diameter, the cross-sectional area will be 625 circular mils (25 x 25). The circular mil area of a stranded wire is the sum of the circular mil areas of each individual strand.

Gauge numbers for wire are inverse to the circular areas. The chart of Table 3.1 shows this relationship, as well as the relationship to the diameter of a solid wire. The description of a cable will include the gauge number and the number of individually insulated wires, such as 14-2 (meaning a pair of 14 gauge wires).

American Wire Gauge (AWG) for Copper Conductor			
AWG Number	Diameter Inches	Circular Mils	Diameter Milimeters
000000	0.5800	336,000.0	14.73
00000	0.5165	267,000.0	13.12
0000	0.4600	212,000.0	11.68
000	0.4096	168,000.0	10.40
00	0.3648	133,000.0	9.266
0	0.3249	106,000.0	8.251
1	0.2893	83,700.0	7.348
2	0.2576	66,400.0	6.544
3	0.2294	52,600.0	5.827
4	0.2043	41,700.0	5.189
5	0.1819	33,100.0	4.621
6	0.1620	26,300.0	4.115
7	0.1443	20,800.0	3.665
8	0.1285	16,500.0	3.264
9	0.1144	13,100.0	2.906
10	0.1019	10,400.0	2.588
11	0.09074	8,230.0	2.305
12	0.08081	6,530.0	2.053
13	0.07196	5,180.0	1.828
14	0.06408	4,110.0	1.628
15	0.05707	3,260.0	1.450
16	0.5082	2,580.0	1.291
17	0.04526	2,050.0	1.149
18	0.04030	1,620.0	1.024
19	0.03589	1,290.0	0.9116
20	0.03196	1,020.0	0.8118
21	0.02846	810.0	0.7229
22	0.02535	642.0	0.6438
23	0.02257	509.0	0.5733
24	0.02010	404.0	0.5106
25	0.01790	320.0	0.4547
26	0.01594	254.0	0.4049
27	0.01420	202.0	0.3606
28	0.01264	160.0	0.3211
29	0.01126	127.0	0.2859
30	0.01002	101.0	0.2546
31	0.008928	79.7	0.2268
32	0.007950	63.2	0.2019
33	0.007080	50.1	0.1798
34	0.006305	39.8	0.1601
35	0.005615	31.5	0.1426
36	0.005000	25.0	0.1270
37	0.004453	19.8	0.1131
38	0.003965	15.7	0.1007
39	0.003530	12.5	0.08969
40	0.003144	9.9	0.07986

Table 3.1 Wire gauge sizes

Chapter 3 Basic Electrical Principles

Ampacity

Wire size is selected for an application based on two criteria; the resistance of the wire and the "ampacity". The resistance determines the amount of voltage drop that will be experienced over a given length, and is related to the gauge of the wire. Ampacity is the amount of current a wire can carry without creating excessive heat that would be damaging to the insulation. The ampacity of a wire is based on its gauge and the type of insulation.

Table 3.2 describes the insulation types and designations for most common wires and cables.

Insulation Type				Resistance To						
Outer Cover	Wires	Temp Range	Flexibility	Abrasion	Water	Solvent	Acid	Base	Flame	Humidity
PVC	PVC	–40 to 220 °F	Exc	Good	Good	Fair	Good	Good	Good	Good
FEP	FEP	–320 to 390 °F	Good	Exc	Exc	Exc	Exc	Exc	Exc	Exc
PFA Kapton	PFA	–450 to 500 °F	Good	Exc	Exc	Exc	Exc	Exc	Exc	Exc
Kapton	Kapton	–450 to 600 °F	Good	Exc	Good	Good	Good	Good	Good	Exc
Glass Braid	PFA	–100 to 500 °F	Good	Good	Exc	Exc	Exc	Exc	Exc	Exc
Glass Braid	Glass Braid	–100 to 900 °F	Good	Poor	Poor	Exc	Exc	Exc	Exc	Fair
HT Glass Braid	HT Glass Braid	–100 to 1300 °F	Good	Poor	Poor	Exc	Exc	Exc	Exc	Fair
Nextel Braid	Nextel Braid	0 to 2200 °F	Good	Fair	Poor	Exc	Good	Good	Exc	Fair
Silica Braid	Silica Braid	0 to 1900 °F	Good	Good	Exc	Exc	Good	Poor	Exc	Fair

Table 3.2 Wire insulation types and their performance characteristics

Table 3.3 lists a number of common wire sizes, the resistance per linear foot and the ampacity rating of each size based on the type of insulation.

AWG No.	Ω per 1000 ft.	Ampacity		
		PVC @ 176 °F	Nylon @ 221 °F	Kapton @ 392 °F
2	0.159	170	200	240
4	0.253	125	145	180
6	0.403	95	105	135
8	0.641	65	75	100
10	1.02	47	58	75
12	1.62	36	45	55
14	2.04	27	33	45
16	4.09	19	24	32
18	6.51	15	18	24
20	10.0	10	13	17
22	16.5	8	10	13
24	26.2	6	7	10
26	41.6	4	5	7
28	66.2	3	4	6
30	105	2	3	4

Table 3.3 Resistance and ampacity of common wire sizes and insulation

Chapter 3 Basic Electrical Principles

Printed Circuitry

A printed circuit is one that consists of thin ribbons of conducting material that have been deposited onto an insulated substrate. Many different types of miniaturized electronic components and semiconductors can then be installed onto the substrate, interconnected by the conducting ribbons, to form complete circuits or sub- -circuits. These types of circuits are called hybrid integrated circuits, and can be as simple or as complex as the task they are designed for. The hybrid designation comes from the combining of miniaturized integrated circuitry components and more conventional components such as transistors and diodes.

The ribbons of conducting material are very thin, often less than one mil (0.001 inches) thick, and only as wide as necessary to pass the power required. The resulting integrated circuit is a very low-power device that operates on very low voltage, such as 10 or 20 volts, and very low current. The current used in integrated circuits is usually measured in milliamperes (one milliampere equals 0.001 amperes).

The advantages of hybrid integrated circuitry is that it is very compact, and utilizes very little power. It therefore does not generate any appreciable heat. The circuits, or sub-circuits, are self contained on a substrate, or card, and are best serviced by replacing the entire circuit. Although this can be more expensive than replacing a single component, it considerably reduces the time spent on troubleshooting.

The disadvantages of this type of circuitry is sensitivity to environmental conditions (humidity, temperature, water, vibrations, etc.) and the low electric power that these conductors can pass.

Closed Loop vs. Open Loop Circuitry

The most common type of electric, and hydraulic, system is one referred to as an open loop circuit. In an open loop circuit, control functions are applied to the input in an attempt to obtain a desired output. An example of an open loop hydraulic circuit is the positioning of a valve to obtain a desired motor speed. An example of an open loop electric circuit is the regulation of an amplifier to obtain a predetermined voltage to position a proportional valve (Figure 3.47).

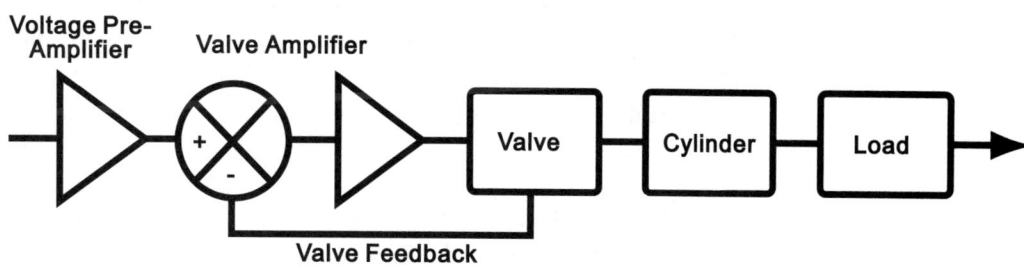

Figure 3.47 Open loop system with an "internal" closed loop

Note that in Figure 3.47, a feedback branch is shown from the valve spool position to the amplifier. This is an *internal* feedback in the system, assuring that the valve spool attains the desired position. It is not a true closed loop system, however.

In the above examples, the *output* of the system can be affected by other things which are not accounted for at the input. The hydraulic motor speed can be affected by fluid temperature and internal leakage, as well as by the output load which affects input pressure and further affects leakage. The *input* to the system does not change, but the output changes because of these other factors.

In the electric example, the *output* of the system can also be affected by temperature, causing both the solenoid and the valve to perform differently. The output can also be affected by a sticking valve spool, causing the solenoid to behave differently even though the system *input*, the amplifier setting, has not changed.

A closed loop system regulates the input to the system based on the desired output. It measures the actual output, compares it to the desired output, and corrects the input accordingly. The correction signal that is sent to the input control is the true feedback signal. The difference between the input signal and the feedback signal is called the "error". The purpose of a closed loop system is to eliminate the error between the input and feedback signals so that the desired output is reached.

Figure 3.48 diagrammatically describes open loop and closed loop systems.

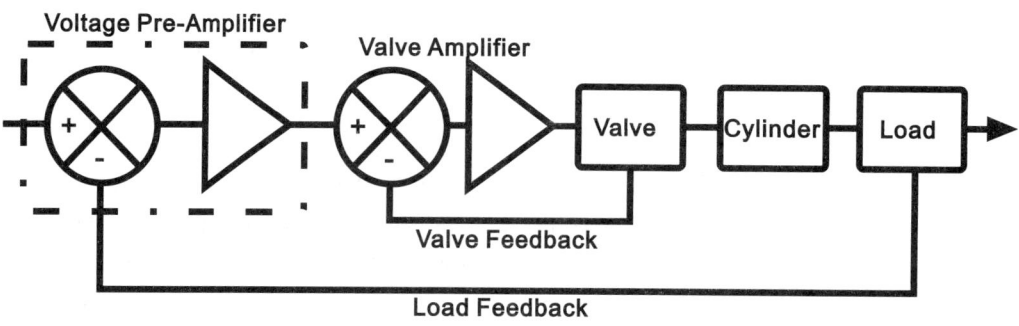

Figure 3.48 Internal and total closed loop

Closed loop systems require electrically operated control valves, such as solenoid, proportional or servo valves, and the associated electric control mechanisms such as operational amplifiers. In addition, the systems require some form of output measurement devices that will convert the desired output parameter into an electric signal that can be used for feedback to the valve control mechanism.

Closed loop systems are categorized into three types; position control, velocity control and force control. Any one application may incorporate one, two or all three of these types. Obviously, the more sophisticated the control requirements are, the more complex and expensive the system will become.

An example of a position control system is one that regulates the height of the blade on a grader. The output (blade position) is monitored by some form of position sensor, such as a laser beam, and any error in the position is transmitted to the blade control for correction.

An example of a velocity control system is one that regulates the rotation speed of a combine header reel relative to the ground speed. In this type of application, ground speed is monitored, constantly compared to the reel speed, and any error in the relationship is transmitted to the reel control to adjust the speed.

Force control is used on farm tractors to regulate the draft (pull) created by ground preparation equipment. For example, if draft forces become too great while pulling a plow, the force information is transmitted to the hitch control valves, and the valves adjust the plowing depth to relieve the load.

In all of these examples, some form of output measurement device is required to perform the feedback function.

Tranducers

The term "transducer" is used to generically describe a family of devices that convert some physical property into an electric signal. The electric signal can then be used to convert the physical information into a readable form, or a form that can be recorded, to sound alarms if predefined limits have been reached, to shut systems down to prevent failure or accidents, and to transmit feedback signals to electrohydraulic closed loop control systems.

Common types of transducers are:

Pressure transducer - these devices are usually of the piezoelectric or the strain gauge type, which will transmit or modify an electric signal in direct proportion to forces applied to them by a pressure. Piezoelectric crystals will generate a small electric charge that can be amplified to a useable current level. Strain gauges are small resistors that, when altered by the application of a force, will change their resistance value accordingly. The change in resistance will cause a change in current in a monitoring circuit, and the change can be amplified to a useable level.

Force transducers - the most common method of measuring forces is with strain gauges. Linear or rotational forces can be measured in this way, by monitoring the resistance change in strain gauges caused by deformation of the material they are applied to. Transducers of this type are also called "load cells" (linear tension or compression) and "torque meters" (rotary).

Angular velocity transducer - the most common type of angular velocity (RPM) transducer is a tach-generator, basically a small DC generator that produces a voltage output in proportion to running speed. A less common but more accurate and more universally applicable type utilizes an inductive sensor positioned near a rotating gear. As the gear teeth pass the sensor, a small voltage pulse is generated that can be counted and integrated by measuring devices to provide RPM. This same device can be used to determine rotary position or the number of revolutions a shaft has made, making it much more flexible than a tach-generator.

Flow transducers - using the same concept as the inductive sensor type of angular velocity transducer, the RPM of turbine blades inserted in a circuit can be integrated by measurement devices to provide flow readings. Instead of registering the number of gear teeth passing the sensor, the number of turbine blades passing would be in direct proportion to the flow across the turbine.

Linear displacement transducer - the simplest form of linear displacement measurement is a linear potentiometer; the wiper is connected to the moving object, and a voltage measurement across the potentiometer would be in proportion to the position of the wiper. This type is very inexpensive, but it is also subject to wear, accuracy and hysteresis problems. A more sophisticated type is called an LVDT (which stands for Linear Variable Differential Transformer). It is a non-contact, linear form of transformer. A known high frequency AC voltage is applied to the primary coil, and the secondary coil voltage indicates the position of the center core. The secondary coil voltage is rectified so that a DC voltage proportional to the center core position is obtained.

Angular displacement transducer - the same two concepts as used on the linear displacement transducers are available for rotary displacement (see also the inductive sensor type of angular velocity transducer). A rotary potentiometer is the simplest form, but subject to the same wear, accuracy and hysteresis problems as its linear counterpart. A rotary form of the LVDT, called an RVDT (for Rotary Variable Displacement Transducer) is a non-contact, rotary type of transformer. The RVDT is limited in angular movement to about 60°.

There are many other types of transducers available to measure properties such as vibration, acceleration, contamination, etc. They all have one principle in common; they provide a voltage or current output in proportion to the property being measured.

CHAPTER 4 .. Linear Actuators

Actuator is the general term used for the output device of hydraulic systems. Two broad categories are rotary actuators, that deliver their power in a rotating or circular motion, and linear actuators that deliver their power in a straight line. This chapter will concentrate on linear actuators; Chapter 5 will cover rotary actuators.

Hydraulic cylinder is the most common term for linear actuators, although other terms such as "ram", "jack", or "stroker" are frequently used. These other terms often have application specific meanings, so cylinder or hydraulic cylinder will be used to describe the majority of linear actuators.

Power in a hydraulic system is generated initially from a rotating device, such as an IC engine, and converted to fluid flow by a pump. The flow is directed by and through the system to the actuators, where it is converted to rotary power by motors, or into linear power by cylinders. It can be said that without actuators, there is no reason for hydraulic systems to exist.

LINEAR ACTUATORS

Construction

The construction of a typical linear actuator is shown in Figure 4.1. The major parts are the body, the rod end head, the cap end head, the piston, piston rod, piston seals and rod seals. There are many additional parts unique to different types of cylinders and different options. These will be discussed throughout this chapter.

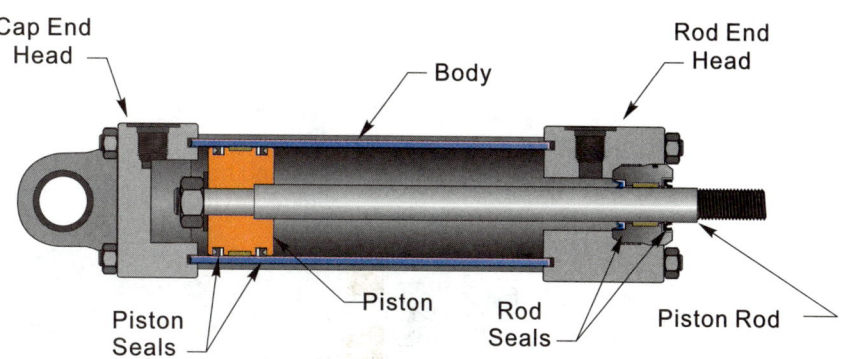

Figure 4.1 Typical linear actuator construction

Chapter 4 Linear Actuators

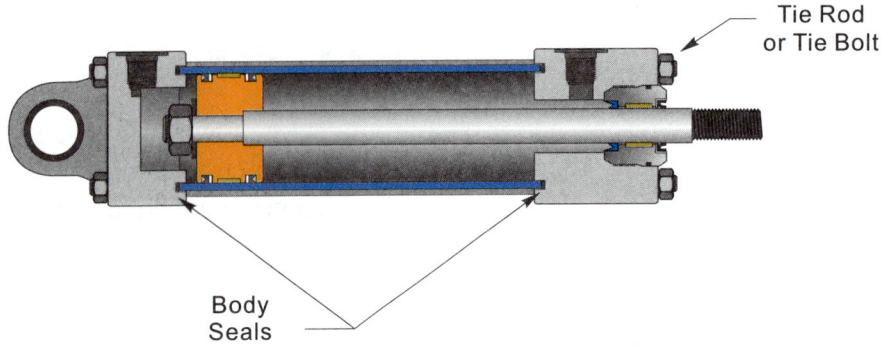

Figure 4.2 Tie bolt construction cylinder

There are two broad categories of construction in common use: the welded construction, and the tie rod, or tie bolt, construction. These two types are illustrated in Figures 4.2 and 4.3. Furthermore, Figure 4.3 illustrates two types of welded construction: a repairable and a non-repairable design. All three types of cylinders are traditionally used on mobile equipment.

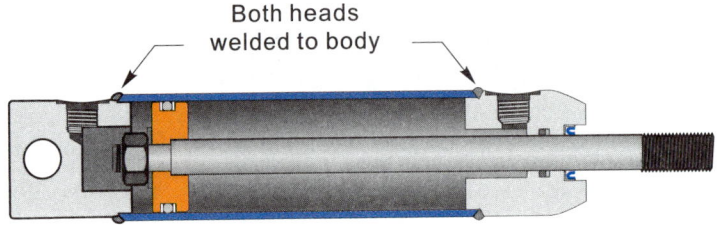

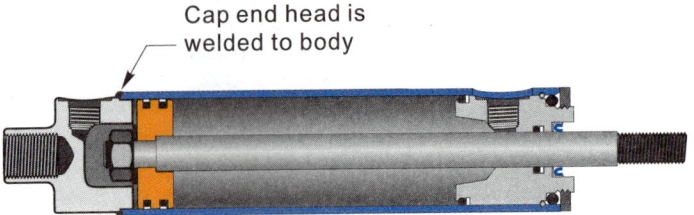

Figure 4.3 Welded cylinder construction

The most common type of cylinder used on mobile equipment is of welded construction. The non-repairable type is common in small diameters, about 2-1/2 inches and less, and is usually used in low stress applications where longer life can be expected. Where larger size cylinders are used, or where cylinders are used in more rigorous applications, a repairable design is generally preferred. Repair of welded cylinders is usually limited to replacement of rod and piston seals, unless special honing equipment is available.

Tie bolt constructed cylinders are occasionally used on mobile equipment, often when special internal valving is required. Tie bolt cylinders are more readily repaired, including honing of the body, than welded cylinders.

Seals

Most cylinders have two locations where fluid must be sealed: across the piston and around the rod. Tie bolt cylinders must also be sealed between the body and the two heads. There are many different types of seals for each of these purposes depending on the fluid being used, the integrity of the sealing required and the desired service life.

Rod Seals

Sealing around the piston rod must be done in two directions; the hydraulic fluid must be sealed within the cylinder, and foreign material must be sealed from entering the cylinder from the outside. These are almost always two different seals (Figure 4.4), referred to as the "rod seal" and the "wiper seal", and may be replaced individually for service. Integrity of the rod seal is very important as leakage from this area of a cylinder is to the outside causing loss of fluid to the system and resulting in a dirty and/or slippery machine. Multiple rod seals are frequently used to provide additional sealing integrity, as shown in Figure 4.5.

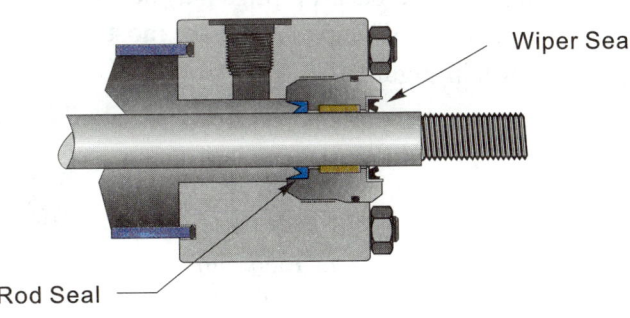

Figure 4.4 Rod and wiper seals

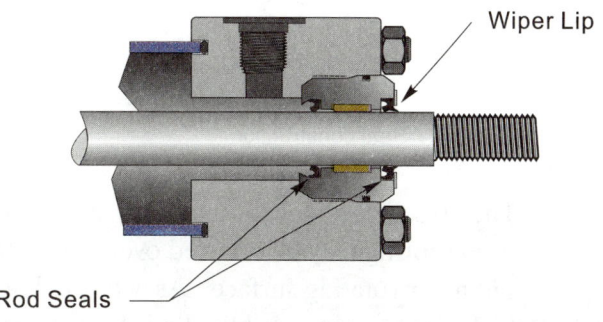

Figure 4.5 Double rod seal and wiper

Two rod seals are also occasionally used to provide a "gland drain" option. A small tube or hose is connected to the rod end head between two seals, draining off any fluid that may pass into this chamber (Figure 4.6). This becomes additional assurance against any fluid leakage to the outside of the system.

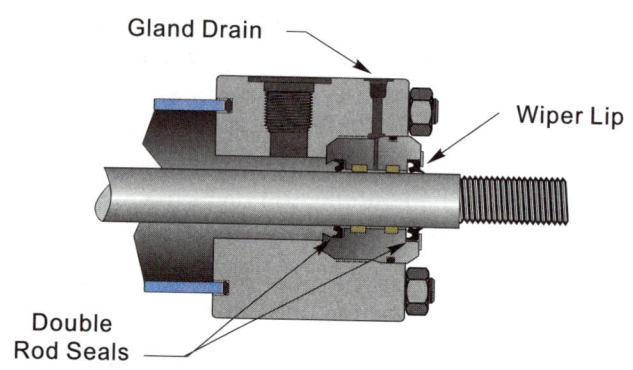

Figure 4.6 Double seal with gland drain option

The seals may be installed directly into the rod end head, as shown in Figure 4.3, but more commonly are installed in a seal cartridge assembly, or gland assembly, which is then installed in the rod end head as shown in Figure 4.2. This latter approach is for manufacturing convenience as well as facilitating service and seal replacement.

Rod seals are a flexible material that is held against the rod surface by a combination of initial compression (the seal inside diameter is slightly smaller than the rod outside diameter) and hydraulic pressure acting against it. A simple "O" ring seal with back-up rings may be used, and a lip-type seal is a popular design. "U"cup or "V" cup packing are most commonly used. Typical seal designs are shown in Figure 4.7.

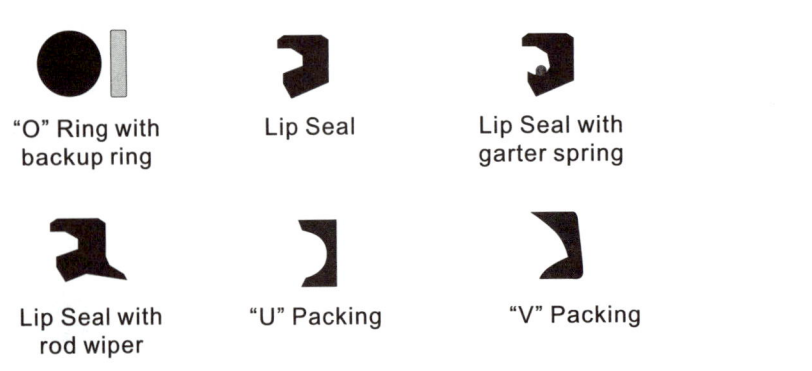

Figure 4.7 Seal construction

The lip seal is a molded material, usually molded onto a metal or hard plastic frame. A coil spring may be inserted over the lip to provide initial contact of the lip to the sliding or rotating surface. As with the U-cup or V-cup, the concave side of the seal faces the pressure, and the lip is forced against the sealing surface by pressure to create a tight seal.

Piston Seals

Depending upon the application, piston seals may or may not require the same leakage integrity as the rod seals. There are many different designs and materials for piston seals, resulting in a wide range of durability and sealing integrity.

The most durable piston seal is made of cast iron, steel or chrome plated steel. An unfortunate characteristic of metal seal rings, however, is a tendency for greater leakage flow than other types. They remain popular due to their durability and their compatibility with high temperatures, and are very satisfactory in applications where a small piston seal leakage is of no concern.

A more popular design utilizes the basic "O" ring, installed in a groove in the piston and supported by one or two backup rings. The backup rings prevent the "O" ring from extruding out of the groove under pressure; a single-acting cylinder normally uses one backup ring, and a double acting cylinder uses two, one on each side of the "O" ring. If two "O" rings are used on the piston, two backup rings are used; one on the low pressure side of each "O" ring. Backup rings are usually made of a plastic compound or an impregnated fiber material. Figure 4.8 illustrates the methods of installing "O" ring seals.

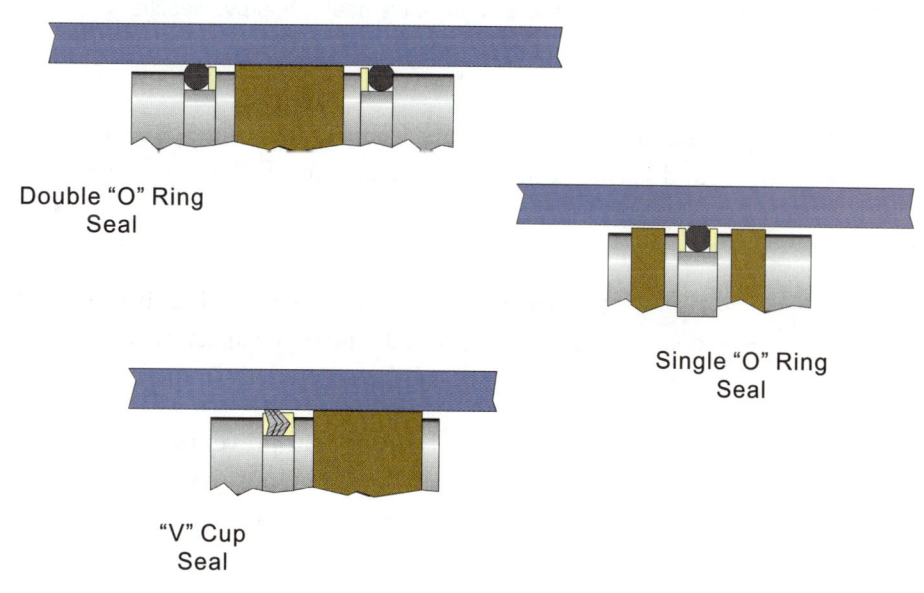

Figure 4.8 Seal installations

U-cup and V-Cup packings are frequently used as piston seals, and are very effective in preventing leakage flow. The seals are installed in sets, consisting of several U-shaped or V-shaped seals held in place by a gland so that the open face of the seals face the pressure side of the piston. This way, pressure forces the lip of the seal against the body wall, creating a tight seal as shown in Figure 4.9. In a double-acting cylinder, two sets of packing are used, each set facing a pressure side of the piston.

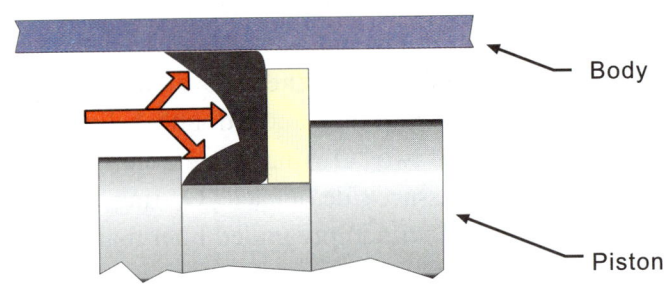

Figure 4.9 Pressure tightens the seal to improve sealing

Seal Material

Materials used for seals are usually synthetic rubber, although rubber compounds and plastic compounds are also used. The main criteria for material selection is compatibility with the fluid being used, wear resistance and temperature conformance.

Seal wear depends a great deal on factors other than the material used; lubrication qualities and cleanliness of the fluid in contact with the seal is by far the most important. Also, for the seal to be lubricated properly, it must be kept "wet" by the fluid.

A "perfect" seal would be one that prevents all leakage. In practice, however, a minute amount of lubrication film must be present for the seal to slide easily over the mating surfaces. In most applications, a seal is considered effective if there is no obviously detected quantity of fluid passing it.

Chapter 4 Linear Actuators

Types of Cylinders

There are many types of cylinders, each having its own advantages and typical uses. The general categories are ram, single-acting, double-acting and telescopic.

Ram (Figure 4.10): Perhaps the simplest actuator is the ram. It has only one fluid chamber and exerts force in only one direction. Most are mounted or used vertically and retract by the force of gravity. Practical for long strokes, rams can be found in "bottle" jacks and automobile hoists.

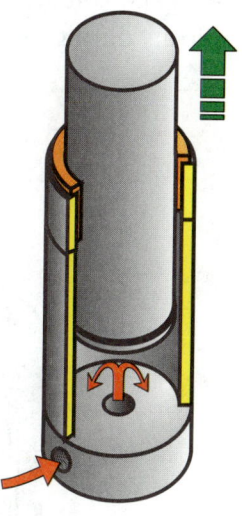

Figure 4.10 Single acting ram

Single-acting cylinders Figure 4.11: Operating much the same as rams, single acting cylinders apply a force in one direction, relying on gravity or a counter-force to retract. The primary difference between a single-acting cylinder and a ram is the single acting cylinder uses a piston, and leakage flow past the piston is ported to the reservoir to minimize external leakage. Single-acting cylinders are typically used for truck hoists and crane booms.

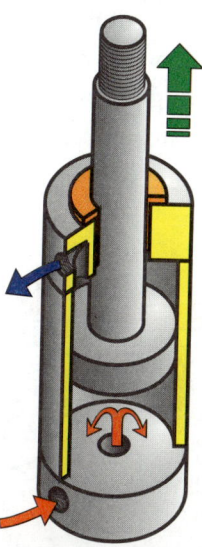

Figure 4.11 Single acting cylinder

Double-acting cylinder Figure 4.12: Perhaps the most popular type of cylinder on mobile equipment, the double-acting cylinder exerts force in both directions, extending and retracting. To extend, fluid is ported into the cap end of the cylinder and the rod end port is vented to the reservoir. During retraction, fluid is ported into the rod end of the cylinder, and the cap end port is vented to the reservoir. Double-acting cylinders are also called differential cylinders because the effective area, and therefore volume, of each end is different by virtue of the space taken up by the rod area and volume. This differential area and volume causes a different force and velocity during extension and retraction.

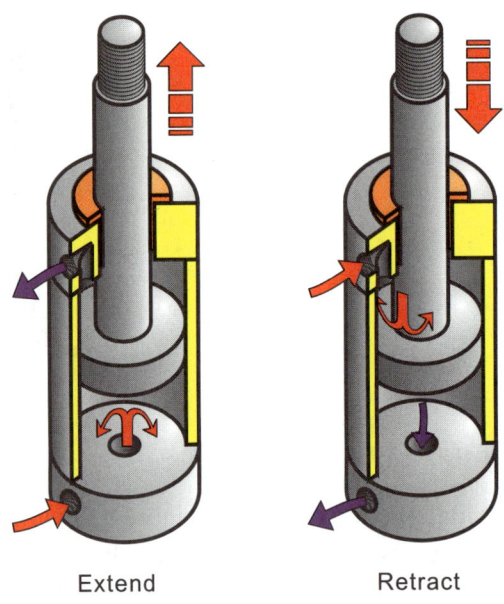

Figure 4.12 Double acting cylinder

A variation of the double-acting cylinder is the double-rod cylinder. In this version, the cylinder rod extends through both end caps Figure 4.13, thus equalizing the area and volume between both ends of the cylinder. This equalizes the forces and velocities during extension and retraction. A typical use for double-rod cylinders is in power steering applications.

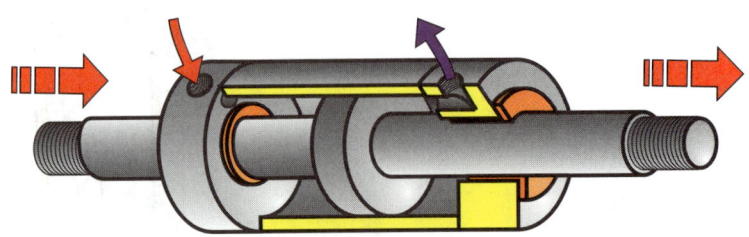

Figure 4.13 Double rod, double acting cylinder

Telescopic cylinders Figure 4.14: Most telescopic cylinders are single-acting. Telescopic cylinders consist of a series of nested tubular rod segments called sleeves. Each sleeve extends individually during extension. There may be two, perhaps three, and possibly up to five sleeves in a cylinder. A long working stroke and a short collapsed length result, making them ideal for applications such as industrial lift trucks and large tilt bed or dump trucks. An inherent feature of telescopic cylinders, due to the sequencing smaller diameters of sleeves, is a reduction of force capability and an increase of velocity during each succeeding stage.

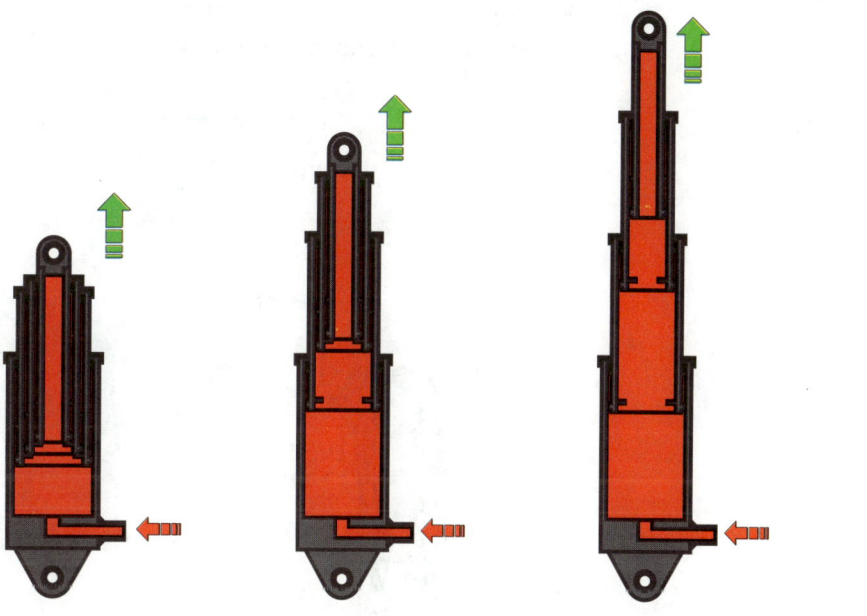

Figure 4.14 Four stage telescopic cylinder

Telescopic cylinders may also be double-acting, see Figure 4.15, although these are not too common. Because of the very small areas involved in the retracting phase, retracting forces are quite low. Double-acting telescopic cylinders are typically used in refuse haulers as high compaction forces and a long stroke are required, but retraction force requirements are very small.

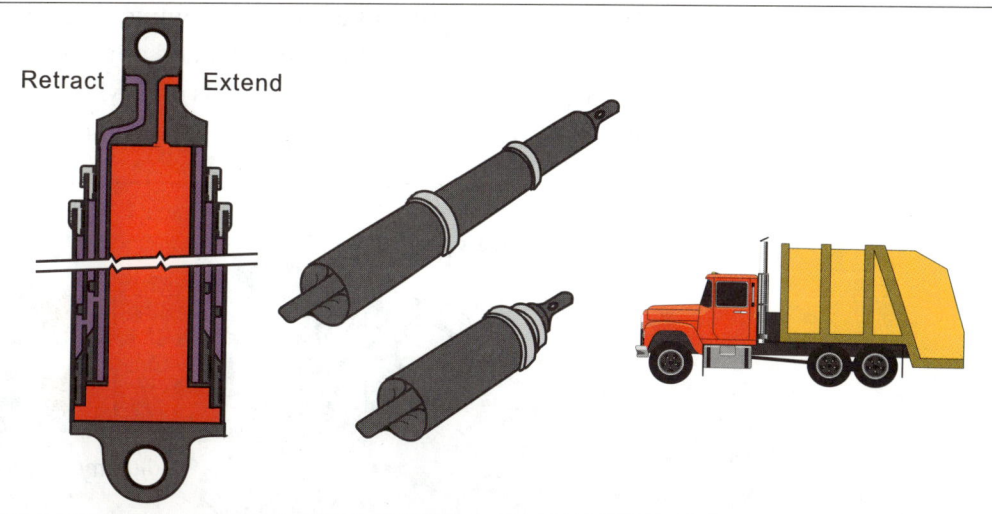

Figure 4.15 Double acting telescopic cylinder

Chapter 4 Linear Actuators

Special Features

There are many special configurations of linear actuators that customize them for particular applications. Some of these are proprietary to specific machine manufacturers, and are not typical beyond those applications. Others, though, are more generally available and may be considered standard options.

Integral Valving - It frequently can be advantageous to incorporate some specific valve requirements within the cylinder, usually into the cap end head. The popularity of screw-in cartridge valves (see Chapter 8) has made some of these features relatively easy and inexpensive to accomplish. Although the counterbalance valve Figure 4.16 is probably the most popular type of integral valve, other types such as directional control, flow control and sequence valves can be found.

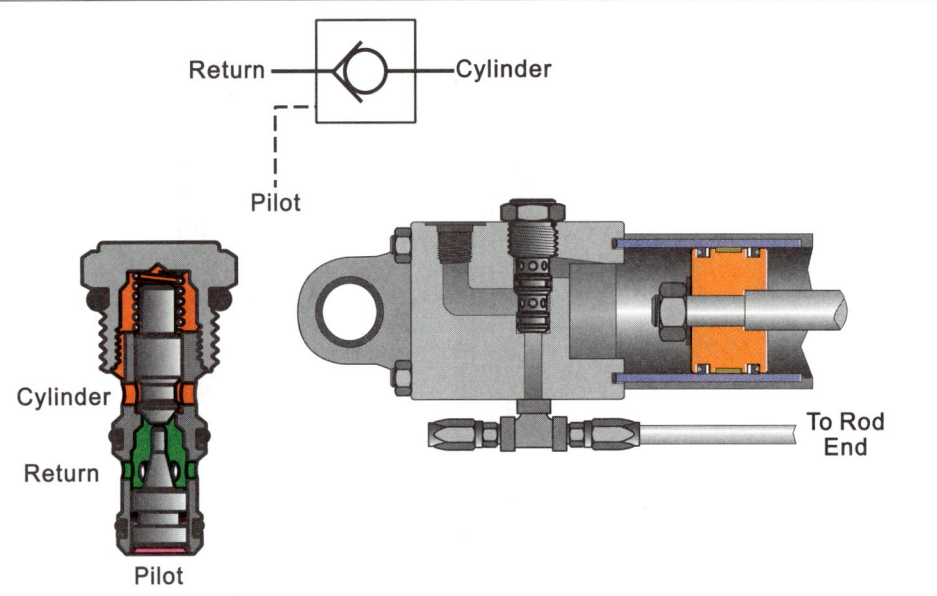

Figure 4.16 Cylinder with integrated counterbalance cartridge

Cushions - Frequent and abrupt end-of-stroke stops can damage a cylinder; cylinders that extend and/or retract at high speed can fail catastrophically in just a few strokes. Cylinder cushions are a fairly common feature on mobile equipment that helps to slow the piston down near the end of its stroke and reduce the impact. Cushions may be found on one or both ends of a cylinder to act as hydraulic brakes during cycling. Figure 4.17 shows one type of cushion used on the retraction stroke.

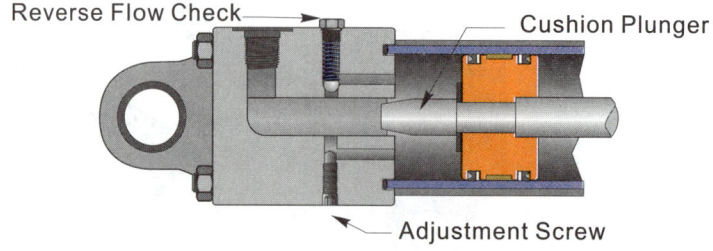

Figure 4.17 Cylinder with retraction cushion

Stroke limiting - Although a simple stop tube (Figure 4.18) can be used to limit the travel of a cylinder, it is frequently desirable to externally adjust the stroke travel. This is done with a stroke control valve similar to that shown in Figure 4.19. The stroke is adjusted by locating the stop flange on the cylinder rod to activate the stop valve at different retraction positions. Reversing the flow direction will compress the valve spring and allow the cylinder to extend.

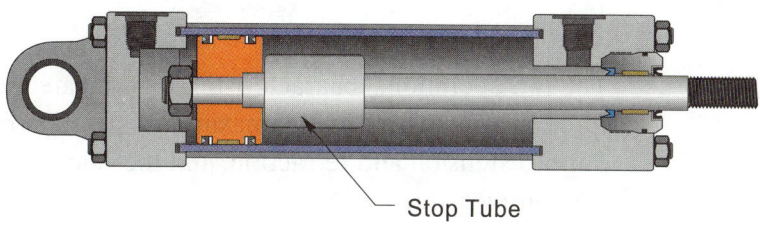

Figure 4.18 Cylinder with stroke limiting stop tube

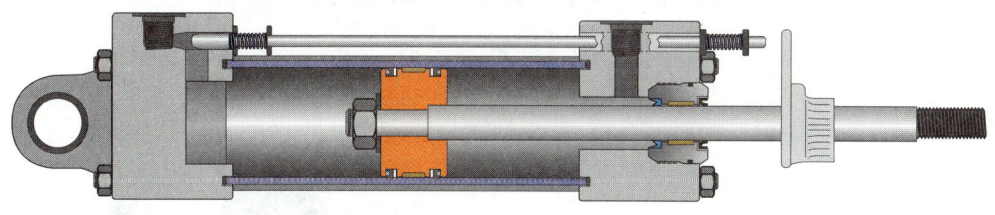

Figure 4.19 Cylinder with adjustable stop valve

Thermal relief valves - Cylinders that have cooled down, and are then subjected to heat from the sun (or other source) may be damaged by the resulting fluid expansion. Extremely high pressures can be developed as the fluid expands, unless there is a way to relieve the pressure and drain off a small amount of the fluid. A small integral relief valve or cartridge (Figure 4.20), set much higher than system pressure, will accomplish this task and prevent damage to the cylinder.

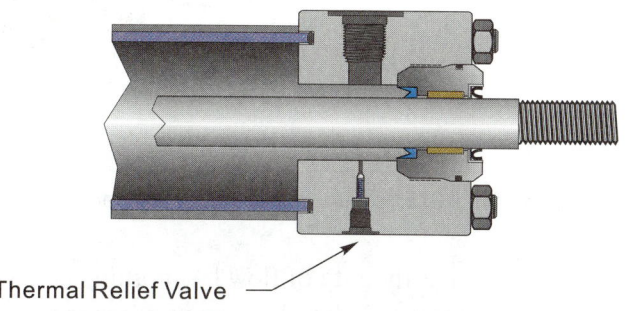

Figure 4.20 Cylinder with thermal relief valve in rod end head

Chapter 4 Linear Actuators

Cylinder Ratings

The ratings of a cylinder include its size and pressure capability. Principle size features are bore (piston diameter or inside body diameter), piston rod diameter and stroke length. The pressure rating is based on the size, design and materials used, and is established by the manufacturer. Refer to the cylinder nameplate or the manufacturer's catalog for this information.

OPERATING PARAMETERS

The two operational characteristics that dictate the selection of a linear actuator are force and speed; that is, the maximum force that can be exerted by the cylinder during extension and retraction, and the velocity of the piston rod during extension and during retraction.

Figure 4.21 illustrates a single-acting cylinder being extended by porting fluid into the cap end. As fluid enters, the piston will rise and push up a load. The magnitude of the load and the area of the piston will dictate the fluid pressure in the cap end of the cylinder. The speed at which the load will rise is dependent upon the fluid flow rate into the cap end and the area of the piston.

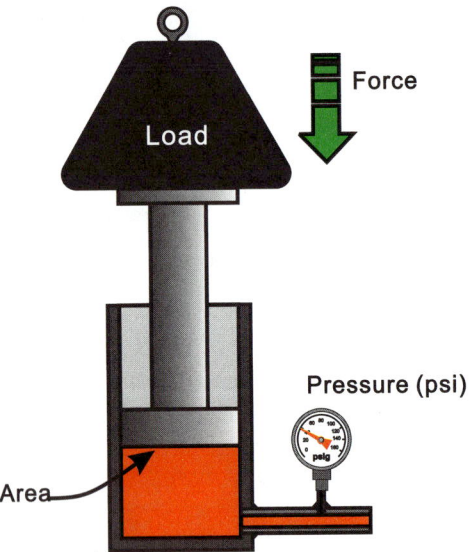

Figure 4.21 Pressure is created by the load and the area of the cylinder

Note that the rate of flow has nothing to do with the pressure that occurs in the cap end of the cylinder. Inlet pressure is strictly determined by the load against the piston, and the area of the piston.

94

Two basic formulae are used to determine the operating characteristics of linear actuators (reference chapter 2):

Formula 4-1

Force = Pressure x Area

Where:

Force = lbs
Pressure = lbs/in² (PSI)
Area = in²

Formula 4-2

Velocity = Flow ÷ Area

Where:

Velocity = in/min
Flow = in³/min
Area = in²

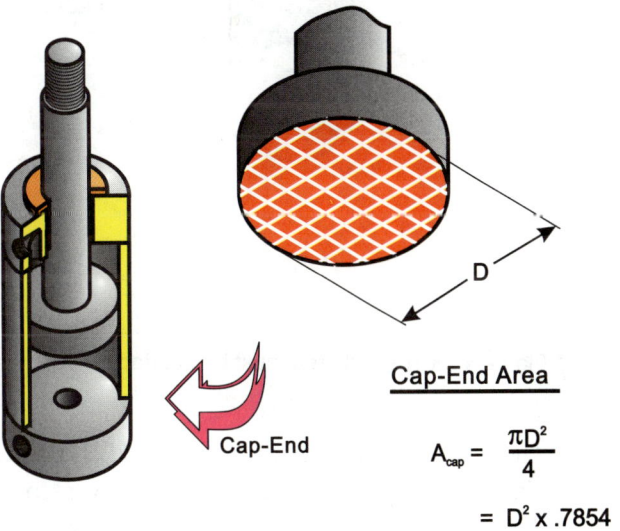

Figure 4.22 The effective area of the cap end of a cylinder is based on the bore diameter

The areas used in formulas 4-1 and 4-2 are the effective areas of the cylinder piston. Figure 4.22 illustrates the calculation of area for any given piston diameter (bore diameter) as:

Formula 4-3

Area = Diameter² x 0.7854

Where

Area = in²
Diameter = in

For the rod end of a double acting cylinder, the effective area is less because the area of the rod is not effective. As Figure 4.23 illustrates, The rod end area is the difference between the bore area and the rod area:

Formula 4-4

Rod End Area = Bore Area - - Rod Area

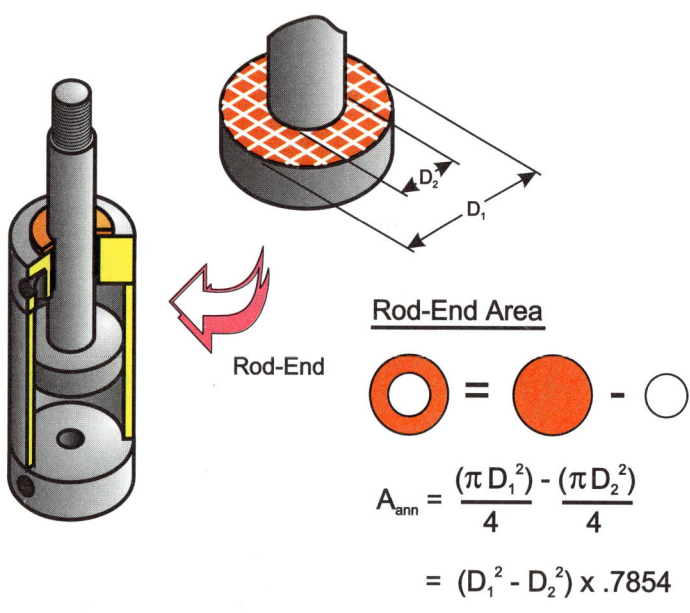

$$A_{ann} = \frac{(\pi D_1^2) - (\pi D_2^2)}{4}$$

$$= (D_1^2 - D_2^2) \times .7854$$

Figure 4.23 The rod area must be subtracted from the bore area to get annulus area

Therefore, the effective area of the rod end of a cylinder will always be less than the effective area of the cap end. The only way to overcome this differential is to utilize a double rod cylinder, which reduces the effective area of both ends of the cylinder equally.

Reviewing formulas 4-1 and 4-2 shows that, if the area of the rod end of the cylinder is less than the area of the cap end, there will be less force available when retracting the cylinder than when extending (pressure being equal). Furthermore, the cylinder will retract faster (higher velocity) than it will extend (flow being equal).

These differential operation characteristics of double-acting, single rod cylinders are normal, and must be either accommodated (accepted) or dealt with by the use of a double rod cylinder or by metering flow by using a flow control.

The differential operation is very predictable by comparing the areas of the rod end and the cap end of the cylinder. A ratio factor can be established by dividing one area by the other, and multiplying or dividing the operating characteristic of one end of the cylinder to predict the characteristic of the other end.

Figure 4.24 illustrates this approach as follows:

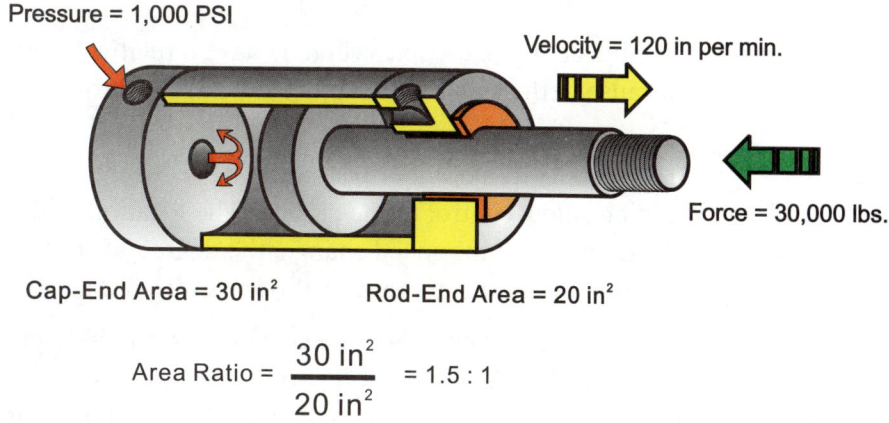

Figure 4.24 Area ratio is very useful for cylinder calculations

The cylinder has a cap end area of 30 in², and a rod end area of 20 in². The area ratio factor for this cylinder is:

R = A$_{cap}$ ÷ A$_{rod}$ = 1.5

Using the area ratio factor, one can now calculate the differential operating characteristics for the cylinder:

Pressure differential - the load on the cylinder has created a pressure of 1000 psi in the cap end of the cylinder. The pressure in the rod end caused by the same load acting in the opposite direction will be:

P$_{rod}$ = P$_{cap}$ x 1.5 = 1500 psi

Velocity differential - The extension velocity of the piston rod is 120 inches per minute. The retraction velocity will be:

v$_{ret}$ = v$_{ext}$ x 1.5 = 180 inches per minute

Volume differential - The volume of fluid required to extend the cylinder full stroke is the area of the cap end times the stroke length, equal to 600 in³. The volume to retract the cylinder will be:

V$_{ret}$ = V$_{ext}$ ÷ 1.5 = 400 in³

Telescopic cylinders exhibit differential operating characteristics of a different sort; the effective area of each succeeding stage is smaller so the maximum force of the cylinder is reduced at each stage for any given pressure. Alternatively, for any given load on the cylinder, the pressure required to raise the load will be higher at each succeeding stage, and the velocity will increase at each step.

Chapter 4 Linear Actuators

Most telescopic cylinders are single-acting; retraction is caused by the weight of the load. For any given load, the pressure in the smallest diameter stage will be higher, so the smallest stage will retract first. Each successive stage will retract in order as the diameter increases and the pressure reduces.

Double-acting telescopic cylinders have a relatively low retraction force capability because of the small area subject to pressure during the return stroke.

Metering Cylinder Flow

The velocity of a cylinder piston can be reduced in either or both directions by adding a flow control to the circuit. The location of the flow control will determine the direction of the speed change, the degree of control over the cylinder operation and may affect the pressure that occurs within the cylinder.

Flow controls can be a simple needle valve or a more sophisticated pressure compensated flow control as discussed in Chapter 9. Either type can be installed on the inlet of the cylinder causing reduced flow and lower velocity of the piston in the metering direction. Flow controls can also be installed on the outlet of the cylinder to have the same effect. The selection of "meter-in" or "meter-out" will have different effects under different circumstances, so it is important to understand what the effects are.

Figure 4.25 shows a meter-in application with the load on the cylinder pulling down, referred to as an over-running load. This approach to metering will control the cylinder speed, but because the load is causing a vacuum in the cap end of the cylinder, cavitation can occur. This can cause imprecise positioning of the cylinder due to a "bouncing" or "spongy" action. Damage to the cylinder components can also occur when the cavitation bubbles collapse.

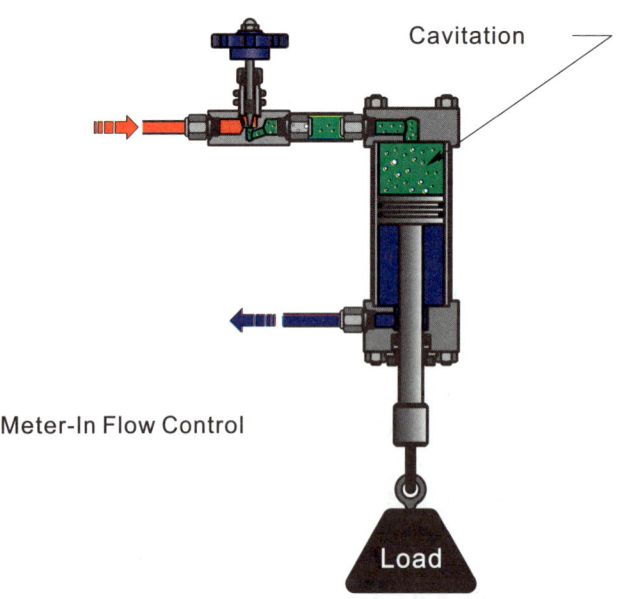

Figure 4.25 Meter-in flow control with an over-running load

Figure 4.26 shows an alternative location for the flow control, metering the fluid out of the cylinder (meter-out) instead of in. Because both the cap end and the rod end of the cylinder are under pressure, cavitation and the associated conditions will not occur. What does occur in this illustration, however, is pressure intensification.

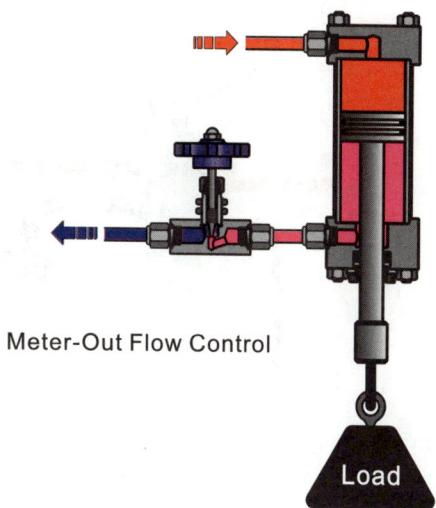

Figure 4.26 Meter out flow control with an over-running load

By metering flow out of the rod end of the cylinder, pressure in the cap end will rise to system pressure. This occurs because the portion of pump flow that cannot go into the cylinder must go over the system relief valve. Figure 4.27 illustrates the magnitude of pressure that can then result in the rod end of the cylinder, caused by the area differential in the cylinder and the weight of the load.

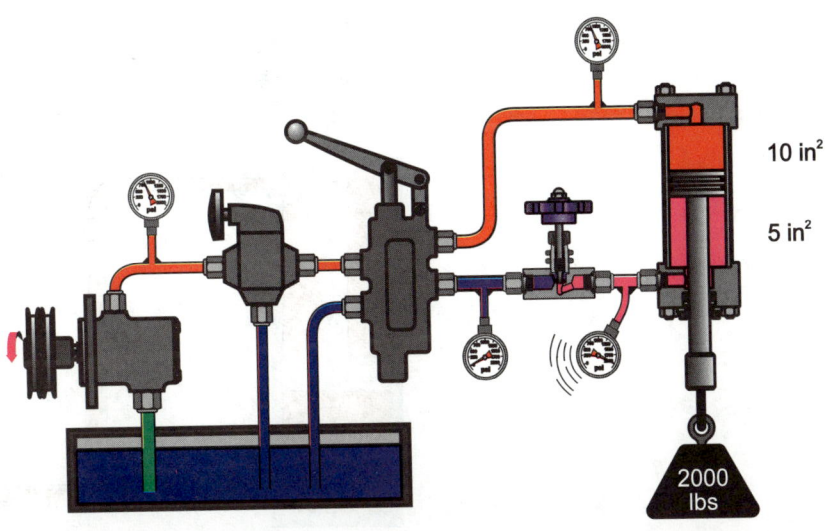

Figure 4.27 Pressure intensification cause by a meter-out circuit

Chapter 4 Linear Actuators

Frequently, hydraulic cylinders will be applied in a situation that will result in both a conventional load and an over-running load during operation (Figure 4.28). Whether metering to control piston velocity is done by a directional control valve or by a flow control, the meter-in or meter-out issue must be addressed, and the effects must be understood to prevent damage to the cylinder or to the machine.

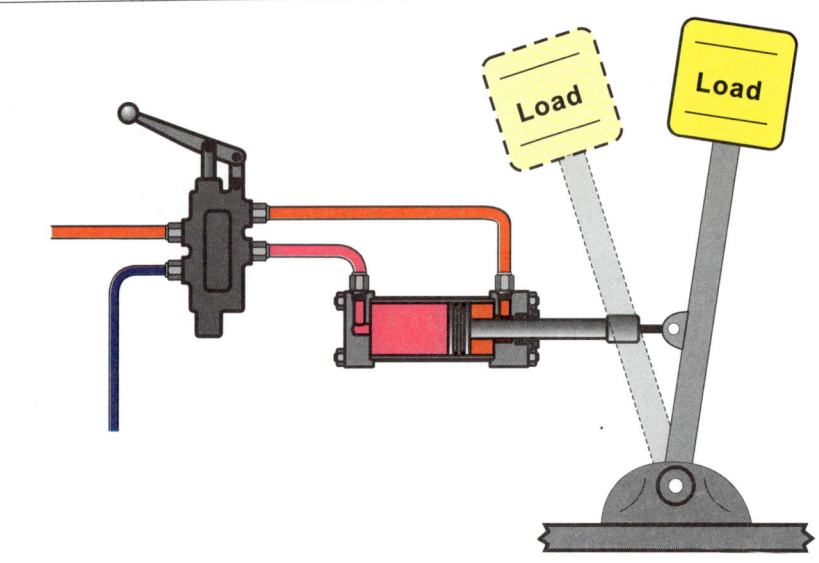

Figure 4.28 Cylinder subjected to conventional and overrunning loads

Regeneration

The area differential of a double-acting cylinder will potentially cause a higher pressure in the rod end of the cylinder when fluid under pressure is applied to the cap end. Because fluid will flow from a high pressure source to a lower pressure area, it is conceivable to connect the output of the rod end to the cap end as shown in Figure 4.29. The fluid from the rod end is then said to be "regenerated" and used to fill part of the volume of the cap end.

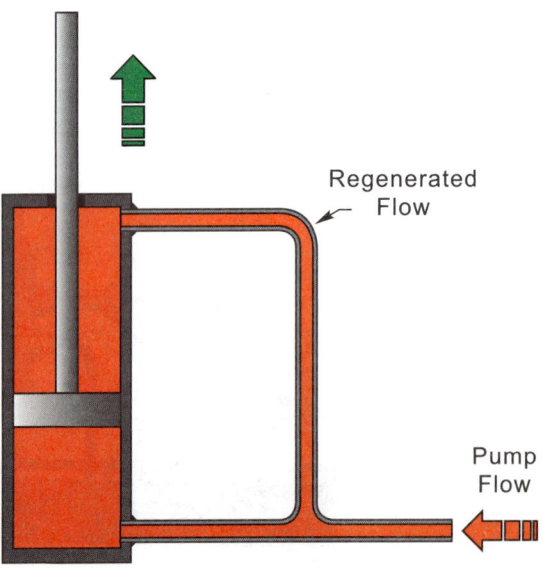

Figure 4.29 Cylinder connected to illustrate regeneration

The volume of fluid flowing from the rod end of the cylinder is equal to the volume in the cap end, except for the volume taken up by the rod. Stated as a formula:

$V_{rod\ end} = V_{cap\ end} - V_{rod}$

Re-orienting the formula will give:

$V_{cap\ end} - V_{rod\ end} = V_{rod}$

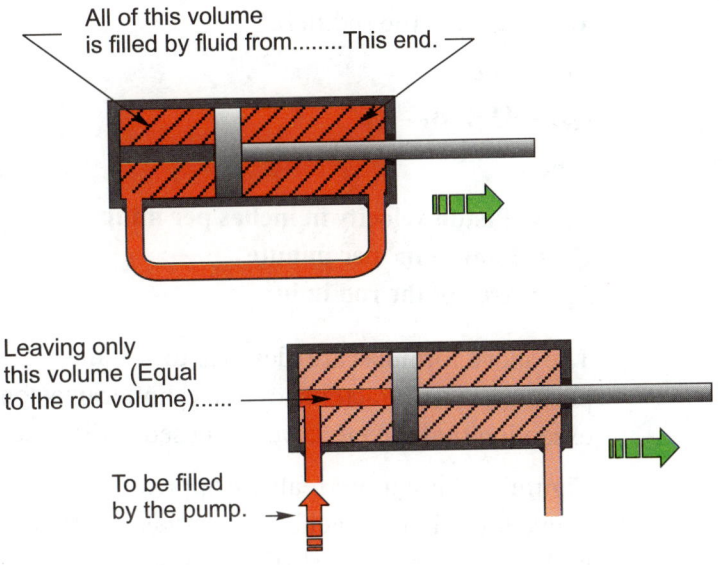

Figure 4.30 During regeneration, the pump displaces only the rod volume

This illustrates that when the fluid from the rod end of the cylinder flows into the cap end, the only volume left to be filled by another source (in this case, a pump) is the volume of the rod, as shown pictorially in Figure 4.30. Furthermore, the only effective area to provide a net force on the piston is the area of the rod, as shown in Figure 4.31.

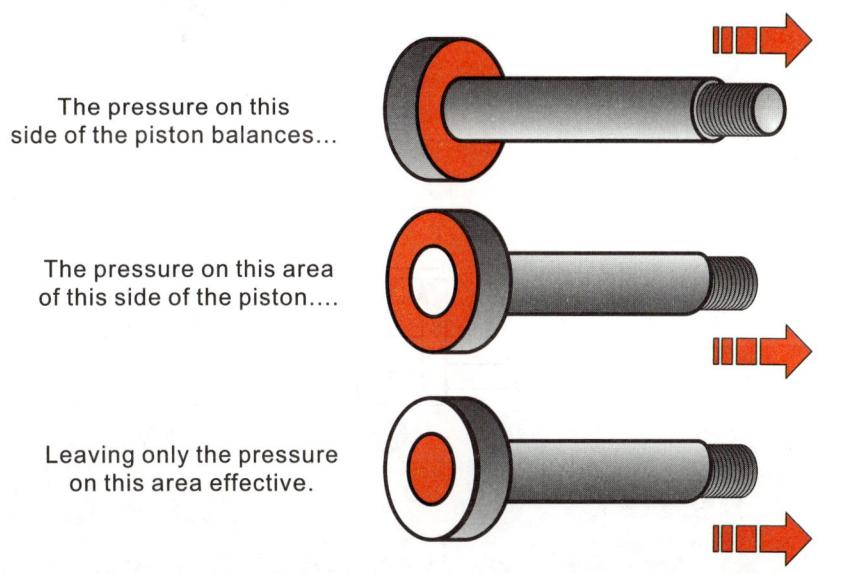

Figure 4.31 Regeneration reduces the effective area that creates extension force

Formulas for calculating force and velocity for cylinders connected in a regeneration mode are:

Formula 4-5

$F_{reg} = P \times A_r$

Where

F_{reg} = Force in lbs of a regenerating cylinder
P = Pressure in psi
A_r = Area of the rod in in^2

Formula 4-6

$v_{reg} = Q \div A_r$

Where

v_{reg} = Piston velocity in inches per minute
Q = Flow in in^3 per minute
A_r = Area of the rod in in^2

The above formulas are identical to 4-1 and 4-2 which were used to calculate the performance characteristics of a cylinder in a conventional connection, with the exception that the effective area used is the area of the rod.

Figure 4.32 is a more realistic application of a cylinder capable of operating in a conventional or a regenerative mode. In this application, using a long cylinder with an area ratio of 2:1 (the cylinder rod area is half the cylinder cap end area), the cylinder can extend at twice the normal velocity while regenerating. With this arrangement, there are three operating conditions:

Position 1 - Low Speed Extend, High Force
Position 2 - High Speed Extend, Low Force
Position 3 - High Speed Retract, Low Force

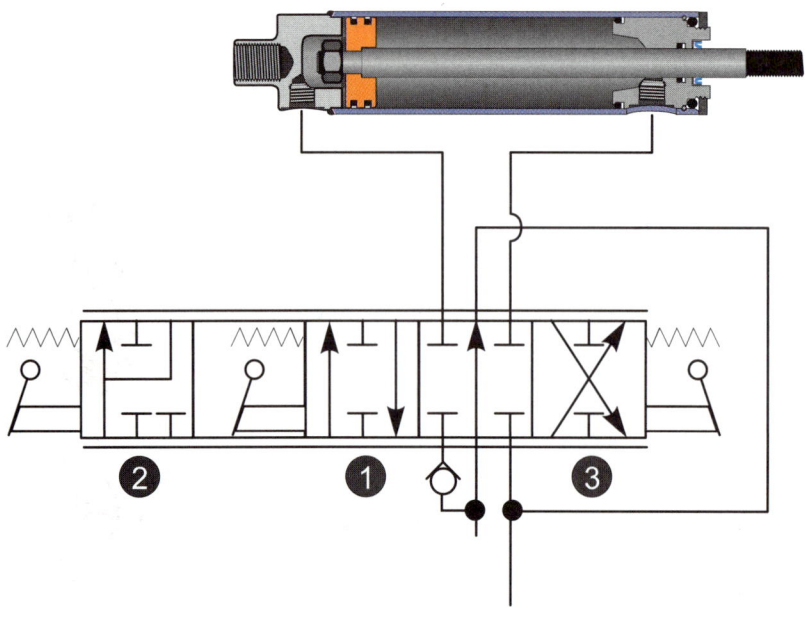

Figure 4.32 Four position valve used for conventional and regenerative operations

Although it is obvious, it needs to be stressed that regeneration can only be applied during cylinder extension, as pressure intensification only occurs in that direction.

Seal Failure

The most common failure on hydraulic cylinders is for one or more seals to begin leaking excessively. Excessive leaking can be considered any leakage that materially affects the cylinder performance, or causes unacceptable quantities of fluid to collect outside the cylinder. The two most common locations for potential seal failure are the rod seal and the piston seal.

Failure of the rod seal (Figure 4.33) is generally quite obvious because it is readily visible. A loss of performance may not be noticed if the leak is small, but problems of appearance, safety, environment and gathering of dirt and dust around the rod need to be considered.

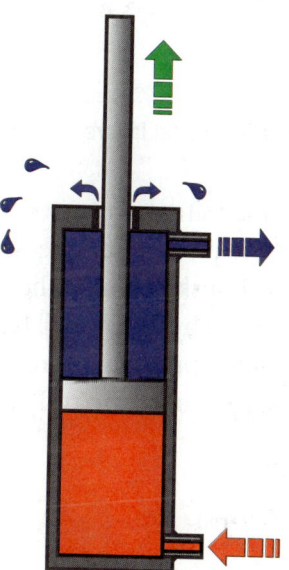

Figure 4.33 Rod seal failure

Failure of the piston seal is not externally visible, but may more readily be identified by reduced performance of the cylinder and machine. As illustrated in Figure 4.34, some fluid will pass the piston seal and not be effective in moving the piston, thus causing slower operation. The larger the leak, the more noticeable it will become.

It is commonly assumed that a piston seal leak will allow the cylinder to collapse while the machine is idle, causing a load to "drift" down. This can only be true if the fluid that leaks past the piston seal has some place to exit the cylinder, such as through the rod seal (external leakage) or past the control valve and/or counterbalance valves to the reservoir. If these two passages are secure, then fluid that passes the piston from the cap end to the rod end will, because of the differential area, be trapped in the rod end and prevent more than a small, perhaps imperceptible, drift downward.

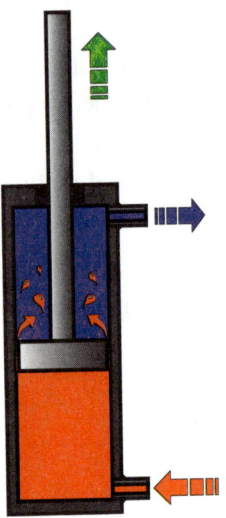

Figure 4.34 Piston seal failure

A very serious situation can occur under these circumstances, however. As fluid leaks past the piston seal, pressure will rises in the rod end of the cylinder. This pressure will push down on the piston, and add to the pressure in the cap end of the cylinder caused by the load. This will cause pressure in the rod end to rise further, aggravating the problem more. This circle of pressure rises will continue until the final pressure in both the rod and cap end of the cylinder become equal to:

Formula 4-7

$$P_{ult} = F \div A_{rod}$$

Where

P_{ult} = The final pressure reached, psi
F = The load on the cylinder, lbs
A_{rod} = The area of the rod, in^2

It can be seen that, depending on the load, cylinder size and rod size, pressures can reach dangerous levels if there is no means of controlling the pressure buildup.

The above pressure intensification problem will not occur if the load is suspended from the cylinder (rod facing down).

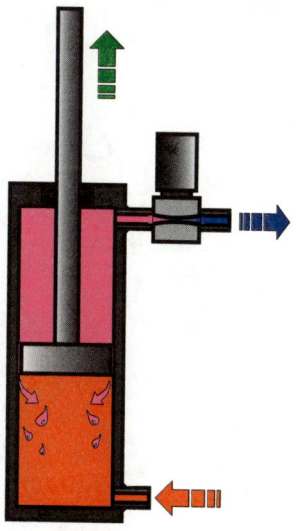

Figure 4.35 Piston seal failure in a meter-out application

Figure 4.35 illustrates another situation that can occur when piston seals leak and the cylinder is used in a meter-out application. Because of pressure intensification that occurs during meter-out control, the fluid in the rod end of the cylinder can pass the seals into the cap end, creating a regeneration effect. Just as if regeneration had been connected outside the cylinder, performance will change noticeably. The piston velocity will increase, and the maximum force that the cylinder can exert will decrease. This change of performance will become more evident as the piston seal leakage rate increases.

Chapter 4 Linear Actuators

CHAPTER 5 .. Rotary Actuators

Just as linear actuators convert fluid power to linear motion, rotary actuators convert fluid power to rotary motion. Fluid is pushed into the inlet of the rotary actuator and causes the output shaft to rotate. Resistance to rotation by an external load creates pressure in the hydraulic circuit and in the inlet of the motor.

Rotary actuators can be divided into two categories (Figure 5.1): continuous rotation and limited rotation. Continuous rotation actuators are usually referred to as motors, and limited rotation actuators are usually referred to as rotary actuators.

Another distinction occurs within the continuous rotation actuators; those that are bi-directional (will operate in either clockwise or counterclockwise rotation) and those that are uni-directional (one direction of operation only). The majority of rotary actuators used in mobile equipment are bi-directional motors.

A third distinction is motors that are designed for high speed, low torque applications and those that are designed for low speed, high torque applications. These are generally designated as either high speed motors or HTLS (high torque, low speed) motors.

And finally, piston motors may be fixed or variable displacement. Variable displacement motors are used occasionally on ground drive, winch or conveyor applications, to name a few. This approach can provide a selection of speed and torque ranges in addition to the basic, or working, range. An example would be in a ground drive installation where a high torque, low speed range would be used for a working range, and a low torque, high speed range would be available for traveling.

The most popular types of motor construction will be reviewed in this chapter: external gear, internal gear, vane, axial piston and radial piston.

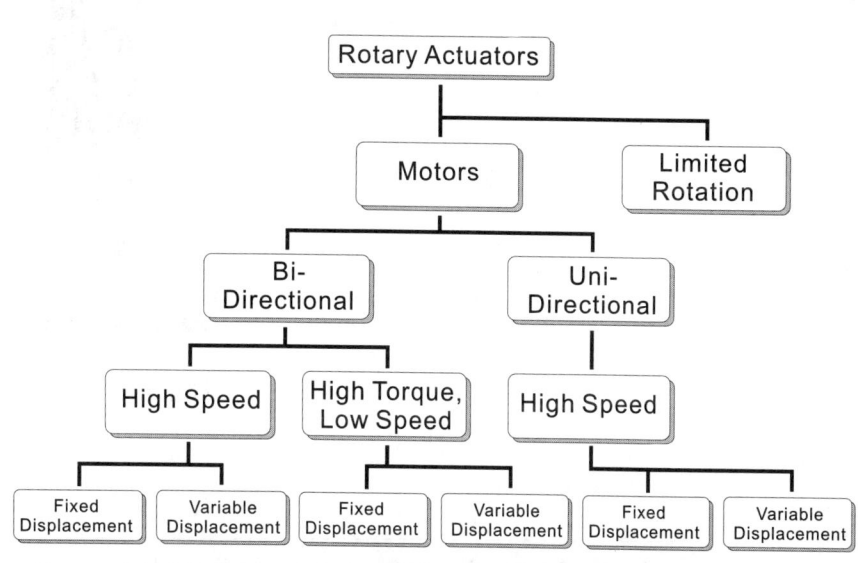

Figure 5.1 Rotary actuator classification

Chapter 5 Rotary Actuators

HYDRAULIC MOTOR CHARACTERISTICS

Hydraulic motors are rated by their torque and speed capabilities, and are usually designated by their displacement. Torque may be defined as a maximum capability in pound-inches (lb-in) or pound-feet (lb-ft), or it may be defined as an operating capability such as pound-inches per 100 psi. Speed is expressed in revolutions-per-minute (rpm), and usually designates the maximum speed capability, but may also express the rpm that will result for a given input flow rate, such as rpm per gallon per minute. Some definitions of typical terms follow.

Displacement: The theoretical volume of fluid that will turn the output shaft one full revolution. The volume is usually expressed in cubic inches or cubic centimeters, and the displacement will usually be designated as CIR or CCR (cubic inches or cubic centimeters per revolution).

Torque: The rotational force that a shaft absorbs or delivers is called torque. Motion is not required to have torque, but if the torque overcomes the resistance to it, rotational motion will result. Figure 5.2 illustrates an application of torque using a wrench, and Figure 5.3 illustrates how that torque can be doubled using either twice the force or twice the lever arm.

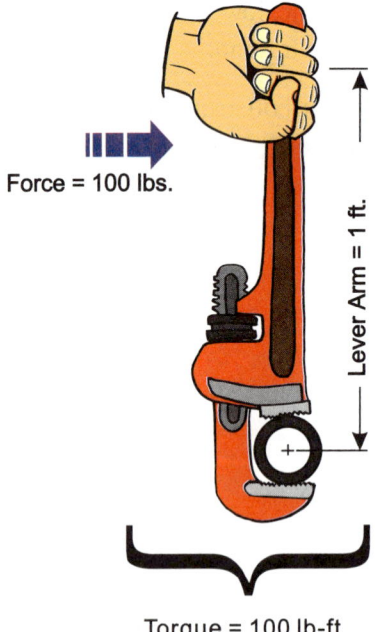

Figure 5.2 Torque is a rotational force determined by multiplying force times distance

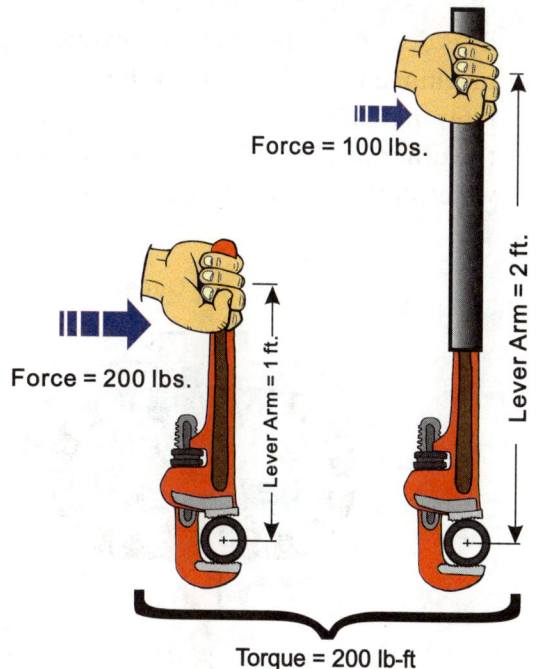

Figure 5.3 Torque doubles by doubling the force or doubling the lever arm (distance).

Starting torque: Friction is the resistance to relative motion between two materials in contact with each other. Static friction is the friction between two materials that have no relative motion, and dynamic friction is the friction between two materials that are moving relative to each other. Because static friction is always higher than dynamic friction, the pressure required to start a hydraulic motor is greater than that required to keep the motor in motion. Starting torque is an important characteristic of hydraulic motors, because it indicates the maximum torque available to begin motion when a mechanism is at rest.

Running torque: Running torque involves dynamic friction, and is therefore a higher value than starting torque. When sizing a motor for an application, both starting and running torque capability must be taken into consideration. One of the distinct advantages of hydraulic motors is that their running torque is constant at any given pressure throughout most of the speed range. The torque/pressure relationship changes only when the running speed becomes very low, and internal friction begins to approach that of starting torque.

Efficiency: Hydraulic motors have two characteristics that prevent them from delivering all of their torque and speed: internal friction and internal leakage. The effect of internal friction causes a torque loss, and is expressed as a mechanical efficiency of the motor. Internal leakage causes a speed loss, and is expressed as a volumetric efficiency. The product of mechanical efficiency and volumetric efficiency is the overall efficiency.

Chapter 5 Rotary Actuators

HIGH SPEED HYDRAULIC MOTORS

External Gear Motors

A cross sectional view of an external gear hydraulic motor is shown in Figure 5.4. This design is referred to as "external gear" design because the gear teeth are machined on the outside of the gears (internal gear designs will be described later in this chapter). One of the gears will be connected to an output shaft, and the other will be an idler gear. Not shown in the illustration are the side plates, which create a sealing wear surface on the sides of the gear set.

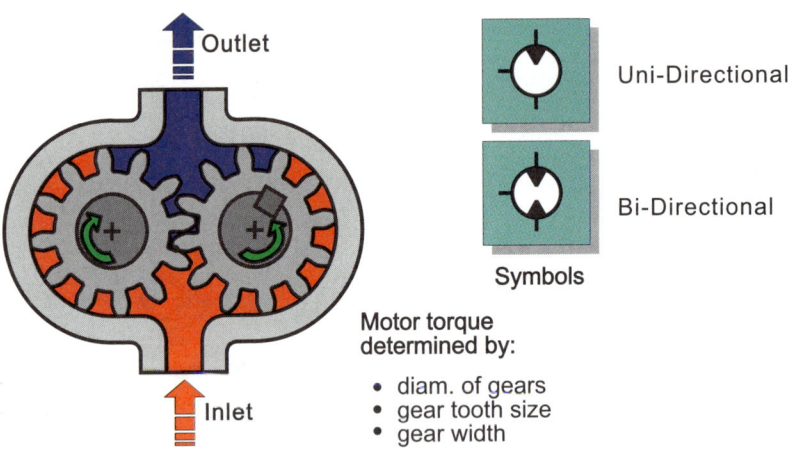

Figure 5.4 External gear motor

Gear motors operate because of a pressure differential, or ΔP, between the motor inlet and outlet. This pressure differential acts across the gear teeth, creating a force that tries to rotate the gear. As the diagram in Figure 5.5 shows, there are two paths around the periphery of the gears, and one path through the intersection of the two gears. In each path, the pressure works against the area of one tooth Figure 5.6, so that there are two tooth forces working in one direction, and one tooth force working in the opposite direction. The net result is the force of one tooth working around the periphery to rotate the output shaft.

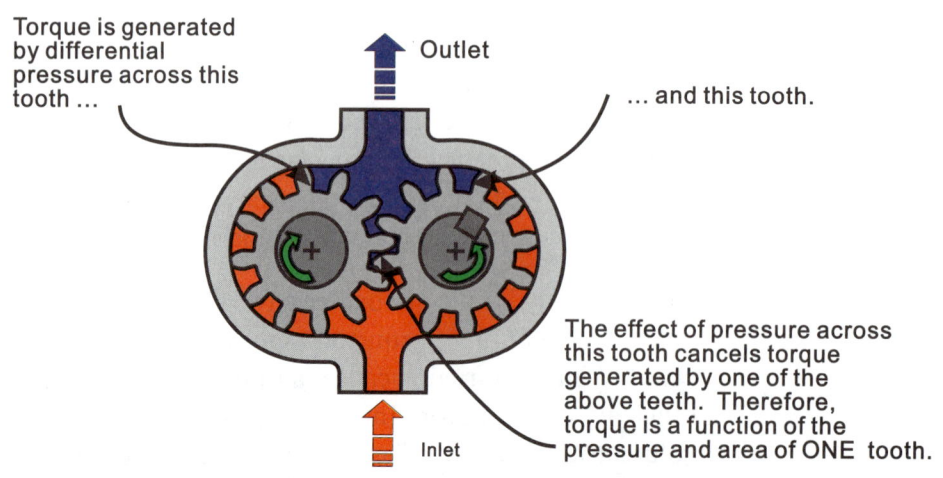

Figure 5.5 Torque generation in an external gear motor

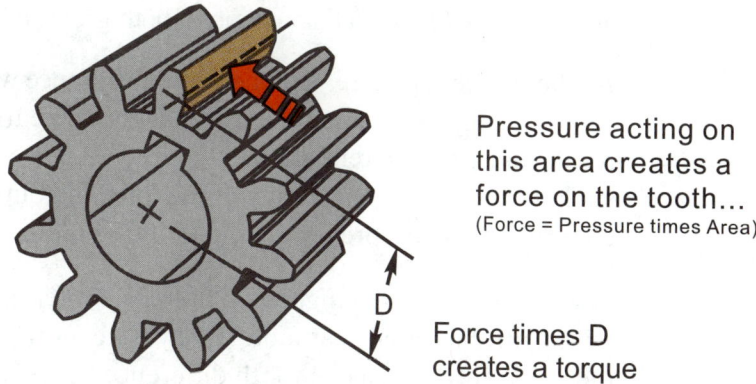

Figure 5.6 Pressure against the area of a tooth creates a force on the tooth

Resistance to rotation of the shaft causes the pressure at the inlet to rise, creating a higher ΔP across the motor and across each tooth. Output torque is created by the resulting tooth force and the distance the tooth is from the center of the shaft. Therefore, a larger tooth (height and depth), a larger gear (tooth center further from the shaft center) and a higher pressure will all create a higher torque output.

Internal Gear Motors

The direct drive internal gear motor is shown in Figure 5.7. The main components are the rotating internal/external gear set, the drive shaft and the housing. Not shown are side plates which create a sealing wear surface on the sides of the gear set. The outer gear of the gear set is considered an "internal gear" because the teeth are cut on the inside rather than the outside. The output drive shaft is connected directly to the center external gear, and the centerline is slightly offset from the centerline of the internal gear.

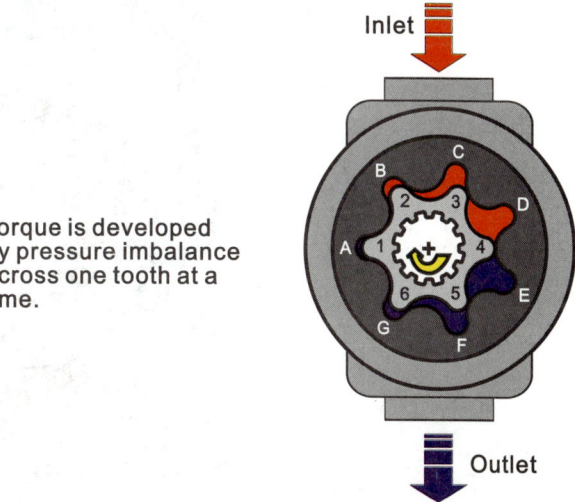

Figure 5.7 Cross section of an internal gear motor

Chapter 5 Rotary Actuators

The internal gear has one more tooth space cut into it than the smaller, external gear. This creates a space for incoming fluid to enter. A pressure differential between the inlet and outlet creates a rotating motion of both gears and a torque output capability.

As the two gears rotate, the one tooth difference will cause the internal gear to rotate slightly slower than the external gear; one tooth per revolution slower. This results in a low differential speed between the two gears, leading to a quiet and smooth operation. The design of the inlet and outlet ports of the gear chamber also contribute to a low pressure "ripple" at the outlet.

The gear set shown in Figure 5.7 has seven internal teeth and six external teeth. Direct drive internal gear motors may have more or fewer teeth on the gears, but there will always be a one tooth difference between the two.

Torque is created by the differential pressure across the gear, the area of the fluid space and the offset distance. Increasing the fluid space and offset distance by increasing the size and thickness of the gear, or increasing pressure, will create a higher torque output.

Vane Motors

A cross sectional view of a balanced vane rotating group is shown in Figure 5.8. The elements shown in the view are the cam (or cam ring), rotor and vanes. The output shaft of the motor is connected to the center of the rotor. The vanes slide in and out of the slots in the rotor so as to make contact with the cam surface. A form of spring, either a spring clip or a small coil spring, is placed under the vane to cause it to stay against the cam surface Figure 5.9. In addition, inlet fluid is also ported under the vanes so as to balance the pressure between the top and bottom and prevent pressure from pushing the vane back into the slot.

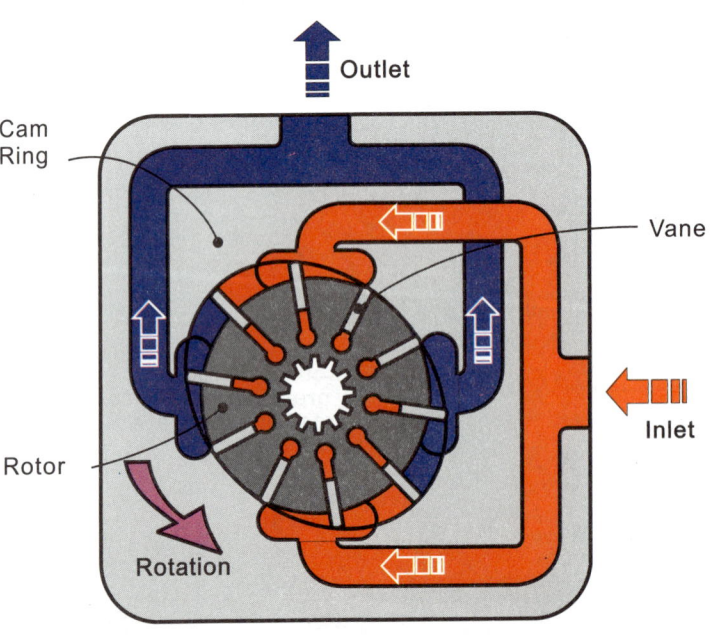

Figure 5.8 Cross section of a balanced vane motor rotating group

Figure 5.9 Springs or spring clips keep the vanes against the cam.

Fluid entering the motor will pressure two opposite sides of the rotor assembly, and return fluid will exit two opposite sides. This way, equal pressures are always opposite each other, balancing forces across the rotor. This relieves any loading on the drive shaft and bearings caused by *internal* pressures and forces.

Figure 5.10 shows how the differential pressure across a vane will create a force on the vane. The amount of vane that is exposed to pressure will determine the magnitude of the force (force equals pressure times area), and the distance from the center of the exposed vane area to the center of the drive shaft will determine the torque that is generated. Therefore, the torque output of a vane motor is dependent upon pressure, size of the vane (height extending above the rotor and width) and radius of the rotor (distance from the centerline of the drive shaft).

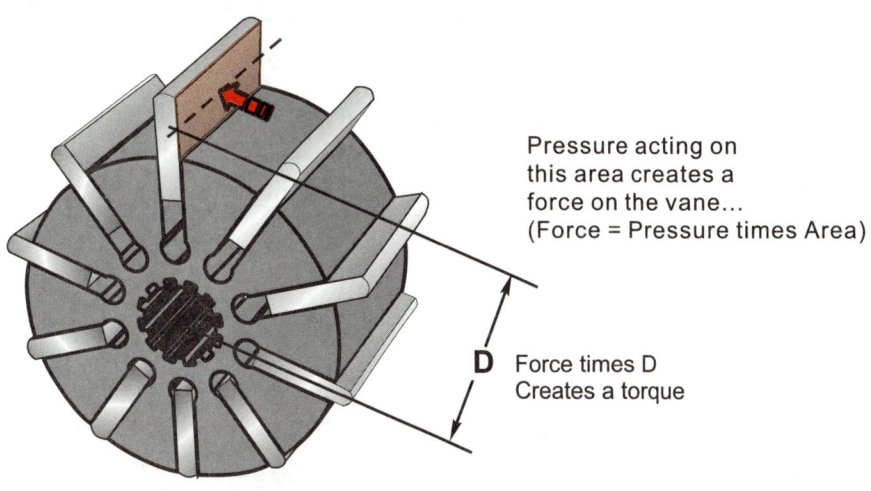

Figure 5.10 Pressure acting on a vane creates torque at the center of the shaft

Chapter 5 Rotary Actuators

In-line Piston Motors

A cutaway view of an in-line piston motor is shown in Figure 5.11. The components that make up a piston motor rotating group are a cylinder block, pistons and shoes, shoe hold-down plate, swash plate, valve plate and drive shaft. The drive shaft is connected by splines to the cylinder block, and the shoes are held down to the swash plate by a hold-down plate.

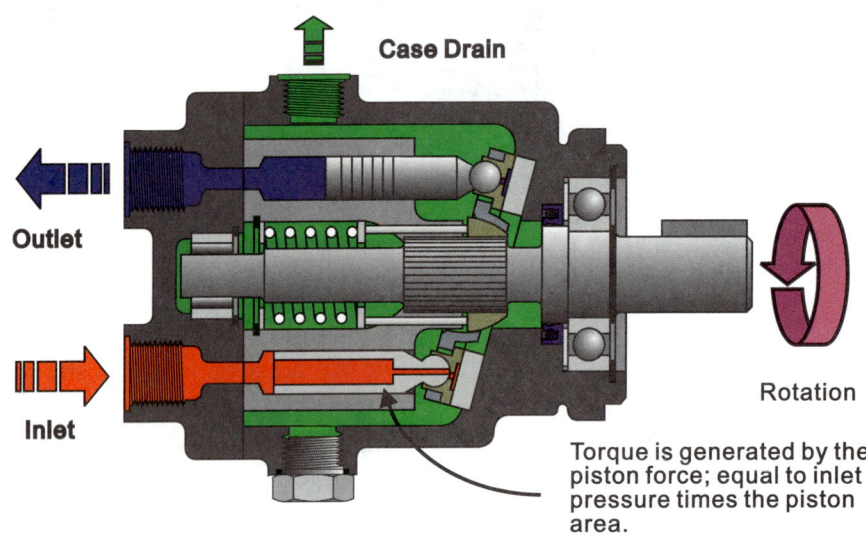

Figure 5.11 Cross section of an in-line piston pump

As fluid is forced through the valve plate into the cylinder block, pistons are forced out of the cylinder block, causing them to slide along the angled swash plate. This causes the cylinder block to rotate along with the pistons, turning the drive shaft. As pistons are forced back into the cylinder block by the swash plate, fluid is forced out through the valve plate and back to the reservoir.

The amount of torque that the motor will deliver is based on the force of the piston (pressure times area), the radius of the piston circle(force times distance) and the angle of the swash plate. The higher the swash plate angle, the greater the torque output for any given pressure.

Bent Axis Piston Motors

A cross sectional view of a bent axis piston motor is shown in Figure 5.12. The main elements are a cylinder block, pistons and shoes, drive shaft and flange, a universal link and a valve plate. The piston shoes are lodged in the drive shaft flange, and the universal link maintains alignment between the cylinder block and the drive shaft so that they turn together.

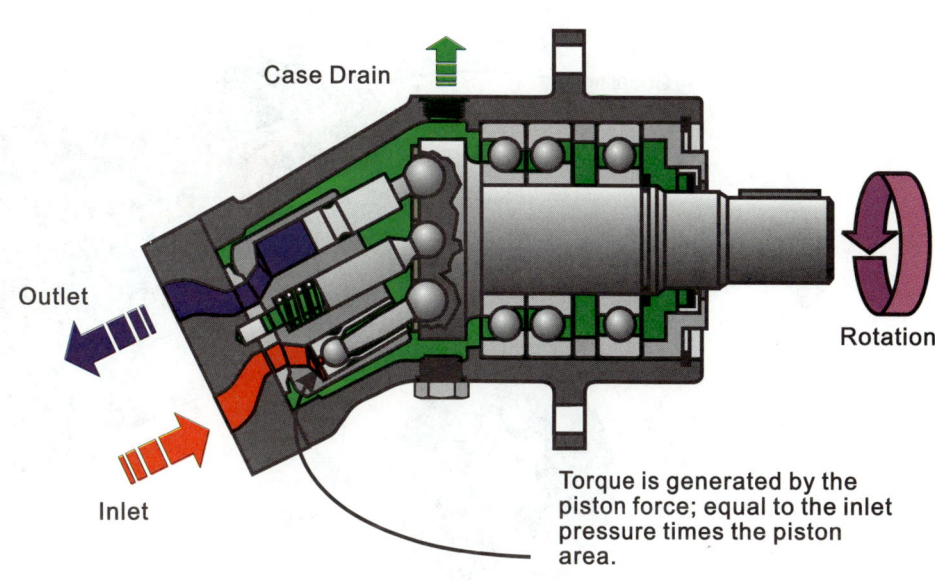

Figure 5.12 Bent axis piston motor

As fluid is forced through the valve plate into the cylinder block, pistons are forced out of the cylinder block, forcing the drive shaft flange to rotate. This causes the drive shaft to rotate along with the cylinder block and pistons. Pistons are forced back into the cylinder block by the drive shaft flange, and fluid is forced out through the valve plate and back to the reservoir. The entire operation is much the same as the in-line piston motor, except that the cylinder block and piston assembly is angled instead of a swash plate.

The amount of torque that a motor will deliver is based on the force of the piston (pressure times the piston cross sectional area), the radius of the drive shaft flange (force times distance) and the angle of the cylinder block. The higher the cylinder block angle, the greater the torque output for any given pressure and piston size.

HIGH TORQUE, LOW SPEED MOTORS

Orbital Internal Gear Motors

A cutaway drawing of an orbiting internal gear motor is shown in Figure 5.13. The major components of this type of motor are an external gear rotor, an internal gear stator, an orbiting drive coupling (or "wobble" shaft), a rotating valve and an output shaft. The drive coupling connects the orbiting internal gear to the output shaft, also causing the rotary valve to rotate.

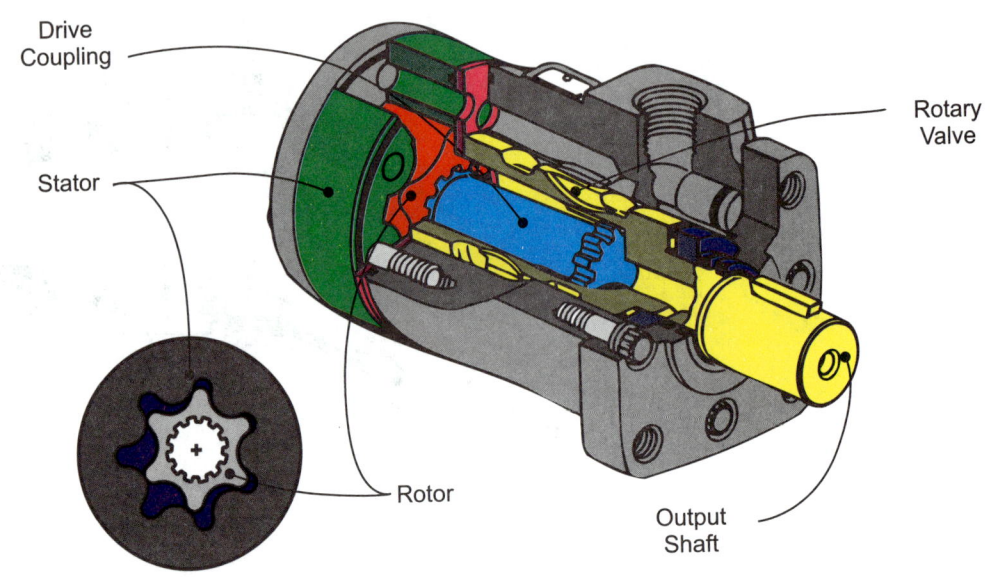

Figure 5.13 An orbiting gerotor motor

The rotating valve may be a cylindrical valve as shown, or it may be a flat plate valve as shown in Figure 5.14. Either one rotates as the orbiting external gear rotates.

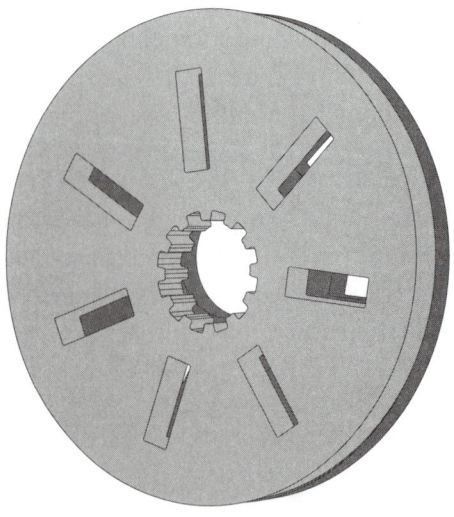

Figure 5.14 Flat plate rotary valve for orbiting gerotor motor

Another variation is to replace the fixed internal gear teeth with rollers as shown in Figure 5.15. Because of very tight clearances between the rotor and stator of an orbit motor, there is high friction between the teeth. The rolling-tooth rollers, or roller vanes, reduce this friction appreciably.

Figure 5.15 Gerotor assembly with rollers in place of internal gear teeth

As fluid enters the motor inlet, it is ported by the rotating valve to the open area between the stator and the rotor. This open area is caused by the fact that the rotor has one less tooth than the stator, and is offset as illustrated in Figure 5.16. As pressure builds up in this area, it forces the rotor across the stator into a new position, which causes it to rotate slightly. This rotating motion also rotates the valve, which ports fluid into the next open cavity and repeats the process. Displaced fluid on the opposite side of the rotor is ported through the valve to the outlet and back to the reservoir.

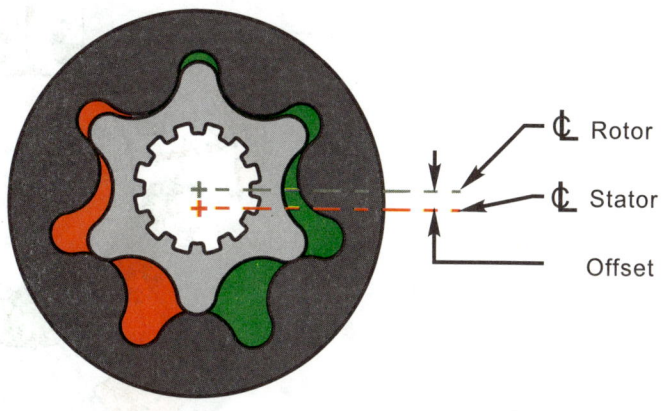

Figure 5.16 The rotor is offset from the stator and has one less tooth

Chapter 5 Rotary Actuators

Figure 5.17 illustrates the complete cycle as fluid is ported successively around the stator. As the figure shows, the rotor has rotated only one tooth during this complete cycle; Because the rotor has six teeth, it will take six fluid cycles to cause it to complete one revolution. Rotation of the orbiting rotor is transmitted to the fixed rotating output shaft via the drive coupling.

Figure 5.17 One complete fluid cycle results in a one tooth rotation of the rotor and output shaft

This six-to-one reduction of speed created by the 7 tooth stator, 6 tooth rotor combination also means a six to one increase in the output torque level, minus the torque lost due to the high friction inherent in this type of design. This is a significant advantage of this type of motor, providing a relatively high torque output and low speed in a compact design.

Torque is generated by the force created by pressure against the side of the rotor (pressure times area). This force, and therefore the output torque, can be increased by increasing the pressure, or by increasing the diameter or the width of the rotor and stator.

Radial Piston Motor

A cross sectional illustration of a radial piston motor is shown in Figure 5.18. Major components are the cylinders, the pistons, the connecting rods, the crankshaft drum and output shaft. Not shown in the drawing is a rotating valve connected to the output shaft on the reverse side of the motor.

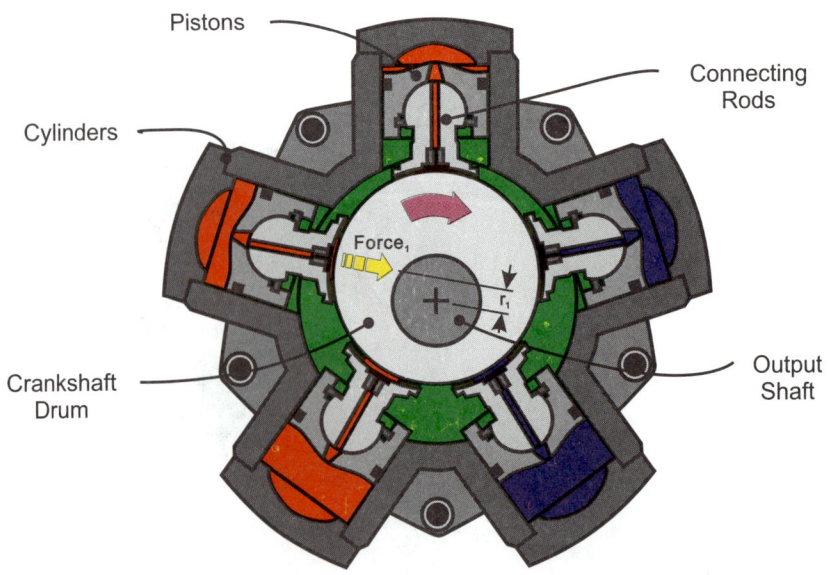

Figure 5.18 Radial piston motor

Figure 5.19 illustrates the location and porting of the rotary control valve.

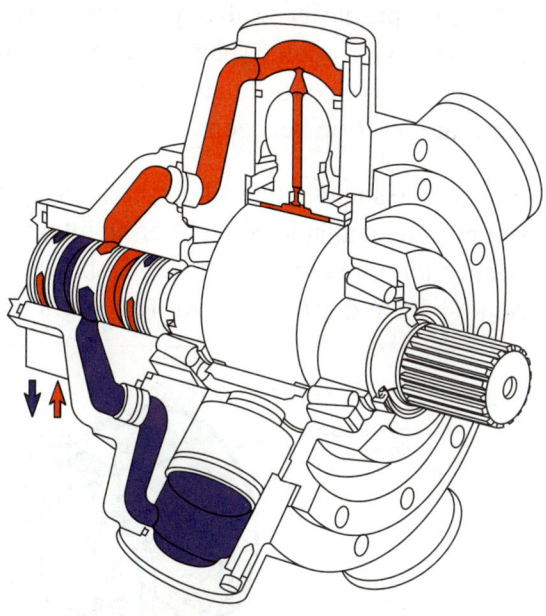

Figure 5.19 Inlet and outlet porting in a radial piston motor

Fluid enters the motor through the rotary valve, which ports it to the pistons. The force created by the area of the piston under fluid pressure, acting against the offset of the eccentric crankshaft drum, creates a rotation of the output shaft as the piston extends in its bore. As the shaft revolves, it rotates the rotary valve, porting fluid into successive pistons and maintaining a continuous rotary motion. Returning fluid exhausted from retracting pistons is ported through the valve and to the reservoir.

The pressurized fluid also travels through a small orifice in the center of the piston and connecting rod to the surface area between the connecting rod and drum. This surface area is fairly large so that the fluid not only acts as a lubricant, but also reduces contact forces between the mating parts.

As shown in the figure, three pistons are pressurized at the same time. This prevents any "dead spots" caused by a single piston being at top dead center, assuring a smooth rotational output.

Output torque is determined by the force of the piston against the drum and by the amount of offset of the drum. Higher pressures, larger pistons and a larger drum (greater offset) will all lead to a higher output torque level. A larger number of pistons will also provide greater torque, and this is accomplished on some very large motors by using a double row of pistons.

Radial piston motors are generally used in applications where a very high torque output level is required, at a low output speed. These motors will normally range from about 12 CIR (200 CCR) up to more than 400 CIR (6,500+ CCR), and require 120 GPM or more to operated at rated speed.

Chapter 5 Rotary Actuators

POWER STEERING MOTORS

A special application of the hydraulic motor is the power steering motor. Although the unit is primarily a valve (perhaps considered an auxiliary valve) that regulates the flow from a source (power steering pump) to a power steering cylinder, the most popular type of control mechanism utilizes an orbiting internal gear motor to maintain position between the manual steering wheel and the vehicle's steer wheels.

Figure 5.20 shows a typical power steering motor without a steering post (these type of units are available with either an integrated steering post or a steering post connection; the connection only is shown in this figure). Port connections are for pump inlet and return and for left / right steering.

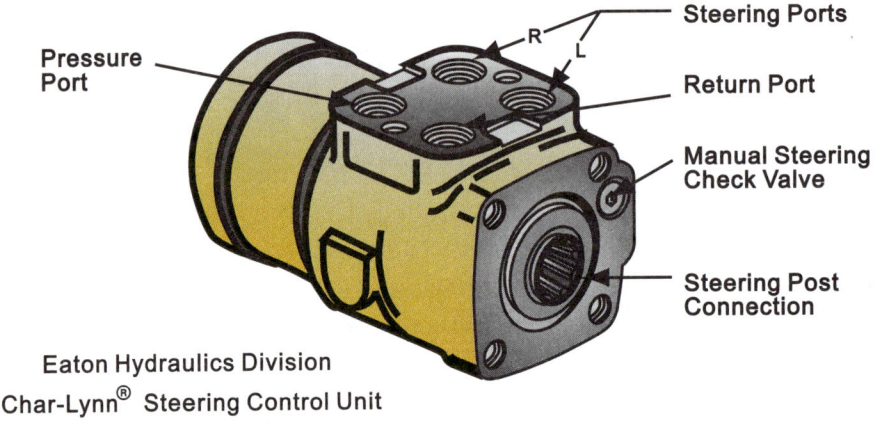

Figure 5.20 Power steering motor

Figure 5.21 is an exploded view of the major components. These consist of the control valve spool and sleeve, a centering pin that fits snugly into the control sleeve but very loosely in the control spool, an orbiting internal gear set and an orbiting drive shaft. When assembled, the drive shaft connects the center gear of the gear set to the centering pin. The steering post connects to the control spool.

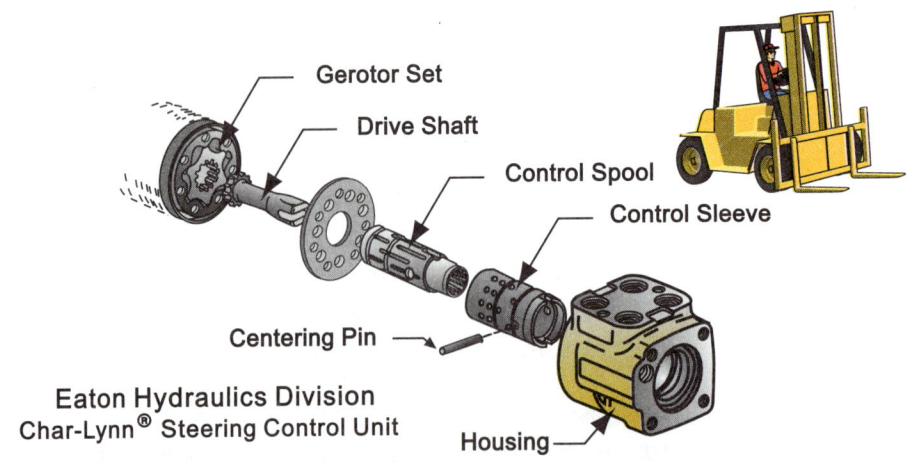

Figure 5.21 Exploded view of power steering motor

When the steering wheel is turned to the right, for example, the control spool is also rotated to the right. This ports fluid from the inlet port to the right outlet port and to the power steering cylinder to effect a right-steer movement. At the same time, fluid is ported to the gear set, causing it to also rotate to the right. As the gear set turns, it turns the drive shaft, which rotates the control sleeve until it catches up to the control spool. When the control spool and the control sleeve are in line, flow to the steering port and to the gear set is blocked, and the steer wheels will stay in the position they have attained. Flow from the return side of the power steering cylinder will proceed through the valve and to the reservoir.

Turning the steering wheel to the left will create the identical action through the left outlet port, and through the gear set to cause a left rotation.

Flow to a power steering motor usually comes from a separate power steering pump or from a priority valve in the main hydraulic system. Therefore, steering flow is always available when the engine is running. If the engine stops, however, steering ability is lost unless some other means is provided. With the power steering motor shown in Figure 5.21, manually turning the steering wheel in a power-off situation will cause the gear set to become a pump, and fluid will be supplied to the steering cylinder. Steering becomes more difficult under these circumstances, but emergency maneuvers can be accomplished.

VARIABLE DISPLACEMENT MOTORS

Axial and radial piston motors can be of a variable displacement design. By reducing the stroke of the pistons, less fluid is required per revolution so that the motor can operate at a higher speed, or the fluid flow rate can be reduced. By reducing the displacement, however, the torque output is also reduced.

Variable displacement motors can be used on applications where a dual range of torque / speed relationship is desirable. An example would be when operating a winch; at heavy loads, maximum motor displacement is desirable to provide maximum torque while winching slowly. During light load or no-load operations, however, motor displacement can be reduced to provide less torque output and higher winch speed.

Motor displacement can be changed in steps, or it can be continuously variable. The more common approach is to shift the motor from maximum to minimum displacement in one step. An alternative is to utilize a compensator that will adjust motor displacement continuously as pressure rises.

> *NOTE: In all cases, a minimum displacement must be established and maintained either through internal stops in the motor or by some external mechanical means. Reducing the displacement of a motor too much will cause over-speeding resulting in serious damage to the motor and/or to the equipment.*

Chapter 5 Rotary Actuators

Variable Displacement In Line Piston Motors

The components of a variable piston motor are very similar to a fixed displacement motor. As shown in Figure 5.22, the primary difference is that the swash plate is mounted on a yoke, and the yoke is free to rotate to change the swash plate angle. The smaller the angle, the smaller the motor displacement.

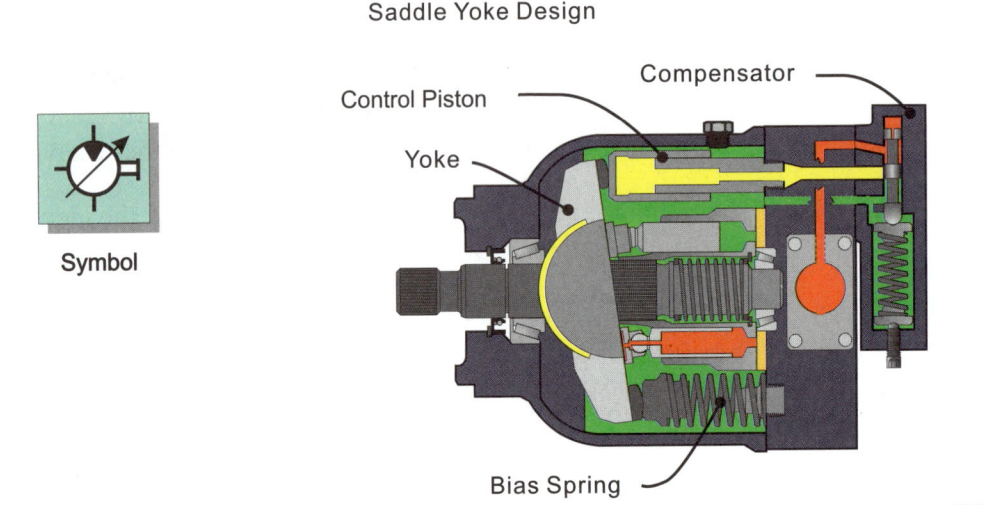

Figure 5.22 Variable displacement in-line piston motor

In Figure 5.22, the yoke is rotated by a control piston, and the control piston is controlled by a compensator. The yoke could also be controlled by a handle connected directly to it, and operated by an external mechanism. Manual operation is also possible, but the internal forces on a yoke caused by pressure and pressure changes is typically too great in motors larger than about 2.5 cubic inches displacement.

The compensator in Figure 5.23 consists of a metering spool and an adjustable spring. The spring is adjusted to a desired pressure level. When that level is achieved in the fluid at the motor inlet, the metering spool moves to open a flow path to the control piston. As the control piston extends, the yoke is rotated to increase motor displacement providing increased output shaft torque and reduced shaft speed. As inlet pressure is reduced, the metering spool moves back and allows the fluid in the control piston to drain to the case, and the yoke will rotate back to reduce motor displacement.

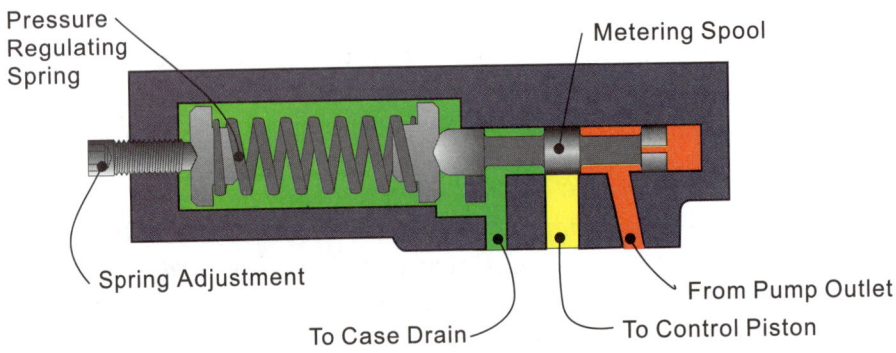

Figure 5.23 Pressure limiting compensator

Chapter 5 Rotary Actuators

Variable Displacement Bent Axis Piston Motor

Varying the displacement of a bent-axis piston motor is similar to that of an in-line piston motor, except that instead of moving a yoke to alter the swash plate angle, the cylinder block / piston assembly is moved to alter the angle between the cylinder block and the drive shaft flange. Figure 5.24 illustrates this method.

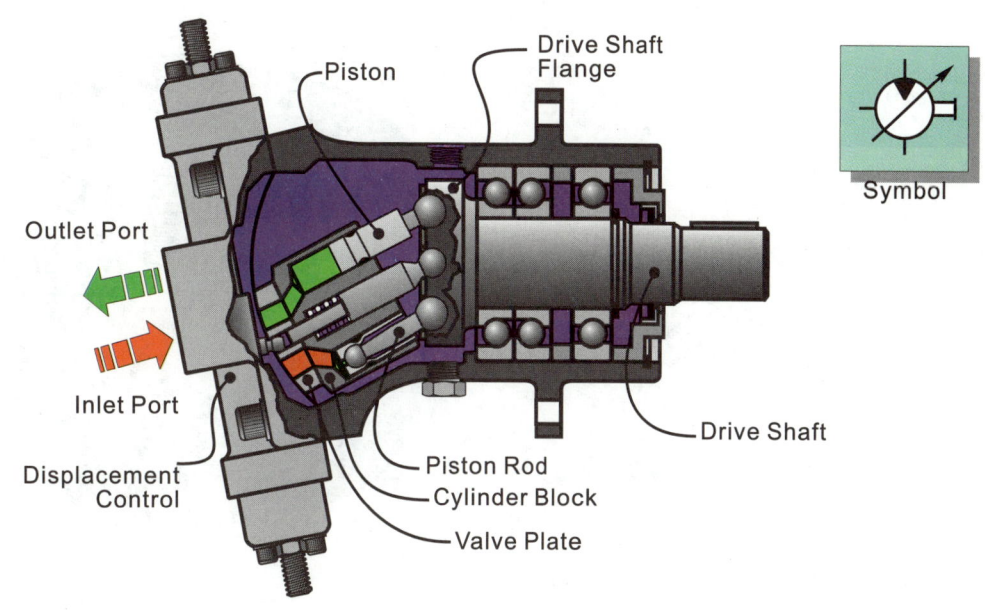

Figure 5.24 Variable displacement bent axis piston motor

Variable Displacement Piston Motor

Radial piston motors can change their displacement by changing the eccentricity of the drum on the crankshaft. This is done by inserting two pistons, one small and one large, inside the drum as shown in Figure 5.25.

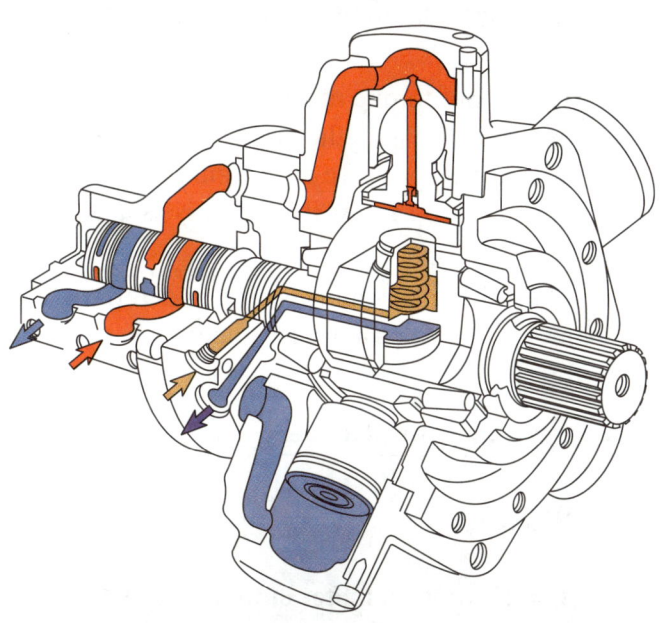

Figure 5.25 Variable displacement radial piston motor

123

The small piston keeps the drum shifted to its maximum displacement, and the large piston shifts the drum to obtain minimum displacement Figure 5.26. This shifting can be done while the motor is in motion, and creates a very smooth change in displacement. An outside pilot pressure is applied to provide the shifting force.

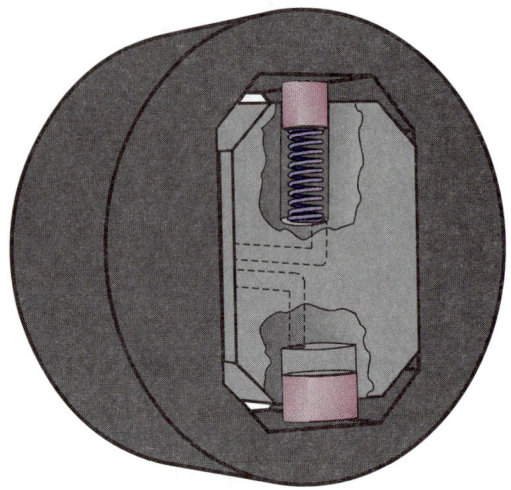

Figure 5.26 Two pistons inside the shaft move the cam to obtain changes in displacement

OPERATING PARAMETERS

The two primary parameters that determine the selection of a hydraulic motor are operating speed and torque capability at operating pressure. A third feature that is very important to the operation is the motor over-all efficiency.

Torque Calculations

The output torque of a hydraulic motor is dependent on the pressure drop across the motor. Where the outlet pressure is usually very close to zero, and there is very little, if any, pressure drop from the inlet of the motor to the rotating group, the pressure drop across the motor is usually assumed to be the inlet pressure.

The following formula is used to determine theoretical torque available from a motor with known displacement:

Formula 5-1

$$T = \frac{P_{in} \times \text{Displacement}}{2\pi}$$

Where:

T = Theoretical motor output torque, lb-in
P_{in} = Inlet pressure, psi
Displacement = CIR

As Figure 5.27 shows, this same formula can be used to determine the pressure required to attain a desired torque:

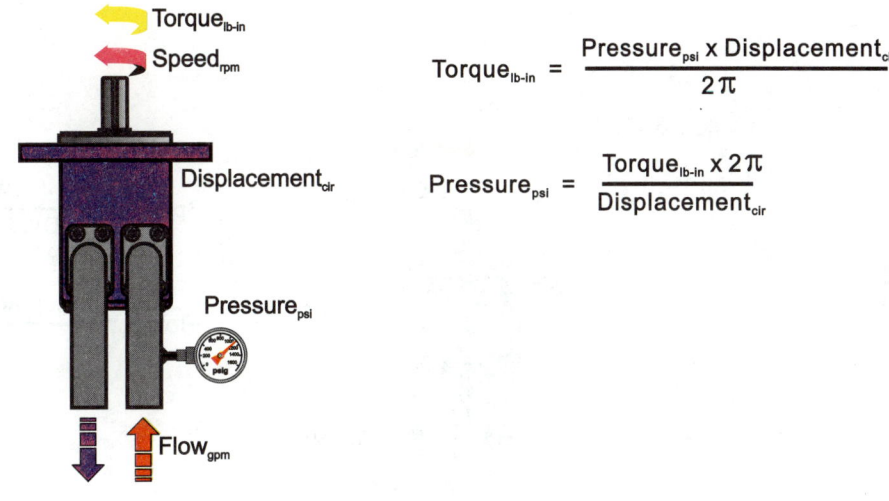

Figure 5.27 Hydraulic motor torque formula

$$P_{in} = \frac{T \times 2\pi}{\text{Displacement}}$$

Speed Calculations

The output speed of a rotary actuator is dependent upon flow rate of the fluid into the actuator. The following formula is used to determine the theoretical speed, in RPM, of a hydraulic motor with a known inlet flow:

Formula 5-2

$$\text{RPM} = \frac{Q \times 231}{\text{Displacement}}$$

Where:

RPM = Motor Output Speed, RPM
Q = Input Flow, GPM
Displacement = CIR

Chapter 5 Rotary Actuators

As Figure 5.28 illustrates, this formula can also be used to determine the theoretical flow required to achieve a desired output speed:

$$Q = \frac{RPM \times Displacement}{231}$$

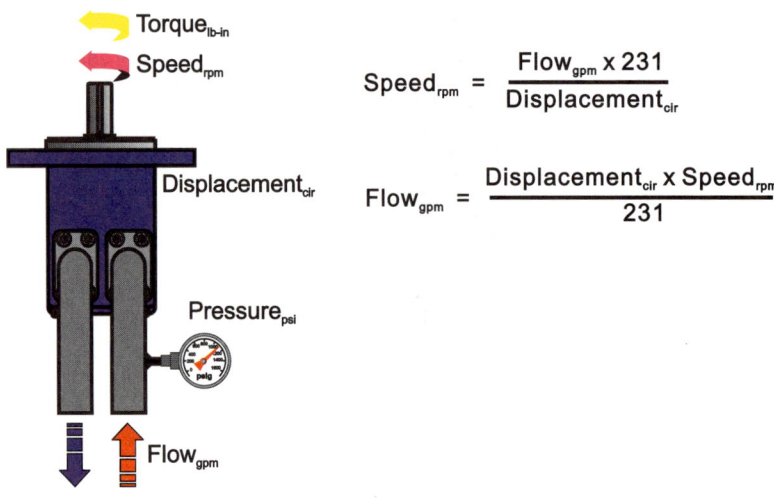

Figure 5.28 Hydraulic motor speed formula

Horsepower Calculations

Output horsepower of a hydraulic motor is based on both the torque and the speed. A high speed motor operating at a low torque level may have the same horsepower output as a motor operating at high torque and low speed. The determination of output horsepower of a hydraulic motor comes from the following formula:

Formula 5-3

$$HP_{OUT} = \frac{T_{lb\text{-}in} \times RPM}{63,025}$$

Where

HP_{out} = Motor output horsepower, HP
$T_{lb.\ in}$ = Output torque, lb-in
RPM = Output speed, RPM

Figure 5.29 illustrates that if motor output torque is measured in pound-feet instead of pound-inches, formula 5-3 becomes;

$$HP_{OUT} = \frac{T_{lb\text{-}ft} \times RPM}{5252}$$

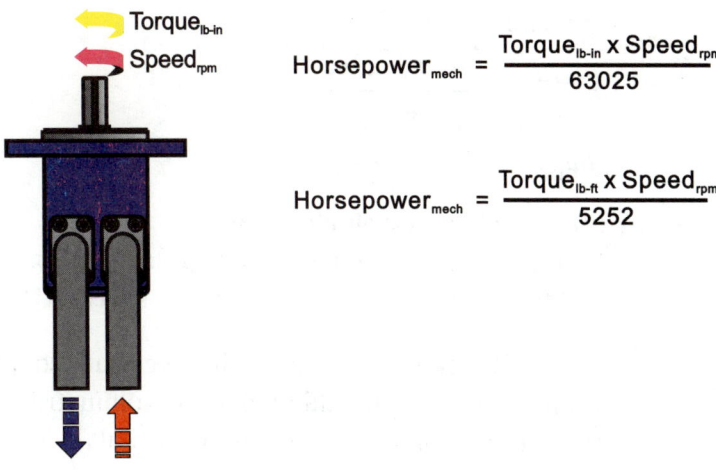

Figure 5.29 Hydraulic motor horsepower formula

Motor input horsepower is also an important factor in motor applications, as that is the power required to actually drive the motor. Input horsepower will always be higher than output horsepower, due to inefficiencies in the motor. Input horsepower can be calculated by the following formula:

Formula 5-4

$$HP_{IN} = \frac{P_{in} \times Q_{in}}{1714}$$

Where:

HP_{in} = Motor input horsepower, HP
P_{in} = Inlet pressure, PSI
Q_{in} = Inlet flow, GPM

Chapter 5 Rotary Actuators

Efficiency Calculations

Volumetric efficiency: As the above formulas illustrate, the speed of a hydraulic motor is totally dependent on the flow rate through the motor, and is independent of the pressure drop. If 100% of the flow was effective, the speed of the motor would be as determined by formula 5-2. Internal leakage occurs, however. Some leakage is intended for lubrication purposes, and some occurs because of clearances between mating parts. Because there is less flow creating motor RPM than is theoretically available, motor speed is somewhat less than calculated. The relationship between actual speed and theoretical speed is called volumetric efficiency, and is expressed in percent (%):

Formula 5-5

$$Eff_v = \frac{RPM_{act} \times 100}{RPM_{theo}}$$

Where:

Eff_v = Volumetric efficiency, %
RPM_{act} = Actual motor output speed, RPM
RPM_{theo} = Theoretical motor output speed, RPM

Theoretical motor speed is the RPM calculated by formula 5-2. Actual motor speed is a measured output, although if the volumetric efficiency is known, the actual motor speed can be determined by formula 5-5.

Mechanical efficiency: As pointed out earlier, friction is the resistance to motion between two contacting materials. Hydraulic motors have many places where materials slide or roll across each other, such as ball bearings or roller bearings, vane tips against cam rings, gear teeth against gear teeth or cam rings, pistons against cylinder block bores, gears against side plates, etc.. These areas of friction are a "drag" on the full torque output of the motor, reducing the output torque to something less than theoretical. The relationship between actual torque and theoretical torque is called mechanical efficiency, and is expressed in percent (%):

Formula 5-6

$$Eff_{mech} = \frac{T_{act} \times 100}{T_{theo}}$$

Where:

Eff_{mech} = Mechanical Efficiency, %
T_{act} = Actual Output Torque, lb-in or lb-ft.
T_{theo} = Theoretical Output Torque, lb-in or lb-ft

Theoretical output torque is the lb-in or lb-ft calculated by formula 5-1. Actual output torque is a measured output, although if the mechanical efficiency is known, the actual output torque can be determined by formula 5-6.

Overall efficiency: Both volumetric and mechanical inefficiencies reduce the overall motor performance, and this is measured in the motor's output horsepower. The relationship between input and output horsepower is the overall motor efficiency as determined by the following formula:

Formula 5-7

$$Eff_{oa} = \frac{HP_{out} \times 100}{HP_{in}}$$

Where:

Eff_{oa} = Overall Motor Efficiency, %
HP_{out} = Output Motor Horsepower, HP
HP_{in} = Input Motor Horsepower, HP

Overall motor efficiency is also the product of volumetric efficiency and mechanical efficiency:

Formula 5-8

$$Eff_{oa} = \left(\frac{Eff_v}{100} \times \frac{Eff_{mech}}{100} \right) \times 100$$

CAUSES OF MOTOR FAILURE

Most hydraulic components fail primarily because of contamination in the system, and motors are no exception. Contamination will induce excessive wear of internal parts leading to a reduction of volumetric efficiency, and it will degrade surface finishes causing increases in friction leading to a reduction of mechanical efficiency. In the extreme case, contamination can cause internal parts to "weld" or "gall" together, causing seizure and catastrophic failure. This is evident by an inability of the motor to rotate.

Another typical mode of failure for hydraulic motors is breakage of the shaft, as shown in Figure 5.30. Motor shafts fail either because of excessive torsional loads (excessive torque) or by excessive side loading. Frequently the failure is caused by a combination of the two.

Figure 5.30 Shaft failure can be analyzed to determine the cause

Examination of the failed shaft may help in the determination of cause. The following two rules will help (reference Figure 5.31):

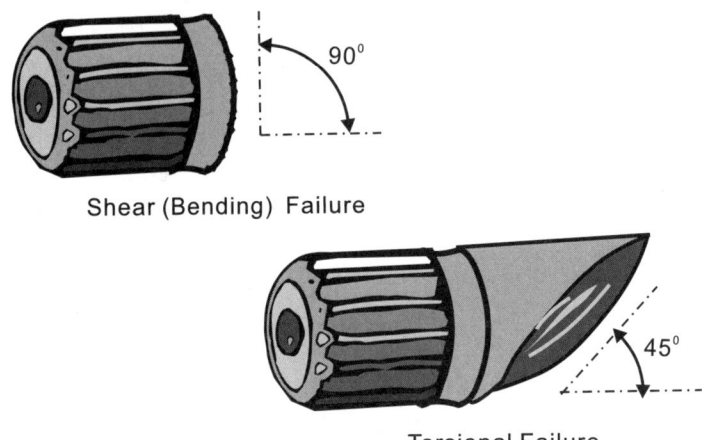

Figure 5.31 Shafts that have failed in pure bending and pure torsion

1. A shaft failing in pure torsion will have a failure line 45° to the longitude of the shaft, and appear spiral in nature.
2. A shaft failing in pure bending (shear) will have a failure line 90° to the longitude of the shaft.

A shaft that has a failure line between 45° and 90° to the axis has failed because of a combination of torsion and bending. The closer to 90° the failure is, the more preponderant bending was to the cause. Likewise, the closer to 45° (spiral) the failure is, the more preponderant torsion was to the failure.

Most often, torsional failures are caused by over-running loads that are not properly accommodated by system circuitry. Bending (shear) failures are usually caused by belt drive or gear drive arrangements that have become overloaded.

CHAPTER 6 .. Directional Control Valves

Directional control valves are the command center of hydraulic systems. They receive input from a machine operator, and accordingly, direct hydraulic fluid in the desired amount to a specific location to perform a desired task. The fluid is conditioned to provide a predetermined level of power, speed and restraint.

Modern variations of mobile equipment use almost every type of directional control valve in existence, including standard enblock and sectional spool-type valves, pilot operated valves, solenoid valves, electrohydraulic proportional valves, servo valves, etc. Furthermore, many of these valves include a variety of functions in addition to the directional role they are usually associated with; functions such as flow control, pressure regulating, pressure limiting, anti-cavitation, load control and more.

Categorizing mobile valves requires consideration of an array of characteristics that would include:

- Construction - Enblock, sectional or hybrid
- Application - Open center or closed center
- Flow Rating
- Pressure Rating
- Number of Controls - Number of spools
- Actuation Control - Manual, mechanical, hydraulic, pneumatic or electric
- Circuitry - Series, parallel or series - parallel
- Circuit Protection - Pressure protection, type and location
- Actuation Type - Tandem, motor, float, single acting

This chapter will examine these characteristics as they relate to each different type of directional control.

Chapter 6 Directional Control Valves

CHECK VALVES

The simplest form of directional control valve is the common check valve. This valve operates as a one-way directional control, allowing flow in one direction, while preventing any flow in the other. Operation of typical in-line check valves is shown in Figure 6.1 and their Symbols in Figure 6.2.

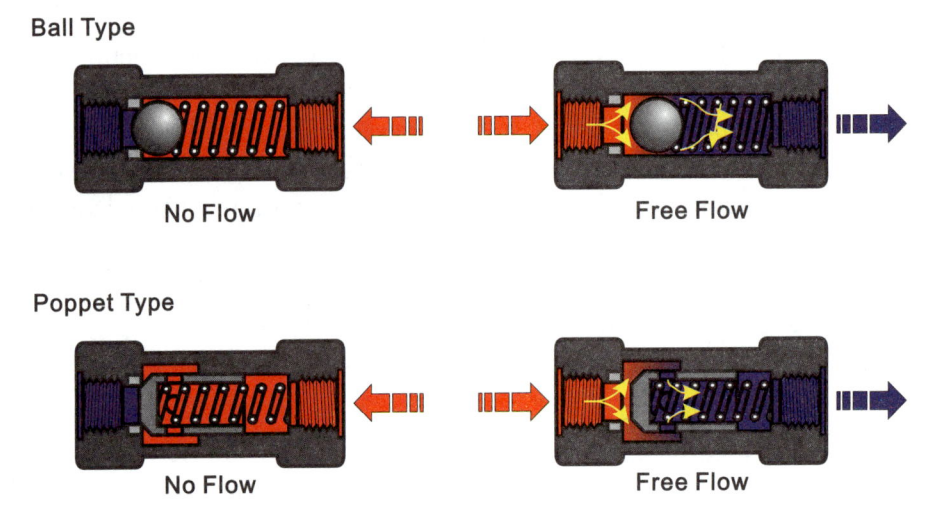

Figure 6.1 Ball and poppet type in-line check valves

The ball or poppet of a check valve is held on its seat by a light spring, usually having an equivalent pressure value of about 5 psi. In the blocked flow direction, pressure keeps the valve closed. In the free flow direction, pressure above the equivalent spring value will open the valve, allowing fluid to pass.

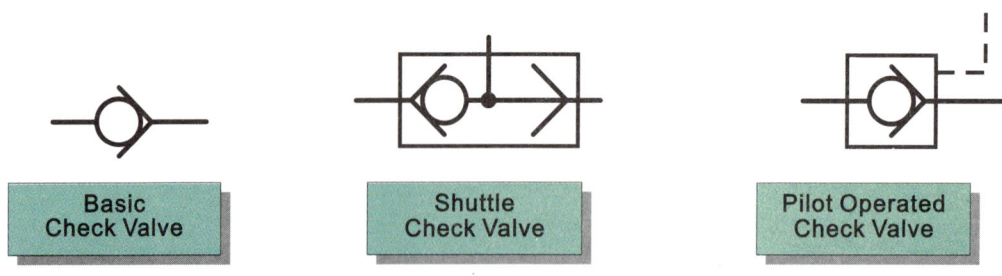

Figure 6.2 Symbols for basic check valve, shuttle valve and pilot operated check valve.

Chapter 6 Directional Control Valves

In-Line Check Valves

To allow fluid to flow straight through, an in-line check valve is installed directly in the hydraulic line. These types of valves are available in a broad range of flow ratings up to approximately 100 GPM. Although in-line valves are convenient to install, making them a popular choice of check valve, they typically exhibit a high pressure drop, particularly as flow increases above the valve rating. To maintain a low pressure drop, the valve must be considerably larger than the hose or tube it is installed in, sometimes negating the cost advantage of this valve design.

In-line check valves are used for safety bypass valves as well, by adjusting the equivalent pressure value of the spring. This type of valve is frequently used as a flow bypass on filters or coolers, such that when the flow through the device becomes restricted, causing the pressure drop to exceed the desired level for the device, flow will take a secondary path through the check valve Figure 6.3.

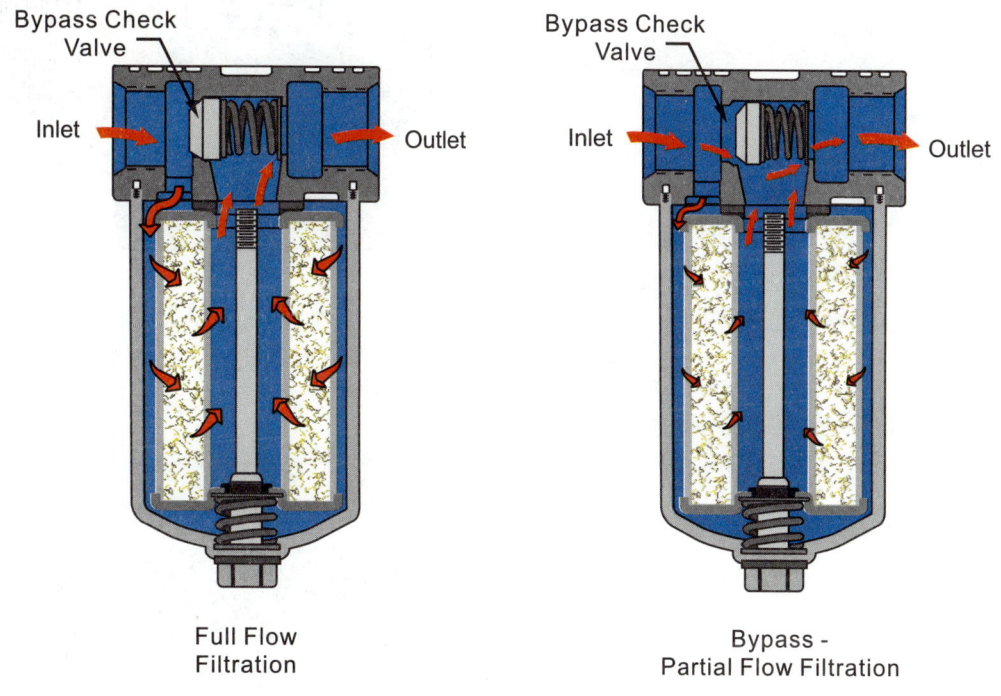

Figure 6.3 Return line filter with a bypass check valve

In-line check valves are also used to intentionally create a low continuous pressure in a circuit. This pressure, usually called back-pressure, may be required for insuring lubrication or sealing on some components, or for generating a pilot pressure to operate controls.

Chapter 6 Directional Control Valves

Right Angle Check Valves

Right angle check valves are designed with the inlet and outlet ports at right angles to each other, as shown in Figure 6.4. This design allows for higher flows with lower pressure drops than the in-line design. Valves rated for flows as high as 300 GPM or more are available with this configuration.

Operation of the valves is the same as the in-line design.

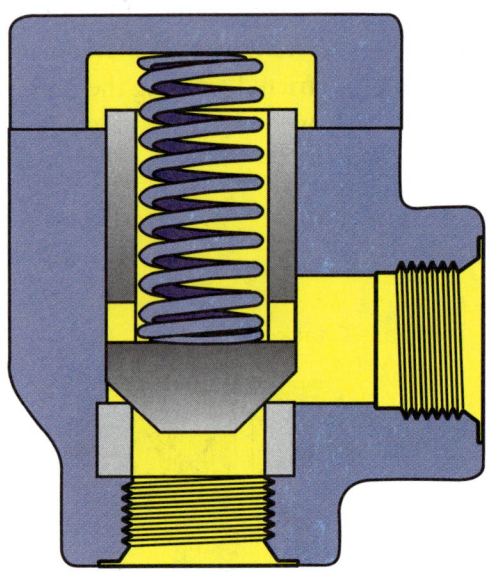

Figure 6.4 Right angle poppet style check valve

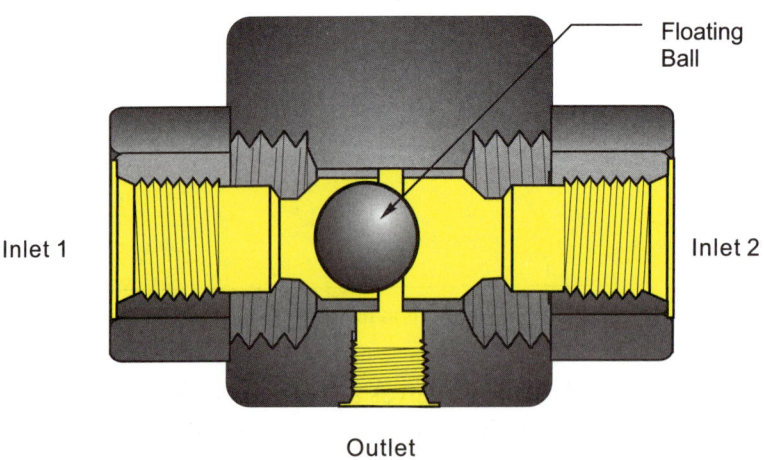

Figure 6.5 A shuttle valve outlet detects the highest of two inlet pressures

Shuttle Check Valves (Shuttle Valves)

A shuttle check valve, or shuttle valve, is constructed as though there are two opposing check valves utilizing the same ball or poppet. There are two inlet ports and one outlet port. Flow at the outlet port will originate at the inlet port having the highest pressure. Figure 6.5 illustrates a shuttle valve. This device is generally used in a circuit where the higher of two pressures is to be sensed. For this reason, they are used in load sensing systems.

Pilot Operated Check Valves

Check valves may also be opened or closed by pilot pressure. These valves are designated "pilot-to-open" and "pilot-to-close" check valves, and have been developed for specific circuit requirements.

Pilot-to-open check valves are used to maintain valve closure until a pilot signal pressure is provided to open the valve to reverse flow. Free flow is still provided in the forward direction, and flow is automatically blocked in the reverse direction, in standard check valve fashion. On command, these valves allow reverse flow. Figure 6.6 shows a basic pilot-to-open configuration.

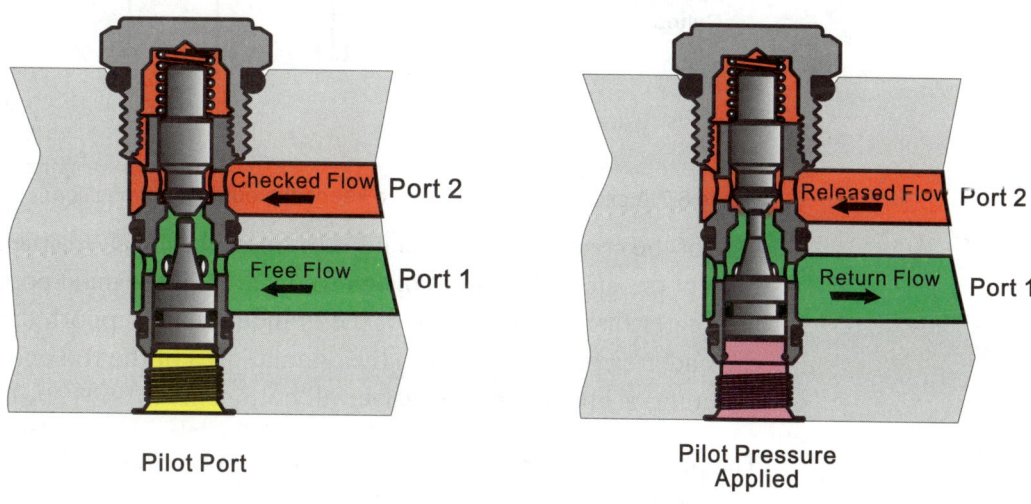

Figure 6.6 Pilot to open check valve

A typical application for this type of valve is to prevent a hydraulic cylinder, such as an outrigger cylinder, from raising until a command pressure is provided. Free flow in the forward direction is allowed to lower the cylinder but reverse flow is blocked, preventing the cylinder from raising until a pilot pressure is applied (see Figure 6.7).

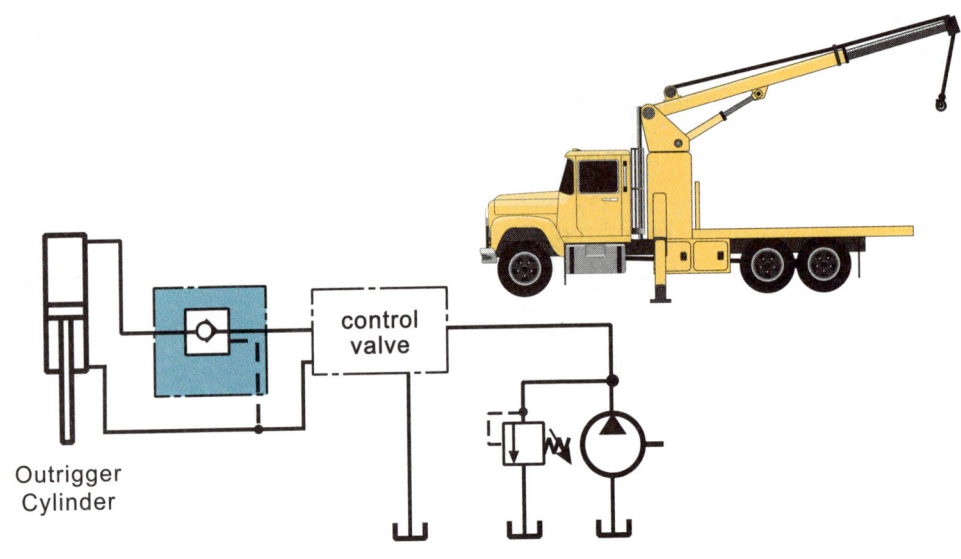

Figure 6.7 A pilot to open check valve keeps the outrigger in position

One of the considerations of a pilot-to-open check valve is the area ratio, i.e., the ratio of the area of the pilot piston to the area of the main poppet. This ratio must be greater than the area ratio of the cylinder being supported, i.e., the ratio of the cap end area of the cylinder to the annular area of the rod end. If the check valve area ratio is not greater, the valve will not open.

The pressure required to release a pilot-to-open check valve supporting a load in this way is determined by the following equation:

Formula 6-1

$$\text{Release pressure} = \frac{[\text{End cap area} \times \text{maximum W.P}]}{\text{Annular (Rod End) Area}} - \text{Load}$$

Where:

W.P. = Working pressure (PSI)
End cap area, rod end Area = in^2
Load = pounds

Pilot-to-close check valves allow free flow in one direction, typical of check valves, but will close off flow entirely with the application of a pilot signal pressure. A valve such as this might be used to relieve the supercharge pressure on a reservoir after the equipment has been shut down. During machine operation, a pilot pressure will keep the valve closed so that reservoir pressure will rise.

Chapter 6 Directional Control Valves

SERIES AND PARALLEL CIRCUITRY

Series and parallel circuitry is described in Chapter 2, and is illustrated again here, using check valves instead of orifices.

Figure 6.8 shows a parallel connection of three check valves, each with a different opening pressure value. It is easy to visualize the flow proceeding over the 100 PSI check valve, in accordance with the rule illustrated in Chapter 2;

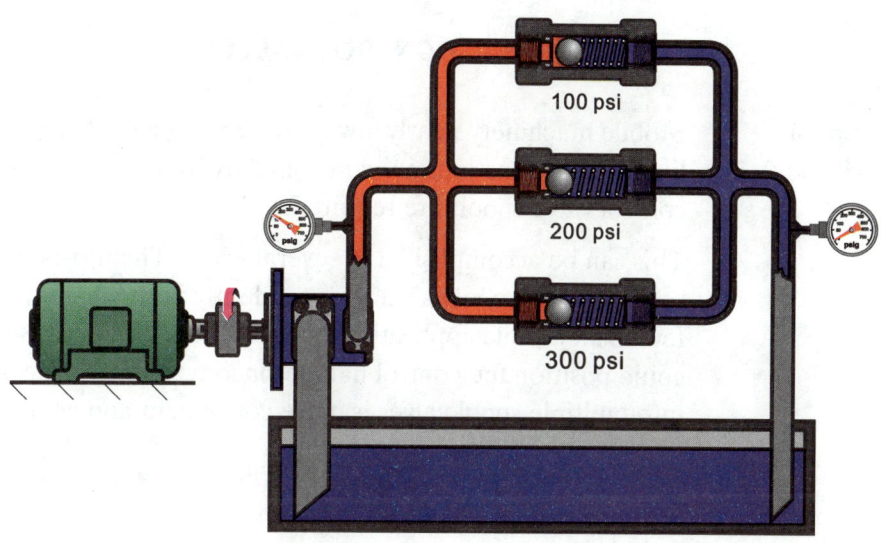

Figure 6.8 Check valves arranged in parallel

In a parallel circuit, flow will take the path of least resistance, and the circuit pressure is the pressure required to force the fluid through this path.

Therefore, the pressure at the pump in Figure 6.8 will be 100 PSI.

In Figure 6.9, the check valves are arranged in a series circuit, meaning that the flow must proceed through all three valves. As illustrated in Chapter 2:

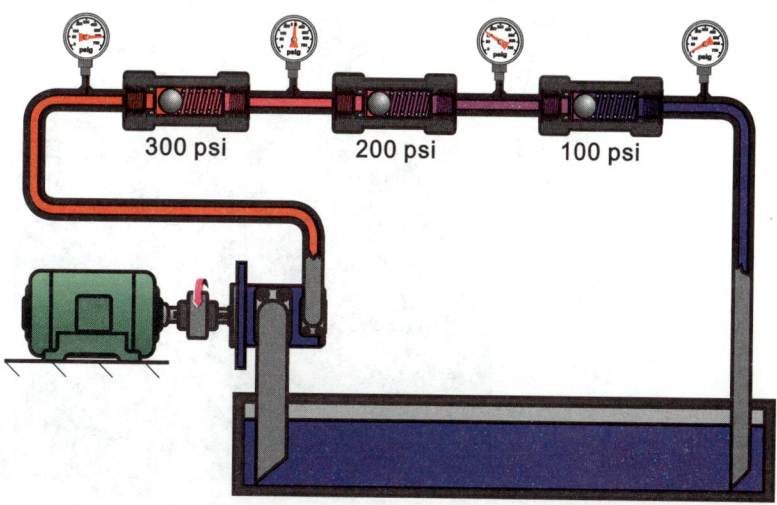

Figure 6.9 Check valves arranged in series

137

In a series circuit, all flow must pass through all branches of the circuit, and the pressure drops across each element of the circuit are additive.

In this instance, the pressure at the pump will be 600 PSI.

Directional control valves have internal construction that is designed to be series construction, parallel construction or a combination series-parallel construction. This will be illustrated later in this chapter.

MOBILE SPOOL TYPE DIRECTIONAL CONTROL VALVES

Multiple Spool Construction

Mobile machinery nearly always requires the control of multiple functions. Very likely, these functions are controlled hydraulically, so that two or more directional control valve spools are required.

This can be accomplished in several ways. The most obvious way is to connect several single spool valves independently in the circuit. Although this is done in a few small mobile applications, it is not common because it is costly, unwieldy, and could position the control handles inconveniently. Packaging several valves together into multiple spool valves is more convenient and economical.

Multiple spool packaging is accomplished in three ways:
Sectional valves
Enblock valves
Combination sectional and enblock valves

Sectional valves are very common and are a convenient, economical and flexible approach to provide a grouping of directional controls. Each spool is contained in a separate valve section, and the required number of sections are bolted together to create a multiple spool valve, as shown in Figure 6.10. Any number of sections can be combined in this way, although more than eight or ten spool combinations in one valve bank is rare.

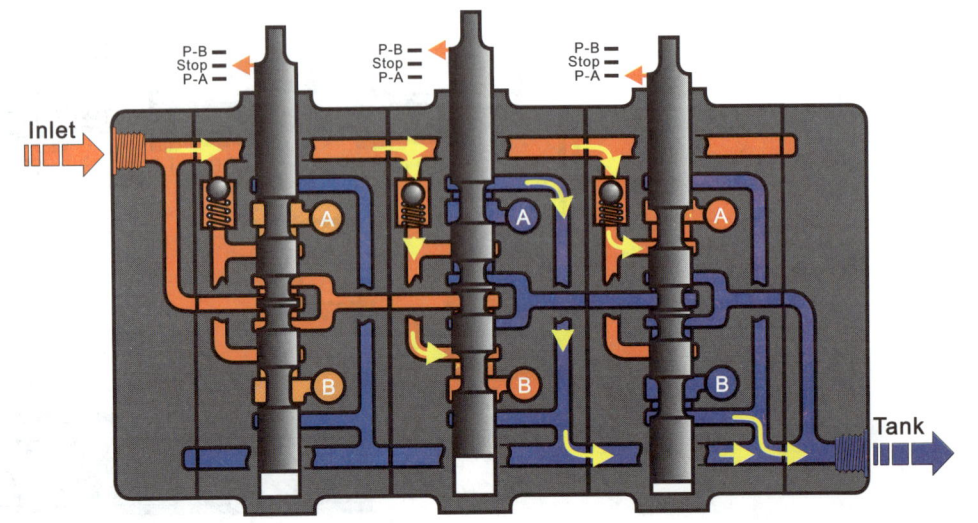

Figure 6.10 Three spool sectional valve

The flexibility of sectional valves makes them a popular choice. A broad variation of spool types and functions can be combined into one valve, and easily modified. Repair can be accomplished on a single spool or section, and sections may be added if additional functions are required at a later date. Multiple sections, however, have potential external leakage points that do not exist on other types of valves. They require positive sealing between sections by use of either O-rings or seal plates inserted between valve sections. It is difficult to achieve leak free assembly of several sections using long "tie bolts" due to a combination of bolt stretch and possible distortion of the valve bank assembly. Furthermore, attempts to keep a compact profile on this type of valve usually result in higher pressure drops than for the equivalent size enblock valve.

Enblock valves, also called monoblock valves, contain all spools in a single valve body. Although this design is much less flexible than the sectional design, it is usually more compact for a given flow size and has fewer potential external leakage points. Body distortion is much less likely to occur, a difficulty with sectional valves which tends to seize spools in their bores. More than four spools in a single body is very uncommon. A typical monoblock valve is shown in Figure 6.11.

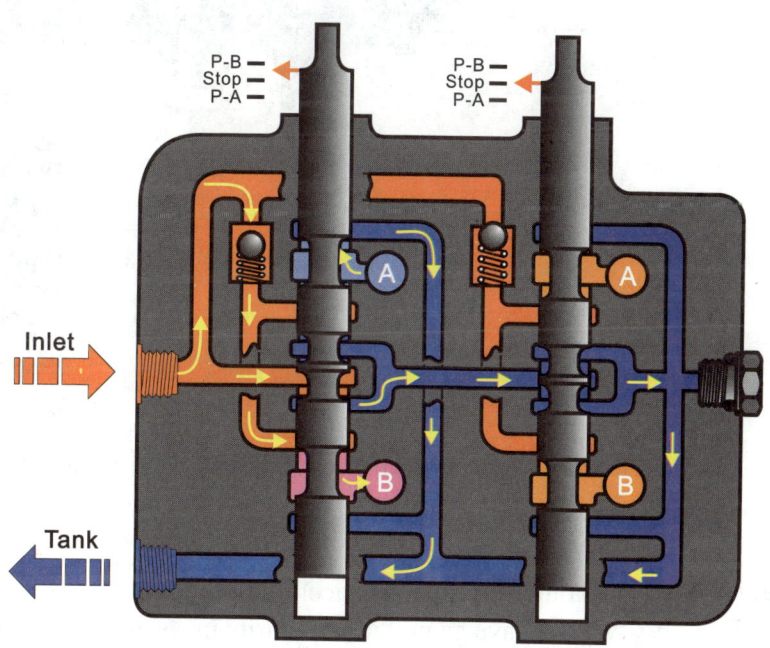

Figure 6.11 Two spool enblock valve

A third type of valve arrangement is a combination of enblock and sectional. This type of valve has become quite common on lift truck applications, where two spools are standard for the lift and tilt functions and an additional one or two spools are frequently required for some combination of auxiliary function, such as side shift, load push, rotate, etc.

Chapter 6 Directional Control Valves

In this hybrid valve, the first two spools are in a monoblock body with one end cap (see Figure 6.12). An additional one or two spool monoblock body is added when needed. Under all conditions, there is only one seam between sections, considerably reducing the potential external leakage paths.

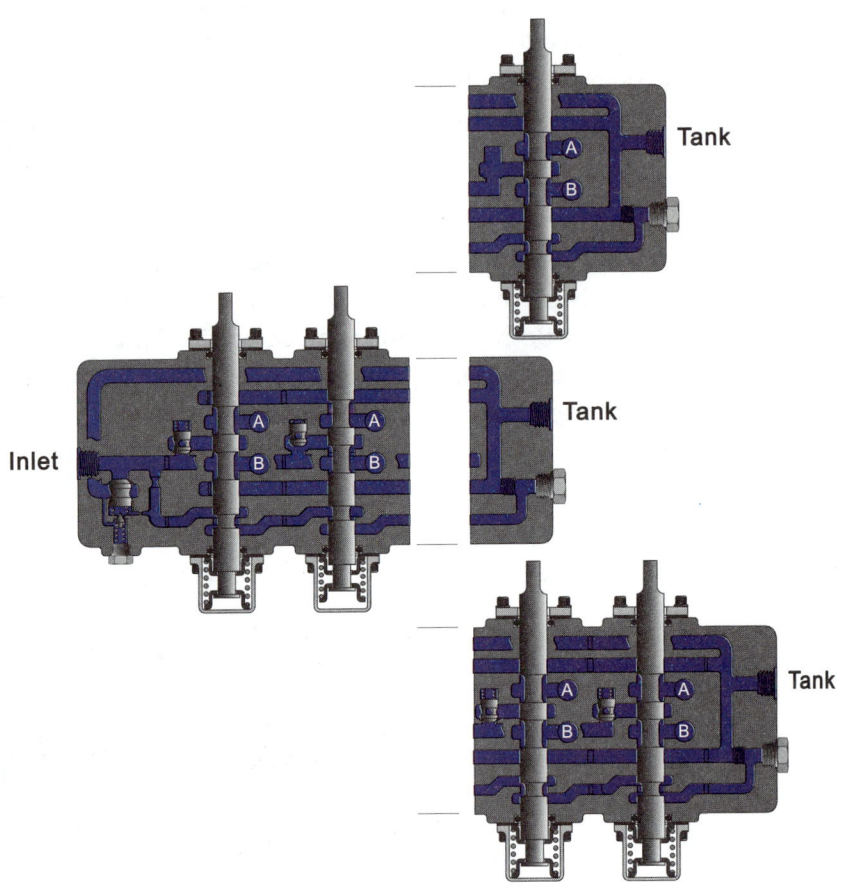

Figure 6.12 Combination mono block and sectional valve

Open Center / Closed Center

Although variable displacement pumps and pressure compensated / load sensing systems have gained in popularity in recent years (see Chapter 13), the vast majority of mobile hydraulic systems incorporate fixed displacement pumps. To conserve power, the directional control valves for these systems are designed with a neutral path through the valve(s) to the reservoir. This is called an "open center" design.

Open center valves have a passage designed in the casting that allows all inlet flow, when the spool is in the neutral, or center, position, to pass through a bypass area at low pressure drop. This flow exits the valve and is either available for another valve connected in series to the first valve, or is free to flow to the reservoir. Therefore, when all valve spools are in the neutral position (no work is being performed), the power consumed by the hydraulic system is at its lowest. Figure 6.13 shows a multiple spool, open center valve with a centrally located bypass passage.

In Figure 6.13 valve section 1 is in the neutral position and the bypass feature is porting flow at the bypass pressure setting to the following valve sections. Valve Section 2 illustrates the bypass operation with the valve shifted out for flow through Port B; and, In Section 3, flow to Port A. Before the flow in either section 2 or section 3 can be useable, the bypass must first be restricted or closed down, or else the fluid will continue through the bypass and do no work (flow takes the path of least resistance). As illustrated in section 2, when the spool is shifted far enough, the bypass will be completely shut off and all flow will then pass to the work port (port "B"). Return flow, coming back through the "A" port, is free to flow to the outlet, or tank, passage to the reservoir. The reverse of this is illustrated in the third section.

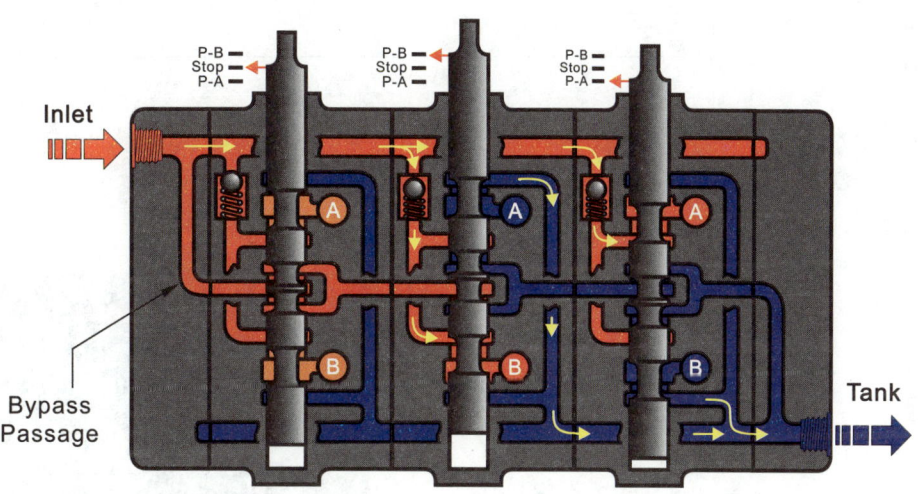

Figure 6.13 Multiple spool valve with center bypass construction

Multiple spool valves are typically either series design or series parallel design. Series design valves are typically the lowest cost and are used in many cost sensitive applications where maximum system pressure is not very high, generally 2000 psi or less. All of the flow is available to each of the valve sections, but at the highest pressure demanded. As such, they are not advantageous in applications where system heat level or energy consumption is critical.

Series parallel designs are the most commonly used type of multiple spool valve. While they permit independent pressure operation for each section, oil flow will follow the path of least resistance and the section with the lowest pressure will tend to have all the flow unless the operator is able to minimize this through the use of metering.

Closed center valves do not have a bypass passage and block all flow through the valve when the spool is in the neutral, or center position. These types of valves are used with variable displacement pumps where system flow in the neutral position does not exist as the pump will be in its "cut off" or "standby" position. Figure 6.14 illustrates a typical closed center mobile spool valve. Closed center valves can also be a combination of inlet spool with outlet poppets; or, inlet poppets with outlet poppets. These designs feature independent control of the valve inlet from the valve outlet with resulting flexibility to the control features of the valve. The Vickers CMX Series valves are an example of a closed center valve typically used in a load sensing system.

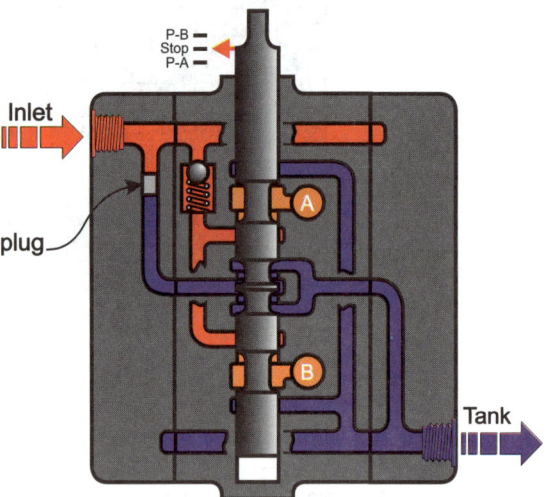

Figure 6.14 Closed Center valve with bypass passage plugged

Figure 6.15 illustrates the CMX valve which is an inlet spool/outlet poppet configuration. CMX valves are covered in more detail later in this chapter.

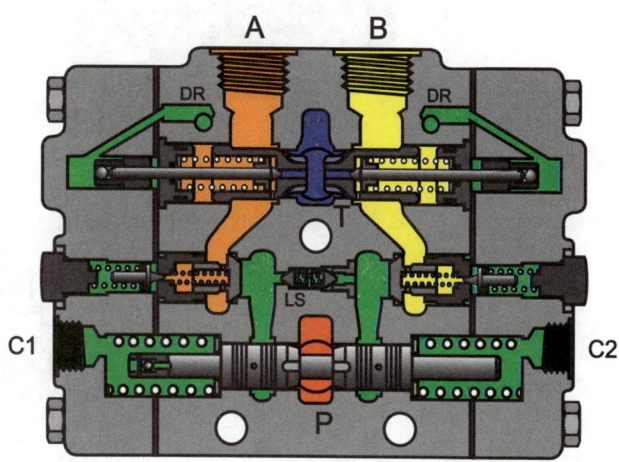

Figure 6.15 Spool and poppet style directional control valve

For safety reasons, each operating section, whether it is open center or closed center, includes a load drop check valve located in the pressure passage to prevent a load from dropping when an operating spool is shifted. The exception to this is those spools that control motors where the use of a load drop check would prevent free directional control of the motor.

Valve Spools

Basic valve spool functions are illustrated in Figure 6.16. Spool valves have a variety of spool types, each for a specific functional control feature. Spool type configurations are:

Tandem
Motor
Closed center
Single acting
Float

Chapter 6 Directional Control Valves

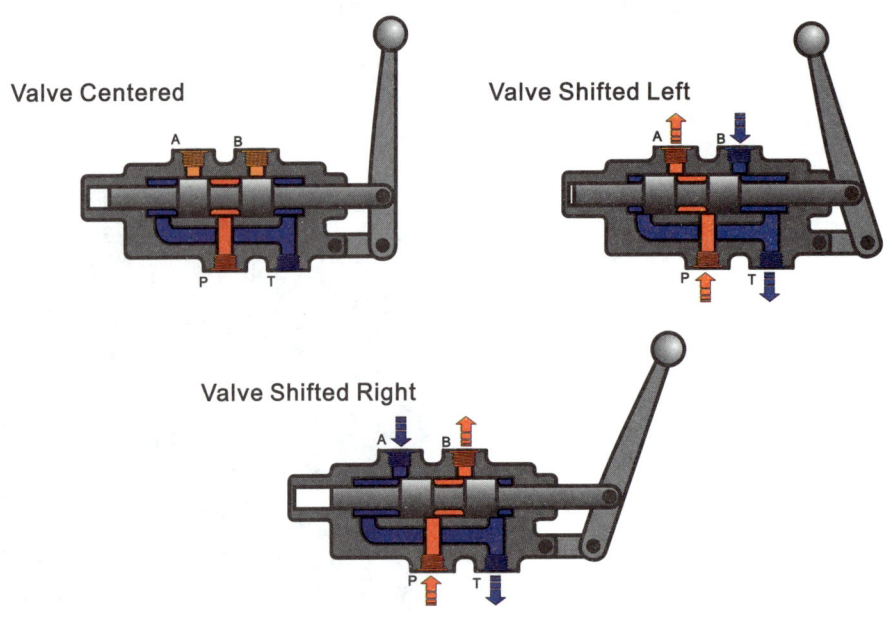

Figure 6.16 Basic valve spool functions

Tandem spools direct flow to either end of a double acting cylinder. Flow from the end not under pressure is returned to tank via internal coring of the valve section. When the spool is in the center position, both cylinder ports are blocked. Typical operation of tandem center spools for cylinder control is illustrated in Figure 6.17.

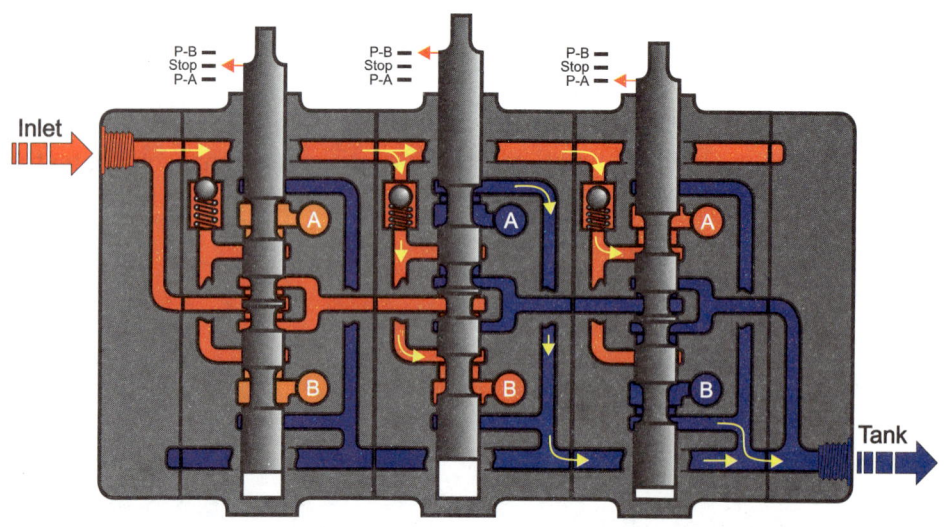

Figure 6.17 Three spool valve with tandem spools

A motor spool is normally used to direct flow to a rotary actuator such as a hydraulic motor. Cylinder ports remain partially open in the neutral position to allow oil flow between motor and reservoir. The motor is not normally provided with a load drop check valve, but for special cylinder applications a load drop check valve can be incorporated. Typical operation of motor spools is illustrated in Figure 6.18.

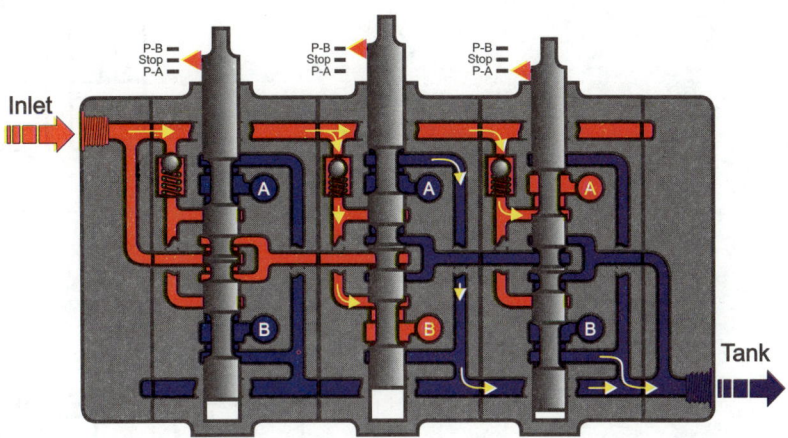

Figure 6.18 Three spool valve with motor spools

Closed center spools are used for either cylinder or motor control in closed center circuits with variable displacement pumps. Figure 6.19 illustrates a closed center cylinder spool. When the spool is in its neutral position flow is blocked and the pump is in its cutoff position. When the spool is shifted, fluid flows through the valve and causes the pump to go into its pumping mode.

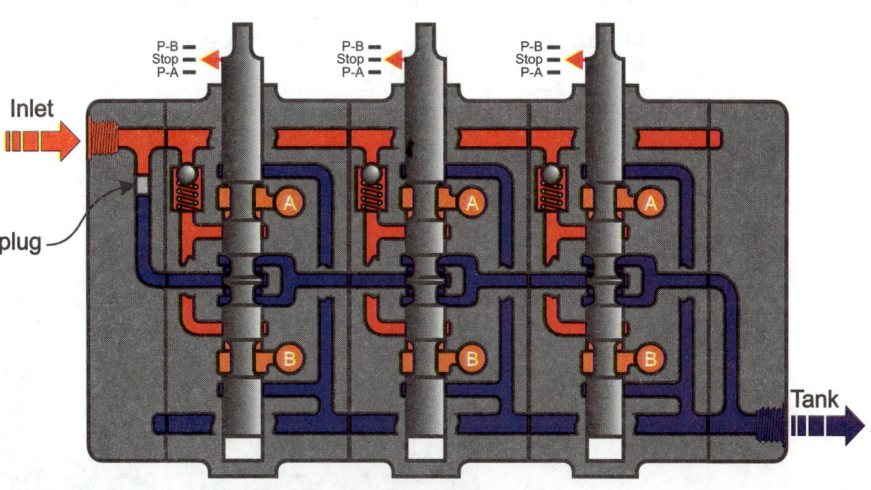

Figure 6.19 Closed center valve with bypass passage plugged

Single acting spools direct flow to only one end of a cylinder, as in the example of a lift mechanism on a lift truck. Return flow is from the same end of the cylinder and relies on gravity or mechanical means. Single acting spools are available for the control of flow to either Port A or to Port B. The application of a single acting spool in a valve bank is illustrated in the basic lift truck circuit schematic shown in Figure 6.20.

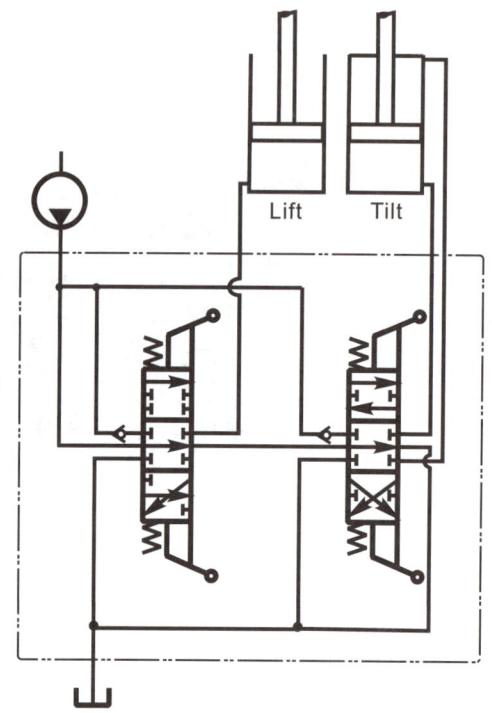

Figure 6.20 Two-spool valve schematic showing single and double acting spool for a lift truck application

Float spools are double acting spools with an additional position that is similar to the center position of a motor spool so that the oil can freely move, or "float" from one port to the other port. Float spools are used in applications such as wheeled loaders or motor graders where it is desirable to have either the bucket or blade follow the ground contour. Typical operation of float spools is illustrated in Figure 6.21.

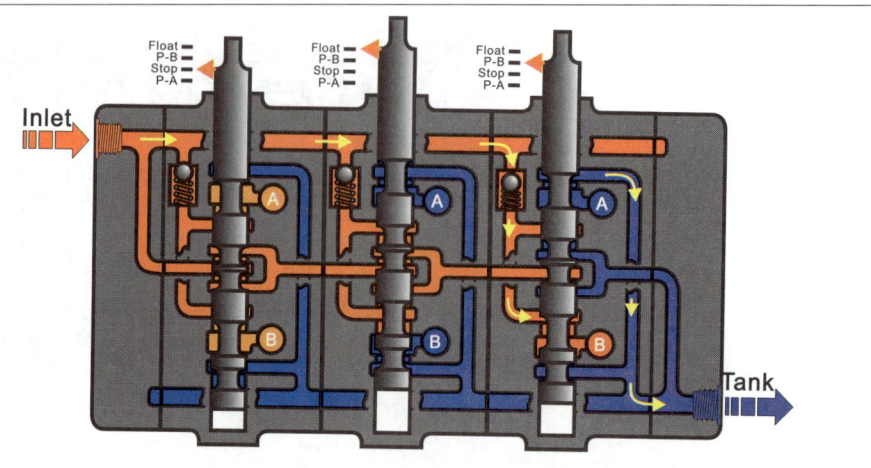

Figure 6.21 Three spool valve with tandem spools including a float position

Valves such as the CMX valve, have similar valve section "spool" functions, but are configured differently due to the separation of inlet control from outlet control.

Flow Forces

Valve spools are subjected to a variable force internally generated that is a function of the flow, pressure drop and spool/valve configuration. This force is typically referred to as a flow force. It is caused by the flow of fluid metering across a spool; it is the result of Bernoulli's Principle (see Chapter 2) of pressure changes with fluid velocity changes.

Figure 6.22 illustrates the marked velocity change that will take place across a spool land while metering flow to an actuator. The pressure change that accompanies the velocity change creates forces on the spool that must be counteracted by the spool return springs. Because there are several metering lands on a single spool, and more than one land can be metering at the same time, the resulting forces are difficult to predict and may act in either direction on the spool. Therefore, the larger the valve size (higher flow rating) and the higher the pressure rating, the greater the return spring force must be.

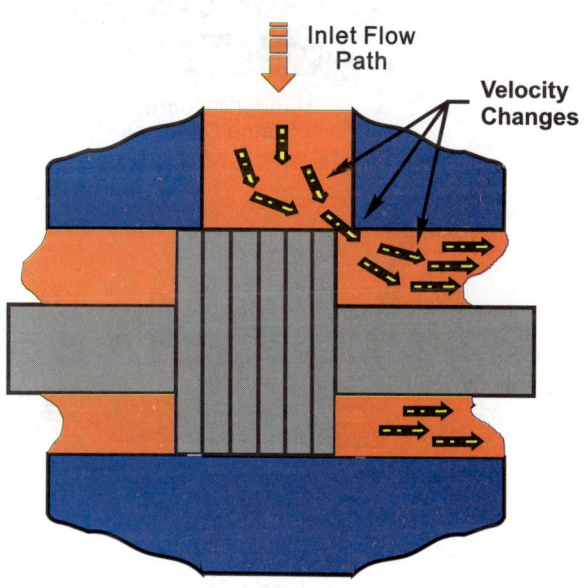

Figure 6.22 Illustration of the dramatic change of fluid velocity during flow metering across a spool

Chapter 6 Directional Control Valves

Spool Actuation

The actuation of mobile valve spools is either manual or pilot operated, with pilot operation being hydraulic, electrohydraulic or pneumatic. Typical valve operators are illustrated in Figure 6.23. Due to the flow forces involved, manual operation is typically limited to flows in the 25 - 35 gpm range with pressures up to 3000 psi. In some instances, where long levers can be used for valve handles, flows for manually operated valves can go up to 70 gpm or higher.

With the trend to higher system operating pressures of 3000 psi or higher; and, the requirement for fine metering control, pilot assisted spool shift control is used. Applications involving pneumatic pilot operation are limited due to the limited force and power available with air pressure, and the unpredictable nature of a compressible fluid for use as a control medium. Hydraulic pilot actuation has been the preferred choice for mobile vehicles, and electrohydraulic pilot operation has increased in popularity. It is projected to be the pilot operated control of choice in the coming years.

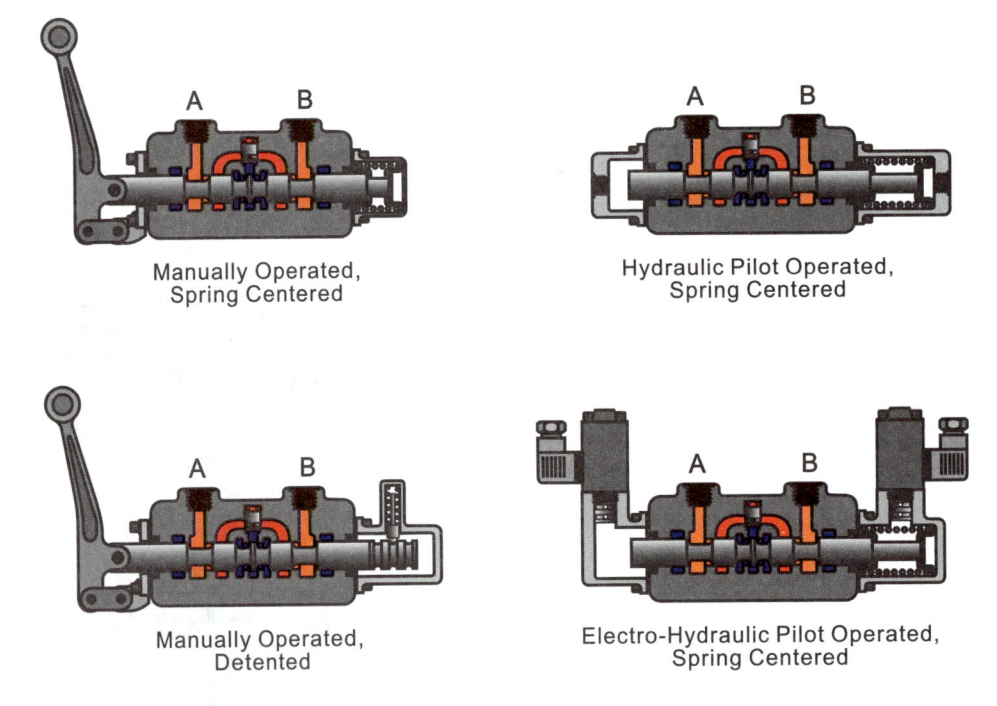

Figure 6.23 Directional control valve operators

Basic Features

Both sectional type valves and enblock type valves have a number of basic features that are common. These are:

Main system relief valves

Cylinder port valving options

Power beyond configuration

Spool end options

Main system relief is provided through a relief valve assembly located just after the inlet to the valve, as shown in Figure 6.24. These relief valves can be either screw-in cartridge type or loose drop-in parts. They are used to provide the main system relief function between the main system pump and the main hydraulic control valve. Cracking pressure is determined by system design requirements and is typically pre-set at the factory.

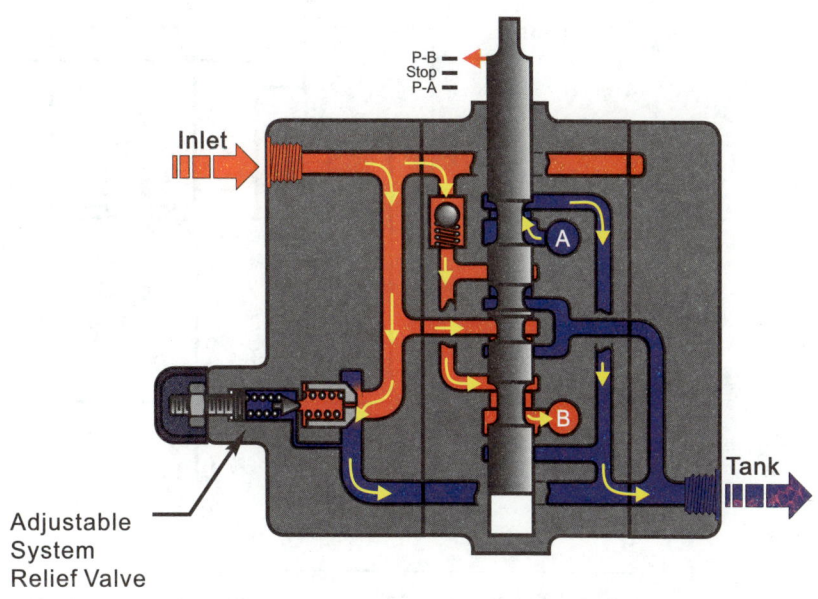

Figure 6.24 Single spool valve with a system relief valve located at the inlet

For multiple spool valve applications, it is sometimes convenient to use an adjustable auxiliary relief valve to control the regulating pressure on the main pilot operated relief valve. Figure 6.25 illustrates the circuit with the two relief valves, the first of which is pilot operated and connected to the main circuit, and the second auxiliary valve is set for the maximum system pressure. At lower operating pressures, the system is controlled by the main relief valve, which will relieve any excess flow at the lower pressure (plus a small pressure value equal to the spring). When system pressure rises to the maximum set by the auxiliary relief valve, the main system relief valve will then be controlled at that pressure. This is particularly helpful on electric lift trucks where it is important to conserve battery life; system pressure is controlled at the working pressure at all times until the maximum system pressure is reached.

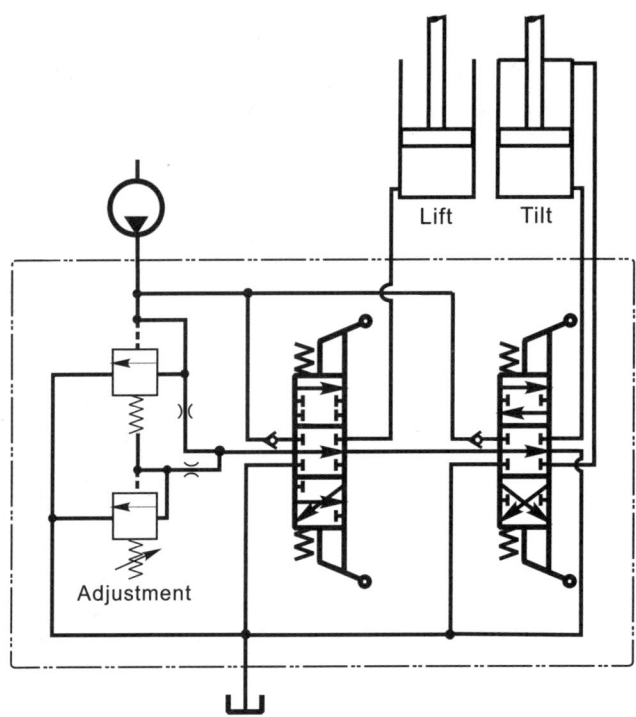

Figure 6.25 A lift truck application using an adjustable relief valve pilot

A second means of energy savings is available through the use of a two pressure valve system particularly for lift truck applications. This arrangement allows the use of lower pressure, lower cost components and hoses for "tilt" and other accessory functions while providing higher pressure for the "lift" function. This arrangement also generates less heat during relief valve operation and gives longer pump and system component life. On electric lift trucks it saves current for longer operation between battery charges.

The two pressure system is obtained by adding a second relief valve cartridge in series with the main relief and the use of a special, single acting spool. This second relief controls the pressure setting for all functions in the valve bank except the lift function. When the lift function is actuated, this second relief is blocked so that it is no longer connected to tank and the main relief valve becomes operative.

Figure 6.26 illustrates the two pressure system.

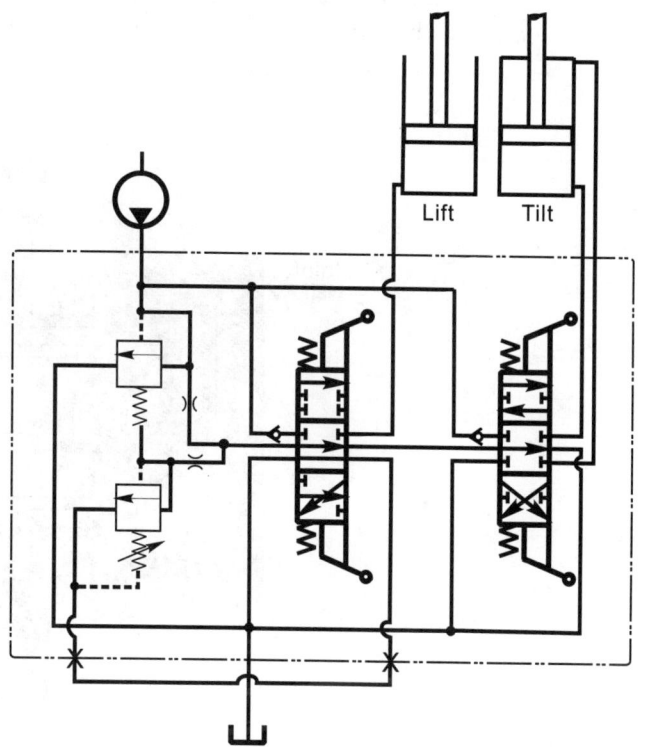

Figure 6.26 A lift truck application using a dual relief system

Cylinder port valve options are located in the valve body between the spool and the external actuator or cylinder port. They include, but are not limited to:

Cylinder port relief valve
Anticavitation valve
Combination cylinder port relief and anticavitation valve

Cylinder port relief valves are located between the valve spool and the cylinder or actuator being controlled. Their purpose is to limit pressure in the hydraulic line between the valve and the cylinder in order to prevent structural damage to the cylinder and inadvertent line rupture. Figure 6.27 illustrates a tandem center valve section with cylinder port relief valves in both port A and port B.

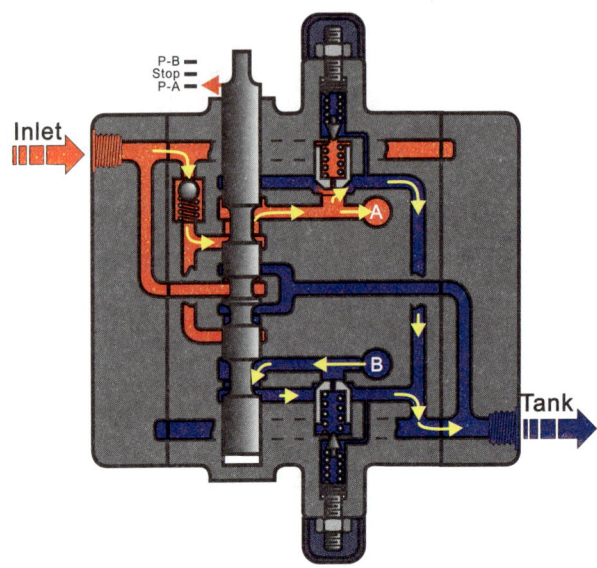

Figure 6.27 Tandem spool valve with dual port reliefs

Anticavitation valves are used to prevent cavitation caused by insufficient oil being available during an overrunning condition. In this situation, tank back pressure is sufficient to overcome the very light spring holding down the anticavitation poppet. This causes the poppet to unseat and enables oil returning from the actuator to be directed from the tank passage in the valve housing to the pressure port to provide makeup flow to the pressure line.

As both the anticavitation valve and the port relief valve fit into the same machined cavity in the valve housing, and typically both are required, it is common to combine these two functions into one valve cartridge assembly. The port valves in Figure 6.27 are combination port relief and anticavitation valves.

Power Beyond Both sectional valves and enblock valves have a "power beyond" option that permits flow through the valve bank to be used for downstream functions. A typical application is with two valve banks connected in series where there is a series connection from the outlet of the first valve bank to the inlet of the second valve bank. There may also be a direct connection from the pump to provide oil to both valves. The second valve bank receives full flow when the spools in the first valve bank are either in neutral or metering in an operating position.

The second valve bank is dependent upon the first valve bank for carryover flow and relief valve protection. Figure 6.28 illustrates the power beyond configuration with the first valve providing flow to the second valve.

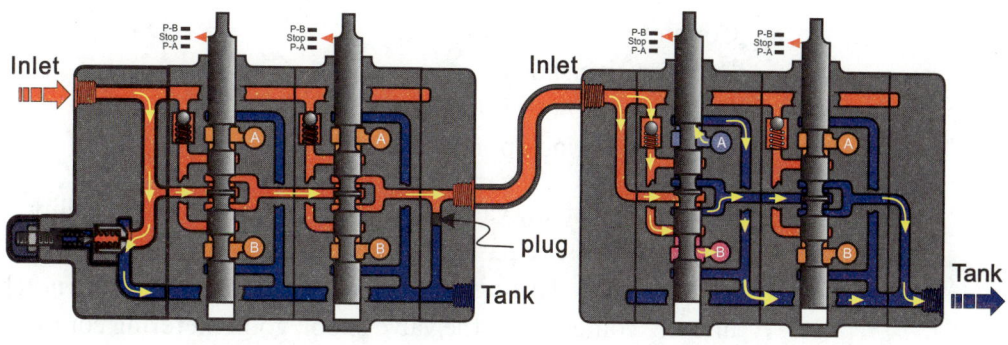

Figure 6.28 Power Beyond feature of first valve provides input flow to the second valve

Spool End Options

There are two basic spool end options: spool detents and electric switches. In normal valve operation the valve spool centering springs return the spool to the neutral position when the operator releases the valve control lever. In some applications it is desirable to provide the means for the operator to remove his hand from the spool shift lever while ensuring the spool is in either the neutral position or in one of the operating positions. An example of this requirement is the use of a valve to control a motor drive on a mixer application. Detents let the operator lock the spool in either neutral or in one of the operating positions. On float spools, a single detent is used to keep the valve in the float position.

Electric switches are used in applications where the hydraulic flow is provided by a pump driven by an electric motor. When the spool is shifted, the initial movement closes the electric circuit and starts the electric motor so that pump flow is instantaneously supplied to the valve. When the spool returns to its neutral position, the electric circuit is broken which then turns off the electric motor and stops the pump. This feature is typically used on electric lift trucks to conserve the battery while ensuring the availability of peak power when required.

Both the detent and electric switch optional features are illustrated in Figure 6.29.

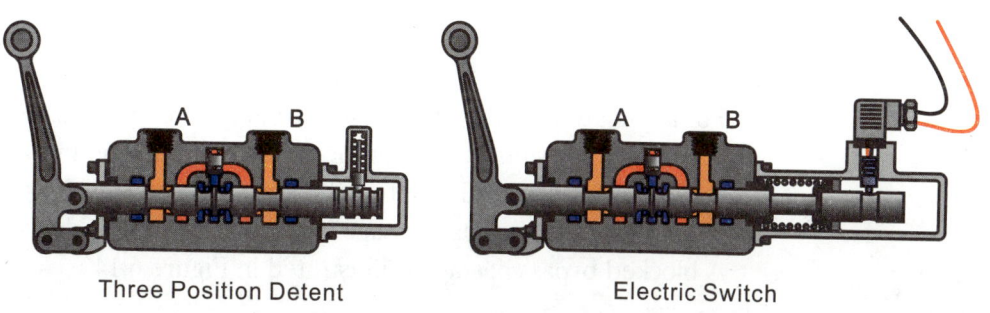

Figure 6.29 Detent and electric switch spool end options

Chapter 6 Directional Control Valves

Specialty Features

There are a number of special features available as options to the basic valve bank configuration. These are:

Flow metering
Closed center valve inlet
Counterbalance spools
Regenerative spools
Controlled flow to cylinder ports
Centering spring options

The success of any valve depends on its metering capability. Metering takes place from the pressure passage to the cylinder ports and from the cylinder ports to the tank passage. Through the use of slots, called metering notches, either machined into or "coined" into the valve spool, good metering control is possible. Figure 6.30 shows some examples of metering notches machined into a valve spool. Alternatively, some valves use metering notches either cast into or machined into the valve body, instead of the spool.

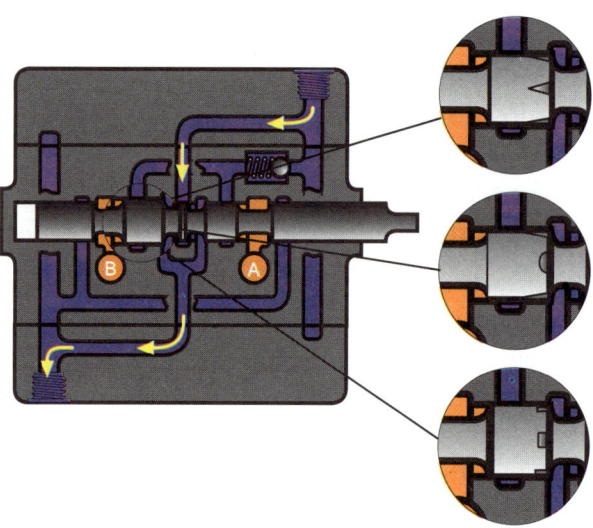

Figure 6.30 Variations in spool metering notches

For applications involving pressure compensated pumps, unloading valve accumulator systems or air motor driven hydraulic pumps a closed center valve inlet section is required. This optional feature typically involves the use of an internal plug to block the bypass passage and integral system relief valve in the valve bank. A blocked bypass passage is illustrated in Figure 6.14.

As a means to prevent overrunning loads, an optional spool configuration is available to provide a counterbalance feature incorporated within the spool. This feature is commonly used in the tilt spools on valves used on lift trucks, as illustrated in Figure 6.31. Operation of a counterbalance valve is discussed in Chapter 9.

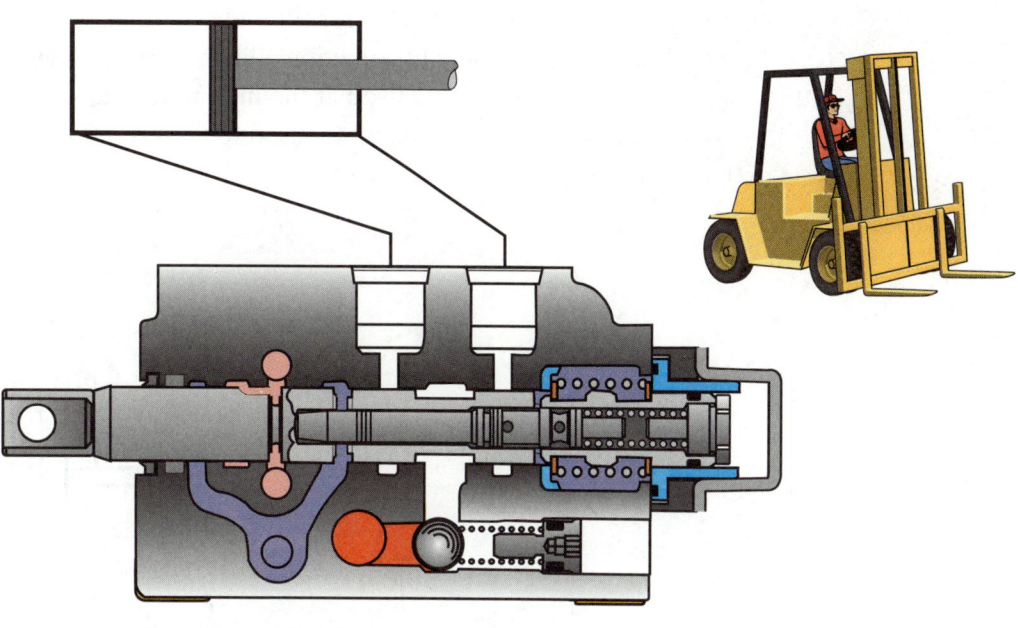

Figure 6.31 Directional control valve with a counterbalance valve located inside the spool

Regenerative spools are used in applications where high cylinder extension velocity at low force would be convenient. An example is an occasional rapid extension of an unloaded crane boom. A regenerative spool allows conventional extension and retraction of a loaded boom during normal operation. Regenerative circuitry is discussed in Chapter 3, and a four position regenerative spool symbol is shown in Figure 6.32.

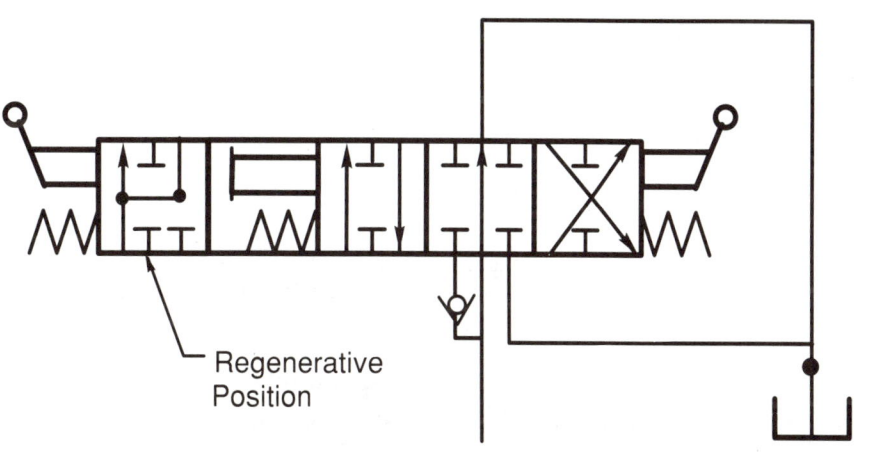

Figure 6.32 Four position valve spool with a regeneration position

Controlled flow for specific functions in a multiple spool valve bank is possible through the use of an orifice located upstream of the spool for which controlled flow is required. The orifice can be either fixed or adjustable. All spool functions downstream of the flow controlled spool section will be limited to the same controlled flow. This feature is quite often used in lift truck circuits for energy savings where the flow required for tilt and other auxiliary functions is much less than that required for the lift function. To conserve energy, a pressure compensated bypass flow control must be used at the inlet to the valve. Figure 6.33 illustrates a circuit schematic where an orifice is used to limit the flow to the tilt section of a lift truck valve bank.

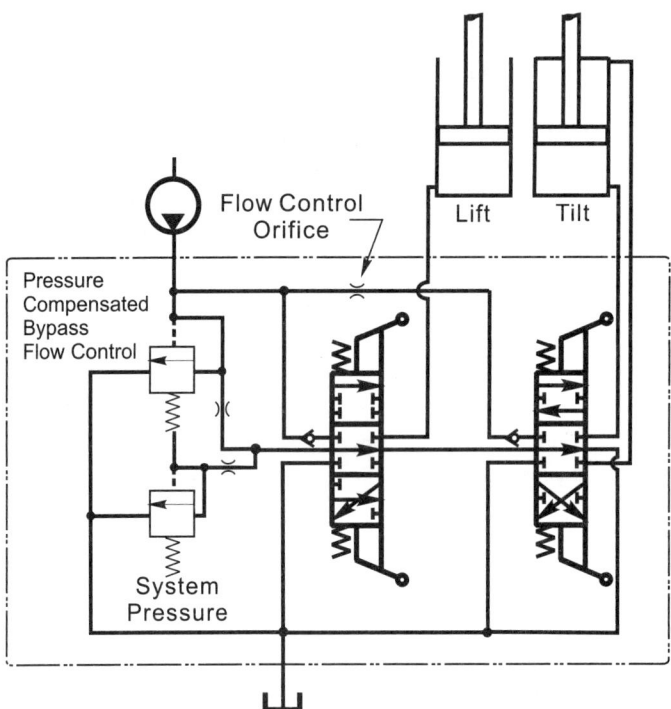

Figure 6.33 A lift truck application with a flow control in the second spool and a pressure compensated bypass flow control and relief valve combination

There are times when users would prefer to have a stiffer or easier force to shift a valve spool. For these occasions there are optional centering springs available that will make the valve harder or easier to shift. Care must be taken, however, that the spring has sufficient strength to compensate for flow forces developed by the fluid.

REMOTE CONTROL OF HYDRAULIC VALVE FUNCTIONS

Remote control of larger directional control valves is used for several reasons:

- Large valves require heavy return springs, which then require high forces to operate the valve spool. Remote control valves can multiply forces, reducing operator fatigue and improving metering.

- Large valves require large conductors and connectors. By keeping the main valves close to the actuator, and using small conductors to connect small remote valves from the operator to the main valves, a more economical system can be devised.

- The high flow that accompanies large directional control valves usually brings substantial heat to the operator's area. Keeping flow low through remote control valves, and keeping the large control valves away from the operator's area, means greater comfort for the operator.

- On many machines, the operator's compartment is quite compact. There is often insufficient room to include large control valves with large conductors and connectors.

- A quieter operator's compartment usually results by removing the high flow components from the vicinity.

There are 3 types of remote control available for the control of individual valve operating functions. These are:

Hydraulic remote control (HRC)
Electric remote control (ERC)
Pneumatic remote control

Pneumatic remote control systems are very limited due to the compressibility of air, and the need for a separate fluid system pump (air compressor). Pneumatic controls tend to be less responsive than hydraulic controls, and somewhat "spongy". For this reason, pneumatic remote controls are seldom used, and will not be covered further.

Chapter 6 Directional Control Valves

The most common types of remote control in use are HRC controls. Figure 6.34 illustrates a typical HRC unit. HRC units are available for hand, foot or mechanical operational control of hydraulic valves. Foot, or pedal, control is quite often used to provide traction control of track type vehicles where individual pedals provide forward, neutral or reverse directional control of the tracks. Pedal controls can also be used to provide an "extra hand" to the operator by allowing him to control "kick in/out" features and/or an additional function beyond those controlled with a hand operated HRC unit. Pedal controls can be either spring loaded or rocker type.

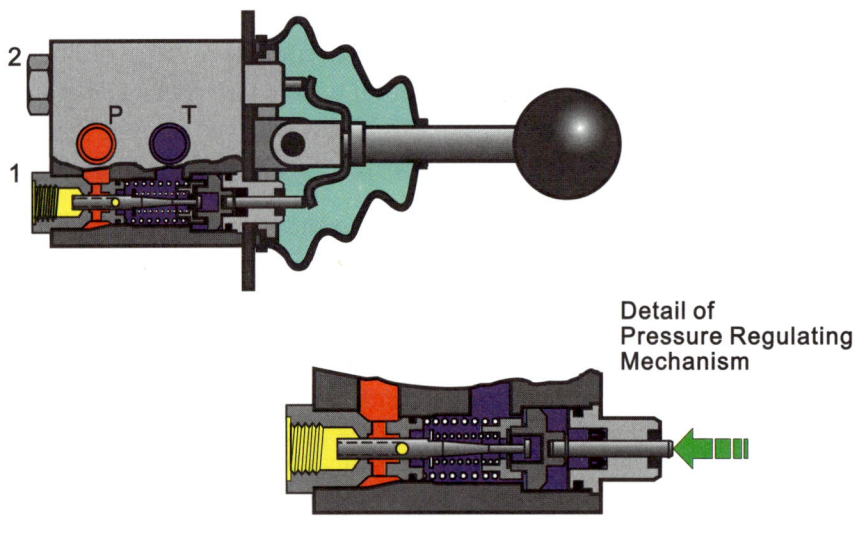

Figure 6.34 Hydraulic remote control valve

Hand operated HRC units are in two basic configurations:

- Single function controls that operate one main system spool. These valves are usually sectional, such that several can be combined in a bank of controls to operate several main system spools.
- Dual function controls in one body. Two main system spools can be controlled with one handle, typically left-right for one spool and forward-back for another. This type of control is commonly called a "joystick" control.

Optional features found in HRC units are: (1) stroke adjustment, (2) 3 position detent, (3) frictional position hold with detent in the neutral position, (4) pull-to-release neutral detent, (5) heavy duty return springs, (6) handle with an electric button control and (7) handle with an electric switch control.

In addition, an HRC can be fitted with one of several spring options that determine the control pressure profile for each end of the main valve spool being controlled. This permits tailoring of a specific control profile for the actuator being controlled.

Figure 6.35 illustrates four pressure profile options.

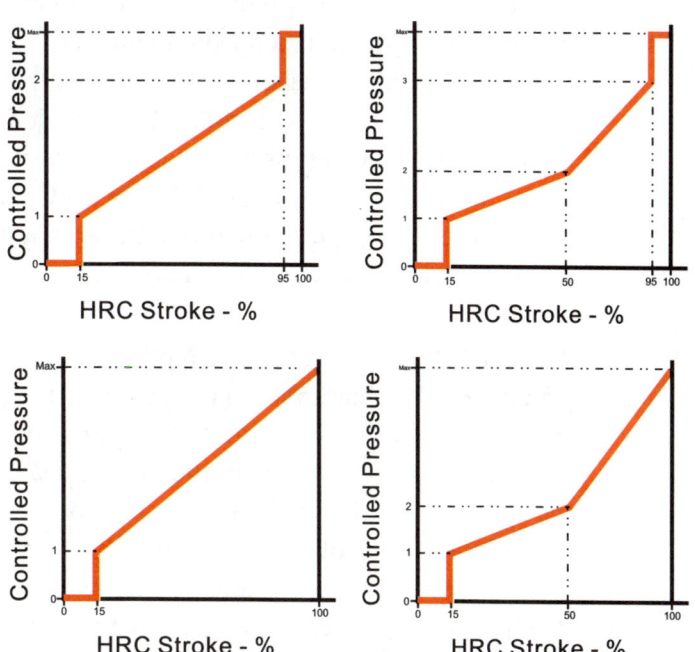

Figure 6.35 Typical pressure profile curves for hydraulic remote control valves

Figure 6.36 illustrates a typical ERC unit. ERC units are more compact in size than HRC units and are more delicate due to their electronics and size. They do not have the sensitivity to ambient temperatures that is common to pneumatic and to some extent HRC units. With the continued trend to miniaturize electronics, a multitude of control features can be built into ERC units that give them increased control precision and features not feasible with HRC units. Because of this there is an increasing trend towards the use of ERC units in newer vehicle designs, particularly those with complex control requirements.

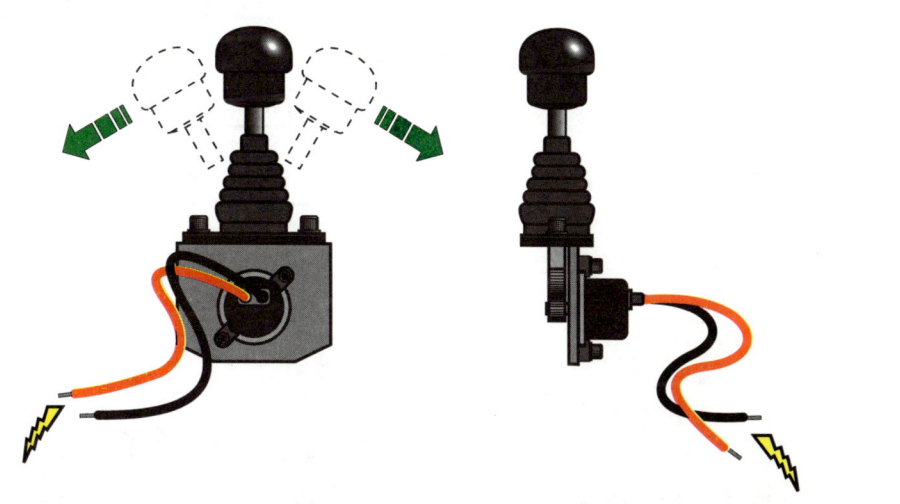

Figure 6.36 Electric remote control (ERC)

Chapter 6 Directional Control Valves

NON-SPOOL DIRECTIONAL CONTROL VALVES

The characteristic feature of non-spool directional control valves is the concept of separate meter-in and meter-out elements. This can be done with a single meter-in element as with the Vickers CMX Series valves or with separate meter-in elements as with the REGO valve introduced in Sweden in the early 1990s. Primary application is in load sensing systems with variable displacement pumps. These valves are controlled by either HRC or ERC units. It is not possible to control them by means of lever operation as there is no physical connection between the meter-in element and the meter-out element.

Design Concepts The CMX Series valves utilize a pilot operated, pressure compensated, proportional sliding spool to control fluid from the pump to the actuator. The REGO valve utilizes separate pilot operated, pressure compensated, proportional poppets to control fluid from the pump separately to each actuator port. In both the REGO and CMX valves, the meter-out elements are pilot controlled metering poppets, and control exhaust fluid from the actuator to tank. Each meter-out poppet functions as a variable orifice between one of the actuator's ports and the tank port with the degree of opening proportional to the pilot signal. As the REGO valve technology has been acquired by Vickers and is not currently in a production configuration, our discussion will be limited to the CMX Series valves.

The basic concept of the CMX valve is illustrated in Figure 6.37. A cross section of the basic valve is shown in Figure 6.38 and the principle of operation for controlled flow to port A and controlled flow from port B is illustrated in Figure 6.39. Pilot control of the meter-in spool and meter-out elements is illustrated in Figure 6.40, illustrating both the HRC or ERC units to control the meter-in and meter-out elements.

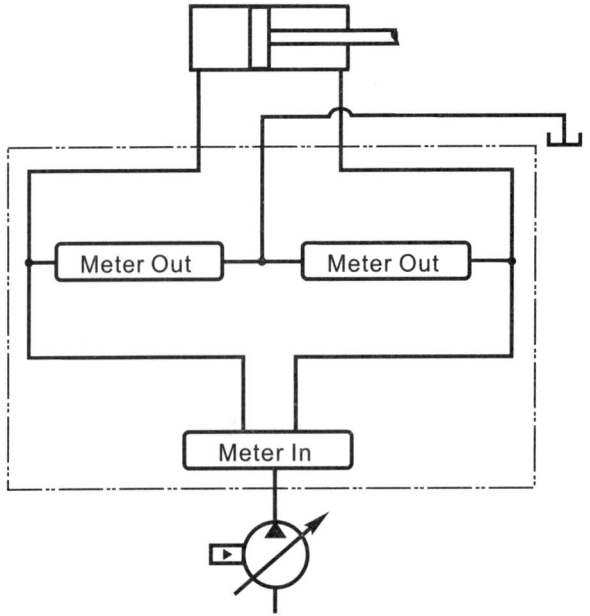

Figure 6.37 Basic non spool valve concept

Chapter 6 Directional Control Valves

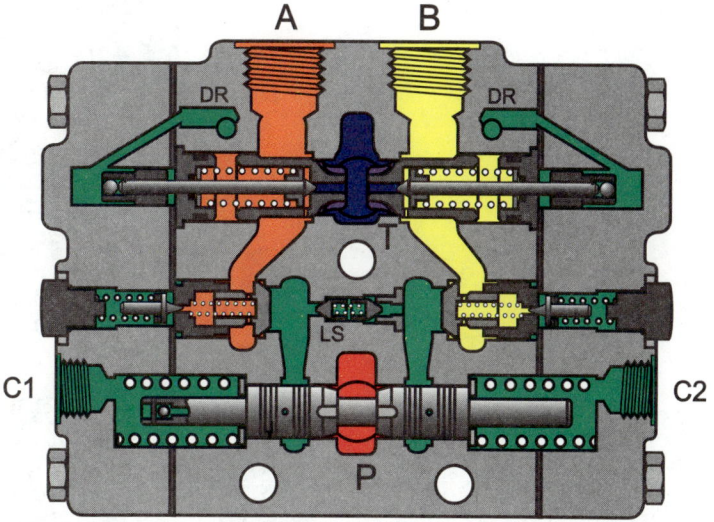

Figure 6.38 Vickers CMX non spool valve construction

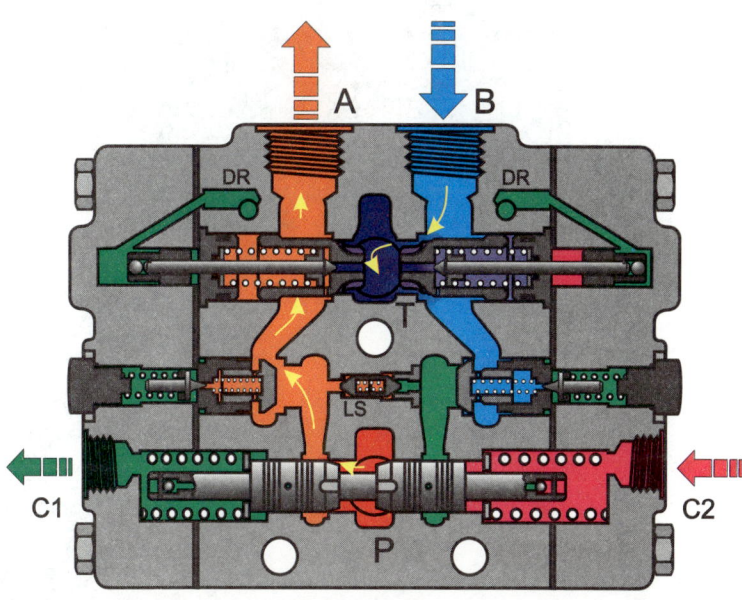

Figure 6.39 Metering fluid to port A and metering return flow from port B

Chapter 6 Directional Control Valves

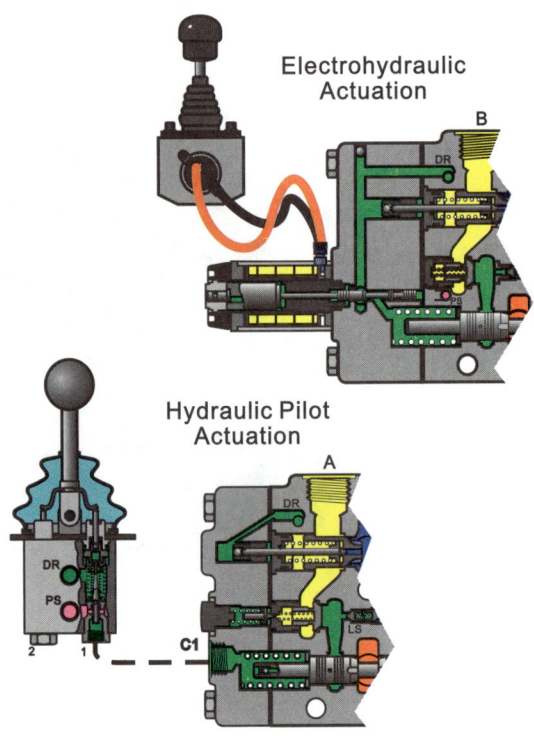

Figure 6.40 Vickers CMX valve can be controlled electrically or hydraulically

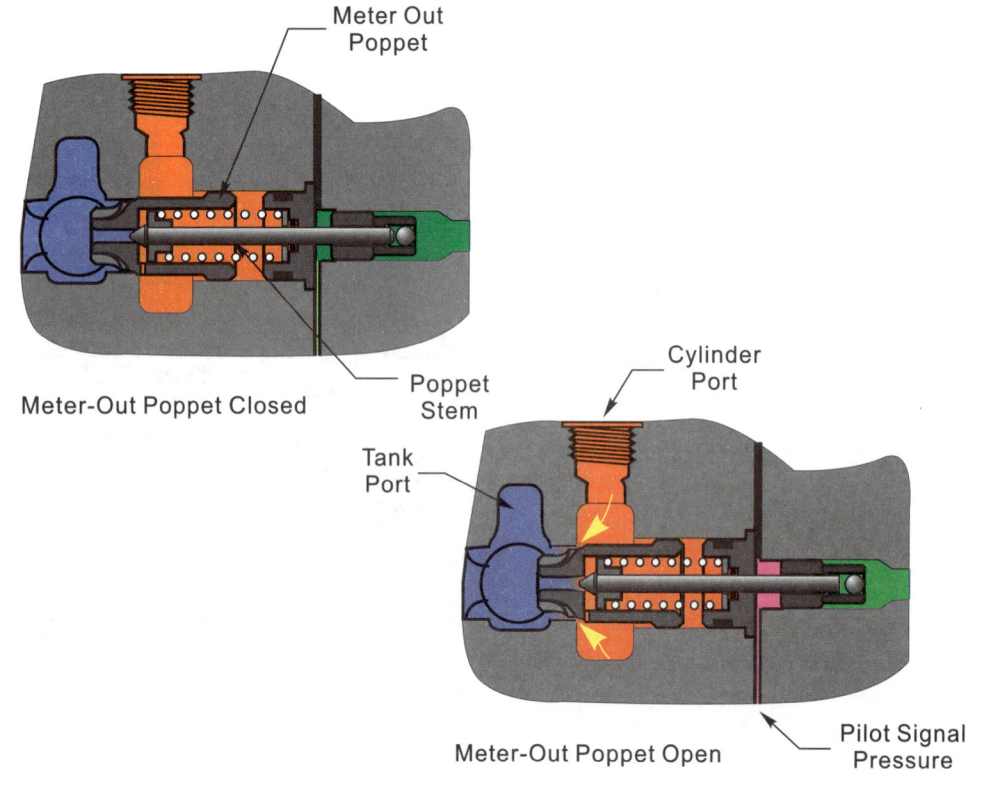

Figure 6.41 Vickers CMX meter out element

162

Chapter 6 Directional Control Valves

Meter-In Control

The meter-in element ports fluid from the valve inlet port to the "A" or "B" meter-in chamber. It is controlled by the pilot pressure signal provide by either an HRC or an ERC unit. The meter-in element is available as either a Flow Control type or a pressure control type.

Meter-Out Control

Meter-out control is achieved by using a pilot poppet stem with a modulating meter-out poppet to form a simple hydro mechanical bleed servo. This is illustrated in Figure 6.41. As shown in the lower view, the meter-out poppet follows the poppet-stem position which is controlled by the pilot signal and the spring it moves against. A small change in position of the meter-out poppet stem, caused by a small flow from the pilot source, results in a large increase in flow passing the meter-out poppet. The ratio of this flow is the gain of the meter-out control.

There are several different meter-out poppets available which provide different area gains. A high gain poppet provides better control when lowering a light load. In contrast, a low gain poppet provides better control when lowering heavy loads. As the meter-out element is controlled by a lower pilot pressure signal than the meter-in element, it is possible to lower a load on a controlled basis without using pump flow, thus providing fuel consumption savings.

Optional Configurations

One of the major benefits to the non-spool type of valve is to quickly modify the configuration for almost any machine and function application. Individual poppets and springs can be modified for different performance on each port, if desired. This is a very difficult undertaking on compound spool type valves. Variations that can be readily accomplished are:

Low flow meter-in spool
Externally vented meter-out poppets
Free coast meter-out poppets
Meter-out spools
Anticavitation option
High flow and regenerative meter-out modules

Chapter 6 Directional Control Valves

ELECTROHYDRAULIC DIRECTIONAL CONTROL VALVES

Proportional Valves

The introduction of proportional solenoid control has greatly enhanced operator performance and productivity through the ability to provide precise, smooth and rapid control of both small and large machines. This is particularly true as vehicles have become more and more complex and the demands upon the operator have increased for more precise work output. This development is reflected in the increasing trend for electrical remote control being preferred over hydraulic remote control.

Figure 6.42 illustrates the mounting of proportional solenoid valves as the valve spool operator on a typical mobile directional valve. The mounting of proportional solenoid control to a CMX valve is illustrate in Figure 6.40. By means of an electrical input signal, precise, accurate and effortless control is made on spool movement.

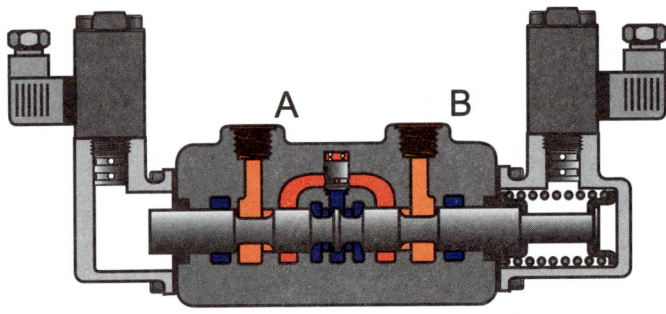

Figure 6.42 Electro-hydraulic pilot operated, spring centered directional control valve

An alternative form of electrical proportional control is illustrated in Figure 6.43. This valve is similar in operation to the Vickers CMX valve, except it uses a force motor and pilot valve to control input pilot pressure. The advantages of the force motor are two wire control versus the four wires required for two proportional solenoids, and usually lower electrical power requirements.

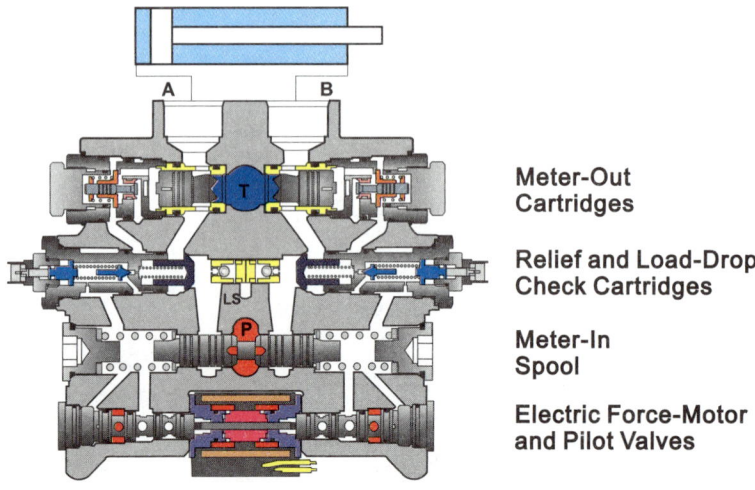

Figure 6.43 Electrically proportional pilot operated mobile valve

The most common type of control used with these types of valves is the ERC. These provide the most basic system, and provide the familiar form of control that operators are accustomed to. The "feedback" in this type of system, that is, the movement of the actuator relative to the movement of the control, is typical hand-eye coordination by the operator.

Technological advances in electronics are providing considerably more sophistication in controls (reference Chapters 3, 7 and 12) such that feedback can be incorporated into the initial control function, and the initial control function can be programmed to supplement or override an operator's command. Control units that resemble personal computers can be programmed to accept position-speed-force feedback information and combine it with operator's input commands to send result oriented information, rather than task oriented information, to the actuators. Wireless transmission of commands and feedback also continues to grow in popularity.

Servo Valves

A servo mechanism is one that produces a large amount of force in relation to a small amount of input, and contains a self correcting feature. A popular example of a servo mechanism is a power steering device, where a small input force at the steering wheel will result in a high force at the wheels, and the angle of the wheels will follow the steering wheel input, self correcting through a mechanical feedback mechanism.

Servo valves perform in the same manner; a small amount of electric power input will produce a relatively large amount of hydraulic power output. In addition, the position of the valve spool is maintained through a self correcting mechanical feedback mechanism. Figure 6.44 illustrates the basic features of a typical servo valve.

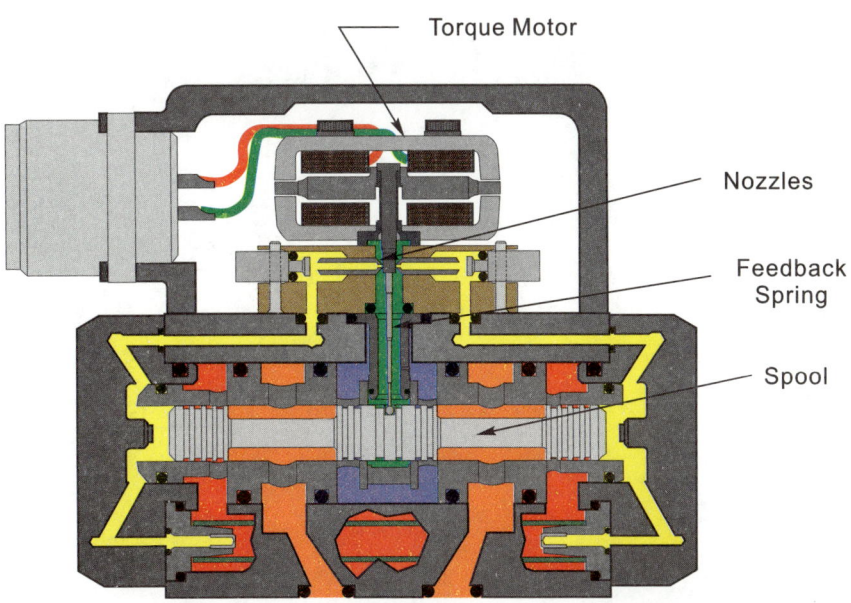

Figure 6.44 Servo valve

A very small amount of electric current, a few milliamperes, will cause the "flapper" to move, which will change the pressure balance between the two nozzles. A difference in pressure between the two nozzles will translate to a difference in pressure at the two ends of the spool, causing the spool to move, providing fluid flow at one of the valve ports. Thus, a relatively large amount of hydraulic power output from a small amount of electric power input.

The feedback is created by the feedback spring, which is moved when the spool moves. This movement will adjust the pressure balance across the nozzles and maintain the spool position determined by the electric input. Note that the feedback is strictly on spool position, not valve output (flow, pressure). Output feedback must still be provided external to the valve with transducers and the associated electronics.

Servo valves are occasionally used in mobile systems. Generally, however, the functions can be dealt with using proportional valves which are less costly. External feedback is required in both instances.

CHAPTER 7 .. Electronic Controls

The integration of electricity and hydraulics, relative to mobile equipment, has come the full range of technology from simple on off solenoids, to microprocessor controlled "smart actuator" systems. Although systems with a high degree of internal intelligence are in low usage at this writing, development labs at OEMs and universities are experimenting with many variations of this approach, and this technology will become commonplace within a few years.

Therefore, it is necessary to become familiar with the types of equipment and systems being developed, understand how they will be used, and be able to differentiate between the various hydraulic products that integrate with these systems.

This chapter is intended to be conceptual relative to the electronic equipment in these intelligent systems; first, because this is a hydraulics manual, not an electronics manual; second, because adequate treatment of electronics would probably double the size of this manual; and third, because electronic equipment and the associated software change very rapidly and software is usually customized for a given application. Hydraulic hardware changes much more slowly and is more evolutionary.

The hydraulic hardware involved in electrohydraulic systems is defined in greater detail in chapters covering the individual components.

BASIC CONTROL CONCEPTS

Control in hydraulic circuits means the control of pressure, flow and direction of fluid. Ultimately, this control leads to the operation of a linear or rotary actuator that is expected to apply a force and do work at a speed and magnitude the operator commands.

Control Parameters

System pressure is a function of the actuator requirement, that is, the linear or rotary force the actuator is subjected to, and the velocity at which the actuator is allowed to operate. Velocity is a function of the flow reaching the actuator, and reducing the flow by restricting it can create higher pressures at the pump to overcome the restriction. "Throttling" or metering a directional control valve leads to higher pressures at the pump than the actuator force might otherwise dictate.

Typical electrohydraulic systems consist of a variable displacement pump with a load sensing compensator (see Chapter 12). This type of pump arrangement will adjust its displacement, and its flow output, to maintain a low pressure drop across the directional control valve, regardless of the metering position. The pressure at the pump, and therefore the system pressure, is strictly governed by the actuator force requirement. In these types of systems, the only two control parameters are flow and direction. Both are dictated, with few exceptions, by the directional control valve (exceptions are discussed in Chapter 12).

Chapter 7 Electronic Controls

Valve Actuation

Electrohydraulic directional control valves are actuated by either solenoids, torque motors or force motors (see Chapter 3). These types of actuation have somewhat different characteristics, such as the amount of electrical power required, proportionality and hysteresis.

Solenoids, by themselves, are a linear, non-proportional, unidirectional device used in on-off control situations. If bidirectional actuation is needed, two solenoids are required. If proportionality is required, a means of applying a proportional signal is needed. Because solenoid operated valve spools are usually actuated directly by the solenoids, higher electric power for control is required (about 25 watts or more). Figure 7.1 illustrates the direct action of a control valve solenoid.

Figure 7.1 Three position, four-way solenoid operated directional valve with direct operating solenoids

Force motors, like solenoids, are a linear force device and are used only with small spools controlling small flows because of their low force capability. Their advantage, however, is that they can be operated in two directions by reversing the current flow. Because of their low force capability, force motors are normally used to control the pilot section of pilot operated valves. Figure 7.2 shows a pilot operated valve activated by a force motor.

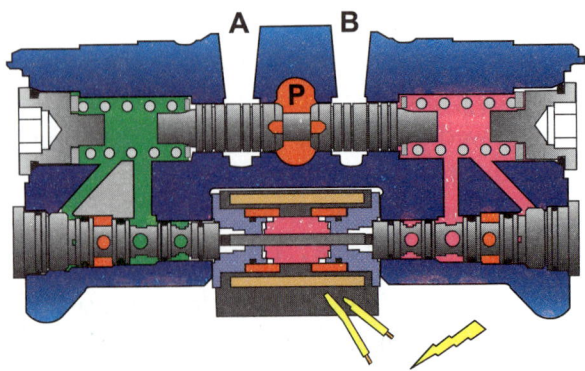

Figure 7.2 Force motors generally operate pilot valves

Torque motors, like force motors, can be operated in two directions by reversing the current flow. They are a rotary force device, but by using only a very small arc of the output movement a nearly linear motion is achieved.

They are used in servo valves, as illustrated in Figure 7.3, which are designed to accept this slight curvilinear motion. Furthermore, the torque motor does not directly move a spool, so the power requirements are exceptionally low.

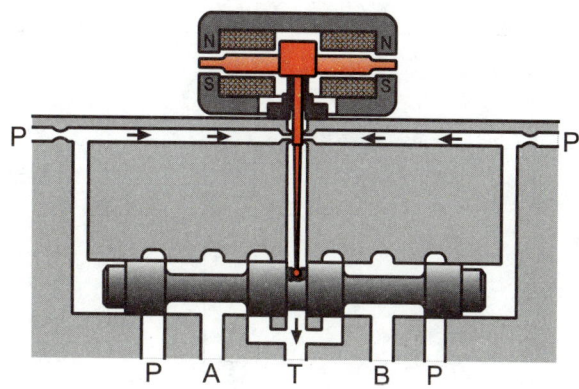

Figure 7.3 Servo valve concept

Pilot Operated Valves

Solenoids, torque motors and force motors used on pilot operated directional control valves require much less power to operate. They do not act on the main spool directly, but regulate a pilot flow which is directed to the ends of the main spool. The pilot flow causes the spool to shift. Less electrical power is required for these types of directional control valves. Pilot operated valves, using solenoids or force motors, require about 15 watts of power to shift. Servo valves, using torque motors, can operate with as little as 1 watt of power, and sometimes less.

Deadband

The edges, or lands, of a valve spool usually have some degree of overlap with the corresponding edges of the valve body ports as shown in Figure 7.4. This is required for several reasons: 1) overlap helps to prevent flow across the spool when the spool is in neutral, 2) it helps to assure that system flow is zero when the electric signal to the valve is zero, and 3) it helps to assure a degree of safety if the electric power is lost and the spool "springs" back to neutral.

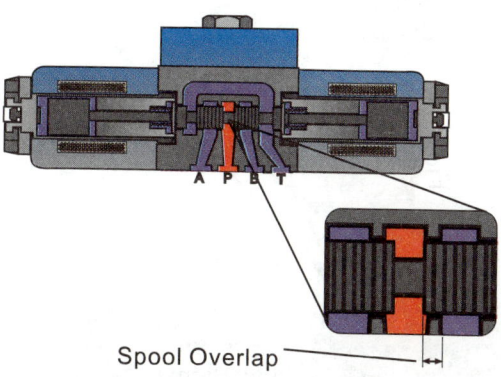

Figure 7.4 Overlap of spool and bore lands leading to a deadband of spool travel

The nature of spool overlap, however, means that there will be a defined movement of the spool in the bore before flow begins to take place. This "dead" movement results in the term "deadband", and refers to the amount of movement the person operating the valve must physically make before any activity (meaning fluid flow) will take place. Figure 7.5 illustrates this characteristic.

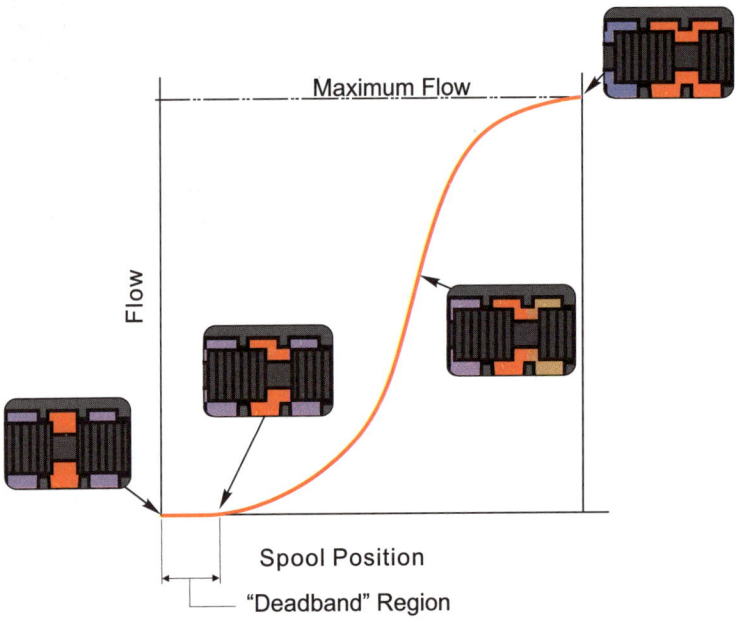

Figure 7.5 Flow vs. spool travel, illustrating the deadband region

The effects of deadband can be compensated for, however, by an electronic circuit called deadband compensation. This circuit automatically increases the gain of the amplifier when the valve spool is near neutral, which allows even a small signal to move the spool across the deadband zone. Deadband can be virtually eliminated in this way. Figure 7.6 shows the effects of deadband compensation.

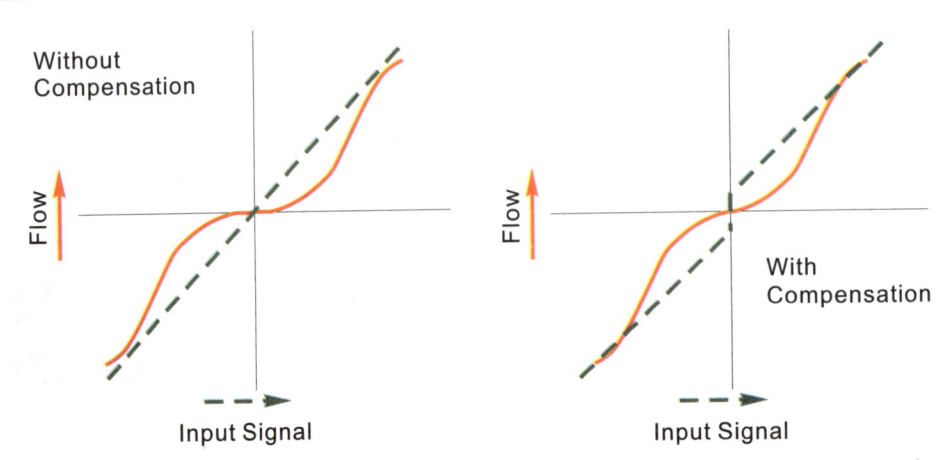

Figure 7.6 Valve spool behavior with and without deadband compensation

Hysteresis

Another characteristic that must be addressed is hysteresis. Sometimes referred to as "static friction" or "stiction", it basically tends to cause a difference in the amount of solenoid current necessary for identical spool position when moving in different directions. Figure 7.7 illustrates the spool position difference that can occur when the spool is moving to increase the flow, versus when it is moving to decrease the flow. At any one point of solenoid current, the flow difference can be in the range of 2% to 8%.

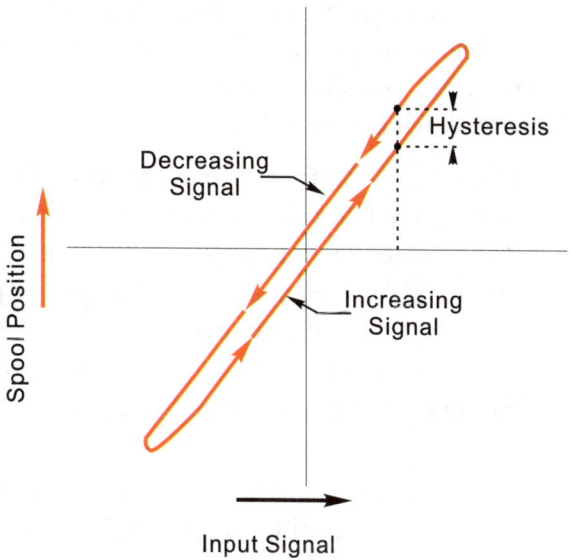

Figure 7.7 The effect of hysteresis on valve spool position

Other factors that contribute to hysteresis are flow forces within the valve (forces caused by pressure drops across the spool due to flow), gravity (affected by valve position), contamination (increasing friction), residual magnetism (between metal parts), inadequate fluid (poor lubrication increasing friction), etc.

Hysteresis can be reduced by incorporating a feature called "dither" in the amplifier. Dither is a low amplitude, high frequency (60 to 100 Hz) signal which is constantly sent to the valve solenoid by the amplifier, superimposed upon the command signal. It causes the spool to vibrate, preventing the spool from becoming stationary in any one position, and reducing "stiction". Figure 7.8 illustrates the appearance of the solenoid signal with and without dither.

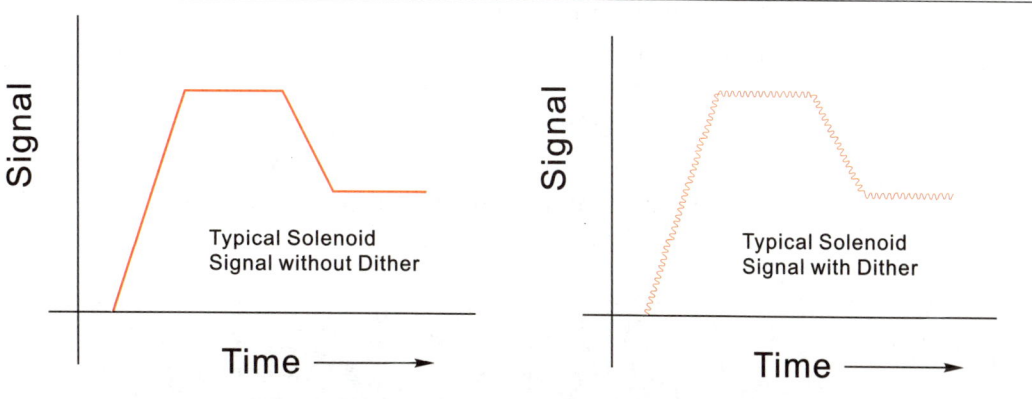

Figure 7.8 Dither is superimposed on the solenoid signal to vibrate the spool and reduce hysteresis

Chapter 7 Electronic Controls

ELECTRONIC CONTROL VARIATIONS

Basic Solenoid Control

Electrical control in mobile hydraulic circuits began very simply, with the on-off type of actuation of basic solenoid valves. This "bang-bang" type of control might be a single solenoid, three-way, two position, spring-return valve used to turn a hydraulic motor on and off (such as a fan motor for crop spraying), or it may be a double solenoid, four-way, three position, spring-centered valve used to raise and lower a cylinder (such as on a snow plow blade). These valves are operated by a basic on-off switch, usually located in the cab of the vehicle. The amount of electrical power required to operate the solenoids is of little concern, provided the size of the wiring and switch is reasonable.

This type of control is also use to control the position of a personnel basket on a small boom truck, for example. This means that the pump flow of the system, and therefore the operating speed of the boom, has to be quite low so that the on and off operation does not cause unacceptable jerking. This may suffice for a small, low-speed machine, but demand for greater productivity (that is, more lift, extend and rotate speed) means that a proportional control would be required to reduce the shock during starting and stopping.

Servo Valves

The application of servo valves in mobile hydraulic systems overcame many of the earlier problems with proportional control of solenoids. The power requirements of servo valves, which are pilot operated by torque motors, are very low and the internal spool-position feedback provides excellent control. This solution is rather expensive, however. Servo valves are much more costly than solenoid valves, and proper operation requires replacing the system's fixed displacement pump with a variable displacement pump. Although this approach provides excellent control, it is not typically cost competitive with manual or solenoid controls.

Proportional Solenoid Control

By today's standards, the early methods of controlling solenoids proportionally were crude, using electrical resistance to control the current flow to a solenoid. It served the purpose, but at the cost of wasted electrical power and the generation of heat (both electrical and hydraulic). The use of electrical amplifiers reduced the power required at the control source (the operator's work station), and improved the proportional performance, but electric and hydraulic heat problems remained.

Significant improvements in the methods of controlling solenoids proportionally has revitalized the interest in this approach. Pulse Width Modulation (PWM), for example (see Chapter 3), provides greatly improved control of solenoid movement (and therefore valve spool movement and valve control) with less power loss and heat generation. PWM has become a very popular technique for controlling proportional valves, and the lower cost of proportional valves makes them very applicable to mobile hydraulic circuits.

Internal Feedback Systems

With expanded use of servo and, more commonly, proportional valves, comes the ability to provide feedback mechanisms within the valve. Servo valves provide a mechanical feedback from spool position to torque motor position, such that the spool position would always be relative to the input signal. The mechanism to accomplish this is detailed in Chapter 6.

Proportional valves can also be equipped with a spool position sensing device, typically an LVDT (linear variable differential transformer). This would provide an electric signal feedback relative to the spool position. By comparing the spool position to the input signal, an error signal is generated which is then amplified and sent to the solenoid. By continuously comparing the input signal to the spool position, a balance will be reached, and correct spool position can be assured.

These approaches, called "internal closed loop" or "internal feedback" control, result in much more predictable control. They significantly reduce the effects of hysteresis, and replace some of the human feedback requirement (eye--hand coordination to accomplish the input-output conformance), allowing a less skilled operator to perform more productively.

Further advancements in control amplifiers allow for tuning the control functions to provide improved control over the valve by conditioning the input provided by the operator. Pre-defined "ramp" functions (a defined slope to the "on" or "off" signal) or control of the "deadband" (the neutral position where control handle movement results in no actuation) are two examples where the amplifier can be designed to replace some of the operator skill. Figure 7.9 shows a typical ramp function created from a stepped input signal.

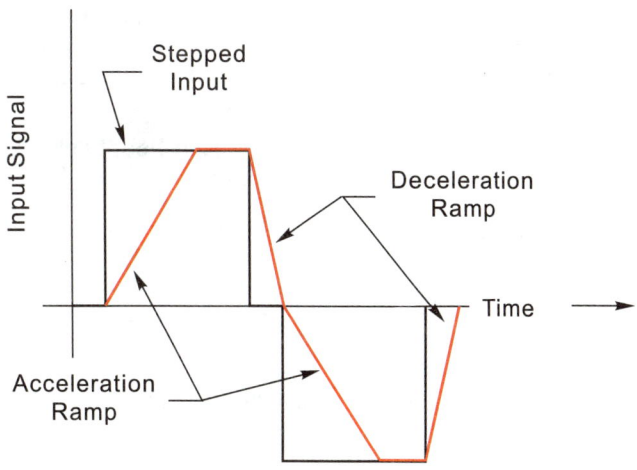

Figure 7.9 A typical ramp function used for controlling acceleration and deceleration

Open Loop Control Circuit

An open loop control system is characterized by the fact that no feedback takes place between the input signal and the output function. The connection is one way; an input signal transmits a command that traverses several steps in the system, and correlation of the output depends entirely on the integrity of each step. There may be some internal feedback elements (such as valve spool position), but these only improve the integrity of an individual step.

Chapter 7 Electronic Controls

Figure 7.10 illustrates the application of an open loop control, attempting to control the speed (or revolutions per minute) that a cement mixer drum will make while traveling to the job site. Communicating the RPM from the dial to the amplifier, valve and motor results in a command that can be influenced by things outside the control system. Things that can effect the actual motor speed are oil viscosity, pump performance, motor efficiency, oil temperature, load (yards of cement), friction in the drive gears and bearing drag in the motor and on the cement drum. With these factors working against the predicted motor speed, actual motor speed will inevitably be different than predicted.

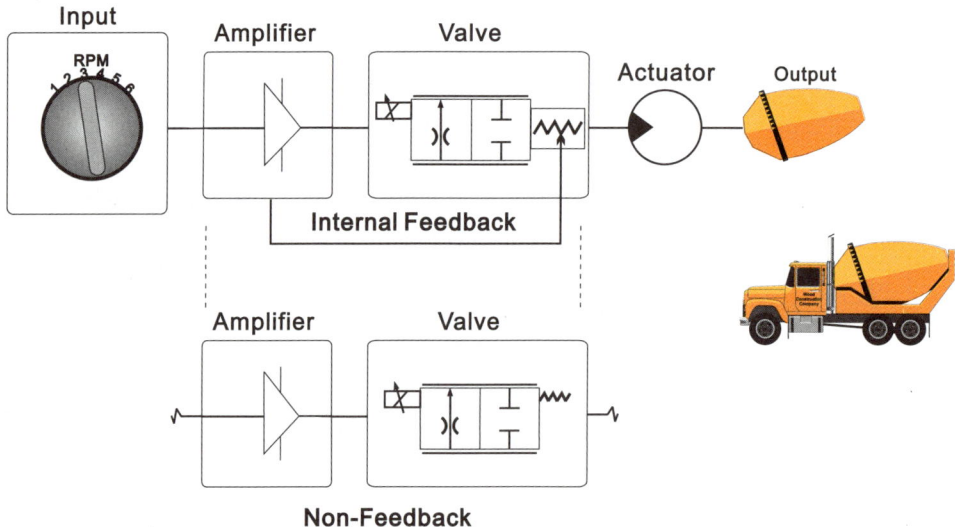

Figure 7.10 Concrete mixer with drum speed control – with and without internal feedback

This illustration also shows the internal feedback of the control valve. This feedback assures that the valve spool is positioned correctly relative to the input command signal. This feedback improves the integrity of the control valve element in the circuit, but all of the external influences can still affect the actual motor speed.

Open loop control circuitry can be a very effective and economical approach to a simple or complex circuit. A designer needs to address the value of an accurate circuit output, such as the regulated speed of the cement drum of Figure 7.10, before pursuing more costly approaches.

Closed Loop Control Circuit

Figure 7.11 shows the same application, but with an "outer loop" feedback circuit added. A transducer added to the mixer drum, provides an electric signal relative to drum speed. This circuit compares the actual motor speed to the intended speed, and causes a correction to be made to the valve spool position (and/or pump flow) in order to attain the intended speed.

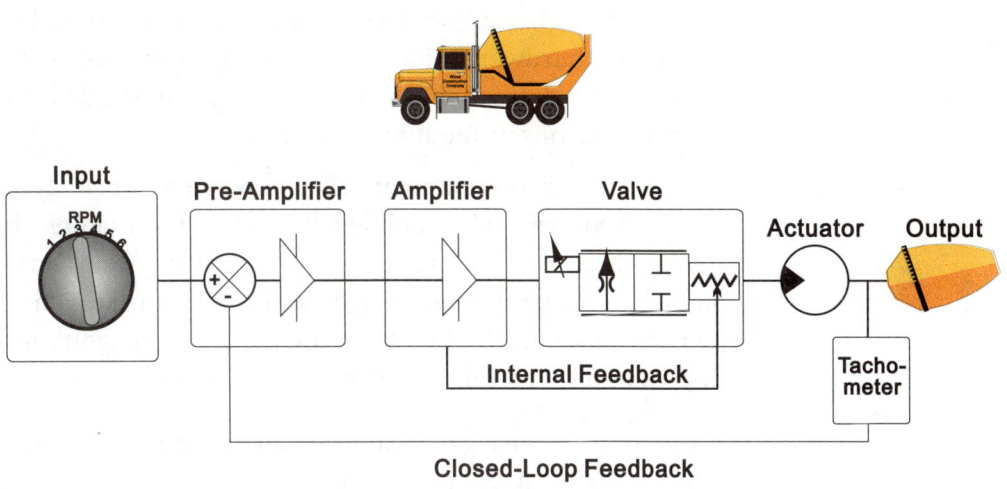

Figure 7.11 Closed loop feedback for a concrete mixer drum speed control

This circuit illustrates a true closed loop control system, in that *actual* output is compared to the input, and a continuous error signal is generated if they do not match. The error signal is amplified, and a correction is made to bring the output and the input signal into line. This is a continuous function, always trying to adjust the system to obtain a "zero" error. The closed loop circuit will compensate for the effects of fluid viscosity, bearing drag and other external factors.

Greater amplification of the error signal can improve response and accuracy of the output. A small error signal, if amplified, can produce a larger correction signal (gain) causing the system to respond more quickly, and to respond to much smaller differences in actual versus intended output. Care must be taken, however, to prevent instability in the system caused by applying too large a signal gain. System designers must determine the correct gain of the amplifier to provide optimum response and accuracy, without creating an oscillation of the actuator caused by "hunting" for the "zero" error point.

Combined Open and Closed Loop Control Circuits

It is fairly typical to combine open-loop circuits and closed-loop circuits on the same application. Using the cement mixer of Figures 7.10 and 7.11 as an example, the drum rotation speed may be controlled by a closed-loop circuit as shown, and the chute position may be controlled by an open-loop circuit, either electrically or manually operated. It is the designer's choice if one element of the machine requires a greater degree of control than others, and whether closed-loop components should be utilized.

Chapter 7 Electronic Controls

CONTROL CONCEPTS

Programmable Logic Control

More recent developments have brought microprocessors (electronically programmable logic control devices) into the control side of mobile systems. Low power electric signals from the operator's control station, or from feedback transducers in the system, or from other sources, can be processed electronically to provide predetermined commands to the directional control valves. These processors are, essentially, miniaturized and specialized computers (hardware) that have been programmed (software) to perform functions in a defined manner upon receipt of signals from one or more of the control locations or feedback sensors.

This opens the door to dramatically improved productivity, energy conservation, safety and cost criteria in mobile equipment. For example, by managing the flow/pressure/direction relationships of an excavator system, sufficient power can be conserved to allow a smaller engine to be used for the same size machine. A smaller engine means a smaller fuel tank, smaller frame, a lighter machine, reduced fuel use (less pollution), less lubricating oil, lower cost and lower operating expense.

Furthermore, feedback mechanisms can provide information to the processor based on the *results* of an input command. Transducers (see Chapter 3) can be used to determine the position, velocity and force magnitude of the machine elements (as opposed to the flow, pressure and direction of the hydraulic fluid). The system will then make control adjustments that coincide with the operator's intention, while still providing an oversight for power management and safety. By providing feedback from the actuator to the control input, a true closed-loop system is attained.

Distributive Control

More advanced electronic control development is beginning to depart from "centralized" control in favor of "distributive" control. Distributive control is a system where the operator's station becomes a "command center", and sends instructions to all actuators at the same time over a single communication link (called a "bus" or "fieldbus"). Each actuator contains its own valve and microprocessor circuitry (and is therefore called a "smart" or "intelligent" actuator) that interprets the bus instruction, and takes action if the instruction pertains to it. It also has its own feedback system to assure that it performs its task correctly, and may also have its own diagnostics and calibration systems. Each actuator maintains communication with the command center, as do other elements of the circuit such as pump controls, engine controls, etc.

Chapter 7 Electronic Controls

Future Trends

Are we developing our way toward machines that do not require an operator? In some respects, we are. There are robotic type machines in existence that perform repetitive functions (automatic storage and retrieval in warehouses) or dangerous functions (land mine sweepers). There are machines that can locate themselves relative to their surroundings (agriculture machines using vision systems for crop row alignment) or geographically (ships that control their course and position using a Global Positioning Satellite). Other machines can select a vegetable and harvest it based on the maturity of its blossom, or pour a concrete curb according to a predefined map.

The trend will continue to move many, and perhaps all, performance control functions from manual or hydraulic to electronic. The intent is to improve reliability In terms of operational safety, economy and consistency.

The operator has not disappeared from the scene, and probably never will. He is, in more and more instances, becoming an on site computer operator and programmer.

Chapter 7 Electronic Controls

CHAPTER 8 .. **Cartridge Valves**

Cartridge valve systems have been in use for many years. Their origin resulted from a combination of: 1) OEMs wanting more flexibility in the design of their hydraulic systems, and 2) a general industry desire for refinements to their "line mounted" hydraulic systems. The primary objective was to simplify the installation appearance of the systems, addressing the multitude of hydraulic lines, fittings, units and mounting brackets while eliminating the associated leakage problems.

Cartridge valves are divided into two basic configurations. One is screw-in cartridge valves, which are commonly used in, but not limited to, low flow systems (up to 120-140 LPM/30-35 GPM). Figure 8.1 illustrates a typical screw-in cartridge valve system installed in a manifold block. The second basic configuration is slip-in cartridge valves, which are used in high flow (150 LPM/40 GPM and up) systems where pressure levels are generally 210 bar/3000 psi and higher. Figure 8.2 illustrates a typical mobile manifold block system using slip-in cartridge valves.

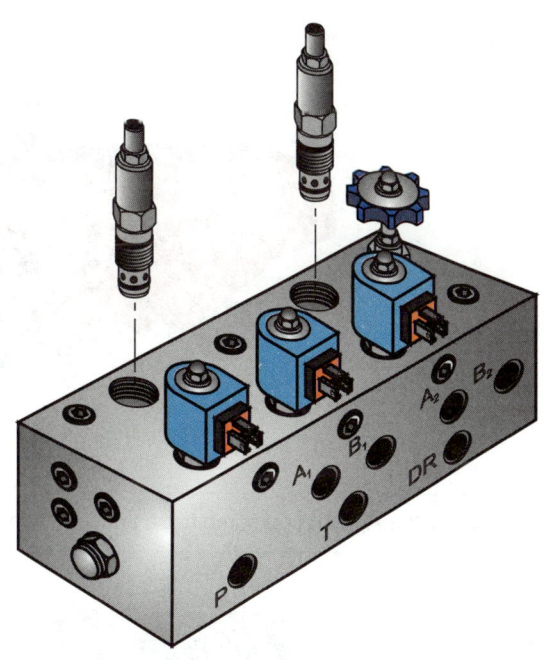

Figure 8.1 A screw-in cartridge valve manifold

Chapter 8 Cartridge Valves

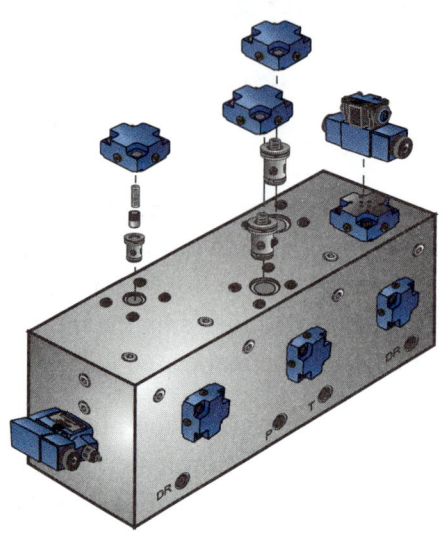

Figure 8.2 Slip-in cartridge valve manifold

Screw-in cartridge valve systems initially gained popularity in agricultural vehicle and implement applications (Figure 8.3). Slip-in cartridge valve systems initially gained acceptance on injection molding machines and steel mill equipment (Figure 8.4). With their high flow and high pressure performance, they are now being adapted into the mobile hydraulics industry as low cost, flexible solutions for some excavator and large wheeled loader requirements.

Figure 8.3 Agricultural combine, a typical machine for screw-in cartridge valve systems

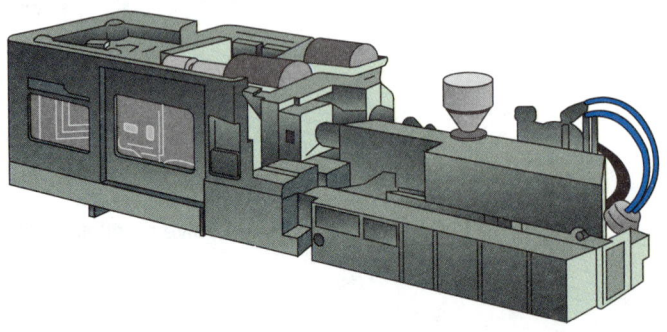

Figure 8.4 Plastic injection molding machine, an early user of slip-in cartridge valves

Chapter 8 Cartridge Valves

CARTRIDGE VALVE CONCEPT

Manifold Block Systems

Manifold block systems are unique solutions to specific hydraulic circuit or system requirements, using one or more cartridge valves installed into either an aluminum or steel block to perform control or work functions. In addition to free standing manifold blocks, blocks may be installed on cylinders or hydraulic motors to form integrated, compact, leak-free actuator packages. Figure 8.5 illustrates a cylinder using a screw-in cartridge valve installed in the cylinder head.

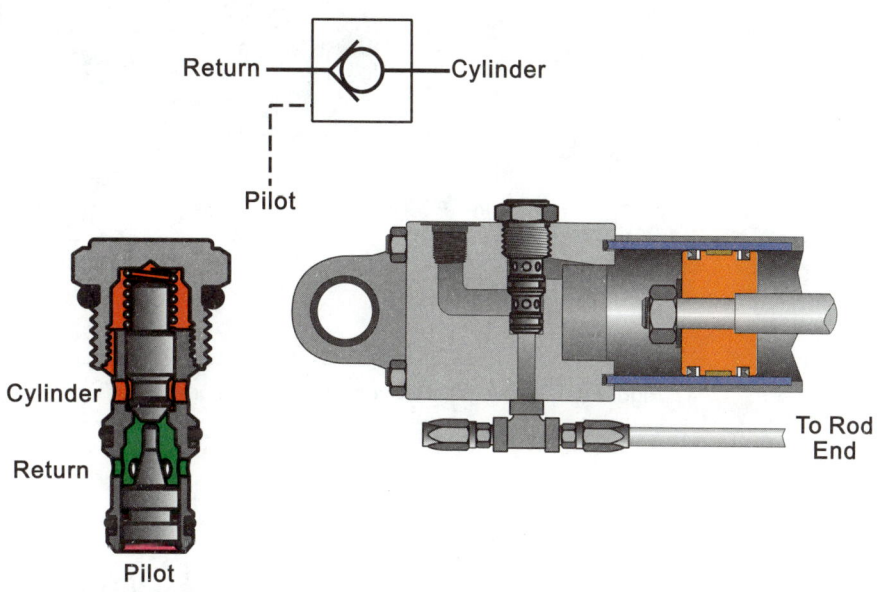

Figure 8.5 Cylinder with integrated screw-in counterbalance cartridge

On mobile applications, manifold block systems typically use screw-in cartridge valves to form an integral circuit for control of one or more system work functions. Quite frequently, as illustrated in Figure 8.6, one or more directional valves, such as the Vickers DG4V-3, are manifold mounted to the basic screw-in cartridge valve block to provide high performance work valve and control valve packages. It is also not unusual to find one or more Vickers' CMX valve sections mounted to screw-in cartridge valve manifold blocks. By combining high flow slip-in cartridge valves with a control circuit of screw-in cartridge valves into one package, it is possible to create low cost, high performance solutions for excavator, wheeled loader, off-highway dump trucks or other similar high flow application requirements (Figure 8.7).

Chapter 8 Cartridge Valves

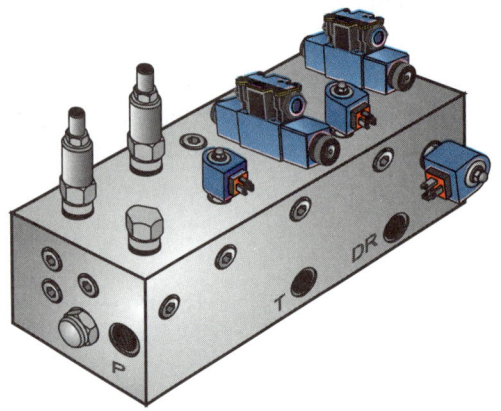

Figure 8.6 Screw-in cartridges and conventional valves on a single manifold

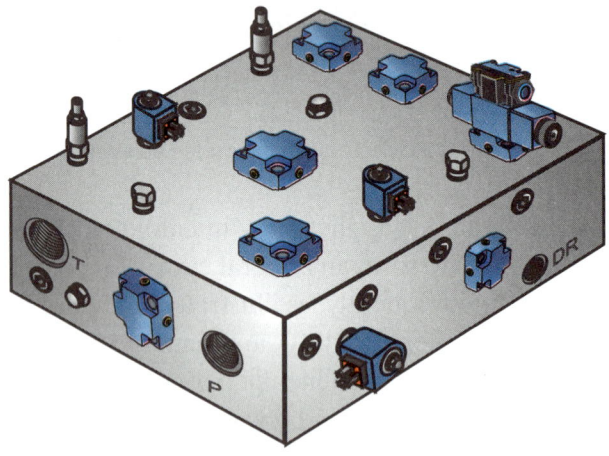

Figure 8.7 A manifold block with a mixture of screw-in and slip-in cartridges

Chapter 8 Cartridge Valves

Consolidating System Segments

As illustrated above, manifold block systems with cartridge valves consolidate system segments by combining valving, plumbing and diagnostics capability into one complete package (Figure 8.8). Valving can be either all screw-in cartridge valves installed into manifold packages, all slip-in cartridge valves, or any combination of screw-in, slip-in and manifold-type valves mounted internally or externally to the manifold block.

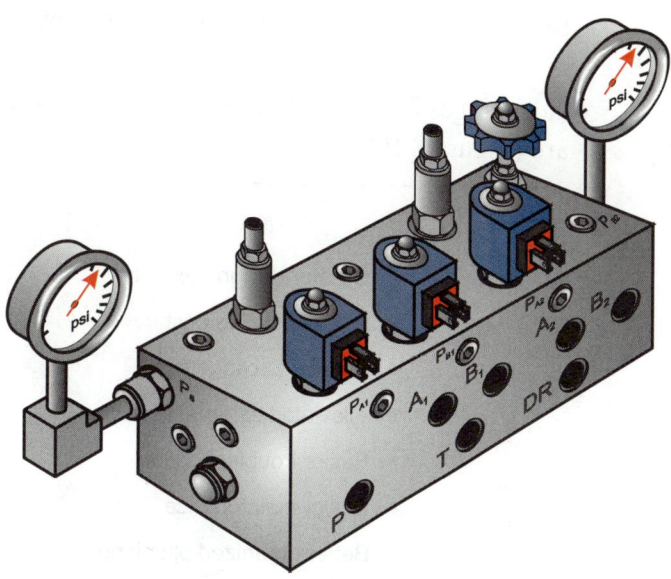

Figure 8.8 Troubleshooting provisions can be incorporated into the manifold block with adequate gauge ports and identification.

A subtle, but significant, difference between the basic types of manifold cartridge valve systems is that in screw-in cartridge valve systems, 100% of the circuit is contained inside the block, as illustrated in Figure 8.9. Slip-in cartridge valve / manifold valve systems typically have only 40% of the system contained inside the manifold block; the balance of the valving is mounted externally.

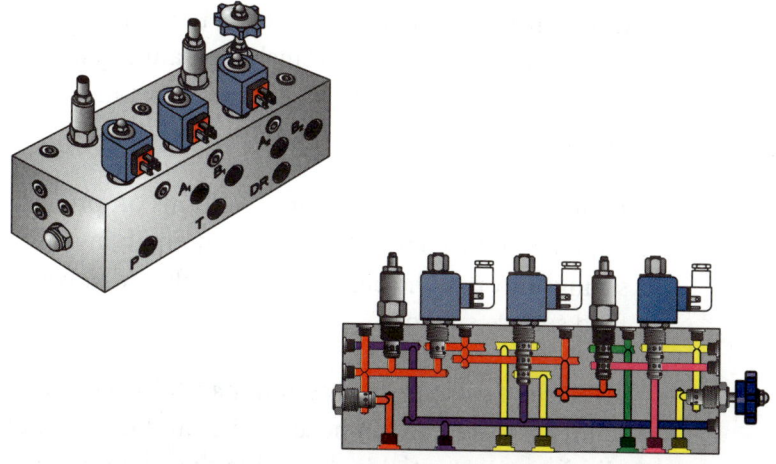

Figure 8.9 Screw-in cartridge systems usually have all the circuit included in the manifold block

Chapter 8 Cartridge Valves

Simplified Installation and Maintenance

Installation and maintenance of cartridge valve manifold blocks is greatly simplified when compared to traditional line mounted valve systems. The necessity of connecting valves by means of hydraulic lines and fittings is greatly reduced by the use of internal passages machined into the manifold block. By combining one or more of the system circuit functions into one or more manifold blocks, the time and cost of installing the hydraulic system on a vehicle or implement is considerably less than traditional line mounted valve systems.

Advantages of Manifold Block Systems

Advantages for an OEM or end user using a cartridge valve manifold block system are (Figure 8.10):

- Greater design flexibility
- Lower installed cost
- Smaller package size
- Alleviated external leakage
- Easier troubleshooting
- Easier maintenance
- Better organized plumbing
- Lower noise levels

Figure 8.10 Advantages of cartridge valve system

Smaller Package Size

The use of internally drilled passages to connect two or more valve functions permits the use of common passages and eliminates the space required for hydraulic fittings. Both of these advantages enable the control valves to be packaged into one compact block, which requires a much smaller volume than if line mounted valves were used.

Design Flexibility

Use of common cavities permits the easy substitution of valves with the same interface, but with either variations on the same functional characteristic or different functional characteristics that provide circuit enhancements or modifications. Also, the vehicle hydraulic system designer can incorporate circuit modifications to enhance vehicle performance by changing only one part (manifold block package).

Lower Cost

Permits significant cost savings in terms of: procurement complexity and time; part numbers to be controlled and maintained; inventory complexity; material and labor savings through the elimination of valve-to-valve hydraulic lines and fittings; and, field service support.

Fewer Leak Points

The use of cartridge valve and manifold systems eliminates many potential leak points associated with hoses, tubes and fittings between each valve. It also eliminates potential leakage due to accidental damage to hoses and tubes connecting individual valve functions.

Chapter 8 Cartridge Valves

Easier to Troubleshoot OEMs and users typically treat screw-in cartridge valves as "throw away" items. As such, if trouble is encountered, it is easier and quicker to remove the old cartridge valve and install a new one. For slip-in cartridge valves, economic reasons typically dictate that the valve would be removed and serviced before a replacement is installed. It is significant to note that for either slip-in or screw-in valves this is normally done without disconnecting hydraulic lines or removing the manifold block from the vehicle. Figure 8.8 illustrates a screw--in cartridge valve manifold block that has been provisioned for troubleshooting.

Easier to Maintain Once installed, the manifold package is typically a "care free" item that requires no tightening of fittings or cleaning of oil leaks. If required, it can be serviced by replacing either an individual cartridge valve or the complete package.

Reduced Noise Level Due to the solid nature of a manifold block system, hydraulic line and valve vibrations are eliminated.

SCREW-IN CARTRIDGE VALVES
General characteristics

Variety Among the first things one notices when looking at a screw-in product catalog is that there is a large variety of basic types of valves. The second thing to be noted is that each basic type has not only the expected spring and/or flow setting options, but also many small modifications that modify the basic function.

Valve design is either spool, poppet or ball check technology, as shown in Figure 8.11. Spool valves are typically either 2-way, 3-way or 4-way in function. Through the use of metering notches or grooves, they provide a very smooth operation and control.

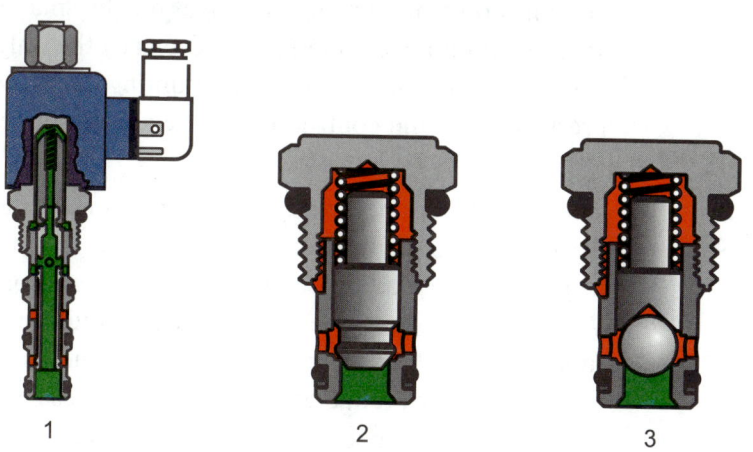

Figure 8.11 Spool type (1) poppet type (2) and ball type (3) cartridge valves

Poppet type valves are 2-way in function, providing leak free operation as an on/off device. Ball type check valves are low cost, simple versions of poppet type valves. Because of the rapid, short stroke, nonmetering nature, these types of valves potentially can introduce shocks into a hydraulic system.

Chapter 8 Cartridge Valves

The attributes of each type of valve make it advantageous for specific applications. While each type is not limited to specific functions, spool valves are typically used for directional, proportional and flow control; poppet valves for directional, pressure, load holding and logic control; and, ball check valves for low cost, simple load holding, shuttle or pressure control.

Actuation of the valves can be by means of a direct acting signal or a pilot signal. These signals can be manual, hydraulic or electrical. Direct acting valves have fast response, although they are generally limited to flow levels of 40 LPM/10 GPM due to flow forces acting on the valve. Pilot operated valves are typically 2 stage valves where a very small amount of flow (from fractional up to 12 LPM/3GPM) is used to control a much larger flow.

Manually operated valves are controlled by either a lever, a knob or a screw. Hydraulically operated valves are controlled by hydraulic pressure, and electrically operated valves are controlled by electric solenoids having either on/off or proportional control characteristics.

By use of one these methods of actuation, valves are designed to provide flow, pressure or directional control. Unique to screw-in cartridge valves is an expansion of the general category of "directional control", to include the categories of "load holding valves" and "logic valves". These will be discussed later in this section.

Compact Designs

Screw-in cartridge valves are compact and require a very small space. While this typically limits the upper flow range to 120-140 LPM/30-35 GPM, it permits the design of very compact, high valve-density manifold block packages.

Screw-in cartridge valves are not attitude sensitive. This permits the block designer great flexibility in locating the valves for the smallest package size to accommodate a specific mounting configuration. Typically the only limitation is the necessity to leave one block surface free for mounting purposes; and to accommodate the pressure, tank and control ports for external connections.

Standardized Valve Cavities

The strength of any cartridge valve program is its ability to be easily used by the designer. This requirement covers not only the product range, but just as important, its ability to be readily used in both simple and complex manifold block systems. While there are proponents of unique cavities for each screw-in cartridge valve function, one of the significant features in the screw-in cartridge valve industry is the use of a standard range of valve cavities. These were first introduced in 1974 by Vickers, and have become the accepted industry standard through their use by most manufacturers today.

Vickers established four standard cavities to service the market. These are the 2-way, 3-way, 3-way short and the 4-way cavities as shown in Figure 8.12. Three way short cavities are similar in design to the standard 3 way cavity with the primary difference being that Port 3 is manufactured as a pilot port and not as a full flow port. This change permits a shorter cavity depth for use with pilot operated 2-way valves.

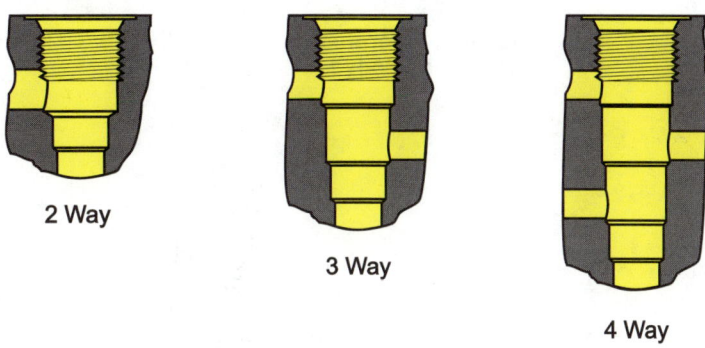

Figure 8.12 Standardized screw-in cartridge valve cavities facilitate design and interchangeability of valves

These 4 basic cavities have provided all the flexibility and freedom necessary to yield both an extensive range of standard valves and a multitude of special valves to satisfy almost any application requirement.

Typical valves used in these cavities are shown in Figures 8.13, 8.14 and 8.15. Figure 8.13 illustrates several 2-way, 2-position valves. Figure 8.14 illustrates several 3-way, 2- position valves, including two 3-way short valves (pilot operated directional valve and pilot operated check valve). Figure 8.15 illustrates several 4-way, 2-position valves.

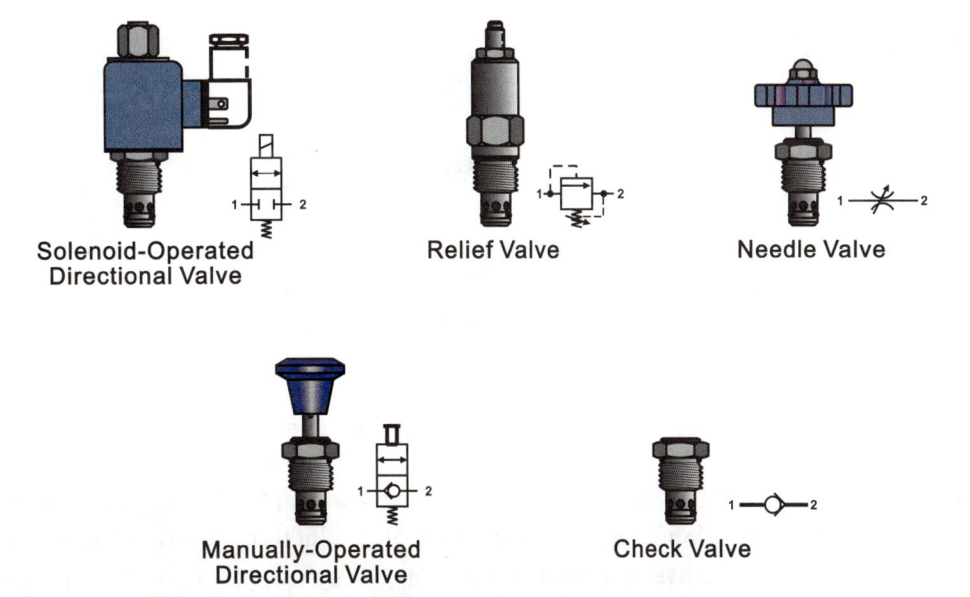

Figure 8.13 A selection of 2-way, 2-port screw-in cartridge valves

187

Chapter 8 Cartridge Valves

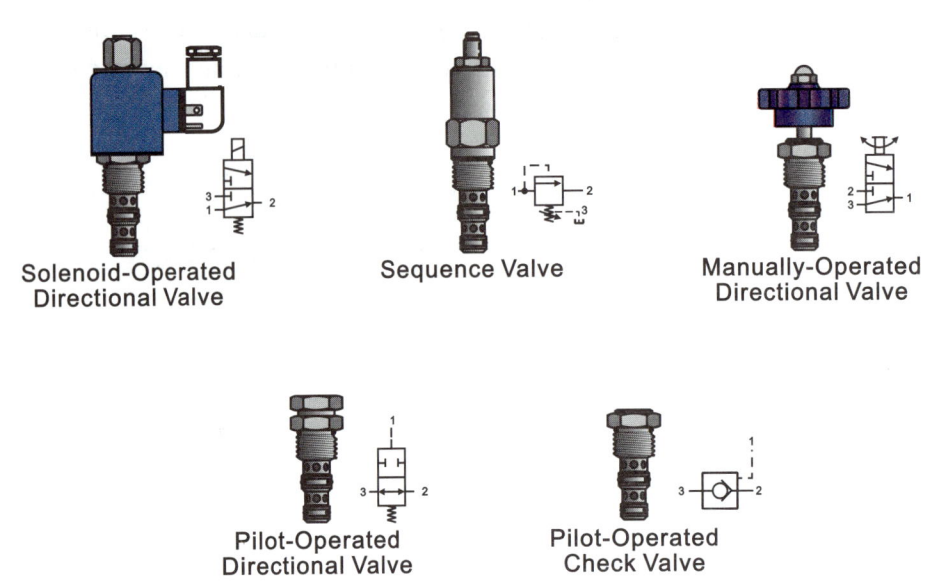

Figure 8.14 A selection of 3-way, 3-port screw-in cartridge valves

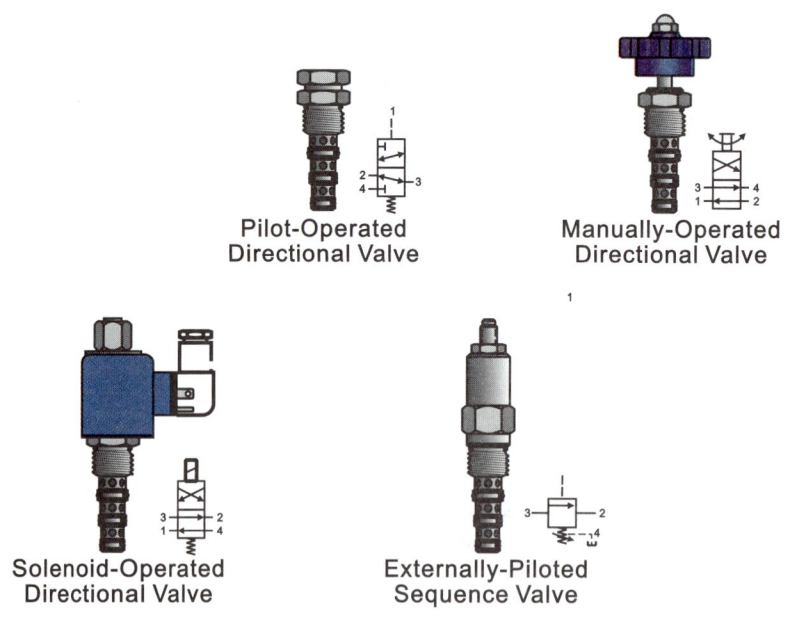

Figure 8.15 A selection of 4-way, 4-port screw-in cartridge valves

The standardized valve cavity concept is based on two simple principles: dependable repeatability of production, and the ability to interchange any 2-way valve in a 2-way cavity, any 3-way valve in a 3-way cavity, any 3-way short valve in a 3-way short cavity and any 4-way valve in a 4-way cavity.

Chapter 8 Cartridge Valves

Functional Characteristics

Directional Controls Directional control valves determine the path, and therefore both the presence and direction of flow in a hydraulic circuit. They are classified as being either direct operated or pilot operated valves.

Direct Operated Directional Control Valves There are 3 different types of *direct operated directional valves*. These are: check valves, shuttle valves and spool valves.

The simplest form of directional control is a check valve where flow is blocked in one direction and permitted in the other when the hydraulic pressure in the line exceeds a predetermined spring force. While check valves can be either ball type or poppet type, the ruggedness, and therefore longer life, of the poppet type makes them the preferred configuration. Figure 8.16 illustrates a typical poppet type check valve.

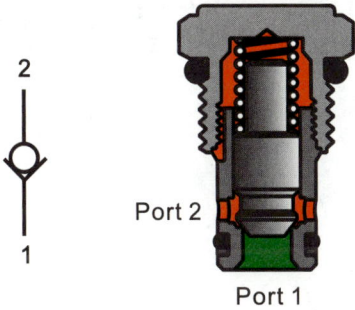

Figure 8.16 A poppet style check valve

Poppet type directional control valves are "On/Off" directional controls. Figure 8.17 illustrates a typical 2-way, 2-position solenoid operated directional valve. Two-way, 2-position solenoid operated directional valves generally are called either normally open or normally closed valves. The valve illustrated in Figure 8.16 is a normally closed valve. Normally open valves permit flow in the nonactuated position while normally closed valves do not permit flow in the nonactuated position.

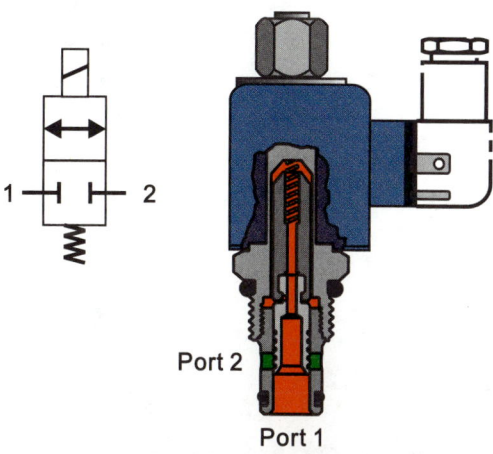

Figure 8.17 2-way, 2-position solenoid operated directional valve

Poppet type directional valves can also be used very specifically for controlling loads. Their use as load holding devices is covered later in this section.

By packaging two 2-way, 2-position poppet valves into a common manifold block, it is possible to create a 3-way, 4-position valve. Figure 8.18 illustrates this valve configuration.

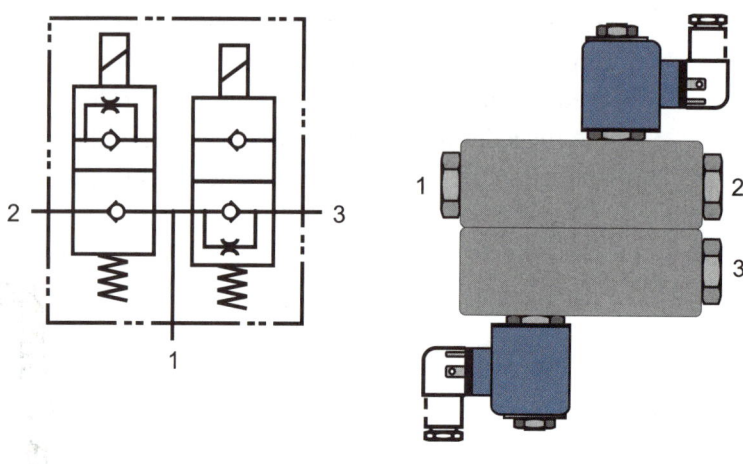

Figure 8.18 3-way, 4-position valve created with two cartridge valves

Shuttle valves are also a simple form of directional control, and use either a ball or a poppet to direct flow from the higher of two flow/pressure sources to a third opening in the valve housing. Figure 8.19 illustrates a typical poppet type shuttle directional control.

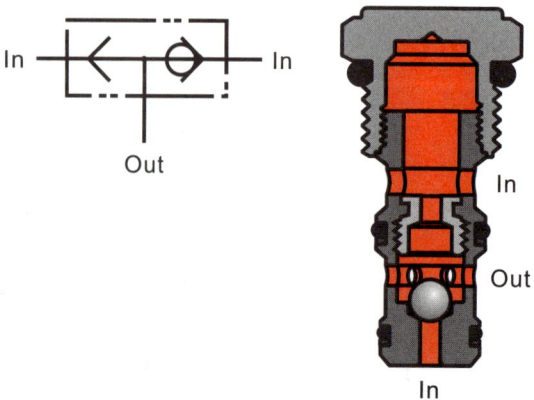

Figure 8.19 Ball type shuttle valve

Spring centered, spool type shuttle valves are typically used as hot oil valves in transmission circuits. They can be either open center or closed center. Figure 8.20 shows the application of a closed center hot oil valve in a circuit.

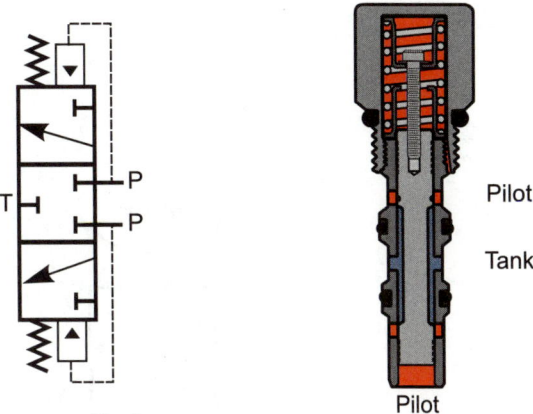

Figure 8.20 Hot oil shuttle valve for hydrostatic transmission circuits

Shuttle valves are also a simple form of logic valve. Logic valves are covered later in this section.

The third type of direct operated directional control is spool valves.

One type of direct acting directional control spool valves is manually operated directional valves. These are either pull to open, push to open, rotary knob or rotary lever operated valves. The push to open and pull to open valves are 2-way, 2-position valves.

Push to open and pull to open directional control valves are "On/Off" and "Off/On" valves that control the entry of flow to a specific branch of a hydraulic circuit. Figure 8.21 illustrates a manually operated pull to open directional valve.

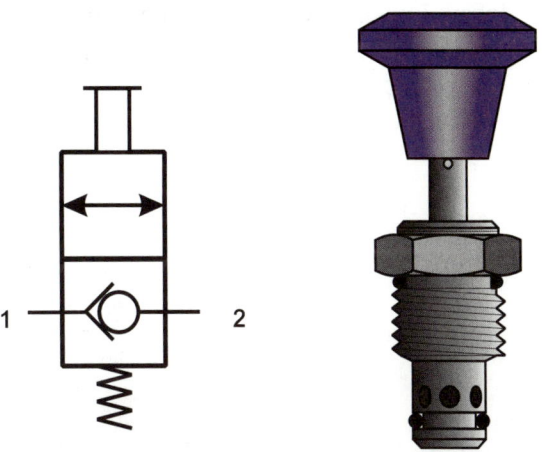

Figure 8.21 Manually operated 2-way directional valve

Manually operated rotary valves are used to direct flow from one branch of a hydraulic circuit to a second branch. Three-way, 2-position and 4-way, 2-position valves are typically knob operated; while 3-way, 3-position and 4-way, 3-position valves are lever operated with a detent. Figure 8.22 illustrates a manually operated 3-way, 2-position rotary directional valve.

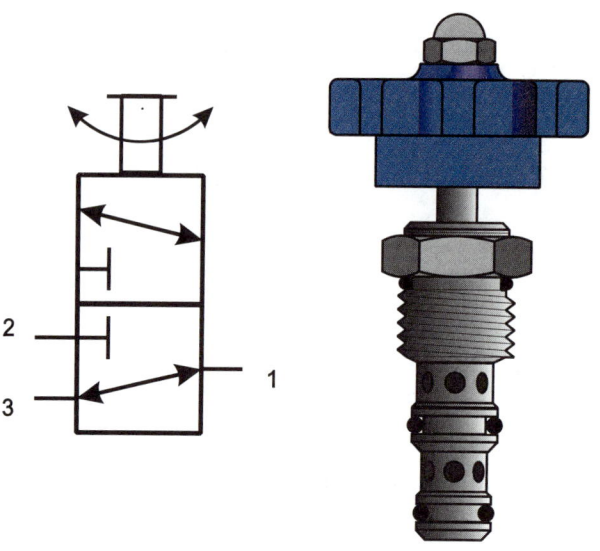

Figure 8.22 Manually operated 3-way directional valve

A second, and more widely used, form of directional control spool valve is the solenoid operated directional valve. These valves enable remote control or electrically programmed control of the direction of flow. On/off directional control is provided by 2-way, 2-position spool valves. Three-way, 2-position spool valves direct flow to and from one branch to a second branch of the hydraulic circuit. Four-way, 2-position and 4-way, 3-position valves have a number of spool configurations that create flexibility in controlling the direction of flow in a branch circuit.

Figure 8.23 illustrates a 4-way, 2-position solenoid operated directional control spool valve, and Figure 8.24 illustrates a 4-way, 3-position solenoid operated directional control spool valve.

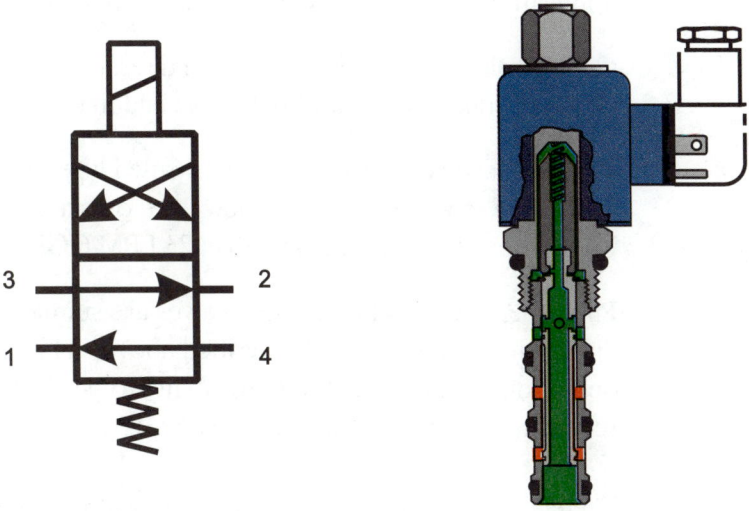

Figure 8.23 4-way, 2 position solenoid operated directional valve

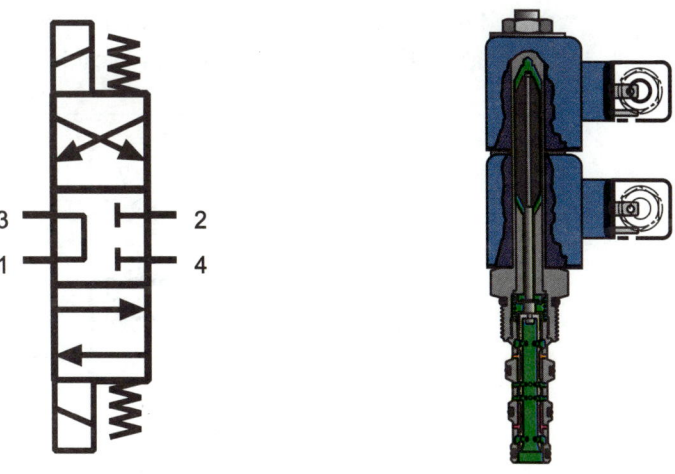

Figure 8.24 3-position, 4-way solenoid operated directional valve

Chapter 8 Cartridge Valves

Pilot Operated Directional Control Valves

There are three basic types of pilot operated directional control valves. These are check valves, counterbalance valves and spool valves.

Check valves can be either ball type or poppet type. Poppet type valves are more rugged and are usually preferred. Pilot operated check valves can be either single acting, or packaged as double acting check valves. A significant number of pilot operated check valve directional control applications are for load holding applications and will be covered later in this section.

Counterbalance valves also control the direction of flow in a hydraulic line. They are better described as a load holding valve and this subject is addressed later in this section.

The third, and perhaps the most widely used type of pilot operated directional control valve is the spool valve. Due to flow force considerations, all spool type directional control valves for application with more than 23 LPM/6 GPM are pilot operated.

Pilot operated directional spool valves are similar in operation to direct acting directional spool valves as described above. The major difference is that the pilot operation feature gives them the ability to perform in systems with flow rate requirements greater than 23 LPM/6 GPM.

Pressure Control

There are 4 basic types of pressure control valves. These are relief valves, pressure reducing valves (sometimes called pressure reducing and relieving), pressure sequence valves and unloading valves. Special purpose valves such as accumulator discharge valves and cross-line pressure relief valves also are part of the pressure control valve family.

Relief valves are 2-ported, pressure limiting devices and can be either direct acting or pilot operated. They can be ball type, poppet type or spool type valves. For durability, ball type relief valves should only be used in low flow (15 LPM, 4 GPM) circuit requirements. Vented versions of the basic relief valve should be used when remote control is desired. Vented relief valves require a 3 port cavity. Figure 8.25 illustrates a typical direct acting relief valve. Figure 8.26 illustrates a typical pilot operated relief valve.

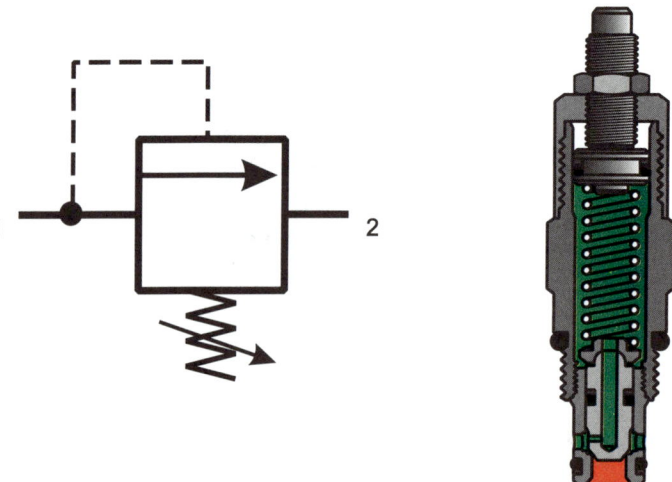

Figure 8.25 Direct acting relief valve

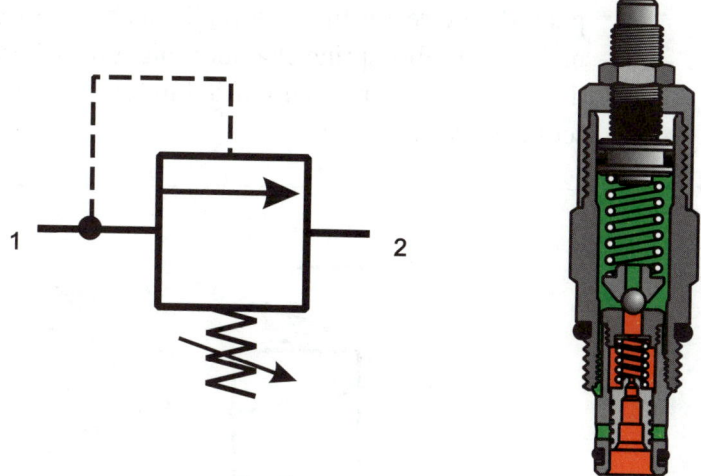

Figure 8.26 Pilot operated relief valve

Pressure reducing and pressure reducing and relieving valves are used to maintain a constant reduced pressure in hydraulic subsystems regardless of pressure variations or high settings in the primary system. This is done by means of either a factory pre-set pressure, or a screw type or knob type adjustable pressure setting.

In addition to the pressure reducing function, there are versions of these valves called pressure reducing and relieving valves, constructed with an internal flow path from the reduced pressure port to the tank port. If pressure in the secondary circuit exceeds the selected pressure, the valve opens this flow path to relieve excess fluid to tank. These valves are 3 ported and can be either direct acting (up to 15 LPM, 4 GPM) or pilot operated. Figure 8.27 illustrates a typical pressure reducing and relieving valve.

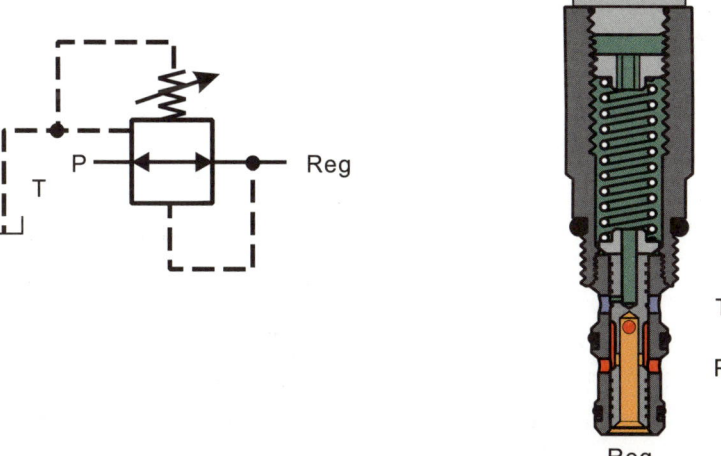

Figure 8.27 Direct acting pressure reducing and relieving valve

Chapter 8 Cartridge Valves

Sequence valves are used to control the sequence of operation for two or more actuators. They typically are 3 port valves, and can be direct acting, externally piloted; or direct acting, internally piloted. For applications where trapped pressure in the adjustable spring chamber might be a factor, 4 ported versions are available to drain this chamber. Figure 8.28 illustrates a typical internally piloted pressure sequence valve.

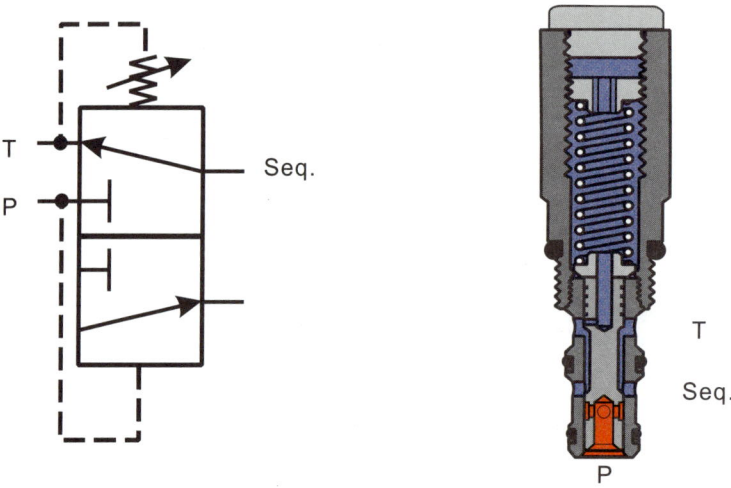

Figure 8.28 Direct acting pressure sequencing valve

Unloading valves are used to provide the basic relief function and the means of automatically loading/unloading a fixed delivery pump according to system demands. They are 3 port, spool valves. Figure 8.29 illustrates a typical unloading valve.

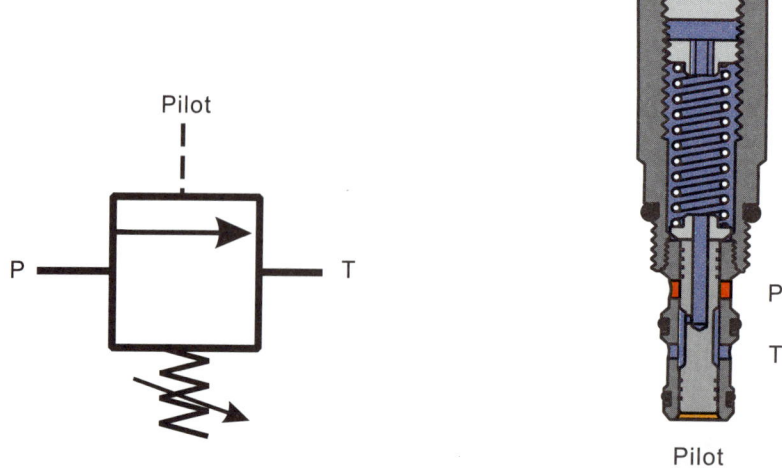

Figure 8.29 Unloading valve

Flow Control

Flow control valves are available in non-pressure compensated or pressure compensated versions.

Typically, the non-pressure compensated flow control is a 2 port valve and is also called a flow restrictor valve. The most basic type is the needle valve, although a slotted spool type valve is also available, with or without an integral check valve to provide unrestricted reverse flow. Adjustment of flow restrictor valves is by means of a knob, lever or adjustment screw. Figure 8.30 illustrates a typical knob type needle valve without unrestricted reverse flow.

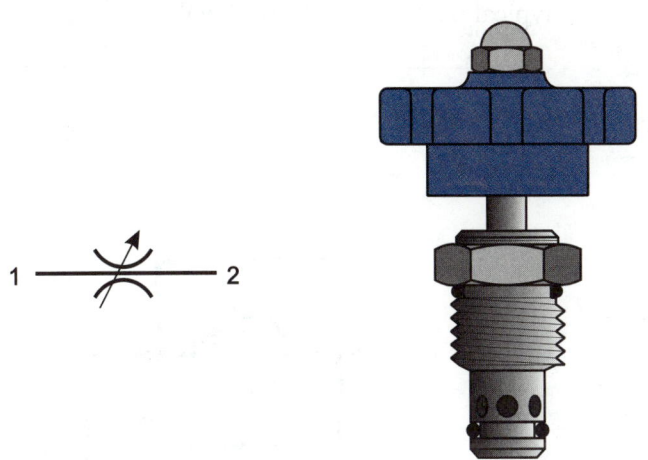

Figure 8.30 Needle valve

Pressure compensated flow control valves are used when it is critical to maintain a constant flow regardless of the hydraulic pressure at the valve inlet or outlet. They are 2 way valves with either pre-set flow or adjustable flow control, and fit into the 2 port cavity. When combined with a priority flow control function, they require a 3 port cavity.

By means of packaging two pressure compensated flow control valves in one housing, either a flow divider/combiner valve or a positive traction transmission valve is created. Figure 8.31 illustrates a typical flow divider/combiner valve.

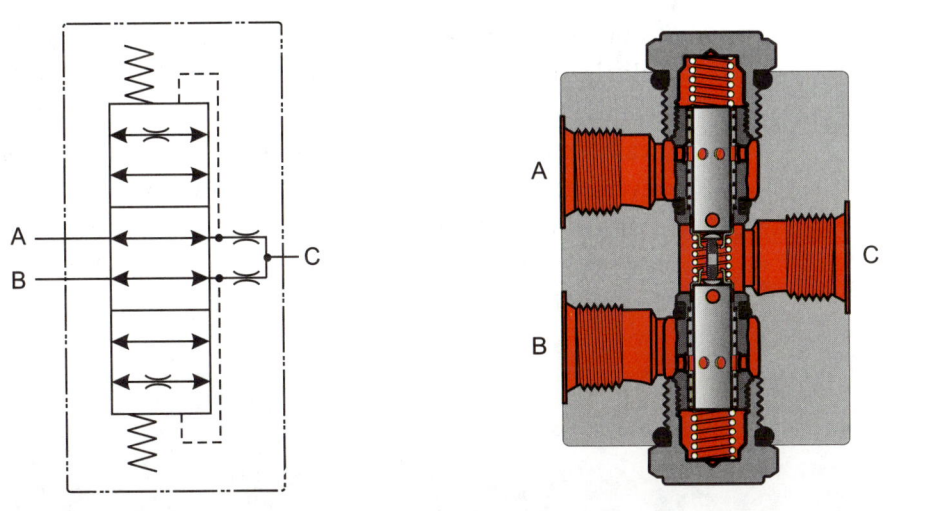

Figure 8.31 Pressure compensated flow divider-combiner valve

Chapter 8 Cartridge Valves

Load Control

Load control valves are used for holding or controlling hydraulic loads by controlling flow. They are important not only for safety, but to prevent the loss of control of actuators.

There are two basic types of load control: pilot operated check valves and counterbalance valves.

Pilot operated check valves are used as a low cost alternative to counterbalance valves when control of overrunning loads and/or load release speed is not required. They should only be used for locking a load into position. Figure 8.32 illustrates a typical pilot operated check valve.

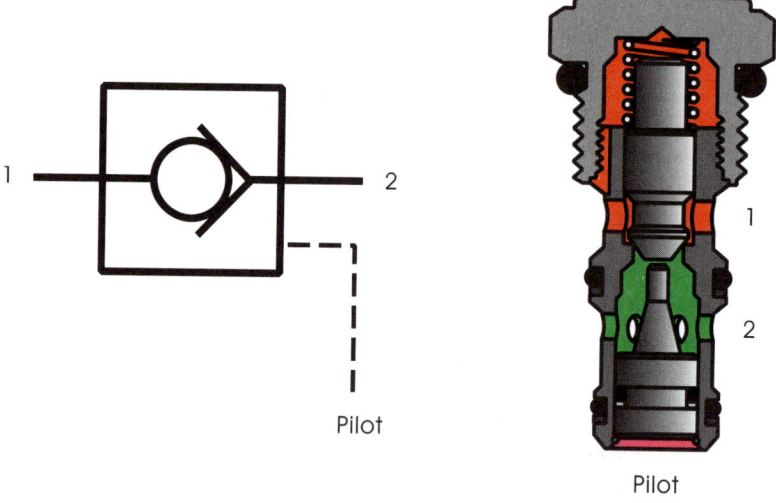

Figure 8.32 Pilot operated check valve

Counterbalance valves are available with or without a spring chamber vent and are called either counterbalance valves or vented counterbalance valves. While specific features differ from manufacturer to manufacturer, both configurations are available with different pilot ratios, adjustable operating pressure settings, built in reverse free flow check valves, different flow ranges and free flow check valve springs to avoid cavitation. Most manufacturers offer a number of standard housings for multiple mounting configurations that offer maximum design flexibility and minimize installation requirements.

Counterbalance valves should be used with an open center, on/off directional valve for precise control of overrunning loads, protection against system cavitation, prevention of actuators running ahead of the pump supply, and to provide load holding and safety in case of hydraulic line failure. Figure 8.33 illustrates a typical counterbalance valve.

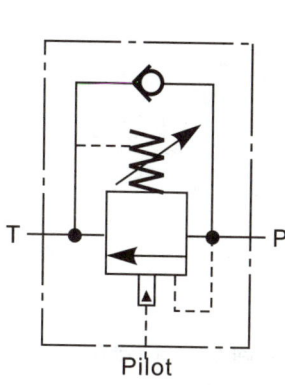 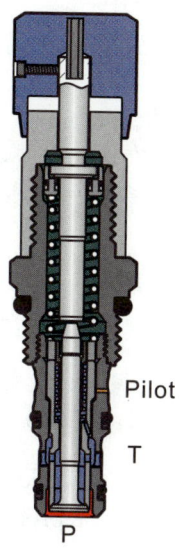

Figure 8.33 Counterbalance valve

Vented counterbalance valves vent the spring chamber in order to ensure stability as a load control regardless of system back pressure. This insensitivity to system back pressure ensures valve performance in proportional, regenerative and meter-out control circuits.

Vented screw-in counterbalance valves are available as a 3 port design or a 4 port design. With the 4th port, any oil remaining in the spring chamber is carried back to the reservoir and not vented to atmosphere. Corrosion of the spring and spring chamber is also prevented by keeping atmosphere out of this area.

Logic Elements

Logic elements are switching devices that use an "On/Off" pressure signal to perform a switching function. There are four basic types of logic elements used throughout the screw-in cartridge valve industry. These are: differential pressure sensing valves, modulating orifice cartridges, 2-way hydrostats and 3-way hydrostats.

Differential pressure sensing valves are used for controlling pressure, flow or direction (including 3-way and 4-way bridge circuits) with the aid of external pilot operators. They are function building blocks which respond to pressure differential signals, providing the capacity to switch or modulate flows and pressure. With logic valves similar to Vickers DPS2-** Series valves, it is possible to obtain five forms of pressure control functions, eight forms of flow control functions, six forms of directional control functions, two forms of 3-way bridge circuits with four basic flow paths, and two forms of 4-way bridge circuits with 13 basic flow paths.

Figure 8.34 illustrates a typical DPS2 Valve.

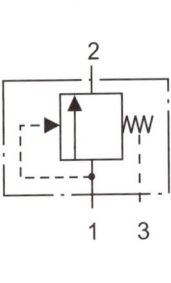
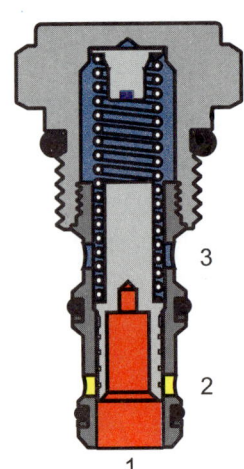

Figure 8.34 Differential pressure sensing cartridge

Modulating orifice cartridges are 4 ported valves. They modulate flow from Port 3 to Port 2 proportional to a back pressure applied to Port 4. When used with a hydrostat and a suitable back pressure pilot valve connected to Port 4, the modulated flow is pressure compensated. Figure 8.35 illustrates a typical modulating orifice cartridge.

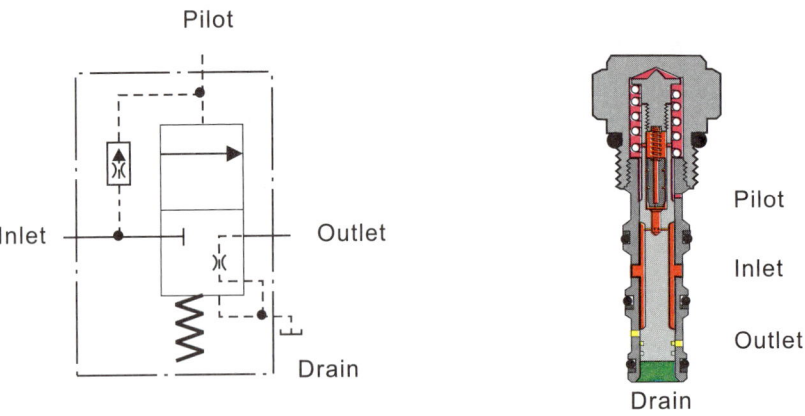

Figure 8.35 Modulating orifice cartridge

Two-way hydrostats are an essential component of a pressure compensated flow control. They are 3 ported cartridges that provide pressure compensation of flow when close coupled in series with a fixed or variable external orifice. A customized housing is required for close coupling the hydrostat and orifice. Excess flow upstream must be diverted through a relief valve to tank. Figure 8.36 illustrates a typical 2-way hydrostat cartridge valve.

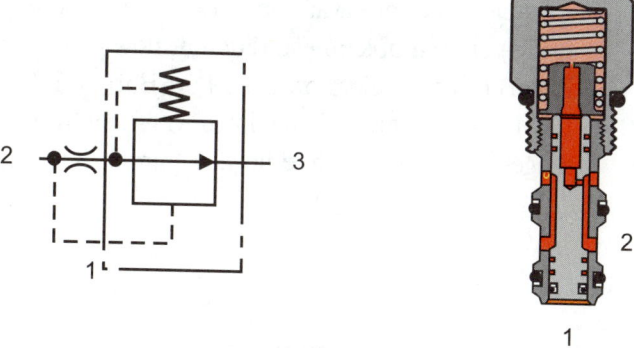

Figure 8.36 Two-way pressure compensated cartridge

Three-way hydrostats with priority are 4 ported cartridges that provide pressure compensation of priority flow from Port 3 while excess input flow is diverted via Port 1 to Port 2. The priority flow rate is controlled by a fixed or variable external orifice close coupled in series with Port 4. A customized housing is required for close coupling the hydrostat and orifice. Excess flow can pass to a secondary circuit or to tank. Figure 8.37 illustrates a typical 3-way hydrostat cartridge valve.

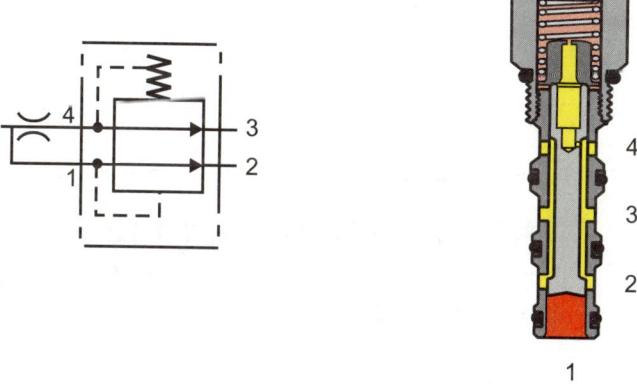

Figure 8.37 Three-way pressure compensated priority cartridge

Valvistor© Screw-in Cartridge Valves

Design Concept Through its patented Valvistor® technology, Vickers is unique in the hydraulic industry with a series of screw-in cartridge valves that provide a new approach to traditional load control problems. Through means of a unique poppet design which incorporates hydraulic feedback, the new series of valves provide low cost, repeatable, high dynamic performance while simplifying inner loop feedback control requirements. The technology is provided by the "hydraulic transistor" or Valvistor® poppet where small flows control much larger flows in the same manner as the electronic transistor uses a very small signal to control a large current. The name "Valvistor®" is derived from the combination of the words "valve" and "transistor".

These valves are 2-way, 2-position, and are electrically controlled by a proportional solenoid. They are available in two sizes. The larger valves (flows up to 160 LPM, 42 GPM) are available in two different flow path configurations: flow from Port A (Port 1) to Port B (Port 2); and, flow from Port B to Port A. The smaller valves (flows up to 30 LPM, 8 GPM) are unidirectional with flow from Port A to Port B. Both valves size ranges permit reverse free flow. Figure 8.38 illustrates an EPV16-A series valve.

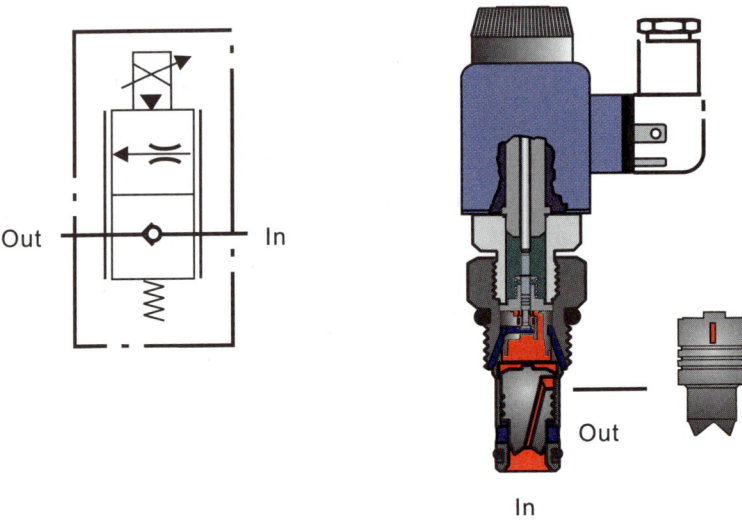

Figure 8.38 Valvistor proportional flow control cartridge

For flows up to 100 LPM (26 GPM) these valves perform as a pressure compensated proportional flow control valve. Above this flow rate, they perform as a proportional throttle valve and a separate hydrostat is required for pressure compensated flow control.

A significant design feature that differs from traditional two stage proportional flow control valves is that a separate hydraulic pilot supply circuit is not required.

Chapter 8 Cartridge Valves

Slip-in Cartridge Valves

Basic Concept

Slip-in cartridge valves, as implied by their name, are cartridge inserts that "slip in" to a cavity in a manifold block. The inserts are retained by a cover which contains one of a number of features to create a specific valve function within the manifold block hydraulic circuit. Figure 8.39 illustrates the basic slip-in cartridge valve configuration of cover and insert.

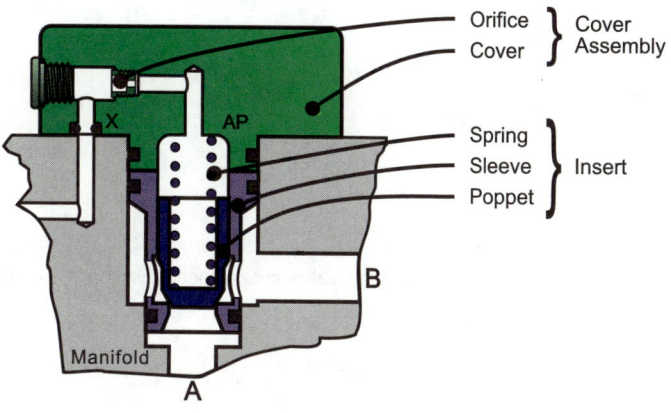

Figure 8.39 Slip-in cartridge valve components

Slip-in cartridge valves are 2-way, 2-position poppet type valves and are commonly known as 2-port check valves. They serve as a one way valve, allowing free flow in one direction while blocking flow in the reverse direction. With some design refinements, these valves can be controlled to overcome the normal blocking action and thus allow control of flow in both directions, i.e., from Port A to Port B or from Port B to Port A. The combination of the 2 port element and the cover with control options into a manifold block system provides many advantages over conventional line or subplate mounted spool type directional, flow and pressure control valves.

As shown in Figure 8.39 the basic slip-in cartridge valve insert consists of a sleeve, spring and poppet. Hydraulic pressure at Port A tends to lift the poppet off the seat and thus allow flow through the valve to Port B. Control pressure at Port AP, plus the spring force, tends to keep the poppet seated, and thus in pressure balance. Through changes in the control pressure applied at Port AP, the pressure balance can be modified to affect the opening and closing of the poppet. The area at the top of the poppet is called Area A_{AP} and the area at the bottom of the poppet (effective area of Port A) is called Area A_A. By changing the ratio between A_{AP} and A_A we can control the action of the poppet for different types of valve functions.

A valve with a ratio of 1:1 is used for pressure control. Figure 8.40 illustrates a typical slip-in cartridge valve insert with a 1:1 ratio. A valve with a ratio of 1:2 is used for directional control and/or flow control. Figure 8.41 illustrates a typical slip-in cartridge valve insert with a 1:2 ratio. A valve with a ratio of 1:1.1 can be used for either directional control or pressure control. Figure 8.42 illustrates four different slip-in cartridge valve insert configurations.

Chapter 8 Cartridge Valves

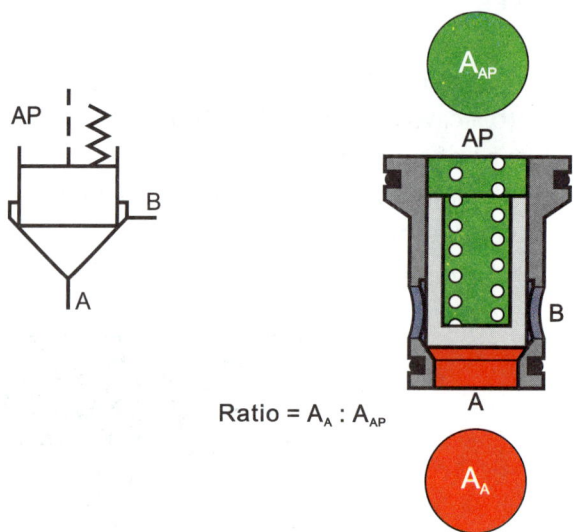

Figure 8.40 A cartridge valve cross section with a 1:1 ratio

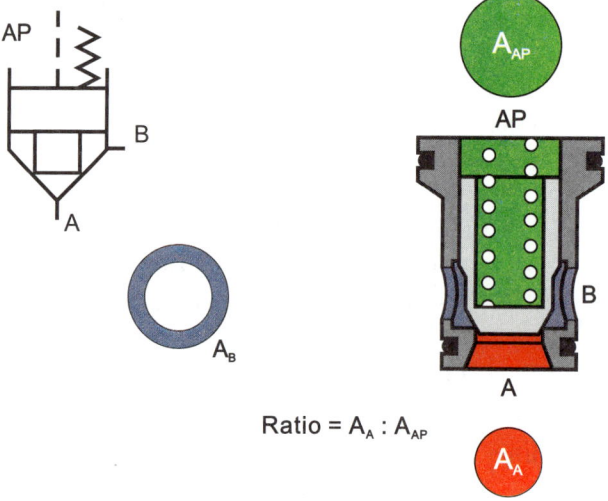

Figure 8.41 A cartridge valve cross section with a 1:2 ratio

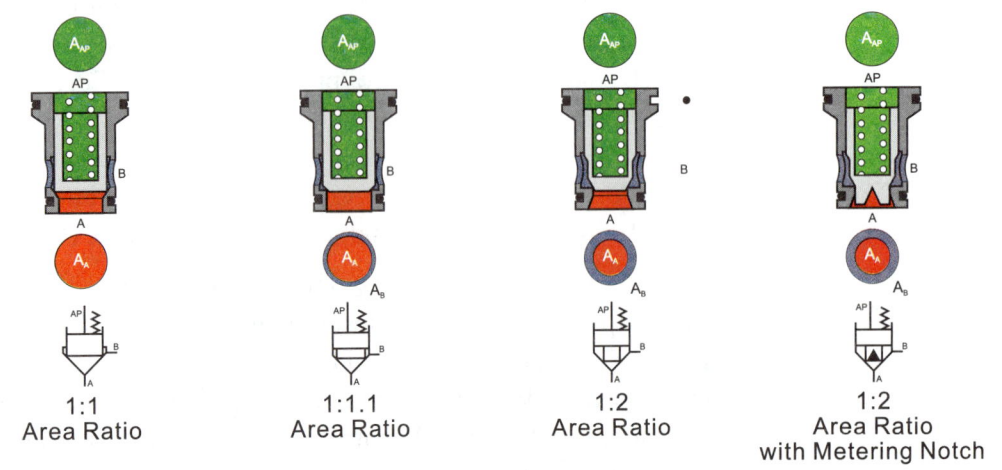

Figure 8.42 Typical slip-in cartridge valve configuration and ratios

Chapter 8 Cartridge Valves

Valve Configurations Available

Slip-in cartridge valve configurations are rated by size in accordance with ISO 7368 (DIN 24342) which gives the nominal size of the valve expressed in millimeters. The nominal size relates to the drilled hole (port size) in the manifold that connects one cartridge to another. Each valve has a nominal flow rate that is defined as the rate of flow with a 5 bar (72 psi) pressure drop. Shown in table 8.1 are typical insert sizes as used throughout the slip-in cartridge valve industry referenced to ISO 7368 (DIN 24342) sizes and the corresponding flow rate. Figure 8.43 illustrates the typical relationships between Size 16 and Size 100 cartridges.

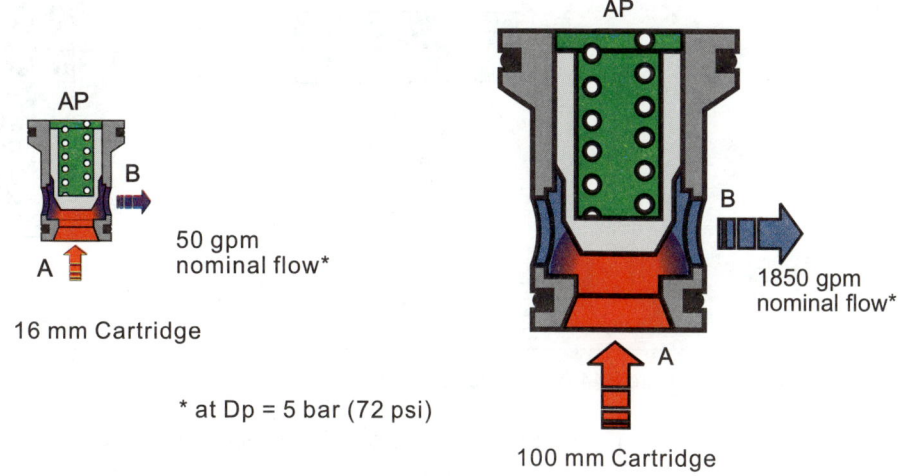

Figure 8.43 Slip-in cartridge valve size range

Size (mm)	ISO 7368	DIN 24342	LPM	GPM
16	06	NG16	200	53
25	08	NG25	450	119
32	09	NG32	700	185
40	10	NG40	1100	291
50	11	NG50	1700	449
63	12	NG63	2800	740
100	13	NG100	7000	1850

Table 8.1 Slip-in cartridge valve sizes and ratings

Chapter 8 Cartridge Valves

Cover Configurations

The basic or standard cover configuration is typically a basic check valve function permitting flow only from Port A to Port B. It contains a pilot pressure passage with an orifice to control the poppet's opening and closing rate.

When the standard cover is used with an internal pilot signal it becomes a pilot operated check valve that is used to open Port A to Port B or to block Port A to Port B. Figure 8.44 illustrates the standard cover configuration used in this manner as a 2-way, 2-position directional valve.

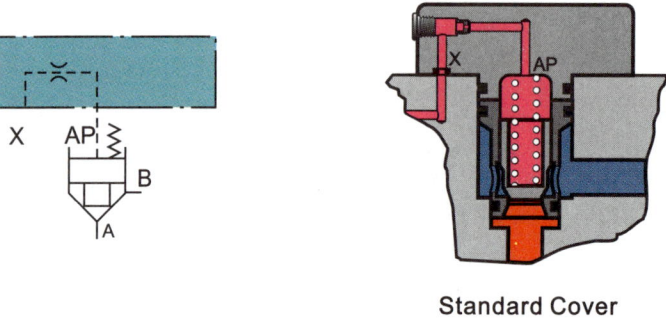

Figure 8.44 A standard cover configuration

When a pilot operated check cover is used with a 1:2 area ratio insert, this performs the function of a 2-way, 2-position directional check valve where flow is permitted between either from Port A to Port B or from Port B to Port A. Flow from Port A to Port B is independent of pilot pressure.

When a shuttle cover is used in combination with a solenoid valve (D03 interface) and a 1:2 ratio insert, the valve acts as a directional valve with flow from either Port A to Port B or from Port B to Port A. When the solenoid is de-energized, the cartridge is shut by the higher of the pressures at Port X or Port Z1 (refer to Figure 8.45).

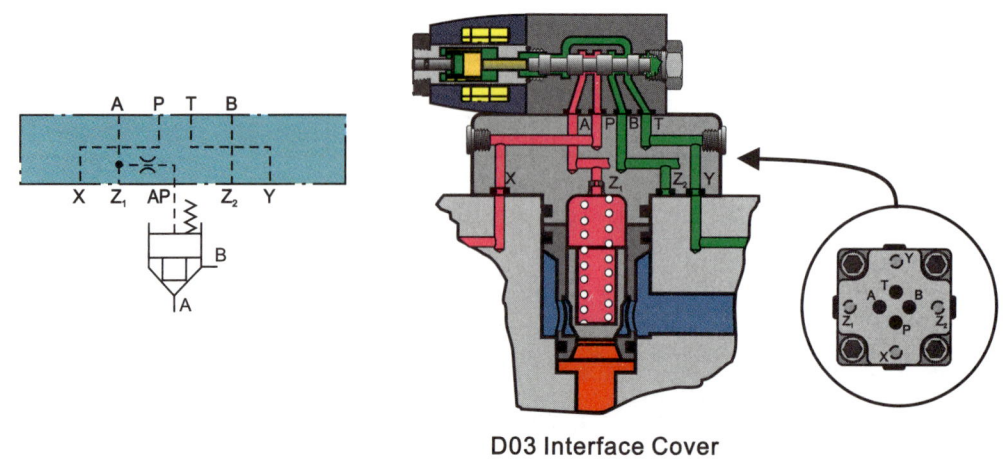

Figure 8.45 A cartridge valve with a D03 interface cover and solenoid valve

When a pressure relief cover is used, the insert acts as a pressure relief function with system pressure being determined by manual pilot adjustment. Figure 8.46 illustrates a typical pilot operated relief cover.

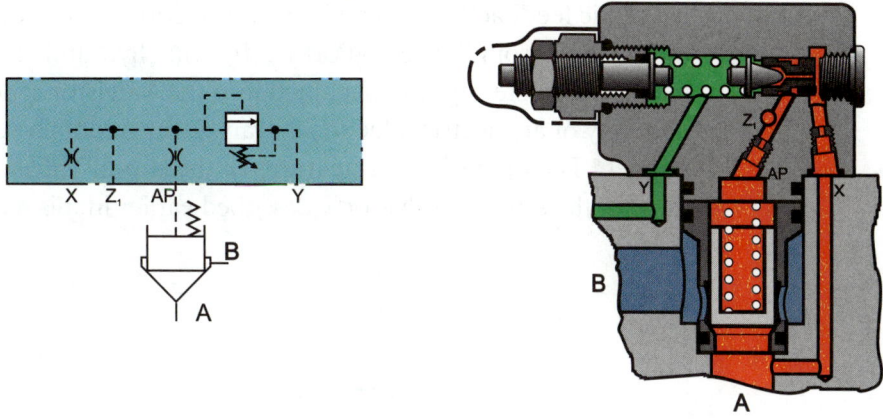

Figure 8.46 Cartridge valve with a pilot operated relief valve cover

When a pressure reducer cover (typically with a size 03 pilot) is used with a pressure reducer insert, it functions as a manually adjusted pressure relief valve to provide a constant outlet pressure below that of the inlet pressure. Unlike other inserts, pressure reducing inserts contain a spool rather than a poppet.

When a stroke adjuster (flow restrictor) cover is used with a flow restrictor 1:2 area ratio insert, the adjustment that limits the insert poppet opening restricts flow in both directions (Port A to Port B or Port B to Port A). Control is from pilot port X and provides both adjustable flow restrictor and directional functions. Figure 8.47 illustrates a typical adjustable flow restrictor and directional slip-in cartridge valve. With this same combination of cover and insert, it is possible to create an adjustable flow restrictor and check function valve. This is accomplished by connecting the pilot signal Port X to Port B.

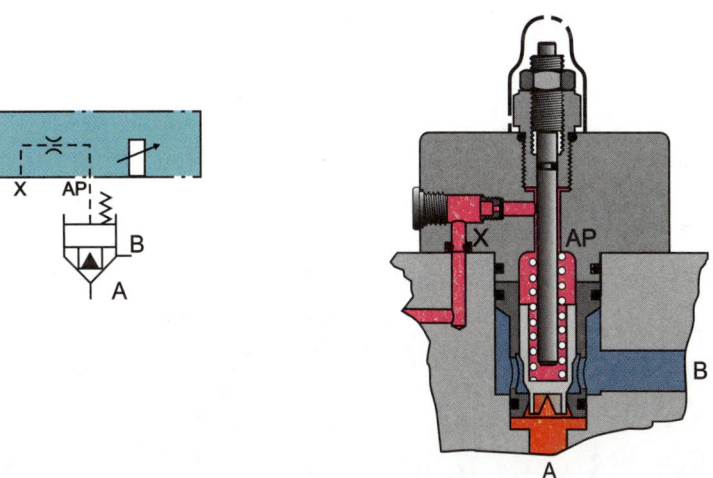

Figure 8.47 Slip-in cartridge valve with flow control cover

Chapter 8 Cartridge Valves

Valvistor® Slip–In Cartridge Valves

Hydraulic proportional throttle control is available from Vickers through use of Valvistor® technology. This uses a directional cover with a single proportional solenoid control and a directional insert with 1:2 area ratio. By using a Valvistor® poppet in place of the standard poppet, proportional throttle control with internal hydraulic feedback is achieved. Flow direction is pre-determined for either Port A to Port B or from Port B to Port A. Internal hydraulic feedback is provided through means of a metering slot machined into the Valvistor® poppet. This eliminates the necessity of an electrical feedback transducer to obtain servo-like control of the poppet. The operation and features of the slip-in Valvistor® control are similar to those of the screw-in Valvistor® described earlier in this section, seen in Figure 8.48.

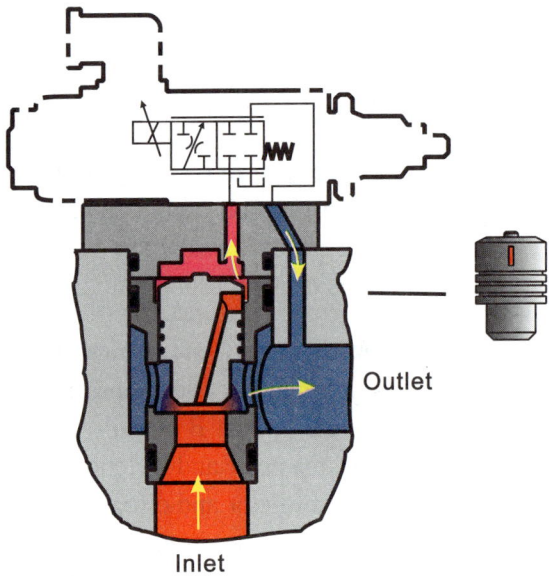

Figure 8.48 Valvistor flow control in a slip-in cartridge valve configuration

Slip-In Cartridge Valve Application Configurations

The number of application configurations for slip-in cartridge valves is limited only by the knowledge and skills of the hydraulic system designer. As slip-in cartridge valves are 2-way, 2-position valves, they can be used alone or in pairs to perform different circuit functions according to specific cover and insert configurations. Typical usage when installed into a manifold block is to use a single valve, a pair of valves or two pairs of valves.

To illustrate the above, consider an application to control a linear actuator. As shown in Figure 8.49, a single valve can be used to provide load holding control. Although this is not a practical application (the cylinder cannot be operated), it illustrates the simplicity of using a slip-in cartridge valve to hold a cylinder's position.

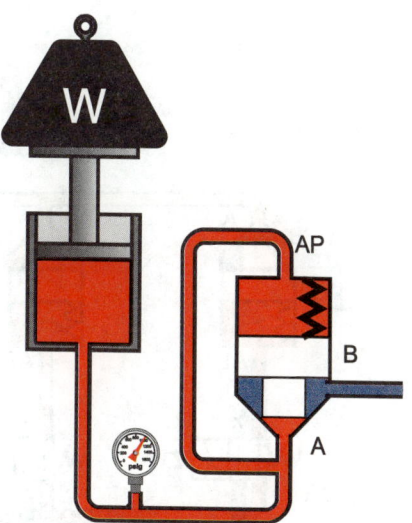

Figure 8.49 Simplistic explanation of how a slip-in cartridge valve will control flow

An alternative solution to controlling reversible directional control is shown in Figure 8.50. This uses a single 4-way, 3-position solenoid valve to control two pairs of slip-in cartridge valves. A relatively small 4-way valve can be used to control the pilot flow and pressure on the AP ports of the cartridge valves, and yet the cartridge valves can control very large flows. Furthermore, because of the differential areas of the cylinder, three different sizes of cartridge valves can be used instead of one.

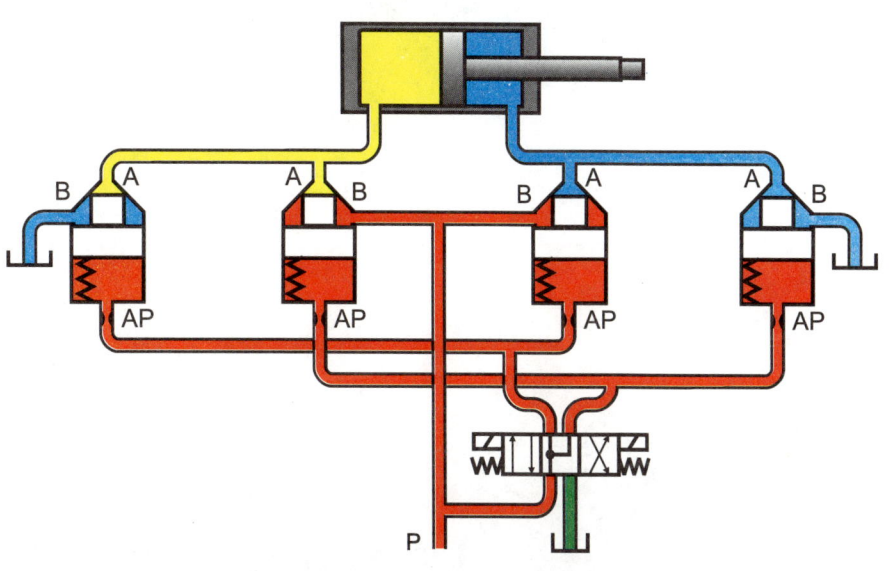

Figure 8.50 Fundamental control of a linear actuator with four slip-in cartridge valves

Chapter 8 Cartridge Valves

Another, more flexible arrangement is shown in Figure 8.51. This circuit uses four small two-way, two-position valves to control the four cartridge valves, providing 16 different combinations of pilot pressure control. Actually, as shown in table 8.2, the 16 combinations result in 11 operating configurations for the cylinder.

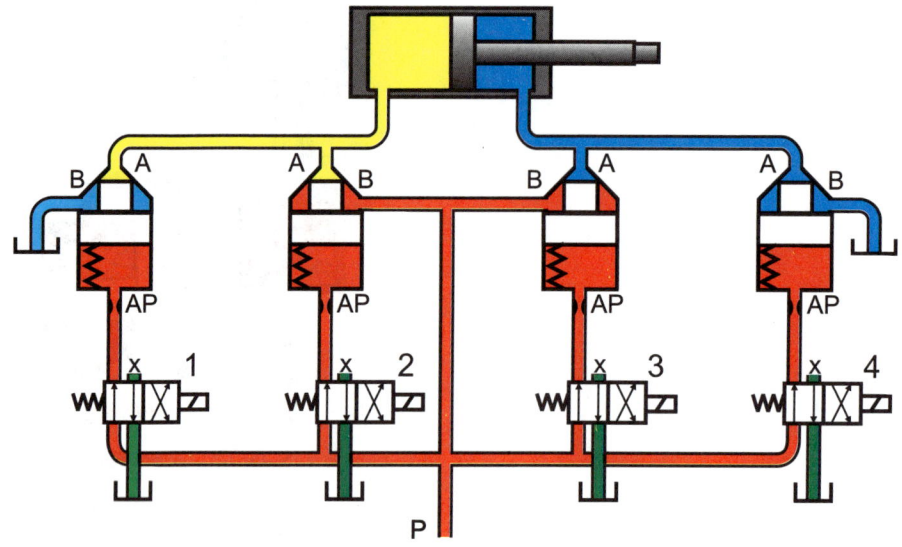

Figure 8.51 Four cartridge valves and four pilot valves will provide 11 combinations of actuator control

Chapter 8 Cartridge Valves

Valve 1 Solenoid	Valve 2 Solenoid	Valve 3 Solenoid	Valve 4 Solenoid	Equivalent Spool Configuration
Off	Off	Off	Off	
On	Off	Off	Off	
Off	On	Off	Off	
Off	Off	On	Off	
Off	Off	Off	On	
On	On	Off	Off	
Off	On	On	Off	
Off	Off	On	On	
On	Off	On	Off	
On	Off	Off	On	
Off	On	Off	On	
On	On	On	Off	
Off	On	On	On	
On	Off	On	On	
On	On	Off	On	
On	On	On	On	

Table 8.2 Four pilot valves will provide 11 different combinations of equivalent spool valve configurations.

Applications

Screw-In Cartridge Valves

Screw-in cartridge valves are used in a broad range of control circuits and low flow work circuits in such diverse markets as agricultural vehicle and implements, aerial work platforms (AWP), construction equipment, forestry, highway maintenance, lift trucks, specialty vehicle and utility truck. Any control or work circuit requirements from 2 lpm/one half gpm up to 120-140 lpm/30-35 gpm can be satisfied by screw-in cartridge valve products in manifold blocks.

By market, some examples of typical screw-in cartridge valve applications are:

Agriculture market: Positioning and/or control of booms on combines, cotton picker baskets, planter wings, implements and spreader motors.

Aerial work platform: Control of boom extend/retract, boom lift/lower, boom right/left and in several applications, propel forward/reverse.

Construction equipment: Work valve pilot control on excavators, cross over relief protection, cylinder anti-cavitation protection, load control and holding, pressure spike protection, outrigger control and cylinder control.

Forestry: Clamping control, winch control and saw control.

On-highway vehicles: Sand and salt spreader motor on/off, snowplow blade and wing blade positioning, screed positioning, vibratory motor on/off and hopper open/close.

Lift truck: Lift/lower control, clamp, tilt and side shift control.

Specialty vehicles: Motor on/off control, boom positioning and outrigger control.

Utility truck: Boom positioning control, cross-port relief protection, tool clamping and positioning and outrigger control.

Screw-in Valvistor® Valves

Screw-in cartridge Valvistor® valves are finding an increasing number of uses where position control requirements demand precise starting and stopping without hydraulic shocks; and where rotary actuator control demands rapid response to changes in vehicle speed. Typical applications include:

Lift truck: Lift/lower control

Salt spreaders: Spreader motor control

Agriculture spreaders: Fertilizer spreader motor control

Forestry: Saw motor control, feed in motor control and feed out motor control

Agriculture planters: Seed motor control

Emergency Vehicles: Generator constant speed control

Slip-in Valvistor® Valves

At the time this section is being written, applications are being evaluated for control of the digging functions on large backhoes and excavators.

Slip-In Cartridge Valves

Potential exists in applications where there are large flow requirements that are not always easily satisfied by conventional mono block or sectional spool type work valves due to the high costs of developing and/or producing valves in relatively small quantities. Examples of these requirements are: drill drive motors on large mobile drill rigs; and, control of the instantaneous large flows involved during boom lower on excavators, wheel loaders and track type loaders.

Chapter 8 Cartridge Valves

CHAPTER 9 .. Auxiliary Valves

If directional control valves are the command center of hydraulic systems, then auxiliary valves are the brains! They determine, and thus *control*, the pressure and flow in a system.

Pressure controls are operated by or control pressure in both the main circuit, and, as required by the system design parameters, in any of its branches or sub-circuits. In the same manner, flow controls control flow in the main circuit and, if required, in branch circuits.

Through individual valve design characteristics, the pressure control valves and flow control valves condition the fluid flowing from a pump, and to or from actuators or directional valves. They provide a predetermined level of power; speed; occasional restraint; and, when required, sequencing of actuator functions. They can be used to limit a vehicle or machine function, or protect components in the circuit. Depending upon valve function, they can provide energy (power consumption) savings and thus reduce operating expenses. They can also reduce the amount of heat going into the system and thus reduce the amount of cooling that is required for long life of the individual components, hoses and seals.

Auxiliary valves are available as standalone valves; or in manifold single function or multi-function blocks using screw-in cartridge valves or slip-in cartridge valves. They are applied as:

- Individual line mounted valves
- Manifold mounted valves on pumps, directional control valves, motors and/or cylinders
- Cartridge valves inserted into cylinder ends or directional control valves

Pressure Controls

By means of either fixed spring settings or adjustable spring settings, pressure control valves control the upper limit of the fluid pressure in either the main circuit, or one of its branch circuits. Pressure controls are classified in the following categories:

- Relief valves - Direct acting or - pilot operated
- Unloading valves
- Pressure reducing valves
- Sequence valves
- Load control valves

Chapter 9 Auxiliary Valves

Relief Valves

The function of a relief valve is to provide protection to a hydraulic system so that the system components do not malfunction, seize or burst and the fluid lines do not burst or leak at their connections. The relief valve performs this function by providing a means for system fluid to be diverted to the reservoir when the valve pressure setting is reached.

The opening of the relief valve is accomplished when pressure of the fluid in the system exceeds a pressure set by a spring force in the relief valve. The spring holds the relief valve in its closed position. As the fluid pressure rises to a level that exceeds the force of this spring, the relief valve opens and creates a flow path to the reservoir. This action results in the "relieving", or limiting, of fluid pressure in the system to the value of the spring force in the relief valve.

Figure 9.1 illustrates a basic direct acting relief valve where either a ball or a poppet is held in the closed position by an adjustable spring. This blocks the flow path to tank.

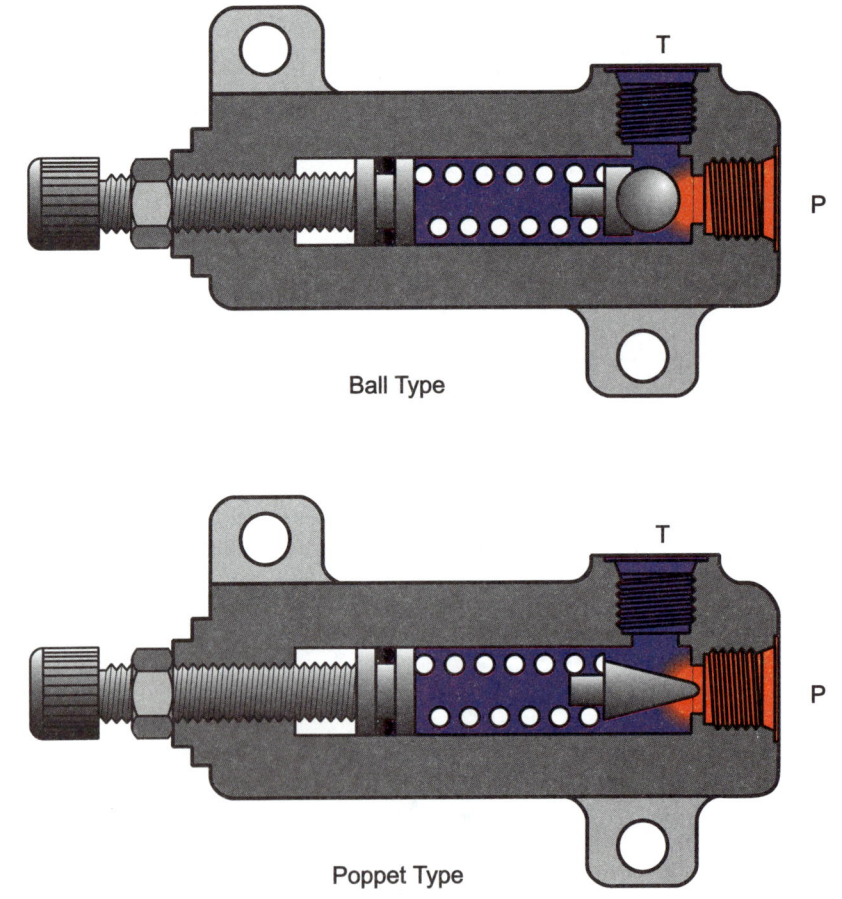

Figure 9.1 Basic direct acting relief valves

A feature of direct acting relief valves is a high pressure override characteristic as shown in the curve in Figure 9.2.

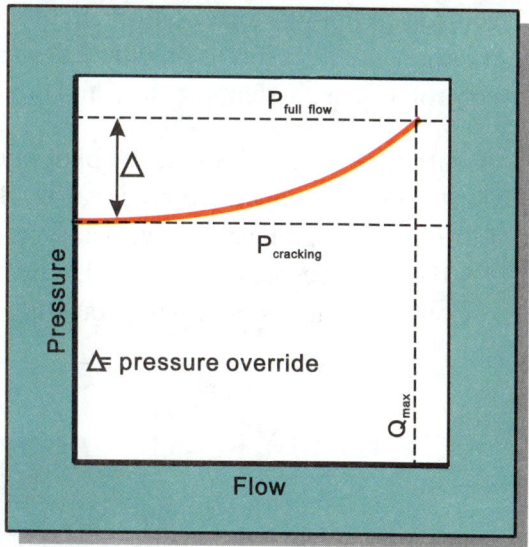

Figure 9.2 Typical override curve

As illustrated in Figure 9.3, relief valves can also be a spring-loaded piston design. Direct acting relief valves are characterized by their simplicity, and corresponding low cost.

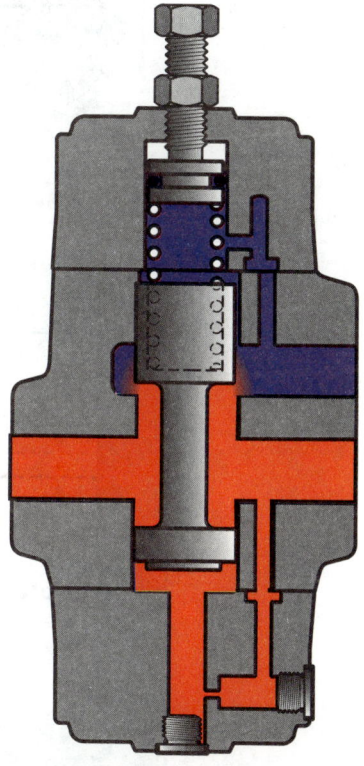

Figure 9.3 Spring loaded piston type relief valve

Chapter 9 Auxiliary Valves

A second type of simple relief valve is called a "hydrostat" which can be either a balanced spool or spring loaded poppet design. Hydrostats are also commonly called pressure compensators and can be found in variable displacement pump controls or in valve logic control circuits. Their use is not so much to provide a flow passage to tank as it is to provide a flow passage to a work function, i.e., in a pump control to sense system pressure and direct flow to the pump stroking piston.

A third type of relief valve is the pilot operated or compound type. Figure 9.4 illustrates a pilot operated relief valve. Because of their design, pilot operated relief valves have a significantly lower pressure override characteristic, as shown in Figure 9.5. For this reason, they are used in hydraulic systems where the amount of pressure override is critical to protecting circuit components and the hydraulic lines.

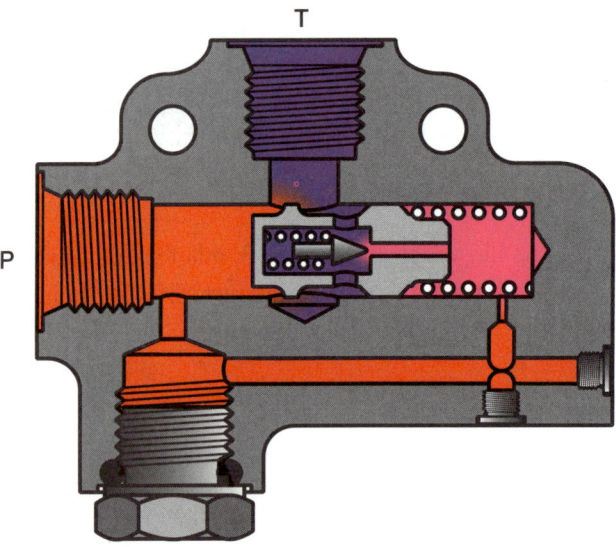

Figure 9.4 Pilot operated relief valve

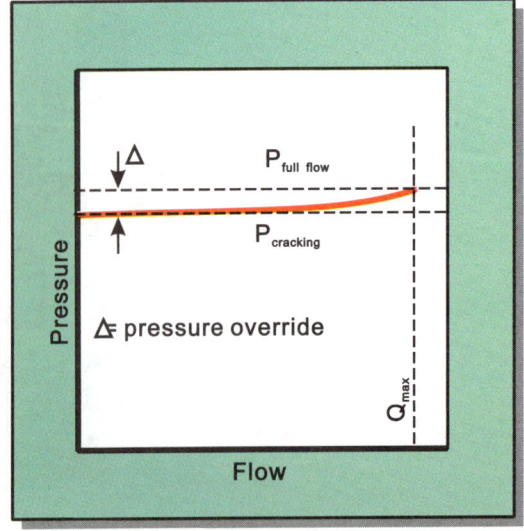

Figure 9.5 Pressure override characteristics of a pilot operated relief valve

Figure 9.6 illustrates the operation of a "balanced piston" type of pilot operated relief valve. Fluid in the main system, passing through the "P" ports, is fed through a small orifice in the main piston, to the small chamber above the piston. A light spring in the piston keeps the piston down on the seat, preventing flow of main system fluid to the tank passage. As system pressure rises, pressure is transmitted through the orifice in the main piston to the upper chamber. When it exceeds the setting of the pilot spring, the control poppet opens. Fluid from the upper chamber then flows past the poppet, down through the main spool and to the tank passage. This limits the pressure in the upper chamber, allowing the main piston to rise if main system pressure rises further. When the main piston rises, system fluid flows freely to the reservoir through the tank passage.

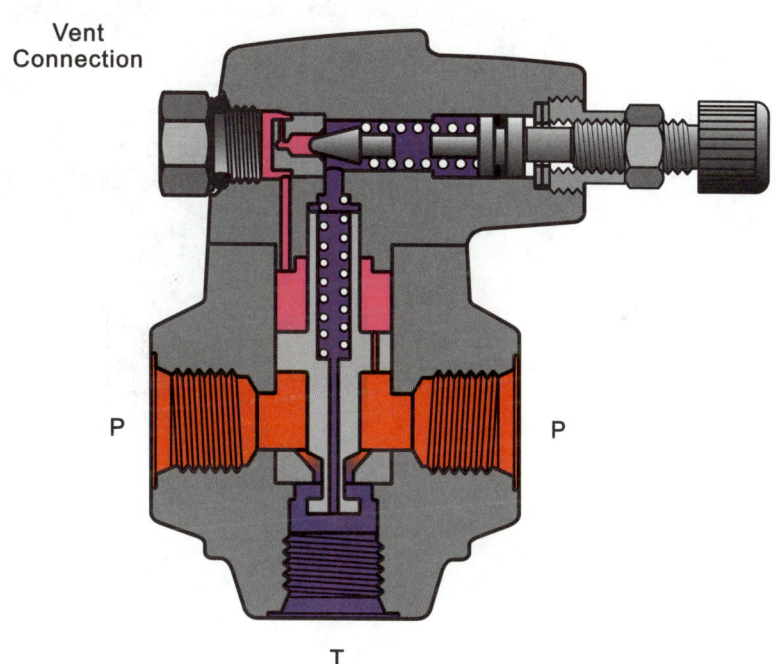

Figure 9.6 Balanced piston type pilot operated relief valve

When system pressure is reduced to less than the pilot spring setting, the control poppet reseats, stopping flow to the reservoir.

Figure 9.7 illustrates the same balanced piston relief valve externally controlled by a remote poppet-type relief valve. By means of adjusting the spring setting on the remote valve, the system pressure in a low flow vent circuit can be used to control the system pressure of a much higher flow passing through the main stage relief valve.

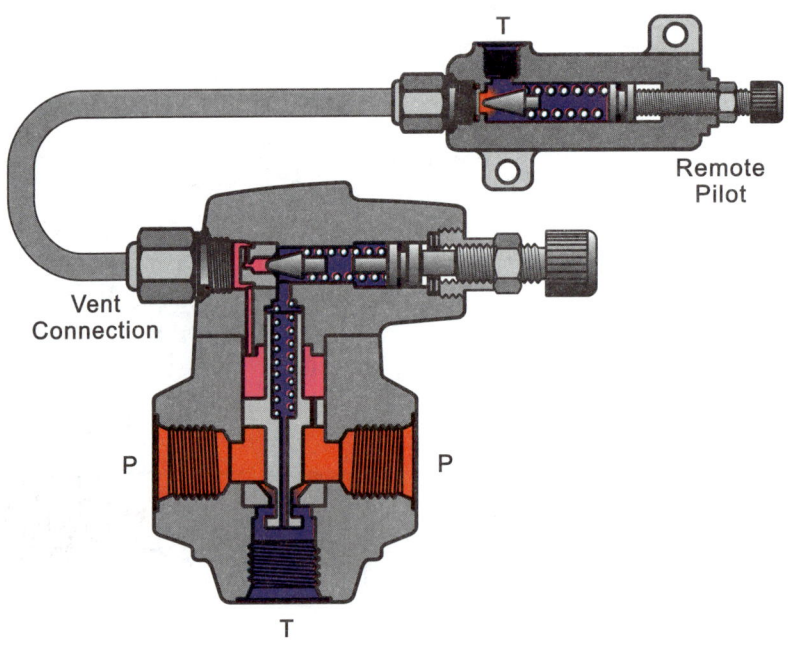

Figure 9.7 Balanced piston type pilot operated relief valve with remote pilot control

Figure 9.8 illustrates the function of a relief valve in a simple fixed pump, directional valve and cylinder hydraulic circuit. In this example the relief valve opens to tank when the circuit pressure overcomes the spring setting of the relief valve, caused by the force acting on the cylinder.

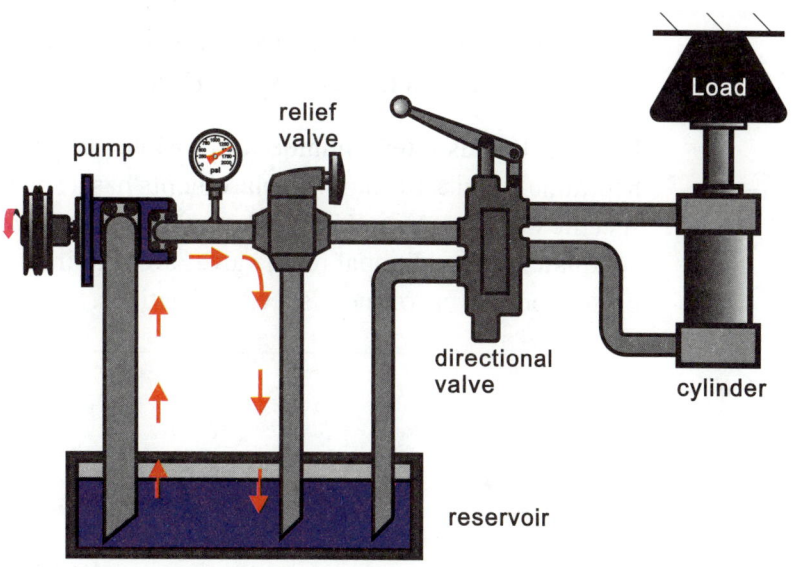

Figure 9.8 Function of a relief valve in a fixed pump circuit

As illustrated in Figure 9.9, fluid passing over the relief valve releases its energy in the form of heat. This could be advantageous on cold winter mornings, when running pump flow over the relief valve may be a convenient means of warming up the hydraulic system before performing work. The disadvantage, however, is that in warm weather, reservoir temperatures can become excessive. High fluid temperatures may cause component malfunction and fluid breakdown (see Chapters 16 and 17).

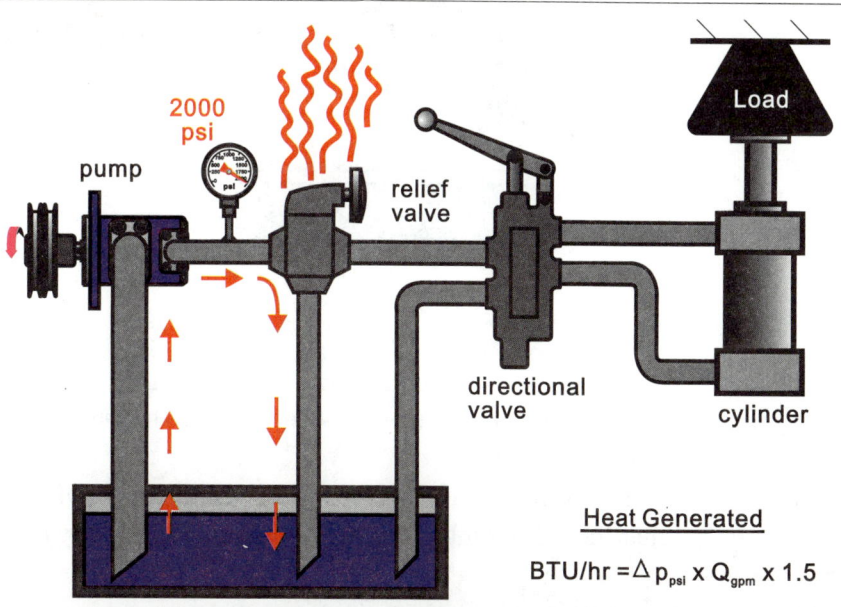

Figure 9.9 Running pump flow over the relief valve releases heat to the reservoir

Heat Generated

$$BTU/hr = \Delta p_{psi} \times Q_{gpm} \times 1.5$$

Heat generation caused by a pressure drop, such as that caused by flow across a relief valve, can be determined by the following formula:

Formula 5-1

BTU/hr = ΔP_{psi} x Q_{gpm} x 1.5

Where:
ΔP = Pressure drop (usually relief valve setting), PSI
Q = Flow through the relief valve, GPM

Figure 9.10 illustrates a number of possible placements for relief valves in a hydraulic circuit. Each of the placements has a specific safety protection for a specific work function. Circuit design determines whether simple relief protection is adequate, or if additional relief valves are required in the circuit to provide protection for one or more specific functions.

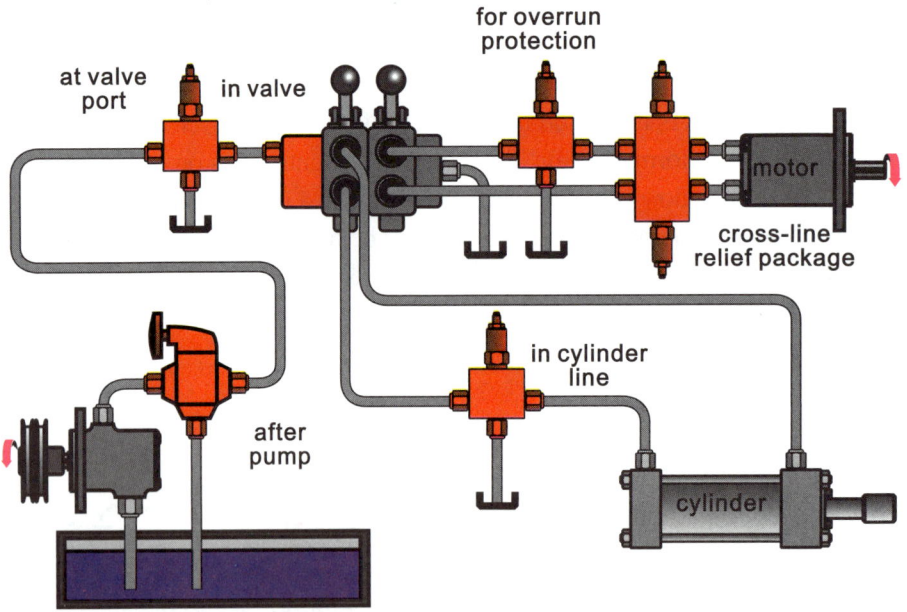

Figure 9.10 Possible placement for relief valves in a hydraulic circuit

When more than one relief valve is used in a circuit, care must be taken to ensure that there is no interaction and resulting instability between relief valves.

As illustrated in Figure 9.11, this can happen when relief valve settings are the same or very close for both the system and the specific work function.

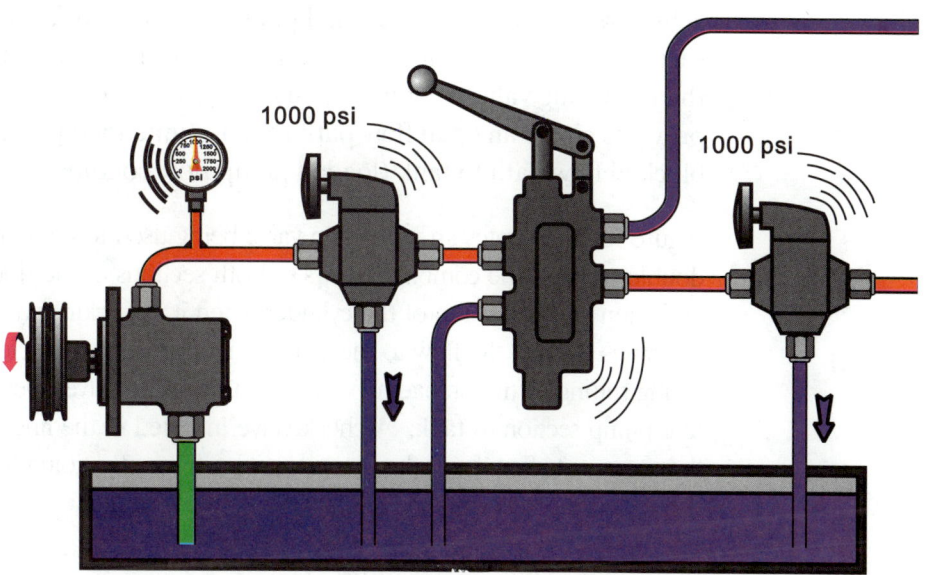

Figure 9.11 Interaction in a system when multiple relief valves are set with the same settings

In this case, the main system relief valve would start to open and relieve fluid to tank at the working pressure of the function being controlled, which is also trying to relieve fluid to tank. The erratic operation occurs as first one relief valve, and then the other, opens and closes alternately at a high rate of speed. To prevent these situations, it is recommended that there always be at least 200 psi differential between the two relief valves.

To illustrate this, if the application were a typical wheel loader, crane or other work circuit, the main relief valve would be set at something on the order of 3000 psi and the work port relief valve set somewhere between 3200 and 3600 psi, dependent upon the pressure capability of the actuator and the need to prevent failure of the hydraulic lines and fittings.

Chapter 9 Auxiliary Valves

Unloading Valves

Unloading valves are typically used in fixed displacement, double pump systems. The unloading valve provides the means of automatically loading/unloading a fixed delivery pump according to system demands. This not only protects the prime mover, but also provides energy savings as flow from the first section of the double pump is relieved to tank at a low pressure, while flow from the second section of the double pump continues to provide flow to the actuator to perform work.

The actuation of the unloading valve on the first section of a double pump is achieved by means of an external pilot signal sensed from the outlet line of the second section. This causes a small pilot piston to act on the main stage piston of the unloading valve to provide "on/off" control of the unloading valve. This provides either an open flow path to tank (unloaded pump flow condition) or a blocked flow path to tank (loaded pump flow condition).

Figure 9.12 illustrates an unloading valve being used to load/unload the first section of a double pump. The combined flows of both sections of the double pump provide rapid extension and retraction of the cylinder when it is working with small loads at lower pressures. When the flow to the cylinder reaches a pressure level predetermined by the spring setting of the unloading valve, the unloading valve opens and directs flow from the first pump section to tank. A check valve installed in the line connecting the two pump flows prevents flow from the second section from also going to tank and ensures this flow is directed through the directional valve to the cylinder.

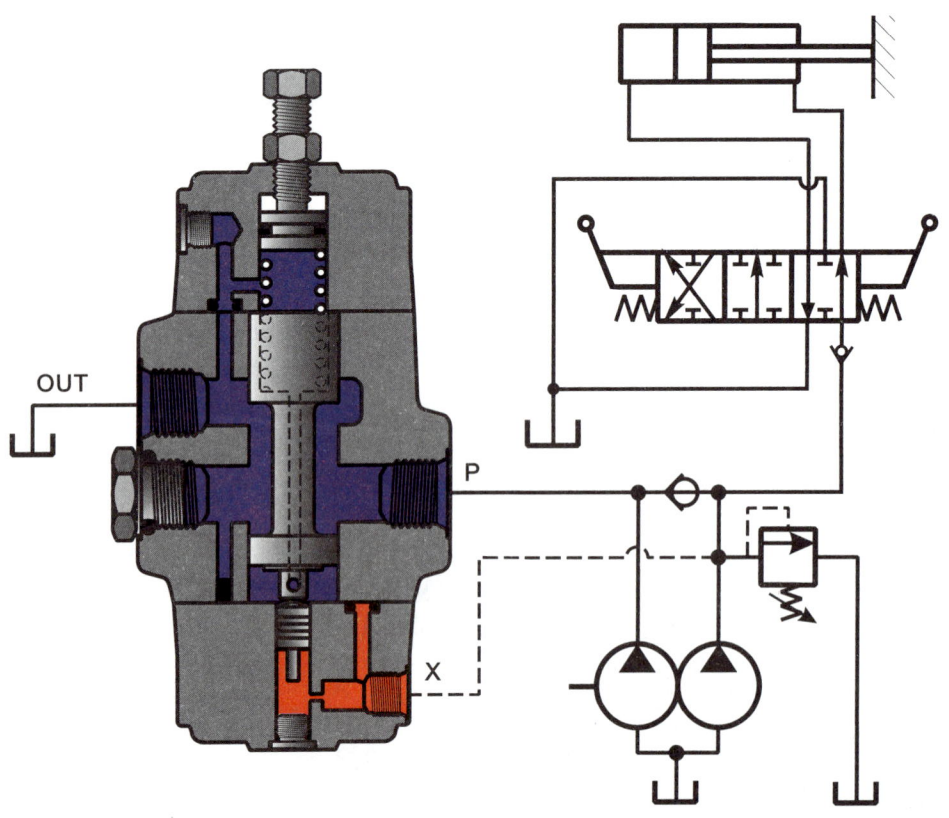

Figure 9.12 Unloading valve used in the first section of a double pump

Typically the unloading valve is set to operate at 1000-1500 psi while the relief valve is set at pressures up to 2500 to 3000 psi. An example of an unloading valve application is on truck mounted cranes, as illustrated in Figure 9.13. Here, the vehicle transmission and power take-off are protected from overloading. Crane operation is limited to low loads at high speeds, and low speeds with heavier loads. These limitations are usually not inconvenient to the operator.

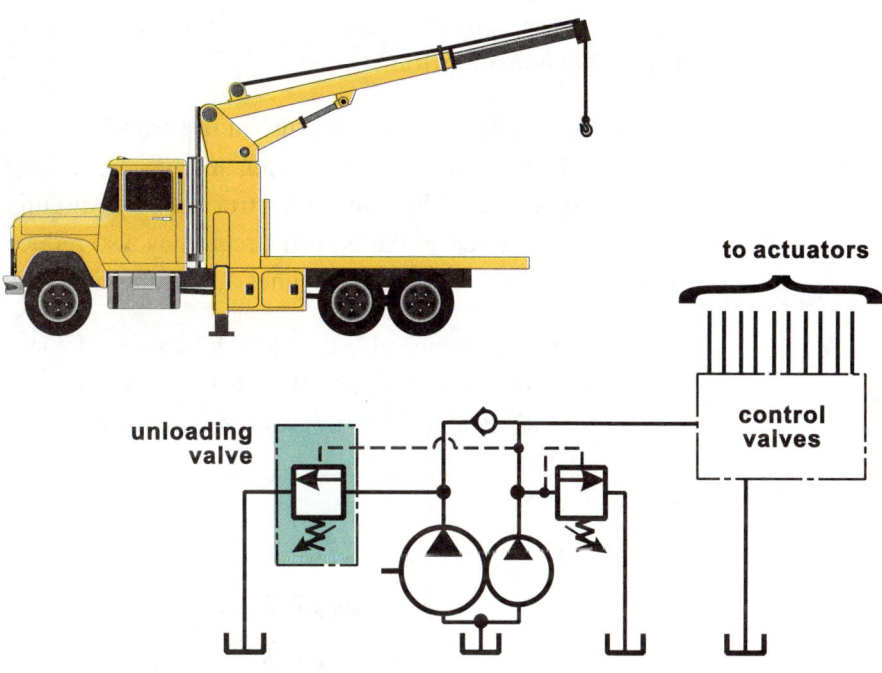

Figure 9.13 Application of unloading valve on truck mounted crane

Pressure Reducing Valves

Pressure reducing valves are used to provide a reduced pressure in a secondary or branch circuit that is lower than the system pressure in the primary circuit. They are inserted between the main circuit and the start of a branch circuit. The ability of these valves to provide a reduced pressure to one or more branch circuits has a number of advantages:

- Controls branch circuit actuator forces

- Permits power savings

- Provides protection to hydraulic units

- Provides protection to lines and fittings

- Provides longer life to the hydraulic units, lines and fittings

- Reduces the potential for leaks in branch circuits

Chapter 9 Auxiliary Valves

Typically, hydraulic circuits on vehicles are required to control a number of different vehicle functions using the same pump. It is rare when all of these functions require the same pressure level for each of the actuators and it thus becomes practical to provide different levels of hydraulic pressure to the various functions. This can not always be done by a relief valve, as the fluid flow would always take the path of least resistance, and this would be through the lowest setting relief valve. To overcome this problem, a pressure reducing valve is used to limit the maximum pressure in a branch circuit, and thus limit the force of any actuator in that branch circuit.

By providing reduced pressure, and therefore reduced power capability to a branch circuit, overall power consumption can be reduced, providing savings through lower fuel consumption.

By means of having a lower pressure in the branch circuit, it is possible to use hydraulic units that have a lower pressure rating and typically a lower cost than those required for the primary circuit. The same holds true for the hydraulic lines and fittings. The use of lower pressure levels in the branch circuit may also reduce the potential for hydraulic leaks in comparison to the units, lines and fittings in the primary circuit.

The pressure reducing valve operates by sensing downstream pressure in the branch circuit to control the flow, and therefore the pressure, of fluid in the branch circuit. Figure 9.14 illustrates the operation of a pressure reducing valve.

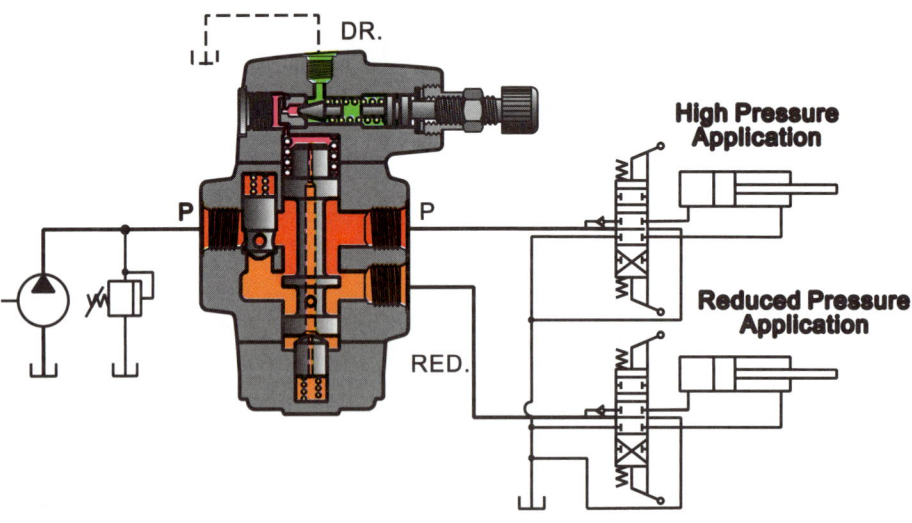

Figure 9.14 Operation of a pressure reducing valve

The value of the reduced pressure setting is determined by means of an adjustable spring loaded poppet, which controls the pressure in the upper chamber. A light spring keeps the piston down in a normally open position. The piston is held in balance as long as the main system pressure and secondary, or branch, circuit pressure are the same and do not exceed the control spring setting.

As pressure in the branch circuit increases to a level exceeding the control spring setting, fluid escapes from the upper chamber past the control poppet. This upsets the force balance, and the piston begins to move upward, closing off flow to the branch. The piston will continue to close down (move upward) until the flow into the branch is sufficient only to maintain the lower pressure setting of the control spring.

A characteristic of pressure reducing valves is their very low pressure over-ride as illustrated in Figure 9.15. Also, good valve design produces pressure reducing valves that maintain a constant branch pressure regardless of pressure variations in the primary circuit.

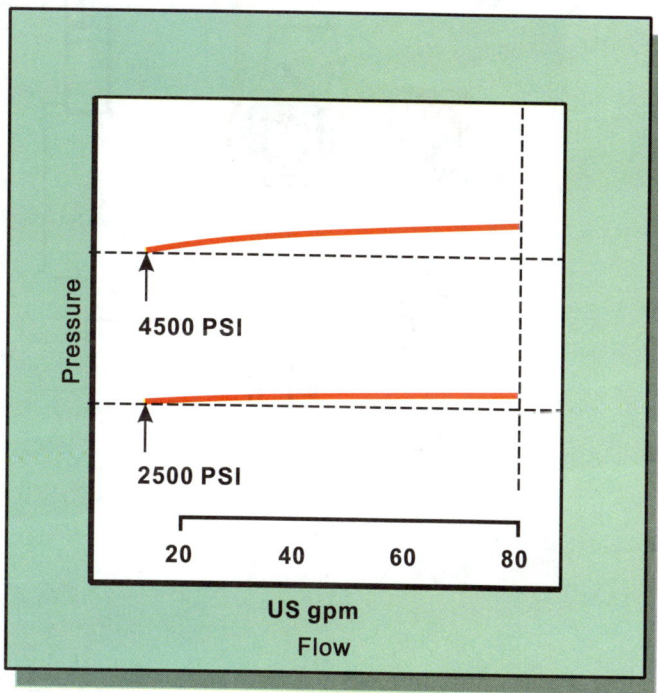

Figure 9.15 Pressure reducing valve override curve

Pressure reducing valves are available with either single or dual reduced pressure level. Dual level valves are controlled by means of either a single solenoid, spring offset solenoid pilot valve or a double solenoid, two position detented pilot valve. By use of a remote electrically modulated proportional pilot, infinitely variable reduced pressure settings and a vent condition can be obtained.

Chapter 9 Auxiliary Valves

Figure 9.16 illustrates the use of pressure reducing valves on a side shift operation of the forks of a lift truck. This provides lower side loading on the forks and upright, as well as a lower potential for leakage at this remote location.

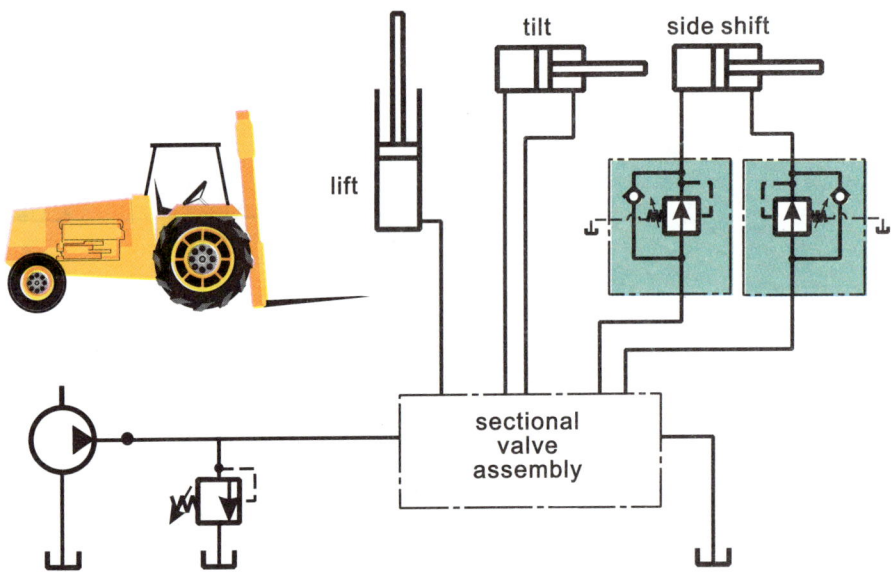

Figure 9.16 Application of pressure reducing valves on a fork lift truck

Sequence Valves

Sequence valves do what the name suggests; they determine the sequence of one function with another. They provide a hydraulic logic so that actuator A will operate before actuator B. They also maintain sufficient pressure for the operation of actuator A (i.e., the extended position of a cylinder) while actuator B is operating.

Figure 9.17 illustrates the use of a sequence valve. In this illustration the normally closed sequence valve initially directs the pump flow to cylinder A until it bottoms out and pressure then builds up in the P port. As this pressure increases, it acts on the small piston located at the bottom of the valve which is held in the closed position by the adjustable spring acting on the top of the spool. When the pressure in the line to cylinder A overcomes the pressure setting acting on the spring side of the spool, the spool shifts and opens up a flow path to cylinder B.

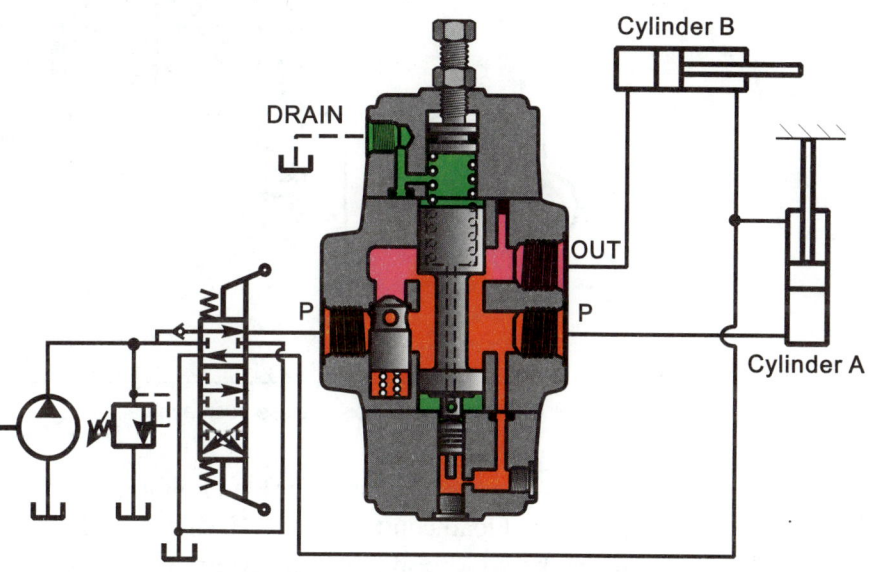

Figure 9.17 Sequencing two cylinders with a sequence valve

Sequence valves have an adjustment feature, typically a screw, cap or knob, which determines the pressure setting at which the valve will open the flow passage to the second actuator. Valve operation can be either internally or externally piloted.

Chapter 9 Auxiliary Valves

Figure 9.18 illustrates the use of a sequence valve on a refuse truck application where the gate must be closed before the compress cylinders are permitted to operate.

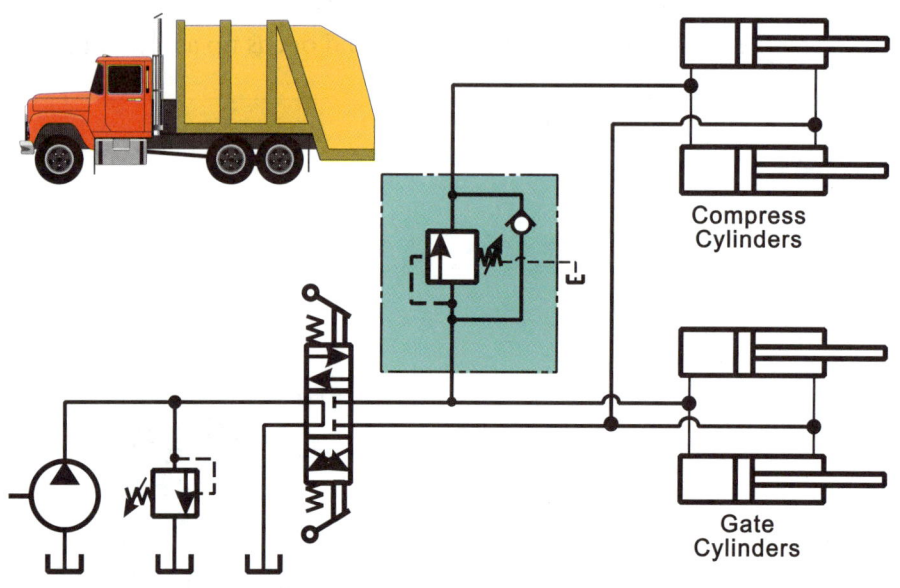

Figure 9.18 A sequence valve on a refuse truck application

Load Control Valves

Load control valves are used for load holding and motion control. There are three basic types of load control valves. Each type or load control has specific features which make it more suitable for certain types of applications. The three types are:

• Counterbalance valves

• Vented counterbalance valves

• Pilot operated check valves

Counterbalance valves are designed for use with an open center on/off directional valve for precise control of overrunning loads; protection from pump cavitation; to prevent an actuator from running ahead of pump supply; and to provide both load holding and hydraulic line failure safety.

Vented counterbalance valves should be used with closed center on/off directional valves and proportional valves for precise control of overrunning loads, regenerative cylinder circuits and meter-out control while providing both load holding and hydraulic line failure safety.

Pilot operated check valves are used as a low cost alternative to counterbalance valves when control of overrunning loads and/or load release speed is not required. They should only be used for locking a load into position.

Figure 9.19 illustrates the use of a counterbalance valve in a circuit to control the lowering of a load by means of a cylinder mounted in the vertical position with the load attached so that the cylinder rod is pushing the load downward. In this situation, the weight of the load would tend to pull the cylinder rod downward in an uncontrolled manner.

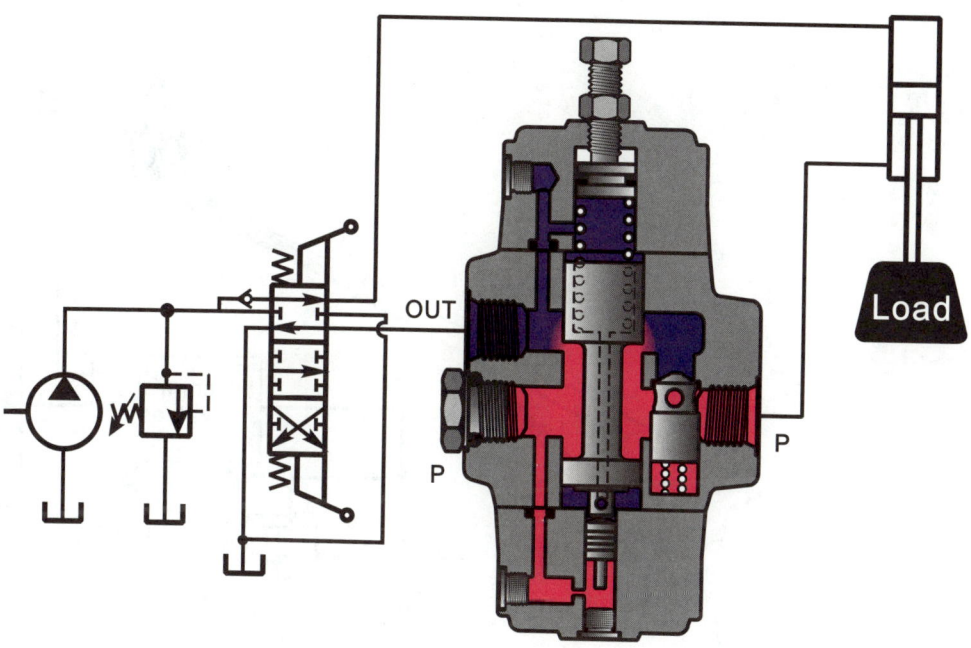

Figure 9.19 A counterbalance valve controls the lowering of a load

The piston of the counterbalance valve is normally held in the closed position by the spring in the upper chamber. This spring will be set to a value higher than the pressure created on the annulus of the piston by the load pulling down on the cylinder rod. Flow from the rod end of this cylinder cannot pass through the counterbalance valve until pressure in the rod end exceeds the control spring value.

When system fluid is directed to the cap end of the cylinder, pressure in the cap end will increase, causing a coincident increase in pressure in the rod end. A sufficient increase in rod end pressure will cause the counterbalance valve piston to open, allowing fluid to exit through the valve. Thus, flow is prevented from escaping the rod end of the cylinder until pressure is applied to the cap end.

When the directional valve is shifted so that fluid flow is to the rod end of the cylinder, the flow path in the counterbalance valve is into the "Out" port, over the check valve and out the "P" port.

It is the pressure of the exhaust fluid from the rod end of the cylinder that provides a "counterbalance" to the downward force of the load, and prevents the load from "running away" from the flow entering the head end of the cylinder. Figure 9.20 illustrates the use of a counterbalance valve on a personnel lift application. Note that in this illustration the rod and head ends of the cylinder are reversed from that shown in Figure 9.19.

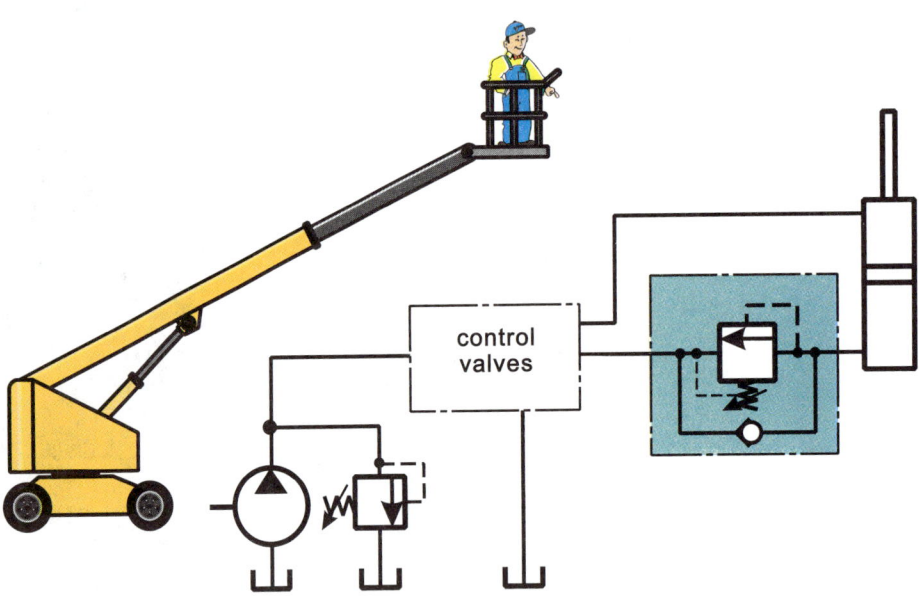

Figure 9.20 A counterbalance valve used on a personnel lift

By installing a counterbalance valve in the circuit between the directional control valve and the head end of the cylinder, the load can be held in place without "drift" while the operator performs his work. Then the platform can be lowered (or raised) at a rate of speed determined by the operator, without danger of the platform dropping in an uncontrolled manner.

A popular type of counterbalance valve is the double valve, frequently called a "holding valve", and applies the counterbalance function to both ends of the cylinder. Figure 9.21 shows a typical design of a holding valve, or double counterbalance valve, typically used on mobile equipment.

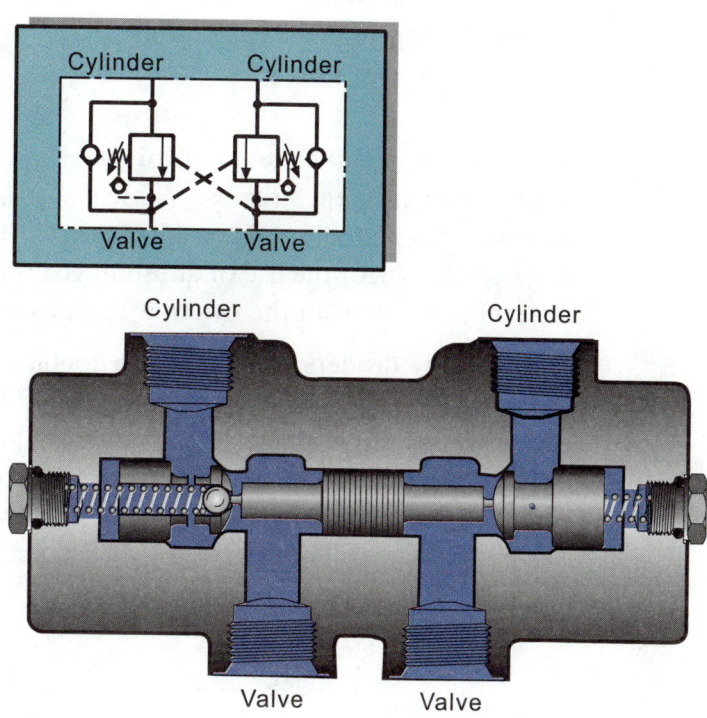

Figure 9.21 Double pilot operated counterbalance valve for cylinders

With the increase in the number of applications of proportional directional control valves, there is a growing trend for the use of vented counterbalance valves. The addition of the vent feature ensures there is no effect of back pressure that may cause hydraulic system instability. This is accomplished by either isolating the counterbalance valve spring chamber from system pressure; or, by venting the spring chamber so that any fluid in the chamber can drain out and is not influenced by any pressure in the line between the counterbalance valve and its directional control valve.

Pilot operated check valves offer a low cost alternative to counterbalance valves for load holding applications. However, they are not suitable for applications where overrunning loads are possible; or where load release speed is critical. They may be suitable for use as outrigger holding valves or stabilizing control on small vehicles. They also lend themselves to use on cost sensitive vehicles such as agriculture vehicles for load holding control of planter wings, 3-point hitch control, combine reel height cylinder, etc.

Chapter 9 Auxiliary Valves

Flow Control

Flow control valves are used to regulate the amount of flow in a branch circuit to a rate below that of the system pump output flow. There are four basic types of flow controls. These are:

- Flow dividers
- Non-compensated flow controls
- Pressure compensated flow controls
- Temperature compensated flow controls

Flow Dividers

Flow dividers are either gear type or spool type. Their function is to divide the flow into two different streams. These two streams typically are equal in flow rate; i.e., each stream is half of the available flow. However, through the use of springs and spool design, or the use of different gear displacements, it is possible to obtain different flow ratios for the two flow streams.

Gear type flow dividers are similar to a double (or triple, or more) gear motor. They use a common inlet and stacked gear sections that are maintained at the same rotational rate by a common shaft (see Figure 9.22). By this means, the incoming flow is divided into a stream of flow for each gear section. It is possible to have more than two sections to produce more than two flow streams. The volume of flow from each stream will be effected by the volumetric efficiency of each gear section.

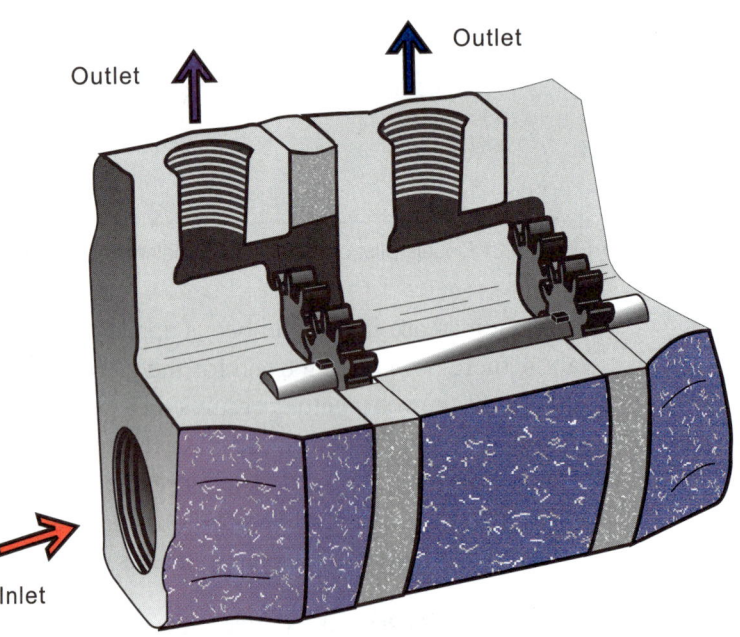

Figure 9.22 Gear type flow divider

Spool type flow dividers are more commonly used as they are typically more convenient to install in the hydraulic circuit, or can be integrated into the outlet cover of a pump. Their design is such that they provide either proportional output flow streams, or a priority output flow stream. When used as a priority valve, the priority flow stream provides a set rate of flow while the secondary stream provides only the flow that is available after the priority stream has been satisfied.

When the accuracy of the output flow streams is critical to the application, a pressure compensation feature can be added to the spool type flow divider in order to prevent changes in output flow rate as inlet or outlet pressure changes. Figure 9.23 is a pressure compensated spool type flow divider.

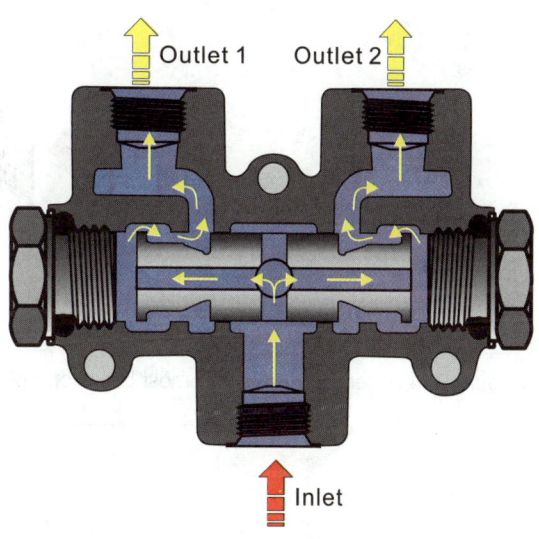

Figure 9.23 Spool type flow divider

Non-Compensated Flow Controls

The simplest type of flow control is the non-compensated flow control. Non-compensated flow control valves are used as restrictors to control the rate of flow into a branch circuit at a given input pressure to the valve. As system pressure increases, flow through the non-compensated flow control will increase with a corresponding increase in pressure drop. If outlet or load pressure increases, flow may decrease proportionally.

The simplest form of a non-compensated flow control is an orifice. An orifice can conveniently be inserted into a hydraulic line as a stand alone valve, or more typically, into a hydraulic fitting as a means of restricting the flow into a branch circuit.

Needle or globe type valves are also commonly used as simple flow restrictions that control the flow into a branch circuit.

Figure 9.24 illustrates a typical needle valve. Figure 9.25 illustrates the use of needle valves as both a meter-in flow control and as a meter-out flow control.

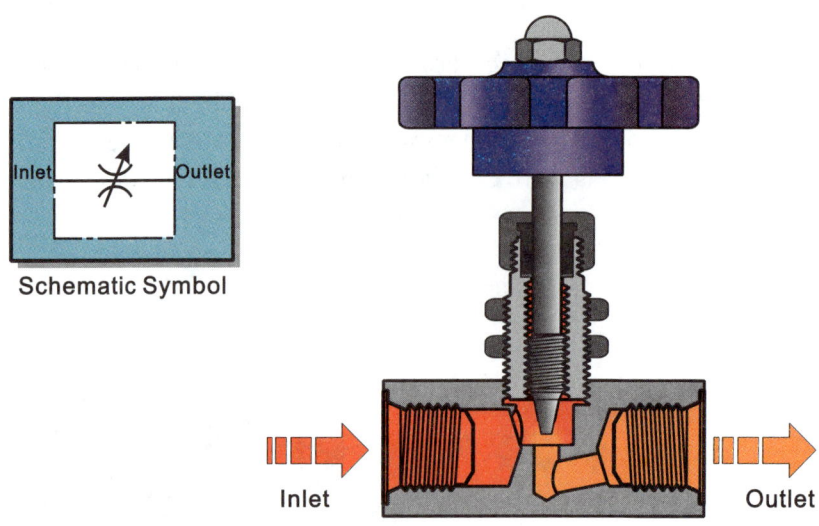

Figure 9.24 A needle valve is the most basic form of flow control

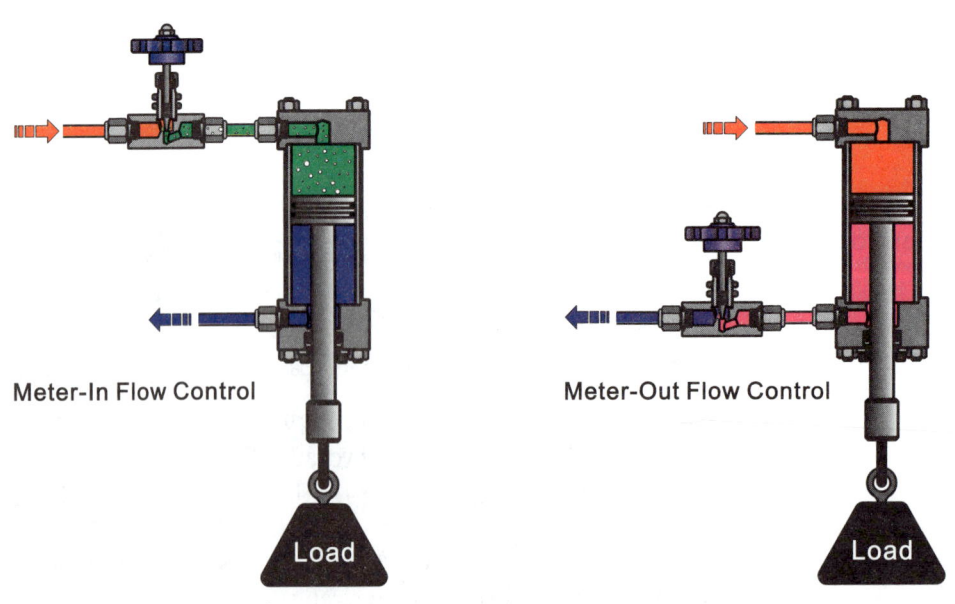

Figure 9.25 Needle valves used as both meter-in and meter-out flow control

Chapter 9 Auxiliary Valves

Pressure Compensated Flow Controls

Pressure compensated flow control valves are used to provide a constant output flow rate into a branch circuit regardless of system input pressure or load pressure. They can be either piston type or spool type. In both cases constant output flow rate is achieved by using springs to maintain a constant pressure drop across the metering orifice. As input pressure increases, output flow is restricted as either the piston or spool is closed off. This prevents an increase of output flow as the increase of input system pressure tries to force more flow through the valve.

Figure 9.26 illustrates a bypass type, pressure compensated flow control with integral relief valve. It appears identical to the pilot operated relief valve of Figure 9.4. The difference is the output flow is controlled by an orifice inserted in the regulated output line. This line maintains a constant flow to a work circuit, all excess flow being diverted to the reservoir through the tank line.

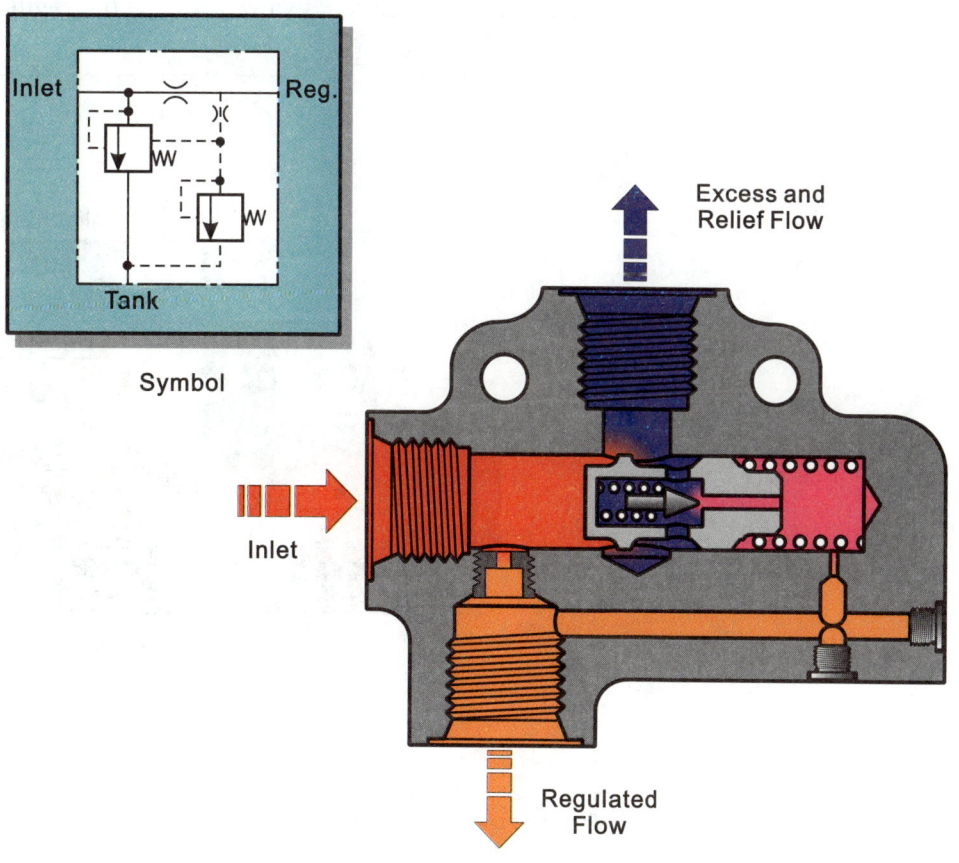

Figure 9.26 Pressure compensated flow control with integral relief valve

237

As the work load increases, pressure in the output line increases, as does the pressure in the spring chamber at the right side of the "hydrostat" piston. Inlet pressure, to the left of the hydrostat piston, will also increase to a level sufficient to overcome the pressure at the right of the hydrostat, plus the value of the spring. At this point, the hydrostat will slide to the right, allowing the excess flow to exit through the tank port. Thus, regardless of the pressure on the right side of the hydrostat, the pressure on the left will be above it by virtue of the spring value. It is this constant pressure drop across the orifice that maintains the pressure compensated flow to the regulated outlet.

If the regulated outlet pressure exceeds the pilot spring setting inside the hydrostat, the pressure in the spring chamber will be limited, and the valve will function as a relief valve.

Figure 9.27 is an alternative design; output flow is controlled by an orifice in the main spool. The spring at the right of the spool maintains a constant pressure drop, and therefore a constant flow across the orifice, and excess flow is diverted to the excess flow port. A separate relief valve protects the regulated flow circuit.

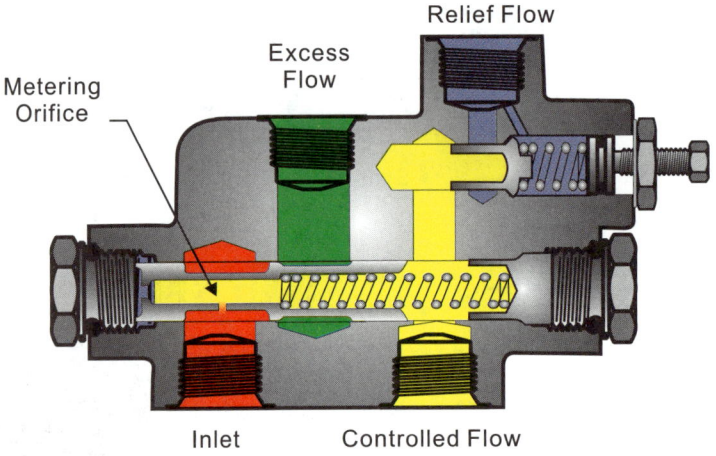

Figure 9.27 Pressure compensated flow control with relief valve

Chapter 9 Auxiliary Valves

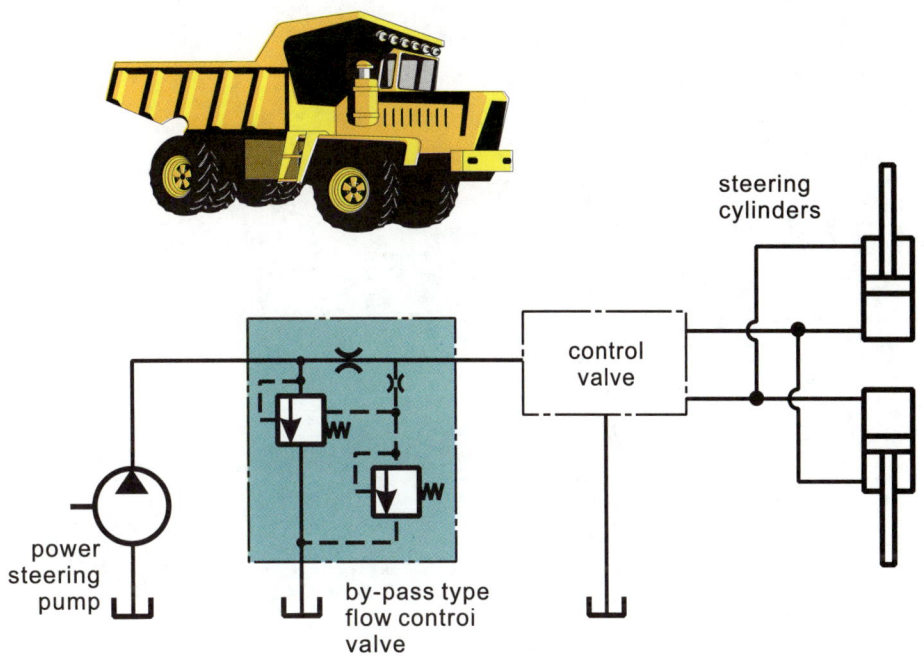

Figure 9.28 Pressure compensated flow control valve controls the rate of steering flow

Figure 9.28 illustrates the use of a bypass type, pressure compensated flow control valve to control the rate of flow into the directional control valve in a steering system on an off-highway dump truck. In this example, the flow control valve would typically be integral to the power steering pump. However, it can also be installed in the circuit as a free standing valve.

Temperature Compensated Flow Controls

Temperature compensated flow control valves are used to maintain a constant output flow rate where a change in the fluid temperature would cause a change. Temperature compensation is usually accomplished through a variable orifice that is controlled by a temperature sensitive, bi-metallic lever or rod.

Chapter 9 Auxiliary Valves

CHAPTER 10 .. Accumulators

Accumulators are devices that can store, or temporarily absorb, hydraulic energy in the form of fluid under pressure. They will accept a quantity of fluid at a predetermined pressure, and retain it until it is needed. Potentially, the fluid will be available to perform a function at that pressure.

The pressure is maintained either by suspended weights, by a spring or by a gas that is compressed. Most common is compressed gas.

As an energy storage device, accumulators are connected to a hydraulic system and are open to accept fluid from the system. The accumulator is pre-loaded; either a pre-set spring, a set number of weights or a pressurized gas is part of the initial setup. When system pressure exceeds this preload, fluid flows into the accumulator. It will remain there until the system pressure is reduced, whereupon, it will flow back into the system.

Hydraulic energy stored in this manner is then available to supplement the system during periods of high demand. When large fluid flows are required in a system infrequently, an accumulator that is filled during periods of low demand can be called upon to discharge its fluid during periods of high demand.

This stored energy and fluid may also be called upon to perform functions when the system has been shut down. The operation of pilot controls or vehicle brakes can be performed by releasing an accumulator's fluid into the system at a time when the system is no longer supplying it.

The energy storage capability also makes accumulators capable of acting as shock absorbers. By absorbing a quantity of fluid above a pre-set pressure, it can help to "cushion" a hydraulic system that is subject to instantaneous pressure surges.

Chapter 10 Accumulators

Accumulator Types

Weight Loaded Accumulator

The weight loaded accumulator is the most basic type, essentially consisting of a piston (a linear actuator) with the rod loaded by weights (Figure 10.1). The area of the piston and the amount of weight determines the pressure required to force fluid into the accumulator.

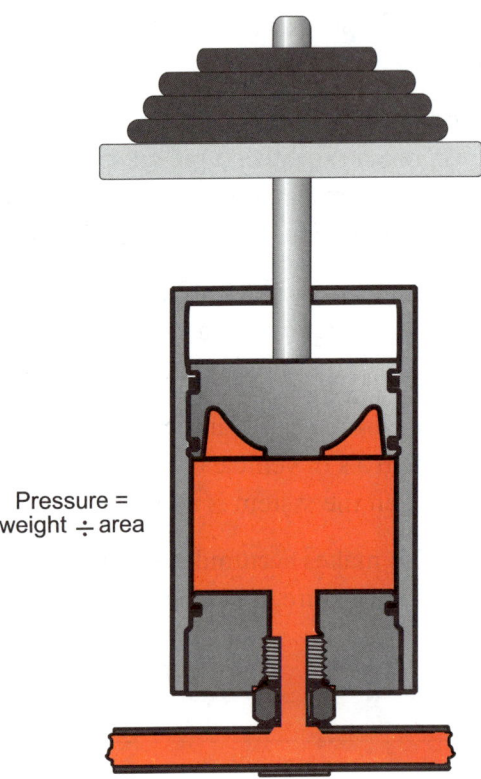

Figure 10.1 Weight loaded accumulator

The primary advantage of weight loaded accumulators is the constant pressure that results as the fluid is discharged. Because the pressure is dependant only on the piston size and the loaded weight, the same pressure will result whether the accumulator is full, partly full or nearly empty. Other factors, such as leakage or temperature do not effect the pressure. This is the only type of accumulator that behaves this way.

On the other hand, the weight loaded accumulator is rather bulky, and is attitude sensitive, making it a poor choice for the restricted confines and typically rolling motions of mobile hydraulic systems.

Chapter 10 Accumulators

Spring Loaded Accumulators

Spring loaded accumulators operate much the same as weight loaded accumulators, the weights merely being replaced by a large spring. The spring provides the force on the piston (see Figure 10.2), and the combination of the force and the piston area determines the pressure on the fluid in the accumulator. The use of a spring instead of weights means the accumulator is not attitude sensitive.

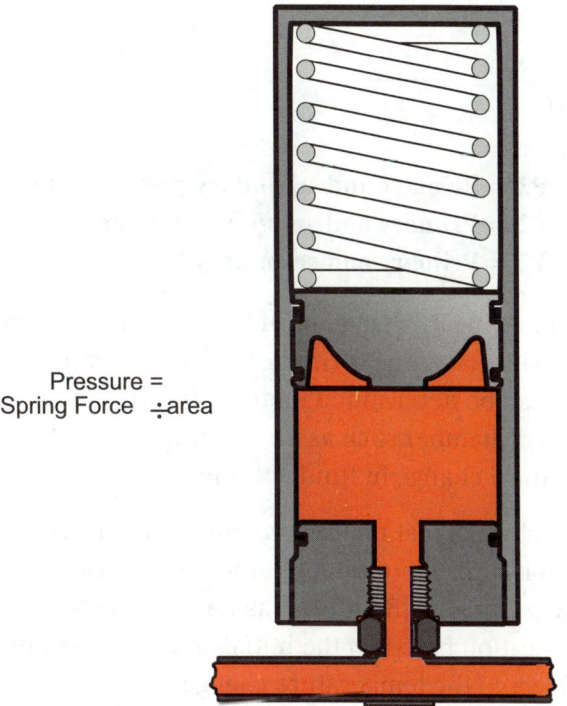

Pressure = Spring Force ÷ area

Figure 10.2 Spring loaded accumulator

To conserve space, the spring must be relatively short, and therefore must have a high spring rate (the unit load required to compress the spring). A high spring rate means that the force resulting from the spring will vary considerably as it is compressed and decompressed. This, in turn, will increase and decrease the pressure in the accumulator fluid, in direct proportion to the spring load.

Spring loaded accumulators are not popular, because of the large size (although they are much smaller than weight loaded accumulators) and the wide variation in pressure between full and empty.

Gas Charged Accumulators

There are several types of gas charged accumulators and all work on the same principle. As fluid is forced into the accumulator, gas that occupies the space is compressed. The compressed gas maintains a pressure on the fluid. The greater the pressure of the gas prior to entry of fluid (precharge pressure), the higher the pressure will be to force fluid into the accumulator. This results in the stored fluid being under higher pressure.

The gas must be a chemically stable one such as dry nitrogen, which is not likely to react with any fluid that is used in the accumulator. Oxygen is never used because of a potentially combustible reaction with some fluids, such as warm oil. For the same reason, air is almost never used because of its oxygen content, although it can safely be used, and occasionally is, with some liquids that will not react with it.

Chapter 10 Accumulators

The behavior of gases under pressure and temperature is very predictable, by virtue of the Ideal Gas Law. This law was developed based on the work of Robert Boyle (1627-1691), who established that at constant temperature, gas volume and pressure are inversely proportional, and Jacques Charles (1746-1823), who established that under constant pressure, gas volume is directly proportional to absolute temperature. The Ideal Gas Law combines the work of these two gentlemen to produce the relationship:

Formula 10-1

$$\frac{P_1 \times V_1}{T_1} = \frac{P_2 \times V_2}{T_2}$$

Where:
P1, P2 = Primary and secondary pressure (absolute)
V1, V2 = Primary and secondary volume
T1, T2 = Primary and secondary temperature (absolute)

By knowing the behavior of the gas in an accumulator, the effect on the fluid is also known. Using the formula 10-1, for example, the change in volume of gas ($V_2 - V_1$) can be determined under a known change of pressure and temperature. In a closed chamber such as an accumulator, a change in gas volume must result in an identical change in fluid volume.

It is obvious that the gas pressure (and therefore the fluid pressure) will change as fluid is forced into or discharged from a gas filled accumulator. It is predictable, however, using the Ideal Gas Law. Accumulators must be sized for a given application based on the initial and final conditions of desired pressure and volume, subject to the temperature changes.

There are three popular types of gas charged accumulators: piston, bladder and diaphragm.

Piston Accumulator

A piston accumulator consists of a free piston separating the gas from the fluid in a cylinder (Figure 10.3). Fluid entering one end of the cylinder will compress the gas in the other end. The gas is normally precharged, and the useable volume of fluid available from an accumulator will be based on the precharge pressure, the pressure of the fluid entering the accumulator and the size of the cylinder. The pressure of any fluid in the accumulator on one side of the piston will be the same as the gas pressure on the other side.

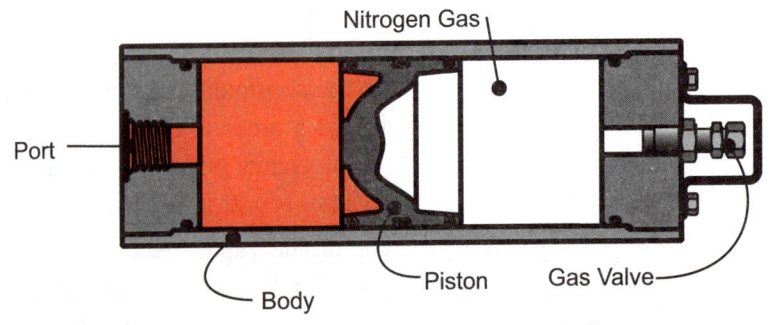

Figure 10.3 Piston type accumulator

Diaphragm and Bladder Accumulators

A diaphragm type of accumulator (Figure 10.4) and bladder type (Figure 10.5) perform quite the same, and also the same as the piston type. In these two designs, gas is separated from the fluid by a synthetic rubber material. Fluid entering the accumulator will compress the gas and perform in the same predictable way as with the piston type of accumulator. A precharge pressure is applied to the gas, and use of the Ideal Gas Law determines the volume and pressure characteristics of the fluid.

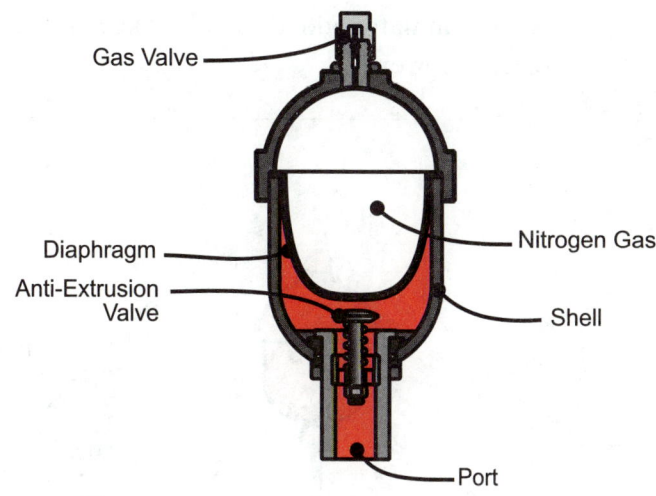

Figure 10.4 Diaphragm type accumulator

Caution should be practiced when alternative types of fluid are used to assure that the bladder or diaphragm material is compatible.

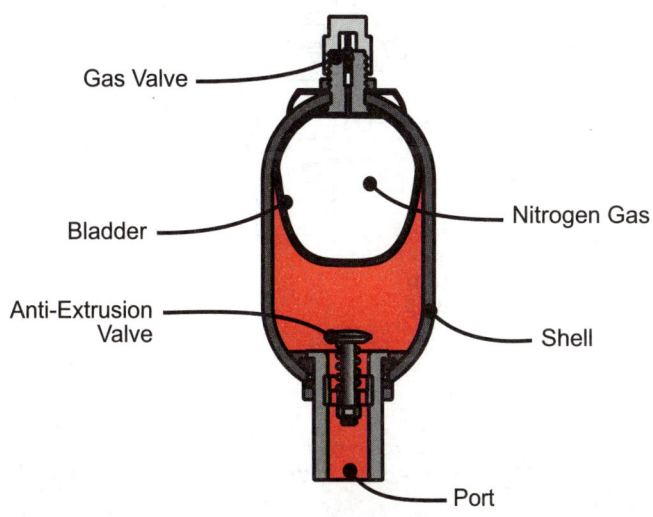

Figure 10.5 Bladder type accumulator

Chapter 10 Accumulators

Accumulator Applications

Mobile equipment utilizes accumulators in a number of applications where hydraulic energy needs to be provided as a supplement to, or in lieu of, the system fluid. It may also be used as a shock absorber, either in the hydraulic system, or as part of a suspension system. The following applications are typical:

Brake Circuit

In many mobile applications using power brakes, the main hydraulic system is relied on to provide fluid to apply the brakes. This means, however, that if the main system is not operating, the brake system will have no source of power.

An accumulator added to the brake system, similar to Figure 10.6, can provide this reserve power.

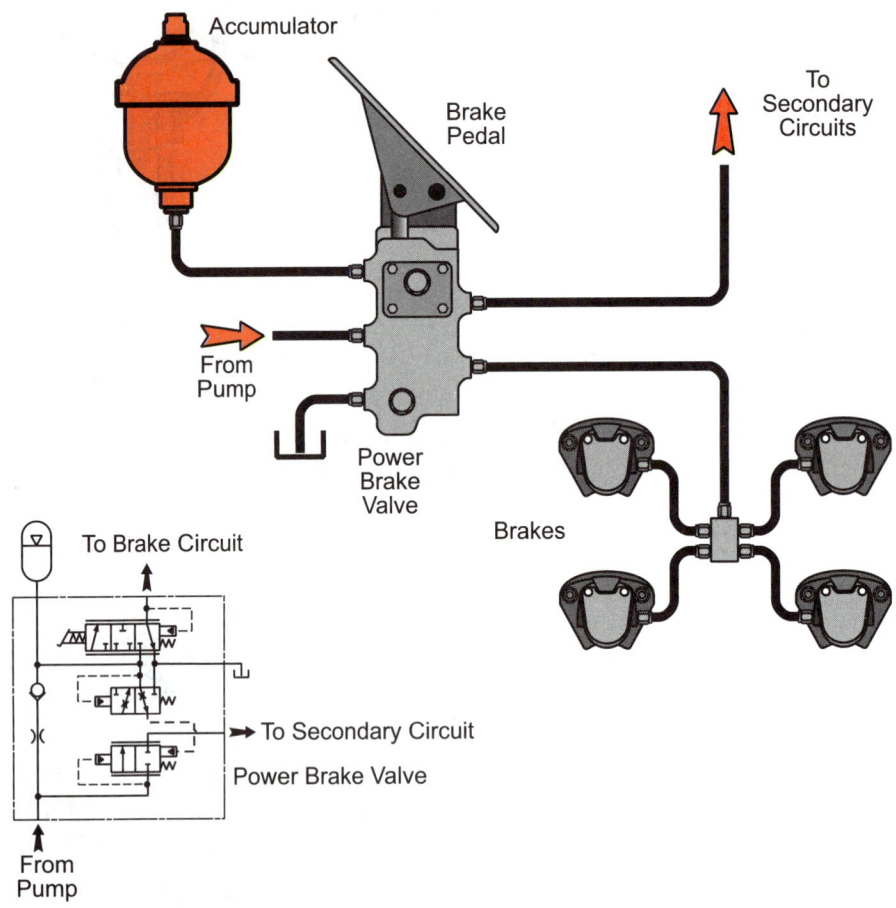

Figure 10.6 Accumulator used in a brake circuit

Chapter 10 Accumulators

Pilot Controls Circuit

When main system valves are pilot operated, and the pilot flow comes from the main system, there is no way to operate the valves when power is off. Normally, this does not present a problem, but there are some instances where it would be advantageous to operate valves when there is no system power.

For example, if a machine operates with one element in a raised position (such as a bucket loader), and the main system power is shut down, there is no method available to lower the bucket unless the valves can be operated (or the machine restarted). An accumulator circuit similar to the one shown in Figure 10.7 would allow limited operation after shutdown.

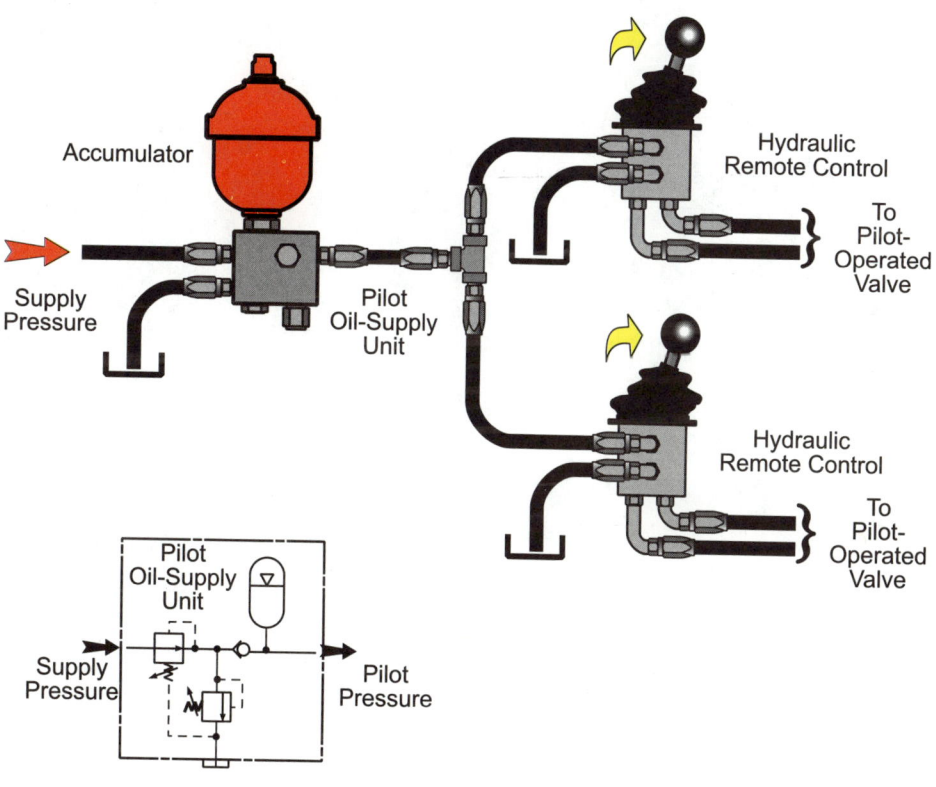

Figure 10.7 Accumulator used in a controls circuit

Chapter 10 Accumulators

CHAPTER 11 .. **Fixed Displacement Pumps**

A pump is one of the most important components in any hydraulic system as it supplies the flow of hydraulic fluid to the system. The pump converts mechanical power supplied to its input shaft into the hydraulic power of fluid flow.

All pumps function by pushing fluid out of the outlet side of the pump. This creates a void within the pump, providing space for fluid to enter the inlet side from the reservoir. Atmospheric pressure in the reservoir forces fluid into the pump inlet.

The pump creates the void by creating an expanding volume in the pumping chambers at the inlet side. A decreasing volume of the pumping chambers at the discharge side is used to push the fluid out. Figure 11.1 is a cross-section view of a vane pump, showing the increasing and decreasing volumes that cause pumping action.

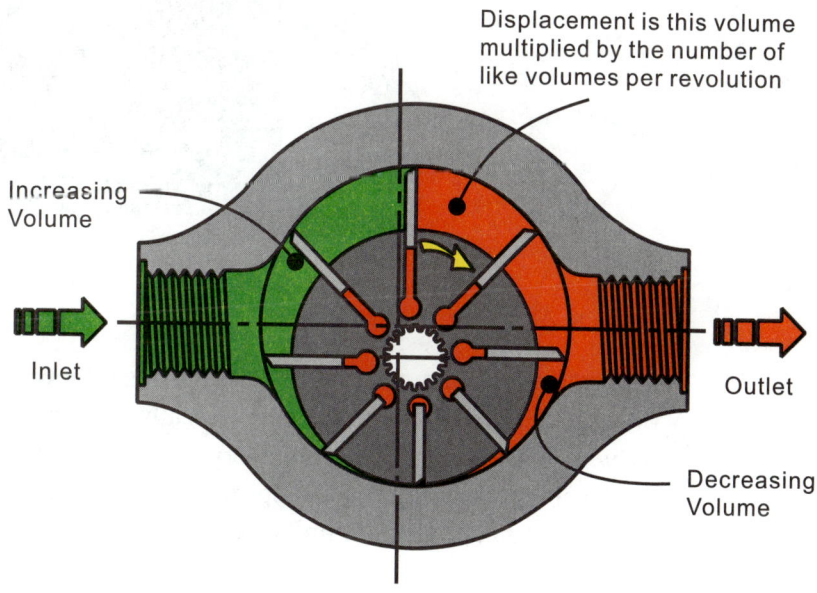

Figure 11.1 A cross-section view of an unbalanced vane pump

249

Chapter 11 Fixed Displacement Pumps

Displacement

With respect to performance, the size or capacity of a pump is determined by the displacement. Displacement is a theoretical volume of liquid that is transferred from the inlet side to the discharge side in one revolution of the pump shaft. If the construction of the pump is a common one, in that multiple chambers transfer fluid from the inlet to the outlet in a single shaft revolution, displacement is equal to the theoretical volume change from inlet to outlet of a single chamber multiplied by the number of chambers. Figure 11.2 also describes this concept. Displacement is usually expressed in cubic inches (in^3) per revolution or in cubic centimeters (cm^3 or cc or ml) per revolution.

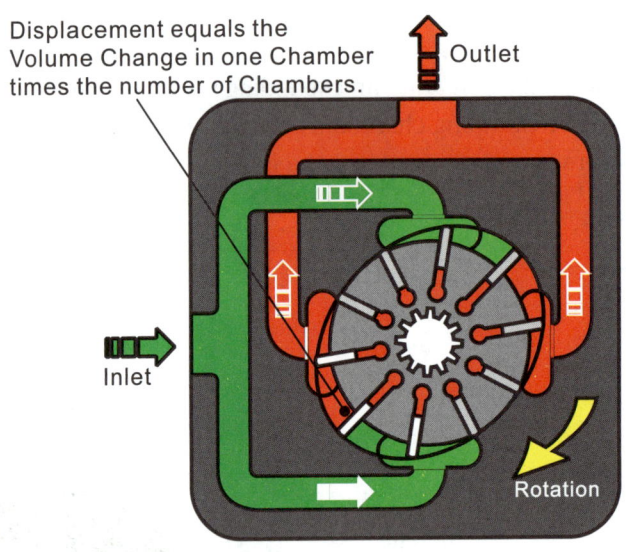

Figure 11.2 Vane pump displacement

Delivery

Alternatively, a pump may be rated in terms of how much fluid is supplied per unit of time. This delivery capability is typically expressed in gallons per minute (GPM) or liters per minute (LPM). Delivery of a pump is found by multiplying the displacement times input shaft speed, taking care to make sure the units agree by dividing by an appropriate unit conversion constant.

Formula 11-1

$$Q = \frac{D \times RPM}{K}$$

Where:

D = Pump Displacement
RPM = Input Shaft Speed in Revolutions per Minute
Q = Pump Output Flow
K = Unit Conversion Constant

If D is expressed as cubic inches per revolution (CIR) and K is 231 cubic inches per gallon, then Q will be calculated as gallons per minute. In metric units, D in cubic centimeters per revolution (CIR) and K being 1000 cubic centimeters per liter produces Q in liters per minute.

Care must be taken in interpreting delivery flow ratings because the delivery flow is proportional to the input shaft speed, and flow will decrease somewhat as discharge pressure increases. Manufacturers typically provide tables or graphs, such as Table 11-1, showing pump deliveries, horsepower requirements, drive speeds, and pressures under specific conditions.

Pump Displacement	Delivery, gpm (lpm) at 1800 rpm			Horsepower (kw) Input at 1800 rpm		
in3 per rev (cc per rev)	0 PSI (0.00 bar) (0.00 kPa)	500 PSI (34.47 bar) (3447.00 kPa)	1000 PSI (68.94 bar) (6894.00 kPa)	0 PSI (0.00 bar) (0.00 kPa)	500 PSI (34.47 bar) (3447.00 kPa)	1000 PSI (68.94 bar) (6894.00 kPa)
0.22 (3.63)	1.8 (6.81)	1.5 (5.68)	1.1 (4.16)	.20 (0.14)	0.9 (0.67)	1.5 (1.11)
0.33 (5.47)	2.7 (10.22)	2.4 (9.08)	2.0 (7.57)	.25 (0.18)	1.2 (0.89)	2.2 (1.64)
0.46 (7.47)	3.7 (14.01)	3.4 (12.87)	3.0 (11.36)	.25 (0.18)	1.4 (1.04)	2.6 (1.93)
0.65 (10.70)	5.3 (20.06)	5.0 (18.93)	4.7 (17.79)	.30 (0.22)	1.9 (1.41)	3.6 (2.68)
1.01 (16.56)	8.2 (31.04)	7.9 (29.90)	7.5 (28.39)	.35 (0.26)	2.8 (2.08)	5.2 (3.87)
1.42 (23.22)	11.5 (43.53)	11.0 (41.64)	10.6 (40.12)	.40 (0.29)	3.7 (2.75)	7.0 (5.22)

Table 11.1 Typical pump performance

Pressure Rating

Besides the rating of a pump on the basis of displacement or flow delivery, it is also important that the pressure rating of the pump be known. The pressure rating is the maximum pressure that should be encountered at the pump discharge port. This rating is specified by the manufacturer based upon reasonable service-life expectancy. Operating at higher pressure may result in reduced pump life or serious damage.

Types of Pumps

There are many different types and designs of pumps. However, different types can be grouped into categories. The most basic way to sort pumps is to group them into positive and non-positive displacement categories. Positive displacement pumps are constructed so that the amount of fluid discharged in a single input shaft revolution is not greatly affected by the discharge and inlet pressures. Non-positive displacement pumps do not have a constant displacement over a wide range of operating conditions.

Pumps can also be categorized according to whether they are fixed displacement or variable displacement. In a variable displacement pump, the geometry of the pumping chambers can be changed, thereby changing the displacement of the pump. This type of pump will be discussed in Chapter 12. Fixed displacement pumps discussed in this chapter have fixed geometries so that the displacement characteristics cannot be changed.

Chapter 11 Fixed Displacement Pumps

Non-positive Displacement Pumps

Most non-positive displacement pumps do not have tight sealing surfaces. They depend upon the centrifugal or inertial forces applied to the fluid being pumped. Figure 11.3 shows a typical version of this type of pump. The turning impeller or propeller pushes on the fluid generating flow.

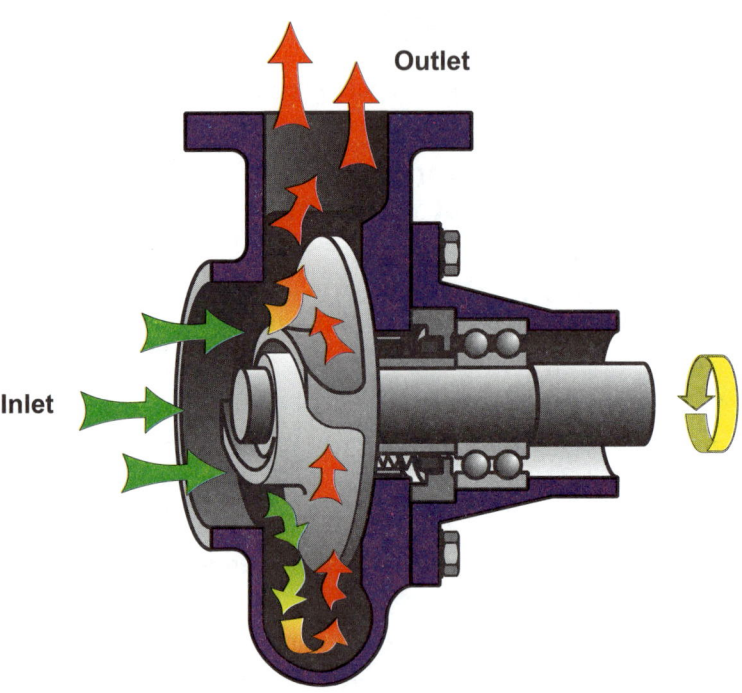

Figure 11.3 A Non-positive displacement pump

Although the output of these pumps is smooth and continuous, increasing resistance to flow at the outlet will cause the flow output to decrease as the impeller is unable to "throw" the fluid past the resistance. In fact, enough resistance will completely stop flow while the pump is running (Figure 11.4). This may be a desirable characteristic in some applications, but it is not acceptable in most power hydraulic systems. With the exception of sometimes being used to supercharge the inlet of a positive displacement main pump, non-positive displacement pumps are seldom used in conventional mobile or industrial hydraulics.

Chapter 11 Fixed Displacement Pumps

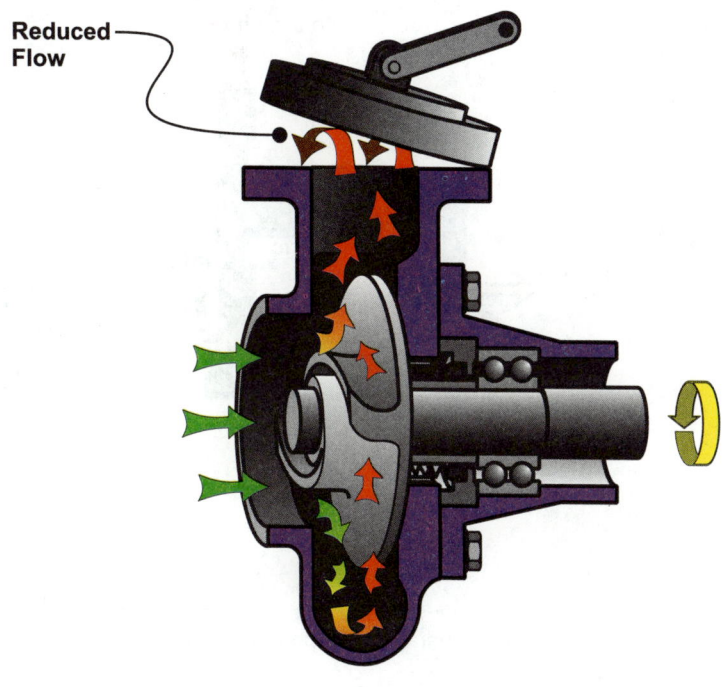

Figure 11.4 Reducing flow in a non-positive displacement pump

Non-positive displacement pumps find more use in lower pressure systems, such as the coolant pump on an automobile engine. There it effectively supplies the needed flow, which varies without damage to the pump when restrictions such as a closed thermostat or a plugged radiator are encountered.

Positive Displacement Pumps

Operation of positive displacement pumps can be seen conceptually by Figure 11.5. This design is typical of hand pumps, such as those used on hydraulic jacks. As the handle is moved to the left, fluid will enter through the lower check valve into the pumping chamber from the reservoir on the lower right. When the handle is then moved to the right, the fluid is forced through the upper check valve to the discharge. The lower check valve prevents fluid from returning to the reservoir.

Chapter 11 Fixed Displacement Pumps

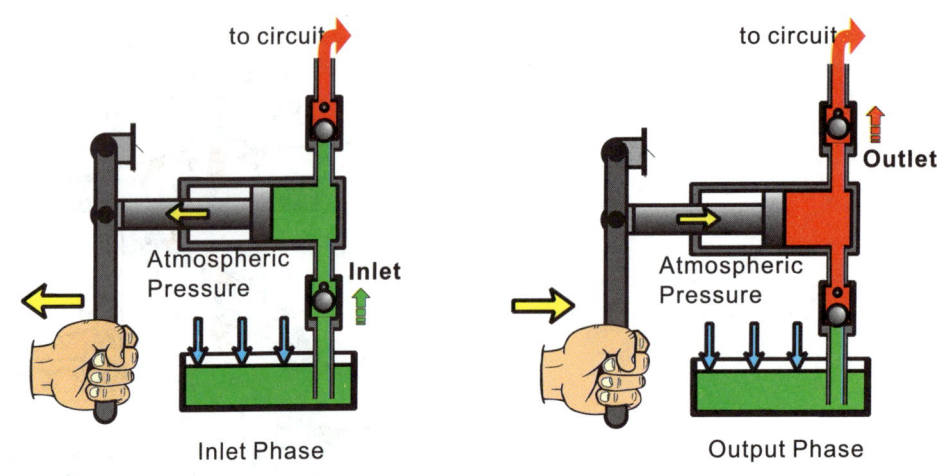

Figure 11.5 Operation of a positive displacement pump

It can be seen that this pump has a fixed and predictable displacement. For example, movement of the handle a certain distance to the right will always generate the corresponding flow out the discharge. Flow per stroke can only be changed by changing the length of the stroke. Delivery in gallons per minute or liters per minute can be changed by changing the length of the stroke or stroke frequency (the number of strokes per minute).

Output of the pump is positive displacement because it is not significantly affected by the resistance to flow. Resistance to flow will determine the discharge pressure and the force required to push the handle. But if the handle is moved, the flow will occur, or the pump will fail.

Positive displacement pumps are available in a variety of designs. This chapter will concentrate on the vane, gear and gerotor, and piston designs which dominate mobile hydraulic applications.

Chapter 11 Fixed Displacement Pumps

Vane Pump Design

Figure 11.6 illustrates a simple vane pump. A rotor with radial slots turns clockwise in the drawing because it is attached to the input shaft. Vanes slide out the of rotor slots following the inner surface of the cam ring as the rotor turns. The vanes are held out by centrifugal force, and pressure under the vanes offsets any pressure above the vane. Pumping chambers are formed between each two vanes, with the rotor, cam ring and two side plates forming the other walls of the chambers.

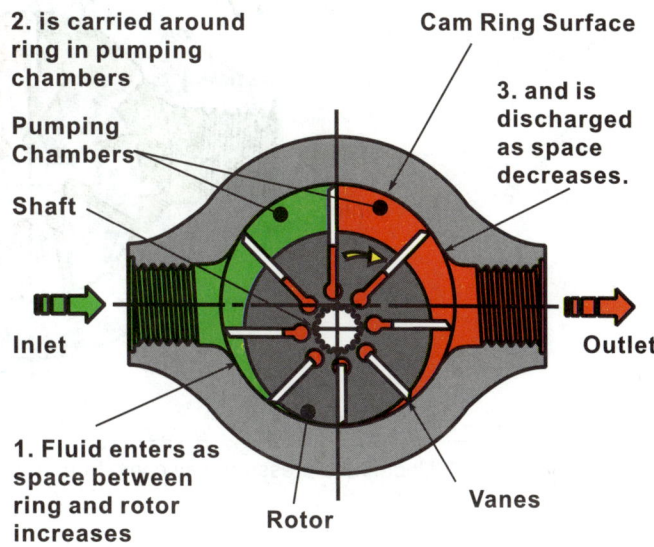

Figure 11.6 Operation of an unbalanced vane pump

Because the cam ring is offset (eccentric) from the rotor and shaft centerline, the chambers change size as the rotor rotates. On the left side of Figure 11.6, the chambers are increasing in size, creating a partial vacuum that allows fluid to be forced into the chambers on the inlet side. The chambers become progressively smaller on the right side of the figure, forcing the fluid out the outlet. Displacement of the pump is determined by the difference in volume of the chambers at the inlet and outlet regions, multiplied by the number of chambers.

Rotation of the rotor, therefore, causes the fluid to be swept from the inlet to the outlet by the vanes. The vanes contact the cam ring surface, and are lubricated by the fluid. The design compensates for vane tip and cam wear by allowing the vanes to extend farther out of the rotor slots.

The vanes are held against the cam ring surface by centrifugal force. The spinning of the rotor causes a centrifugal force which pushes the vanes out of the rotor slots. Effective sealing at the vane-cam ring surface requires that the rotor be turning at a certain minimum speed, depending upon the pump and the operating conditions.

Since centrifugal force is usually insufficient to overcome high pressures at the outer tip of the vane on the outlet side, pressurized oil is also supplied under the vanes. One method of accomplishing this pressure balance is to have a passageway from the rotor surface to the base of the vane, as shown in Figure 11.7. This causes the pressure under the vane to be equal to the pressure on the top. Another method is to supply pump outlet pressure to the bottom of all the vanes, as shown in Figure 11.8.

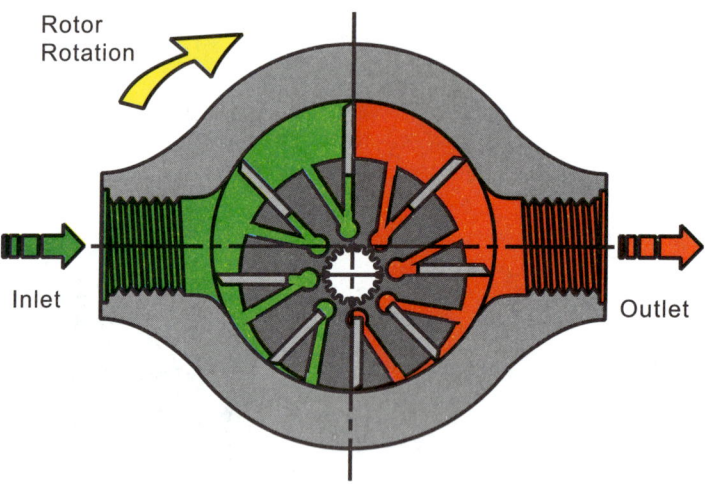

Figure 11.7 Undervane pressure may come from rotor surface

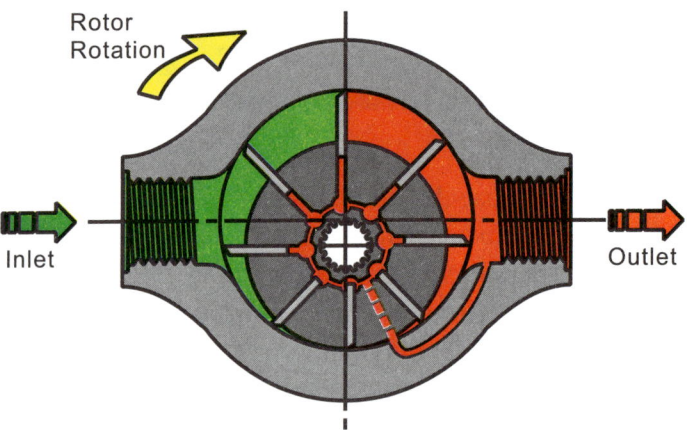

Figure 11.8 Undervane pressure provided from outlet chamber

Outlet pressure under the vanes causes a greater vane tip force against the cam surface at the inlet side of the pump. Certain high-performance pumps seek to optimize the vane extension pressure by using a combination of outlet pressure and surface pressure. The fluid passages in the rotor are designed so that the combination is varied appropriately as the rotor turns.

Chapter 11 Fixed Displacement Pumps

Figure 11.9 shows a high-performance pump in which intra-vanes, or small inserts, are used in the vanes. Full outlet pressure is fed into the intra-vane area continuously to generate enough force to hold the vane tip against the cam ring at all times. Meanwhile, the surface pressure is applied to the underside of the vane. This combination can work better at higher speeds or higher pressures than either outlet pressure or surface pressure by itself.

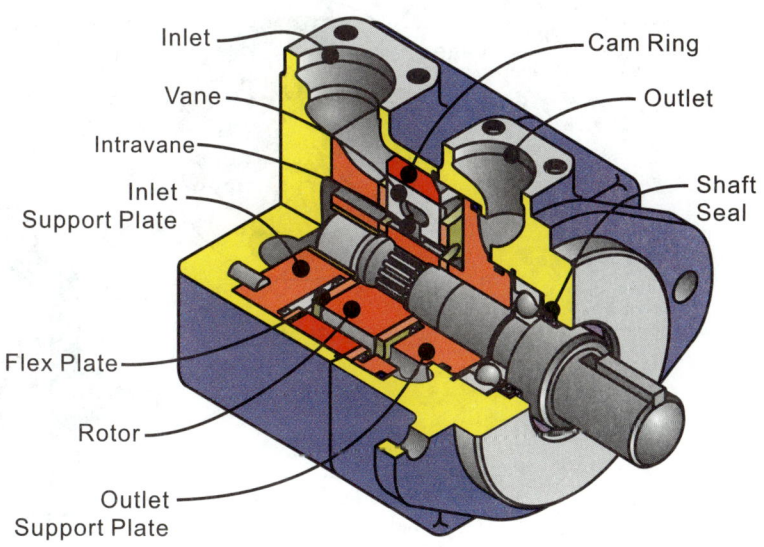

Figure 11.9 High performance intra-vane pump construction

Balanced Vane Pumps

The difference between the low pressure at the inlet and the high pressure at the outlet will cause a significant force to be transmitted to the rotor and therefore to the pump shaft and bearings. In Figure 11.6 this force will be to the left. To overcome this problem, most fixed displacement vane pumps use a balanced design. The circular cam ring of the unbalanced design is replaced with an elliptical cam ring in the balanced design.

Figure 11.10 shows the balanced design principle. This design has opposing sets of inlet and outlet ports. Since the ports are positioned exactly opposite each other, the high forces generated at the outlet ports cancel each other out. This prevents side-loading of the pump shaft and bearings and means that the shaft and bearings only have to carry the torque load and external loads. Since there are two lobes to the cam ring per revolution, the displacement of the pump is equal to twice the amount of fluid which is pumped by the vanes moving from one inlet to its corresponding outlet.

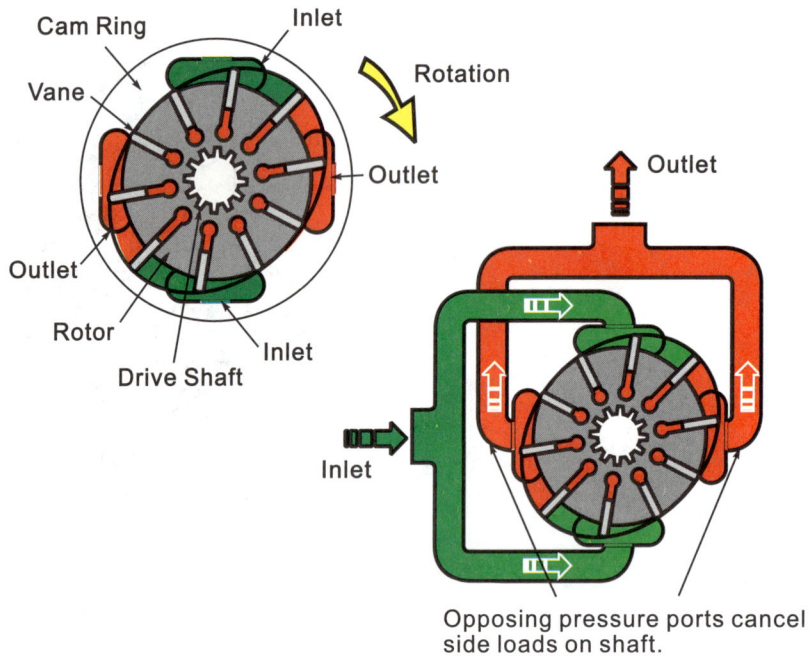

Figure 11.10 Balanced vane pump with elliptical cam ring

For fast replacement of the vanes and cam ring, where the greatest wear occurs, a preassembled cartridge design is often used in high-performance vane pumps, as seen in Figure 11.11. The cartridge consists of a ring, rotor, vanes, vane inserts, outlet support plate, inlet support plate, locating pins, and assembly screws. As shown in Figure 11.12, the direction of shaft rotation can be easily changed by repositioning the cam ring, rotor and vane assembly. The inlet and outlet ports remain the same.

Chapter 11 Fixed Displacement Pumps

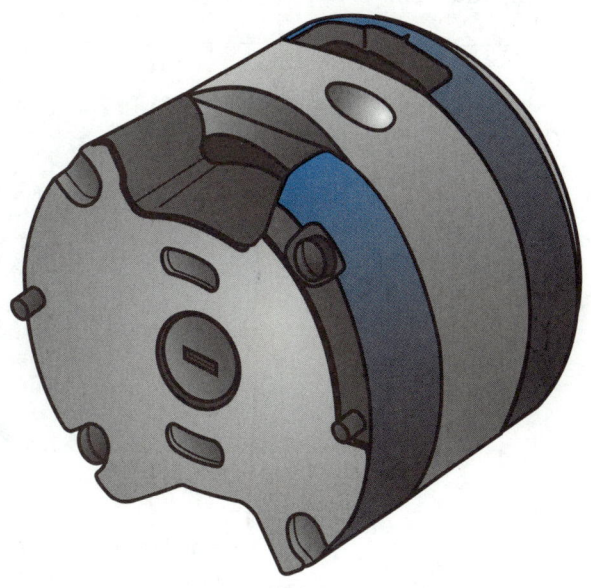

Figure 11.11 Preassembled cartridge

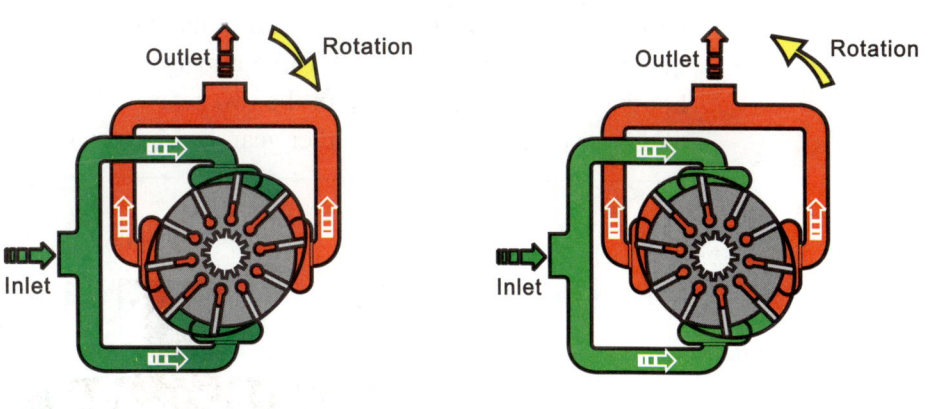

Figure 11.12 Reversing pump rotation

The vanes rub against the cam ring and the inlet and outlet support plates. Despite the materials used in these components and the inherent fluid lubrication, there will be some wear. As shown in Figure 11.10, these pumps often use the combination pressure discussed above to hold the vanes against the cam ring at the appropriate pressure to avoid leakage and to compensate for wear. The side plates are similarly pressurized.

Chapter 11 Fixed Displacement Pumps

The displacement of fixed displacement vane pumps can be changed by either changing the width of the cam ring, rotor and vanes or by using a different cam ring profile. This is shown in Figure 11.13.

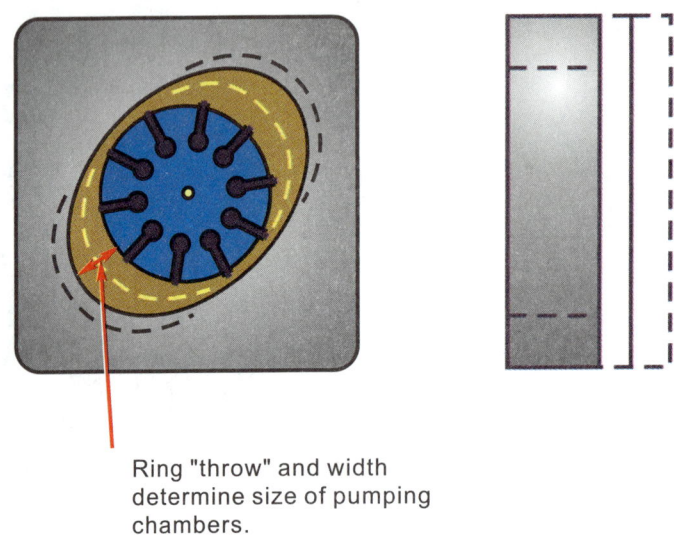

Ring "throw" and width determine size of pumping chambers.

Figure 11.13 Variations in vane pump displacement

Gear Pumps

Another common type of hydraulic pump is the gear pump. The operation of a typical external gear pump (so called because the gear teeth are on the external surface of the hub) is shown in Figure 11.14. A gear pump carries oil from the inlet to the outlet in the spaces between gear teeth. The pumping chamber is formed by the gears, the pump housing, and side plates. One of the two gears, called the drive gear, will be connected to the drive shaft. The other, idler gear, is driven by the drive gear.

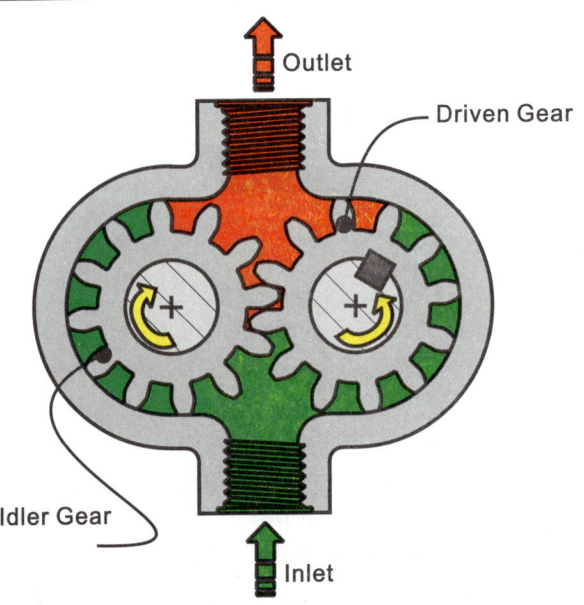

Figure 11.14 External gear pump

As the gear teeth unmesh at the bottom of Figure 11.14, a partial vacuum is created, allowing fluid into the spaces between the teeth. As the gears rotate, the fluid is carried around to the outlet at the top of Figure 11.14. The fluid is expelled from the spaces between the teeth as the gears mesh. The oil cannot return to the inlet because the spaces between the teeth are filled with a meshing gear as the teeth return to the inlet.

The displacement is equal to the size of each space between teeth multiplied by the number of such spaces which pass in a single input shaft revolution. The number of spaces is equal to the number of teeth on each gear multiplied by two since there are two gears. The flow output of such a pump is again equal to the rotational speed (rpm) times the displacement.

Although there can be pressurized wear compensation at the side plates, there is no wear compensation at the cam surface. If the cam ring portion of the housing or the gear teeth wear, the internal leakage of the pump will increase.

External gear pumps are inherently unbalanced. In Figure 11.14, the outlet pressure will create a force trying to push the gears down and apart. There will also be a force generated from the transmission of the power from the drive gear to the idler gear. These loads, combined with the external loads, must be borne by the shafts and bearings. Despite these drawbacks, gear pumps are very popular due to their simplicity and robustness.

Internal Gear Pumps

An internal gear pump (so called because the gear teeth are internal to the gear) replaces the external idler gear with an internal gear, as shown in Figure 11.15. A crescent segment machined into the pump body causes the teeth to unmesh at the lower left of the top figure, thereby allowing fluid into the spaces between the teeth. At the top of the top figure, the teeth mesh again, forcing the fluid out at the outlet.

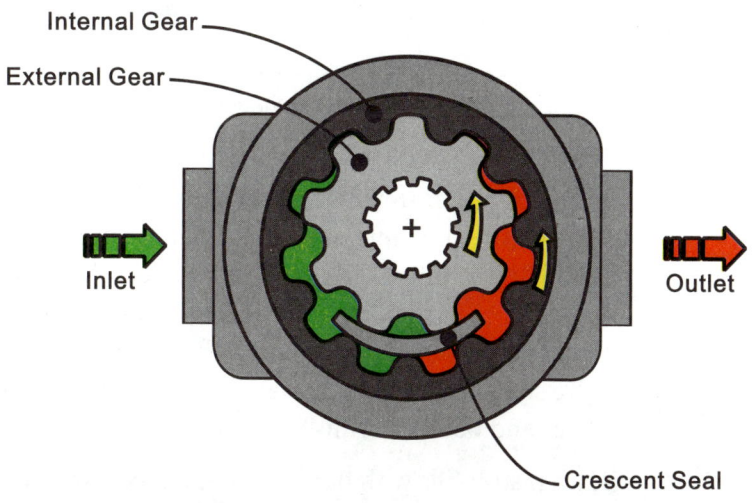

Figure 11.15 Crescent-seal internal gear pump

Again, the size and number of spaces between teeth determine the displacement. The internal gear pump is similar to the external gear pump in terms of wear compensation and unbalanced loading. These designs tend to be quiet. Due to the crescent and the internal gear, the manufacture of these pumps is slightly more difficult than an external gear pump.

Gerotor Pump

Figure 11.16 shows another type of internal gear pump called a gerotor pump. The external (inner) and internal (outer) gears have special teeth shapes. The external gear is mounted on the shaft. It has one less tooth than the larger internal gear. Due to the eccentricity of the centers of the two gears, the spaces between the two gears change. On the bottom of Figure 11.16 the fluid is being forced in to the expanding spaces by atmospheric pressure in the reservoir. On the top it is pushed out of the outlet due to the decrease in the spaces caused by the gears meshing.

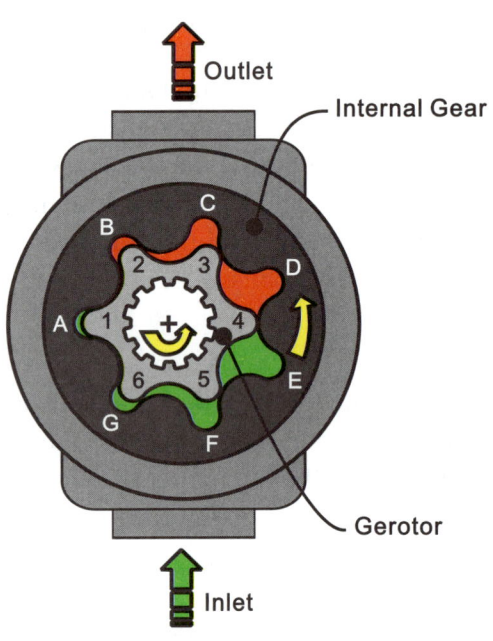

Figure 11.16 Gerotor internal gear pump

Piston Pumps

Piston pumps form another category of popular pump designs. All piston pumps work on the principle of pistons reciprocating in bores. The piston retracts, allowing fluid into the bore, when it is at the inlet. At the outlet, the piston moves forward into the bore, expelling the hydraulic fluid. The two basic designs are axial and radial, depending upon whether the pistons are parallel or perpendicular to the drive shaft respectively. Axial piston pumps are further divided into in-line (swash plate) and bent-axis (link) types.

Each piston fills with fluid at the inlet, carries it to the outlet, and expels it. The displacement of the pump is therefore determined by the diameter of the pistons, the stroke of the pistons, and the number of pistons.

Piston pumps have several sets of surfaces which move with respect to one another, with the piston motion inside the bore being the most obvious. The sliding of piston shoes on the swash plate in axial in-line pumps is another example. Often one of the surfaces will be steel and the other bronze to lower friction. Care is also taken during the design of the pump to insure that lubrication flow is provided to the surfaces which need it. A pump drain, connected to either the reservoir or pump inlet, is provided to carry away the lubrication and leakage flows. This lubrication fluid also helps cool the pump and may become quite heated itself. Because of the closely fitted parts and finely machined surfaces in piston pumps, cleanliness and good quality fluids are vital to long service life.

Although the different types of pumps (vane, gear, and piston) are each available in a wide range of qualities and capabilities, high-quality piston pumps are most popular for applications needing high pressure and high efficiency. Of course, the precision of the many components in a piston pump may be reflected in its cost. Piston pumps are inherently unbalanced. That imbalance, as well as thermal and lubrication management, must be considered in the design of the pumps.

Radial Piston Pumps

Figure 11.17 illustrates the operation of a radial piston pump. The cylinder block rotates on a stationary pintle inside a circular reaction ring or rotor. As the block rotates, centrifugal force, charging pressure, or some form of mechanical action causes the pistons to follow the inner surface of the ring. The center lines of the ring and the cylinder block are offset from one another causing the pistons to move in and out in the cylinder block. Porting in the pintle permits the pistons to take in fluid from the inlet as they move outward and allows the fluid to be discharged to the outlet as the pistons are forced inward.

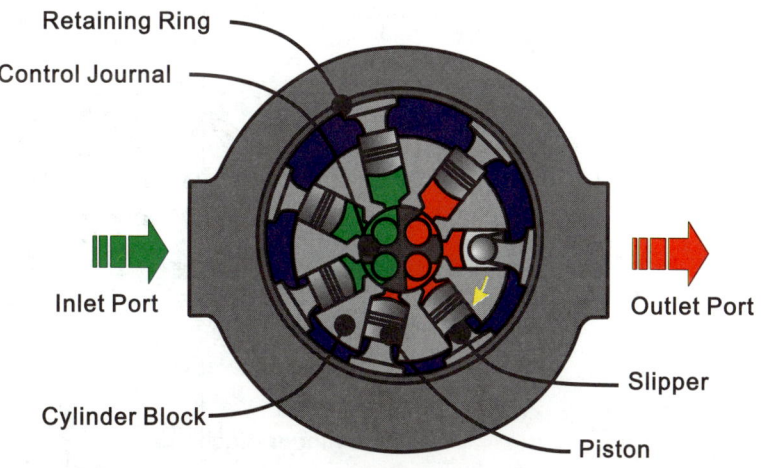

Figure 11.17 A radial piston pump

Chapter 11 Fixed Displacement Pumps

Axial Piston Pumps

The in-line axial piston pump is the most common design. Figure 11.18 shows one design of such a pump. The piston at the bottom has just finished expelling fluid at the outlet and will retract, allowing fluid in as it passes the inlet porting, until it is at the top position. The fluid will then be expelled out the outlet as the piston returns to the bottom of the drawing. The shoes of the pistons slide on an angled swash plate that moves the pistons in and out as the cylinder block rotates. The cylinder block is turned by the drive shaft. The valve plate is designed to couple the passing cylinders to the inlet or outlet at the proper times corresponding to piston retraction and forward motion respectively. Figure 11.19 illustrates the piston motion. The shoe plate (or retractor ring) makes sure that the piston shoes follow the swash plate.

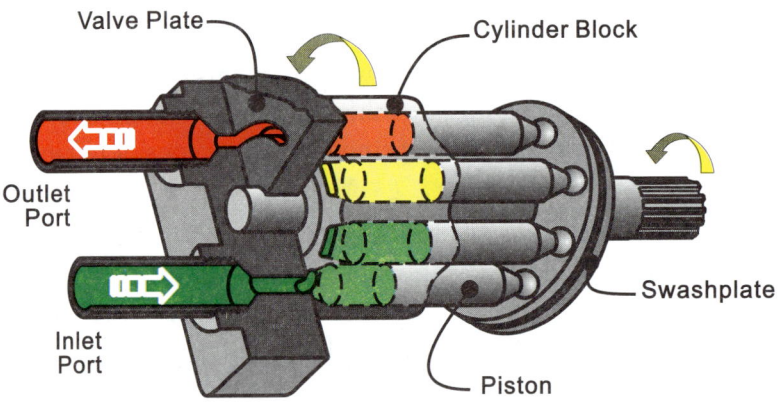

Figure 11.18 Inline design piston pump

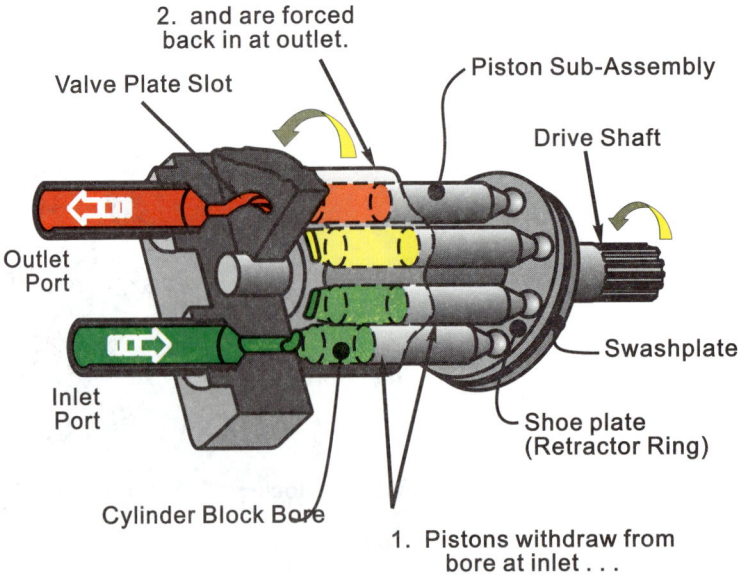

Figure 11.19 Swash plate causes pistons to reciprocate

Bent-axis axial pumps have no swash plate. Instead, the reciprocating motion of the pistons in the cylinder block is caused by an angle in the connection from the drive shaft to the cylinder block. Figure 11.20 illustrates such a pump. The piston rods are attached to the drive shaft flange by ball joints and are forced in and out of their bores as the distance between the drive shaft flange and cylinder block changes. The piston at the bottom is retracted, while the one at the top is fully forward in the bore. It is generally easy to recognize this type of pump due to the distinctive angled shape of the pump housing.

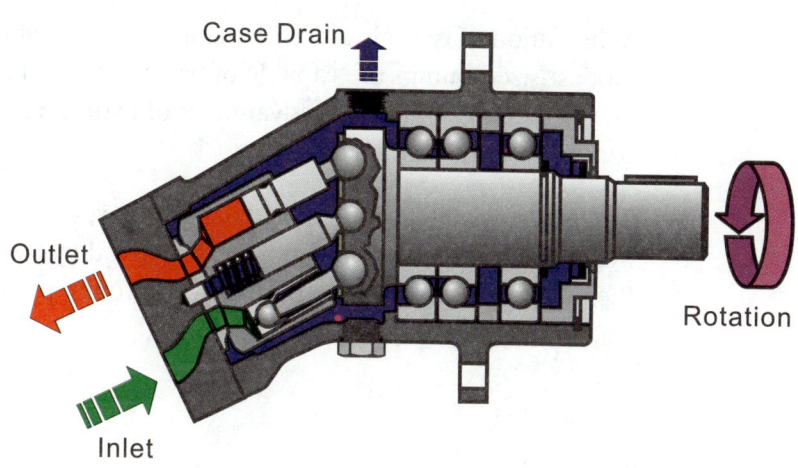

Figure 11.20 A bent-axis piston pump

Operation of Fixed Displacement Pumps

The pumps discussed in this chapter are all fixed, positive displacement. That means (disregarding leakages, compressibility, and similar effects) a constant amount of fluid is pumped per revolution of the drive shaft. Therefore, the displacement per revolution times the speed determines the output flow of the pump. In traditional units, the theoretical (neglecting leakage, etc.,) output in gallons per minute is equal to the displacement in cubic inches per revolution multiplied by the speed in revolutions per minute and divided by 231 cubic inches per gallon. That is:

Formula 11-2

$$Q_{Theo} = \frac{D \times RPM}{231}$$

Where:

Q = Theoretical output flow (GPM),
D = Displacement (in^3 / rev),
231 = In^3 per gallon

The theoretical hydraulic horsepower output of the pump can be found from:

Formula 11-3

$$HP_{out} = \frac{Q \times P}{1714}$$

Where:

HP_{out} = Hydraulic horsepower,
Q = Actual output flow (GPM),
P = Output pressure (psi),
1714 = Constant to resolve units.

Substitution of typical values of Q and P into the above equation shows that even modest-sized pumps are capable of transmitting a large amount of power. This is one of the most important advantages of hydraulics over other power and motion control technologies.

The pump must be supplied mechanical power to be able to generate the output hydraulic horsepower. The power is supplied by the torque and rotational speed of the input shaft. The mechanical power into the pump can be calculated in traditional units as:

Formula 11-4

$$HP_{in} = \frac{T \times RPM}{63025}$$

Where:

HP_{in} = Mechanical input horsepower,
T = Input torque (lb.-in),
63025 = Constant to resolve units.

The input power is usually supplied by connecting the input shaft to an output of a vehicle's engine, either directly or through a gear or belt drive.

Although modern hydraulic pumps are well designed and efficient, they are not 100% efficient in converting all of the mechanical input power into hydraulic output power. The pump efficiency is important. It can be calculated as:

Formula 11-5

$$Eff_{OA} = \frac{Hp_{out}}{HP_{in}}$$

Where:

Eff_{OA} = Overall pump efficiency.

The efficiency calculated from this formula is usually multiplied by 100 so that it is expressed in percent. For example, if 20 input horsepower is required to generate 18 hydraulic output horsepower, then (HP_{out} / HP_{in}) = 0.9. The efficiency would then be 90%.

The 10% of power lost in this example represents the pump losses. Pump losses are generally divided into the categories of mechanical or volumetric (fluid) losses. Some mechanical losses are due to the friction of the moving parts in the pump, such as the friction of the seals and bearings. Others are due to fluid friction and turbulence. These losses are reflected in the increased torque necessary to rotate the pump against a given pressure.

Volumetric losses are deviations in the amount of fluid pumped compared to what might be expected of a perfect pump. Internal leakages reduce the output flow. Compressibility of the fluid and deflections within the pump may have a small effect. In addition, some of the fluid may be used for internal lubrication flow.

The total losses include both the mechanical and volumetric. The relationship between input power and output may therefore be expressed as:

Formula 11-6

$$HP_{out} = HP_{in} \times Eff_{oa}$$

or:

$$HP_{in} = \frac{HP_{out}}{Eff_{oa}}$$

The overall efficiency is the product of the mechanical and volumetric efficiencies according to:

Formula 11-7

$$Eff_{oa} = Eff_v \times Eff_m$$

Where:

Eff_v = Volumetric efficiency,
Eff_m = Mechanical efficiency.

The power lost within the pump is converted to heat energy. This heat must be dissipated directly to the atmosphere, or carried away to the reservoir, typically via an external case drain, where it is then dissipated to the atmosphere.

Chapter 11 Fixed Displacement Pumps

Causes of Pump Failure

If the heat is not carried away or excessively hot fluid is supplied to the pump, the pump may overheat, leading to possible pump failure. There are many other potential causes of pump failure. As the power source, the pump experiences the highest horsepowers and pressures in the hydraulic circuit. It is one of the most complex components with moving parts and sliding surfaces. Given the heavy loading and many places for possible failure, it is not surprising that pump failure can be a problem. And pump failure is not a trivial matter because the pump is often the most expensive single component in a hydraulic circuit.

Some studies point toward contamination as the leading cause of pump failure (see Figure 11.21). Contamination is defined as the inclusion of unwanted debris or material in the hydraulic fluid. The contamination can be divided into solid, gaseous or liquid categories. The solid category can be subdivided into metallic and nonmetallic debris.

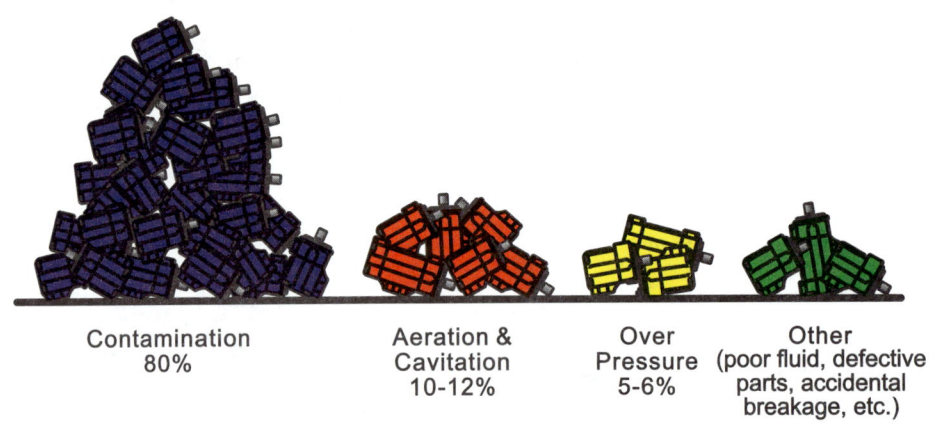

Figure 11.21 Typical causes of pump failure

Metallic debris includes metal chips, filings, and component wear particles. The debris is usually steel and/or bronze, although cast iron particles may be present. The debris may have been externally introduced to the hydraulic system or internally generated by it. It may have been introduced by the pump itself, or by other system components. For example, a failing pump will release debris particles into the fluid. Figure 11.22 illustrates the effect of contamination in a hydraulic pump, and the resulting generation of metallic debris.

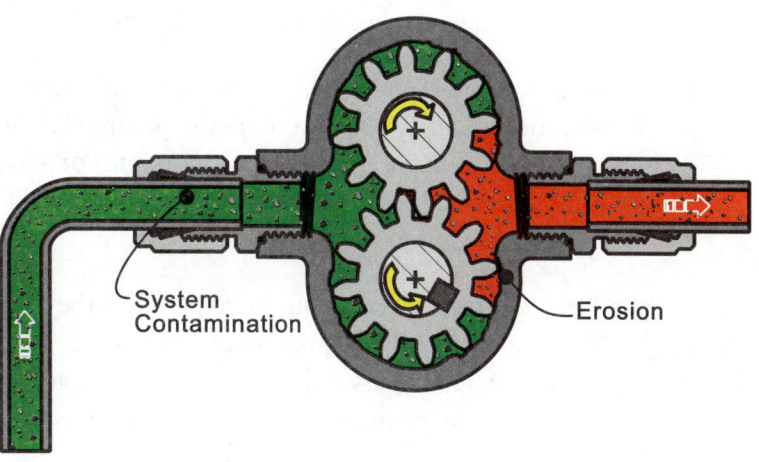

Figure 11.22 Contaminated fluid and contamination induced erosion

Non-metallic debris includes dirt and other non-metal materials. These materials may be internally introduced, such as core sand from component castings or seal debris. But, more often non-metallic debris is externally introduced into the system. Mobile equipment generally works in dirty, dusty environments. The dirt must be kept out of the hydraulic system. The fluid reservoir is a critical location. Clean fluid must be added. Dirt must not be allowed to enter through fill holes or breathers. Dirt can also enter past seals, especially on cylinder rods. Chapter 17 provides a thorough review of contamination and the methods for keeping a hydraulic system clean.

Liquid materials such as water and chemicals may be externally introduced into hydraulic systems through the same openings. Water can also condense from air in the reservoir. These liquids can cause chemical reactions which lead to failure. Water can cause rust, cavitation, or pitting. Other chemicals can also cause such problems as corrosion, sludge, and pitting.

The second leading cause of pump failure is improper installation or application. The pump installation needs to be correct for both the mechanical input and the hydraulic output. The input shaft must be properly aligned with the power source and no excessive axial, transverse, or bending loads must be applied to the shaft. The pump body must be properly supported and aligned. Failure to do proper mechanical installation may result in excessive loads on bearings or other components.

The hydraulic installation should be properly done so that the fluid lines to the inlet, outlet, and drain are big enough and don't have excessive restrictions. The overall system must have adequate cooling and filtration and the proper fluid must be used.

In addition, the proper pump must be selected for each particular application. The application must not impose excessive pressures on the pump, neither as operating system pressures, nor as shock pressures. The pump must not be run at excessive speeds. If the pressures or speeds are excessive, either a more tolerant pump must be selected or the application must be modified.

Aeration and cavitation are two additional categories of pump failure causes. Aeration occurs when air enters the pump inlet along with the hydraulic fluid. The air can enter due to a leak in the inlet lines or a low fluid level in the reservoir. Air bubbles can also be mixed with the hydraulic fluid returning from the hydraulic system or from turbulence in the reservoir.

The low pressure at the pump inlet causes the aeration air bubbles to expand. However, the bubbles quickly collapse as they go through the pump due to the high outlet pressure. This violent collapse, or implosion, causes rapid erosion of the internal pump components. Figure 11.23 illustrates this phenomenon.

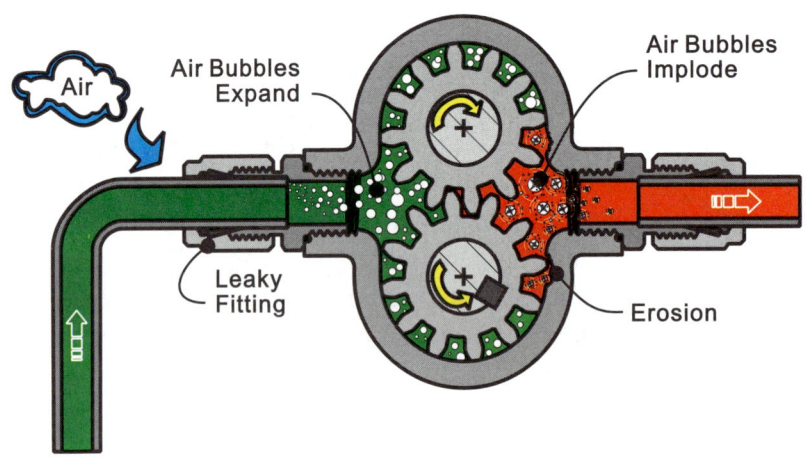

Figure 11.23 Aeration induced hydraulic pump erosion

Cavitation is similar to aeration, except that rather than already having air bubbles as the fluid approaches the inlet, vapor bubbles form in the fluid at the inlet. These bubbles form due to excessively low inlet pressure (an excessive vacuum) for the fluid being used. The bubbles again collapse violently at the outlet causing component erosion and heat.

One category of causes for cavitation includes restrictions in the flow of the hydraulic fluid to the inlet. The inlet hose may be too small, may be bent, or may collapse under vacuum (see Figure 11.24). Excessively clogged inlet screens or filters may not allow fluid to easily reach the pump inlet. Or the fluid may be too viscous to be easily forced into the pump if the fluid temperature is too low or the wrong fluid is used.

Chapter 11 Fixed Displacement Pumps

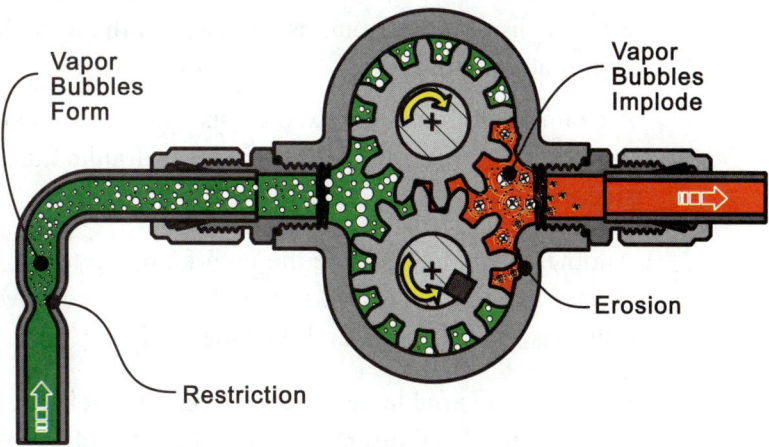

Figure 11.24 Cavitation induced by an inlet restriction, leading to pump erosion.

Cavitation can also occur if the fluid, or part of the fluid, vaporizes too easily. This may happen if the fluid is too hot. Water in the fluid can become water vapor bubbles. Proper choice and care of the fluid is necessary.

Running the pump too fast can also cause cavitation. As Figure 11.25 illustrates, the inlet system may be unable to deliver the large flow rate of fluid which the high speed requires. Often, a combination of several of these causes leads to cavitation. It is usually wise to try to minimize the potential contribution of many of these causes.

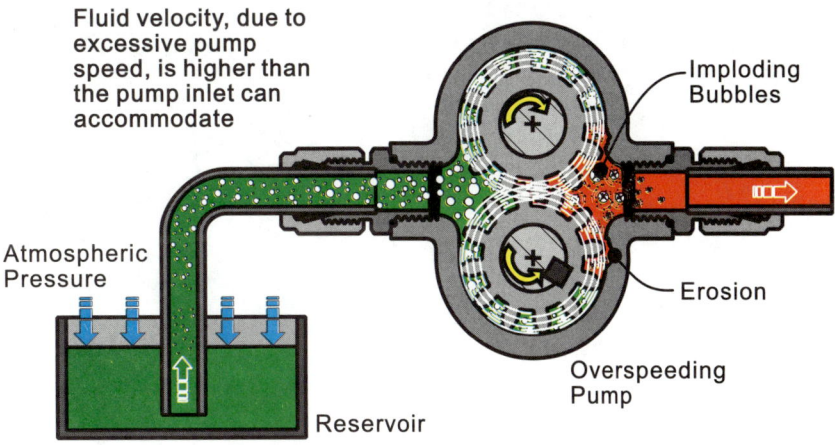

Figure 11.25 Cavitation can be induced by overspeeding the pump.

Contamination, installation/application, and aeration/cavitation failures are generally preventable. Proper hydraulic system design, operation, and maintenance lead to long lives in today's well-designed pumps.

Chapter 11 Fixed Displacement Pumps

Typical Applications

As the supplier of hydraulic power, at least one pump is required in every hydraulic system. Both the fixed displacement pumps discussed in this chapter and the variable displacement pumps discussed in the next chapter are widely used in mobile hydraulic systems.

A common example of a fixed displacement pump application on a mobile vehicle is the power steering pump. It provides hydraulic fluid flow from the reservoir to the steering valve.

Another example would be the use of a pump to supply the flow to valves which control the position of forks on a lift truck. The valves direct pump flow to the cylinders which lift, tilt, and shift the load.

Even when a variable displacement pump is being used, it is common to have a low-pressure, fixed displacement pump to act as a charge pump, which provides fluid from the reservoir to the variable displacement pump's inlet. This helps prevent aeration and cavitation of the expensive main pump. The charge pump also may supply a low-pressure flow for off-line filtration, circuit replenishment and control activities.

CHAPTER 12 .. Variable Displacement Pumps

Fixed displacement pumps discharge a set volume of fluid regardless of the system requirements. This volume can be changed only by changing the drive speed of the pump. If the system requires less fluid than the pump is discharging, the balance of the flow must find an alternate path, which is usually over a relief valve and back to the reservoir.

This amount of excess flow, at the relief valve pressure setting, results in lost energy to the system and heat being added to the fluid and to the reservoir. The amount of heat can be approximated by the formula:

Formula 12-1

BTU/Hr = ΔP x GPM x 1.5

Where:

BTU = British Thermal Unit
ΔP = the relief valve setting
GPM = the flow passing over the relief valve
1.5 = a constant to resolve the units

To conserve energy and to prevent generating heat, variable displacement pumps are used. These pumps provide a means of changing the pump's displacement so that sufficient fluid is provided to the system, and no more. Formula 11-1 shows that if the flow passing over a relief valve can be reduced, the heat generated is reduced. Furthermore, the power required to drive the pump is also reduced, based on the formula:

Formula 12-2

$$HP = \frac{P \times GPM}{1714 \times Eff}$$

Where:

HP = the horsepower to drive the pump
P = the pump output pressure
GPM = the pump output flow
Eff = the pump overall efficiency
1714 = a constant to resolve the units

Reduced input power, reduced energy loss and less heat generation make the concept of a variable displacement pump quite attractive.

Chapter 12 Variable Displacement Pumps

Metering Losses Most mobile equipment requires a variable rate of operation of actuators. Figure 12.1 illustrates two different operations of an excavator that require markedly different rates of flow to the actuators; high speed operation while moving dirt, and very slow operation while setting pipe. A fixed displacement pump that provides sufficient flow for the high speed operation will generate large amounts of lost energy and heat when the actuators are moved very slowly.

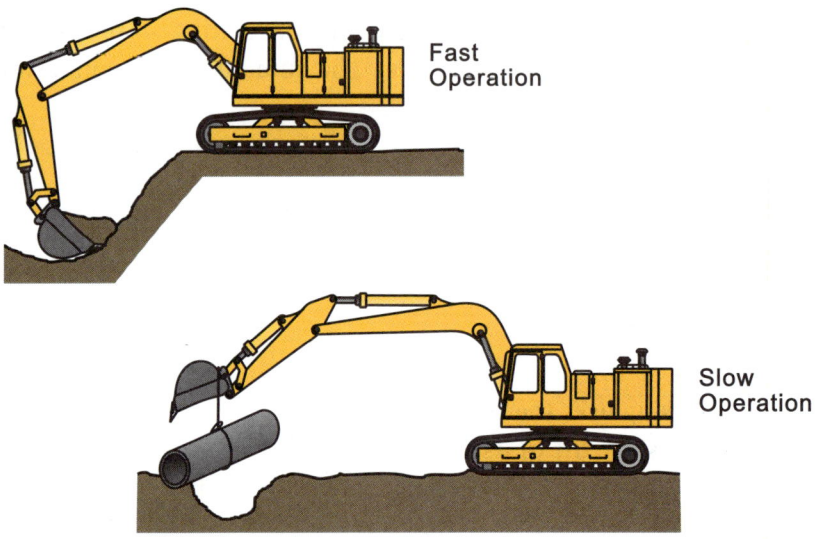

Figure 12.1 Typical fast and slow operations of an excavator

This slow operation is accomplished by the directional control valve, which is called upon to "meter" the flow to the actuators. The directional control valve essentially becomes a non-pressure-compensated flow control valve that restricts the amount of fluid into the actuator circuits. The balance of the flow passes over a relief valve. This "metering" activity can waste large amounts of energy, even though it is a frequent operation in most mobile machinery. An example of the magnitude of energy loss that can occur, from the fuel to the work being done, is illustrated in Figure 12.2.

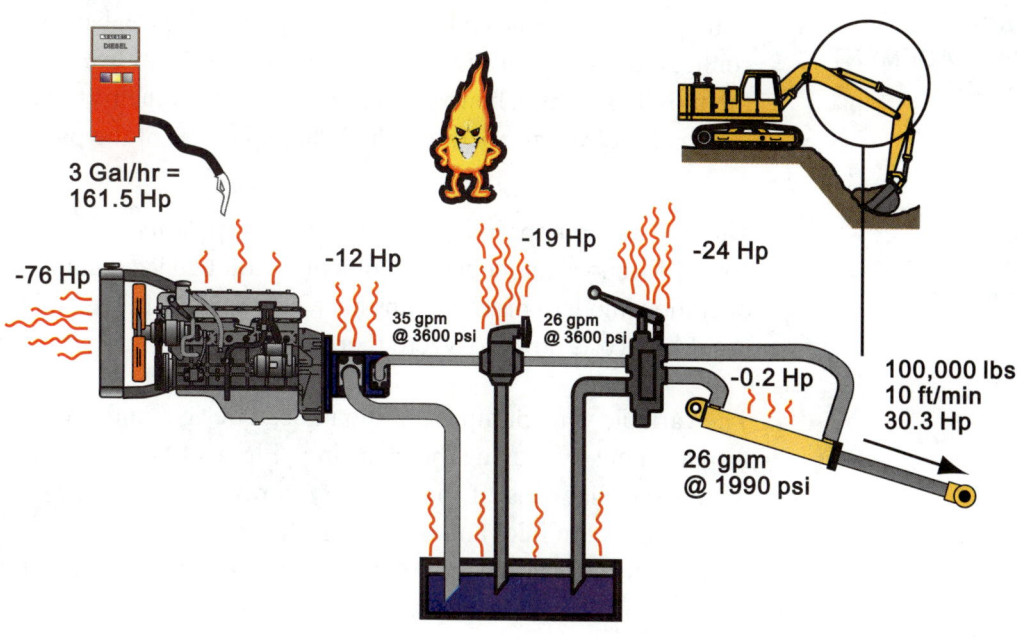

Figure 12.2 Potential energy losses during a metering situation

Figure 12.3 shows the relationship between the amount of energy being used by the hydraulic system (the "corner horsepower" point), to the amount of energy going into work (the "metering" point). All energy above and to the right of the metering point is wasted.

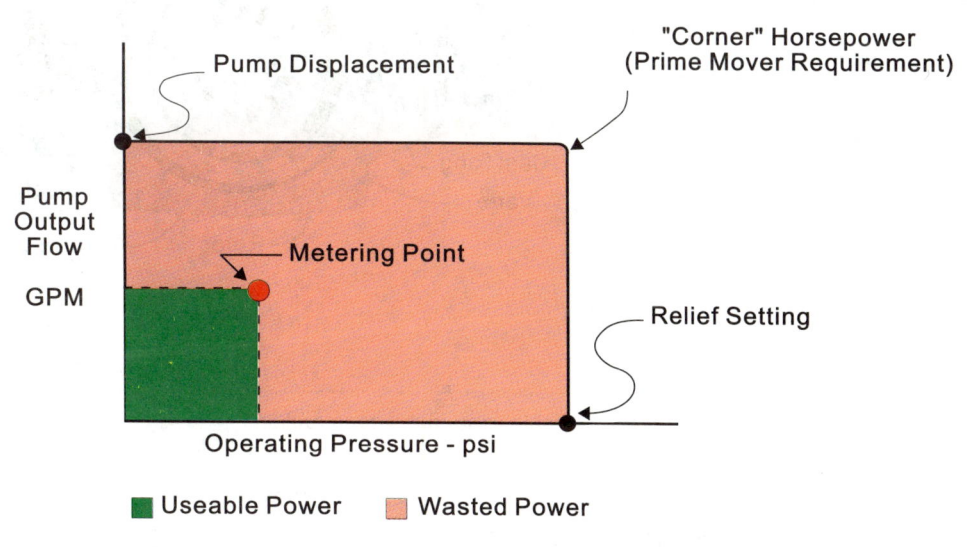

Figure 12.3 Metering with a fixed displacement pump circuit

By controlling displacement, the flow of fluid out of the pump is matched to the system flow demands, and metering losses are significantly reduced.

Chapter 12 Variable Displacement Pumps

VARIABLE DISPLACEMENT PUMPS

Vane pumps and axial and radial piston pumps are available in variable displacement configurations. By far the most popular type is the in-line design axial piston pump. In the most basic sense, however, they all accomplish their varying displacement by altering the change of volume between the inlet and outlet of the pump.

Variable Displacement Vane Pump

Variable vane pumps have some degree of popularity in industrial markets, but few, if any, applications on mobile equipment. Although vane pumps are noted for quiet operation, modern piston pumps have also become quiet, and high production volume has driven costs down to where there is little benefit to the variable vane design.

The variable vane pump is an unbalanced design, and creates the changing displacement by moving the cam ring. Figure 12.4 shows the movable cam ring and the pistons that cause it to move. By moving the cam to the left, the differential volume between inlet and outlet is reduced, because the eccentricity between the cam and rotor is reduced. As the cam is moved back toward the right in Figure 12.4, pump displacement is increased.

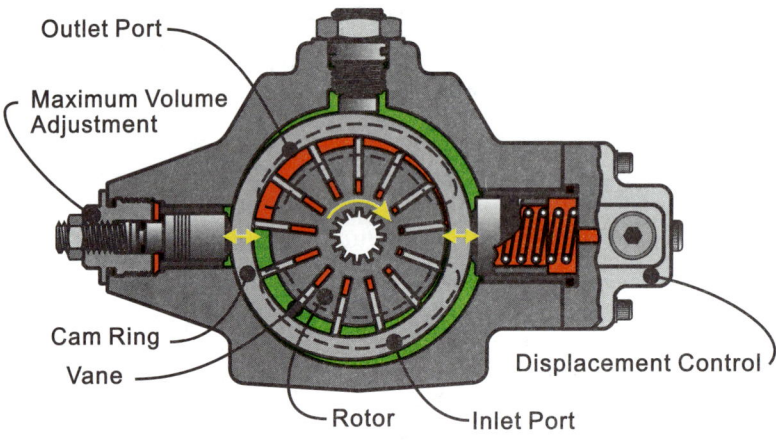

Figure 12.4 Variable vane pump construction

The piston on the left is called the "bias" piston, and attempts to maintain maximum displacement at all times. The piston on the right is called the "control" piston, and has a larger diameter (area) than the bias piston. When system pressure reaches a preset level, the control piston will force the cam to the left; because the control piston is larger, it will overcome the bias piston and the pump displacement will be reduced.

The mechanism that operates the control piston is called a "compensator". There are many versions of compensators, the more common of which will be covered later in this chapter.

Variable Displacement Radial Piston Pump

Figure 12.5 shows a typical radial piston pump that has been modified for variable displacement. It works on the same principle as the variable vane pump, in that the outer cam is moved left or right to change eccentricity with the cylinder block, and change piston stroke which changes displacement. A bias piston on the left, and a larger control piston on the right, move the cam left or right to change displacement.

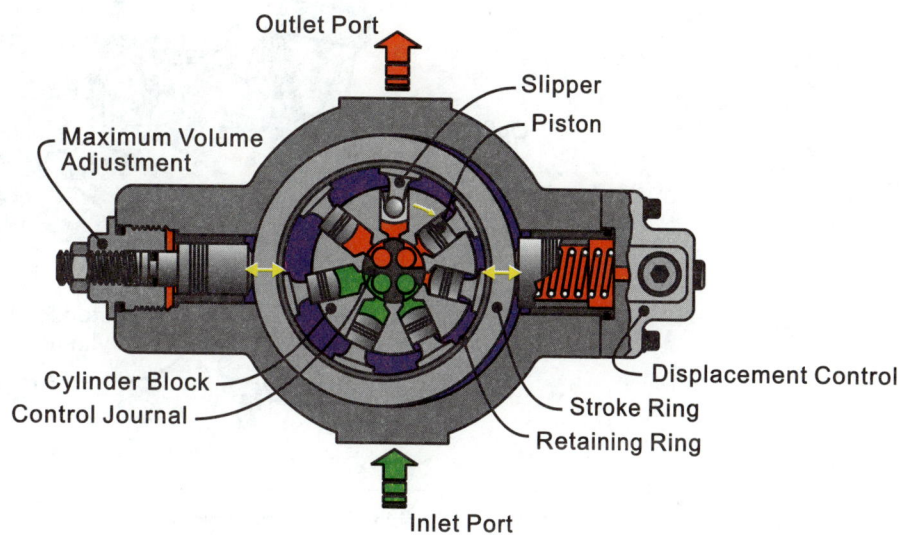

Figure 12.5 Variable displacement radial piston pump construction

Variable displacement radial piston pumps operate very well, and are usually very reliable and durable. They perform equally as well at high pressures and low pressures, although cost usually prohibits their use in low pressure applications.

Chapter 12 Variable Displacement Pumps

Variable Displacement Bent Axis Piston Pump

Figure 12.6 shows a cross-section view of a bent-axis axial piston pump. Rotation of the drive shaft causes the pistons and cylinder block to rotate, which in turn causes the pistons to move in and out of the cylinder bore. As the angle between the cylinder block and drive shaft (Figure 12.7) changes, the piston travel in their respective bores changes, increasing or decreasing the pump displacement.

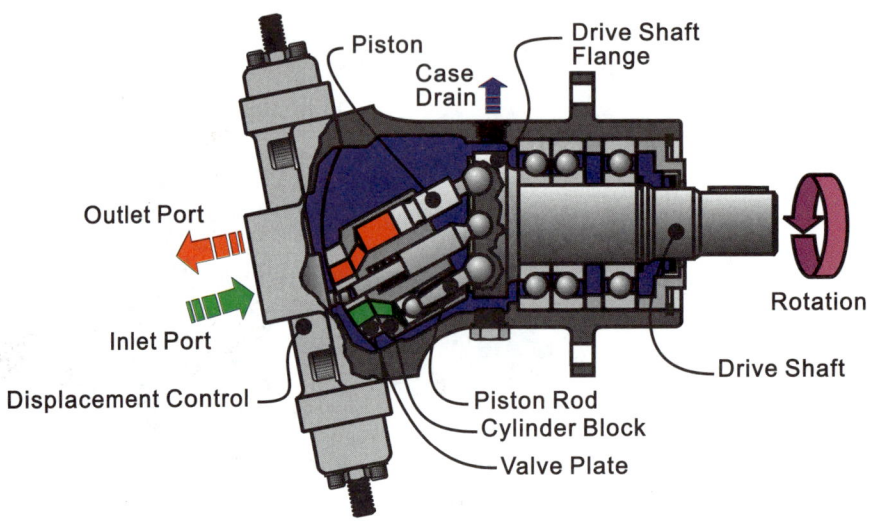

Figure 12.6 Bent axis variable displacement piston pump

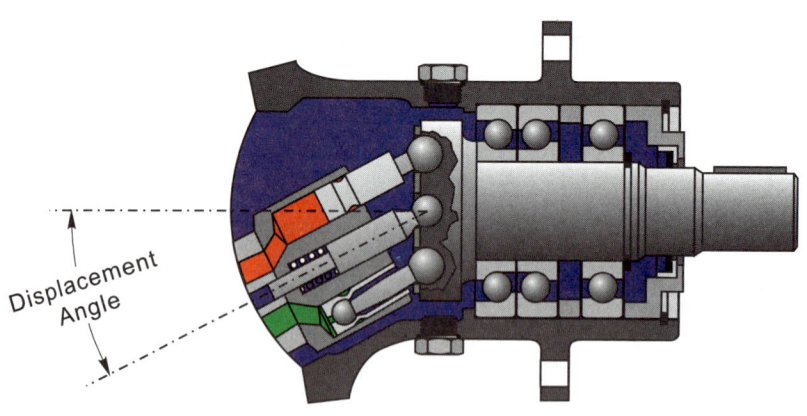

Figure 12.7 Bent axis pump output is in direct proportion to the displacement angle

The displacement angle is controlled by the displacement control valve mounted on the rear of the pump. The displacement control valve consists of a compensator, a bias piston and a control piston. The bias piston attempts to keep the cylinder block at its maximum displacement angle at all times, and the control piston, being of a larger diameter (area), will stroke the cylinder block to a lesser displacement angle as the system requires.

The displacement may also be controlled manually by operating the displacement control valve with a lever. This is more popular when the pump is used in a hydrostatic transmission arrangement, where the handle becomes the forward-reverse and speed control. The displacement angle of the cylinder block is still governed by the control piston, however.

Variable Displacement Inline Piston Pump

The most popular design of piston pump is the inline design, so called because the pistons and cylinder block are in line with the drive shaft. Figure 12.8 shows the individual parts that make up the inline design.

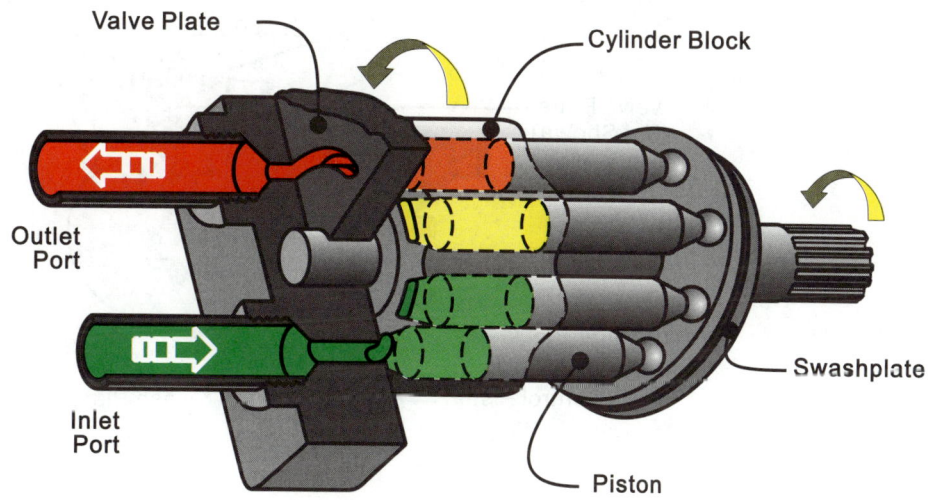

Figure 12.8 Inline design piston pump

Rotation of the drive shaft causes the cylinder block and pistons to rotate, which causes the pistons to move in and out of their bores as they slide over the angled swash plate. The swash plate is supported by a yoke, which pivots on bearings. Two types of bearings are prevalent; either a pair of roller or needle bearings on pintles, or a bushing that supports a saddle-type yoke. These two bearing arrangements are illustrated in Figure 12.9. Pivoting the yoke changes the angle of the swash plate, which changes the stroke of the pistons in and out of their bores. Displacement of the pump decreases with a decrease in the swash plate angle.

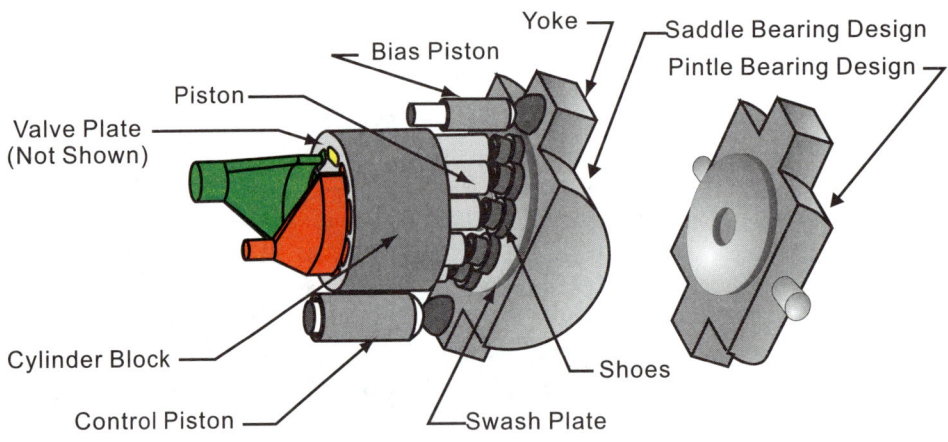

Figure 12.9 Inline design variable displacement piston pump components

The yoke can be pivoted manually or hydraulically, the most common being hydraulic. Only smaller pumps, up to about 2.5 in^3/revolution, can be stroked manually due to the high internal forces on the yoke with larger pumps. Manual operation is popular on smaller hydrostatic transmissions, where the handle becomes the forward-reverse and speed control lever.

Hydraulic stroke control is accomplished with a control piston working against a bias piston or a bias spring. These two arrangements are illustrated in Figure 12.10. The bias spring is quite common, although it is limited to pumps with displacements of about 5 in^3/revolution due to the high yoke forces in larger pumps. Both the bias spring and the bias piston perform the same function, to keep the yoke at maximum angle when the control piston is not forcing it back.

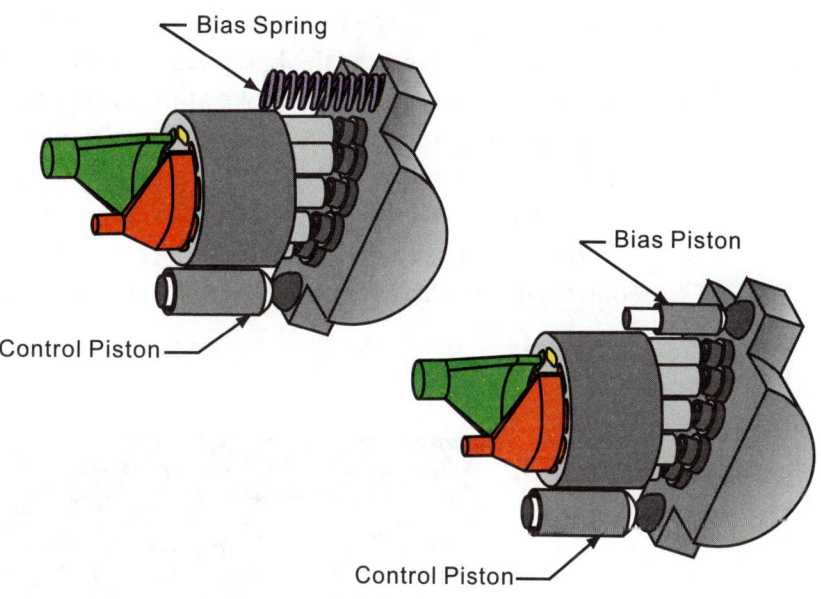

Figure 12.10 Inline piston rotating group showing bias piston and bias spring designs

INLINE PISTON PUMP CONTROLS

Controls for inline piston pumps, called compensators, have one single purpose; to pressurize the control piston and force the yoke to reduce its angle, reducing pump displacement. There are many types of compensators for regulating different parameters of a hydraulic system, but the only control function they perform is reducing pump flow.

Two broad categories of compensators are hydraulic and electrohydraulic. Both types include a valve, usually mounted on the pump, that regulates flow and pressure to the control piston. The hydraulic compensator reacts to hydraulic system parameters such as pressure or flow, and the electrohydraulic compensator reacts to electronic signals that may come from a controller, a transducer, a power amplifier, a control card or computer, etc.

Chapter 12 Variable Displacement Pumps

Pressure Limiting Compensator

The most basic compensator is the pressure limiting compensator, frequently called a pressure control or a pressure compensator. This compensator constantly senses pump outlet pressure. When a preset pressure level is reached, the compensator applies flow and pressure to the control piston to reduce pump displacement. Displacement will continue to reduce, reducing pump output flow until it is at the minimum level required to maintain the preset pressure. If no flow is required by the system, such as when a cylinder has reached the end of its stroke, the pump output flow will be zero.

The operation of a pressure limiting compensator is illustrated in Figure 12.11. The pump rotating group (cylinder block and pistons) discharges fluid to the outlet (1) which is fed constantly to the bias piston (2), the compensator (3) and the system (7). The bias piston holds the yoke at its maximum angle, causing the pump to discharge maximum flow. Any resistance to the flow, due to the valve (7) or the actuator, will create a pressure at the pump outlet (1). This pressure will be transmitted into the pressure limiting spool (4) area, and attempt to move the spool to the right against the adjustable spring (5). The spring can be preset (6) to any equivalent pressure up to the maximum rating of the pump.

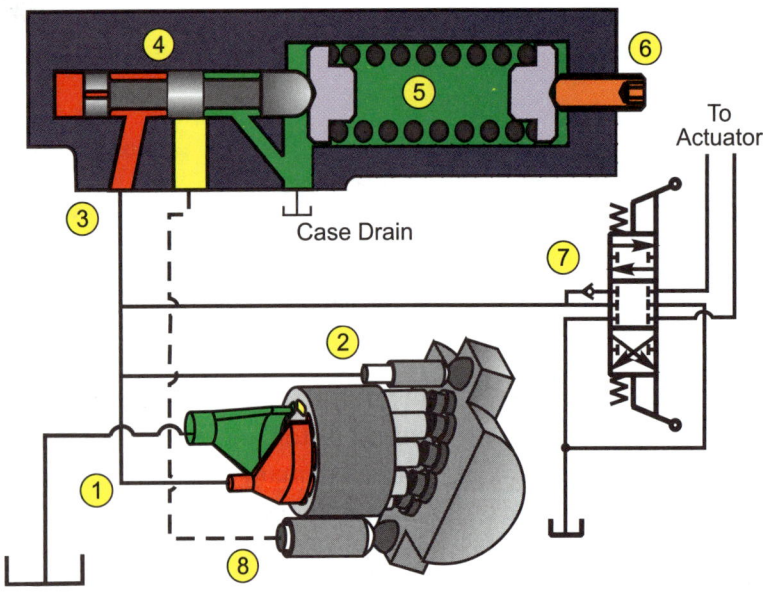

Figure 12.11 Pressure limiting compensator construction and operation

As system pressure rises due to metering of the valve (7) or due to higher pressure requirements at the actuator, increasing pressure will eventually achieve the setting of the adjustable spring (5), and the spool (4) will begin to move to the right, opening the passage to the control piston (8). This will cause the pump to reduce stroke, reducing flow into the system.

As long as system pressure remains above the compensator spring setting, the control piston will continue to force the yoke to a lower angle, reducing flow out of the pump. Flow will continue to reduce until the system pressure is equal to the compensator spring setting. The compensator will then regulate the flow at that level until the system parameters change (that is, until more or less flow is required). Figure 12.12 shows how the system flow has been reduced to the metering point level, and also shows the energy savings that resulted.

If the system flow requirement is reduced to "zero", such as when the control valve is in neutral or the actuator reaches the end of its stroke, the pressure limiting compensator will cause the pump to reduce the stroke to "zero". This can happen very quickly: typical reaction times may range from 25 to 75 milliseconds (.025 to .075 seconds) to stroke from full displacement to "zero" displacement. In practice, the pump does not stroke to a true "zero", as a small amount of flow is required for lubrication and leakage within the pump.

Load Sensing Compensator

Figure 12.12 illustrates the reduction of energy waste by the use of a pressure limiting compensator during a metering operation. There is still a significant amount of lost energy, however, because the pump is still operating against maximum system pressure. A load sensing compensator addresses this element of waste.

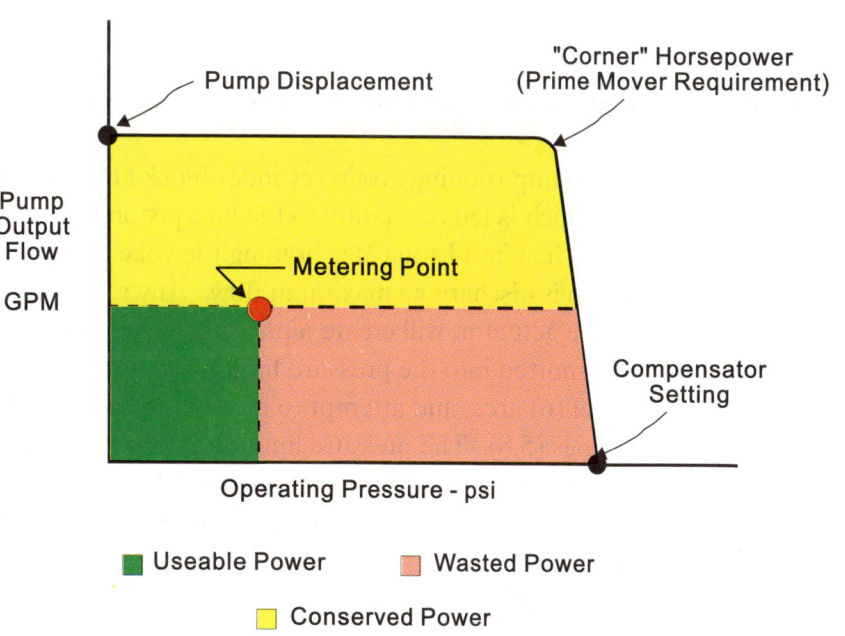

Figure 12.12 Metering with a variable displacement pump and a pressure limiting compensator

Figure 12.13 shows a combination load sensing and pressure limiting compensator. The pressure limiting component, the bottom spool (4), operates the same as described earlier in Figure 12.11. The upper spool (6) is the load sensing component.

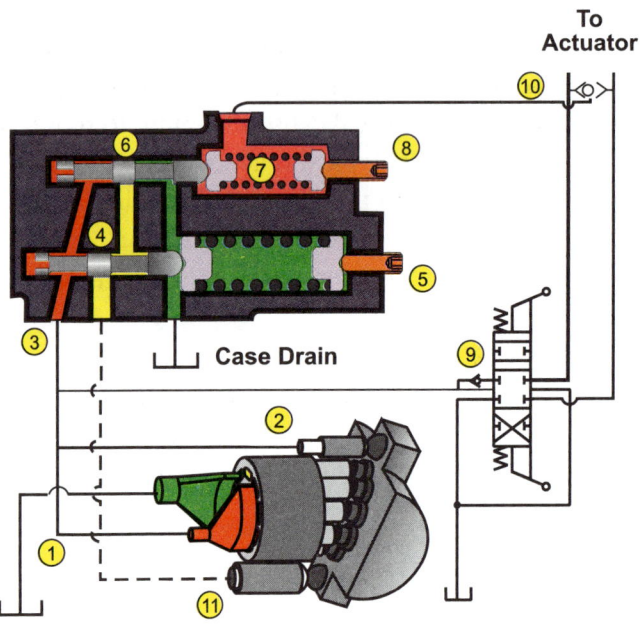

Figure 12.13 Load sensing and pressure limiting compensator construction and operation

The pump rotating group (cylinder block and pistons) discharges fluid to the outlet (1) which is fed constantly to the bias piston (2), the compensator (3) and the system (9). The bias piston (2) is holding the yoke at its maximum angle, such that the pump is discharging maximum flow. Any resistance to the flow, due to the valve (9) or the actuator, will create a pressure at the pump outlet (1). This pressure will be transmitted into the pressure limiting spool (4) area, as well as the load sensing spool (6) area, and attempt to move both spools to the right against their adjustable springs (5,8). The pressure limiting spring (5) can be preset to any equivalent pressure up to the maximum rating of the pump. The load sensing spring (7) can be set to a nominal pressure drop required for proper operation of the directional control valve (9). This "stand-by" pressure is normally factory set in the range of 150-450 psi.

Chapter 12 Variable Displacement Pumps

The work operation may cause the pressure in either of the actuator lines to rise (depending on which direction the actuator is moving). The shuttle valve (10) detects the highest of the two actuator line pressures, and transmits this pressure to the spring chamber (7) of the load sensing spool. This pressure, plus the equivalent pressure of the load sensing spool spring, now acts upon the load sensing spool to regulate system pressure (1) by metering fluid and pressure to the control piston (11). Pump outlet pressure (1) will be regulated at the load pressure (10), plus the value of the load sensing spring (7). The pressure drop across the directional control valve will therefore be held constant at the value of the load sensing spring (7) regardless of changing load pressures.

When there is no work load, and the actuator pressure is zero, there will be no pressure added to the spring chamber (7). The load sensing spool will then regulate the pump at a "standby" pressure governed by the load sensing spring only. This will become a low power standby condition for periods when no work is being performed.

If working pressure exceeds the pressure limiting spring setting (5), the pressure limiting spool will move to the right and the pump will now be regulated at the maximum pressure setting.

By regulating the system pressure to a level slightly above the load pressure requirement, an additional element of energy savings has been accomplished during the metering phase, as shown in Figure 12.14.

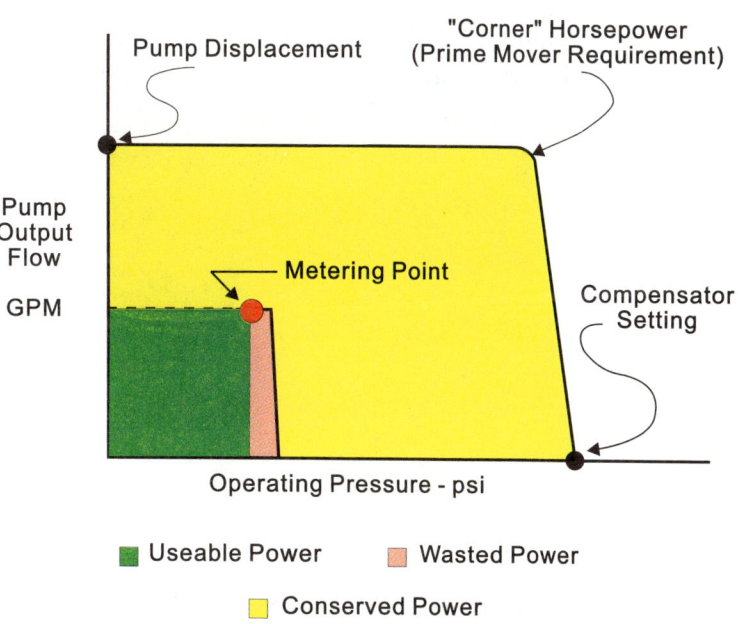

Figure 12.13 Load sensing and pressure limiting compensator construction and operation

Chapter 12 Variable Displacement Pumps

Torque Limiting Compensator

The upper right corner of the curves in Figures 12.3, 12.12 and 12.14 shows the maximum power required by the hydraulic circuit, called "corner horsepower". This is a maximum pressure, maximum flow requirement of the pump for which the prime mover must be sized to prevent stall. The relationship in formula 12-2 illustrates how pressure and flow together result in system horsepower. When both pressure and flow are at their highest level, maximum horsepower is absorbed by the system, and the prime mover must be large enough to deliver this horsepower.

Many types of mobile operations, however, never require maximum flow and maximum pressure at the same time. Examples are:

Excavators: maximum pressure is required for severe digging operations, such as heavy clay, rock, concrete, etc., but this is a slow operation where high hydraulic system flow is not required. On the other hand, moving dirt out of a hole is frequently a high speed operation, requiring high system flow, but lower pressures. The two types of operation never combine to require high pressure and high flow at the same time.

Cranes: high speed operations are frequently convenient during some swing movements, cable payout or boom lifting. When heavy loads requiring high system pressures are involved, however, movements are slow in order to maintain control of the load and prevent equipment damage. It is not only unnecessary to have high flow and high pressure at the same time, it is occasionally unsafe.

A torque limiting compensator is designed to prevent high pressure and high flow occurring at the same time. Either one can occur independently, that is high flow <u>or</u> high pressure, but not simultaneously. The result is a lower "corner horsepower" requirement, a smaller prime mover requirement and protection against stalling of the prime mover.

Figure 12.15 shows a version of a torque limiting, pressure limiting and load sensing compensator. The upper portion of the assembly (8) is the same pressure limiting and load sensing compensator as described earlier (see Figure 12.13) and operates in the same way to provide the energy conservation characteristics shown in Figure 12.14. The load sensing signal (13) is provided from a shuttle valve which selects the highest of the actuator line pressures.

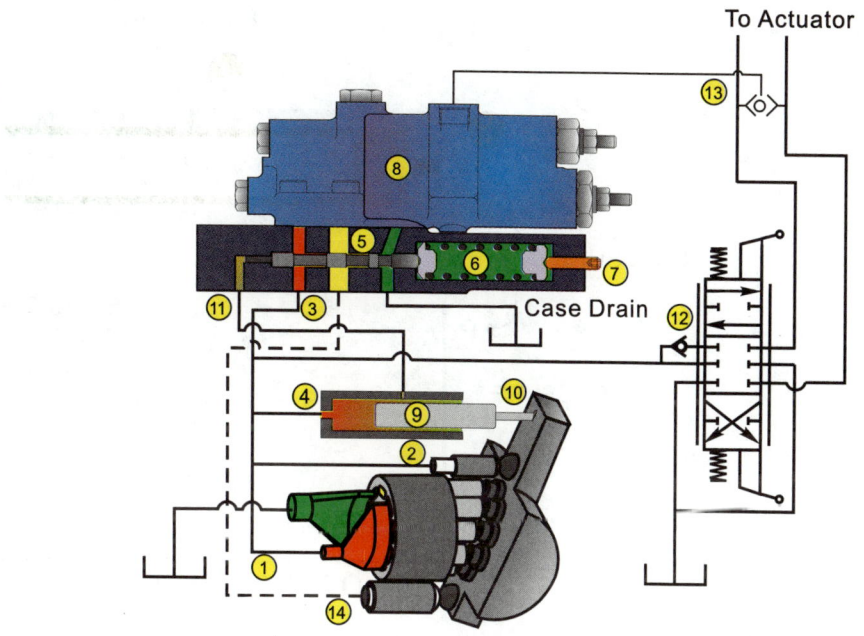

Figure 12.15 Torque limiting compensator with pressure limiting and load sensing

A third spool (5) has been added to the assembly, which is the torque limiting spool.

The pump rotating group (cylinder block and pistons) discharges fluid to the outlet (1) which is fed constantly to the bias piston (2), the compensator (3), the yoke position sensing piston (4) and the system (12). The bias piston (2) is holding the yoke at its maximum angle, such that the pump is discharging maximum flow. Any resistance to the flow, due to the valve (12) or the actuator, will create a pressure at the pump outlet (1). This pressure will be transmitted into the torque limiting spool (5) area, as well as the yoke position sensing spool (4) area. The pressure entering the compensator (3) will also enter the load sensing and pressure limiting spools (8) in the same manner as shown in Figure 12.13.

Chapter 12 Variable Displacement Pumps

Figure 12.16 is a cross section of the yoke position sensing piston (item 9, Figure 12.15). It is attached directly to the pump yoke such that it will move right (out of) its bore as the yoke is rotated to maximum angle, and move left (into) its bore as the yoke is rotated to minimum angle. The pressure being applied to the left side of the piston (1) comes from the outlet side of the pump rotating group. The right side of the piston (3) drains into the pump case, which is drained to the reservoir. This pressure is consistently very close to zero, and is considered zero.

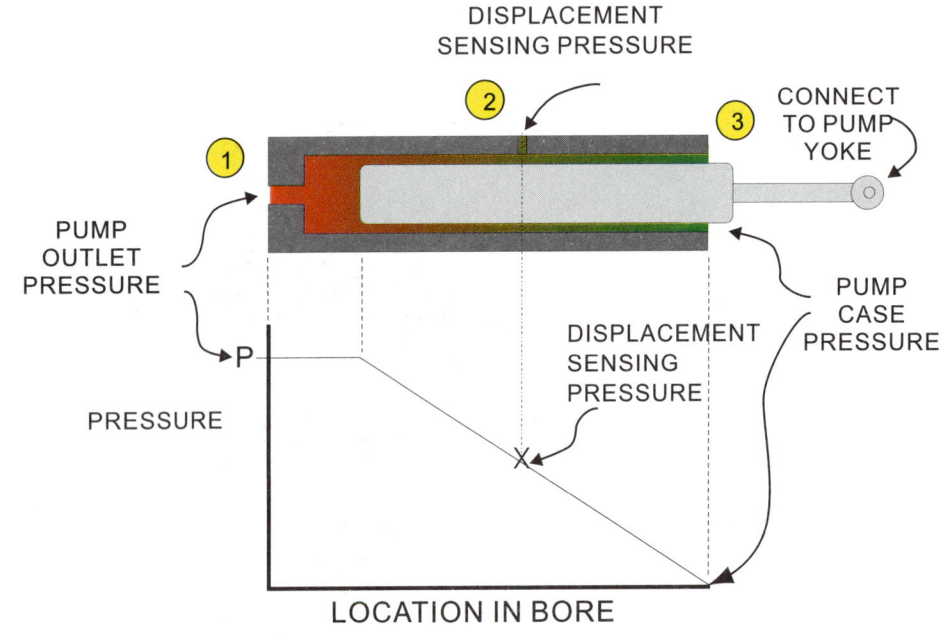

Figure 12.16 Position sensing spool is connected to the yoke, providing a pressure to the pump indicting yoke position.

There is a small clearance between the yoke position sensing piston and the bore, such that there is a gradual decline in pressure from pump outlet pressure (on the left side) to zero (on the right side). The pressure that will be detected at the displacement sensing port (2) will be in direct proportion to the pump outlet pressure and the location of the piston. When the yoke is rotated to its maximum stroke angle, the yoke position sensing piston will be furthest to the right (as in Figure 12.17), providing a pressure at port (2) very close to pump outlet pressure. When the yoke angle is small, the yoke position sensing piston will be furthest to the left, providing a pressure at port (2) closer to zero.

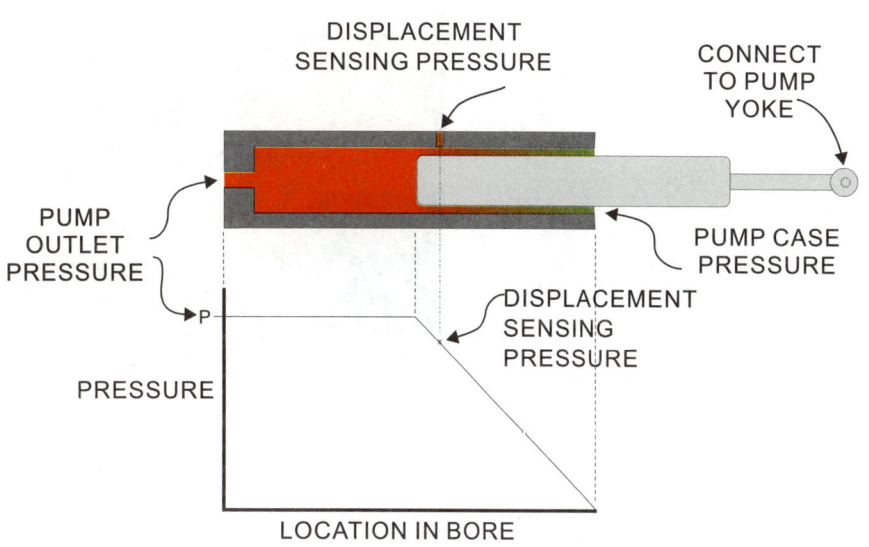

Figure 12.17 Fully angled yoke and fully extended piston results in pump outlet pressure being transmitted to torque limiting spool

The pressure sensed at port (2) is fed to the torque limiting spool (5) through port (11) (Figure 12.15). The intensity of this pressure depends on the position of the yoke and the pump outlet pressure. When the outlet pressure is high and the yoke is at its maximum displacement angle, the pressure at the torque limiting spool (5) will be great enough to force the spool to the right against the adjustable spring (6). Flow will then be metered to the control piston (14), which will reduce the displacement of the pump by rotating the yoke to a lesser angle. As the yoke angle decreases, the pressure from the yoke position sensing piston (9) will reduce until the torque limiting spool returns to the left. The spool (5) will meter flow to the control piston (14) to maintain the reduced displacement position until the pump outlet pressure (1) reduces (due to a drop in load requirement). When pump outlet pressure reduces sufficiently, the torque limiting spool will return to the left, allowing the yoke to rotate to full displacement.

Chapter 12 Variable Displacement Pumps

The pressure at which the torque limiting spool begins to regulate a lower displacement is set by the spring adjustment (7), which would be set based on the peak horsepower that the prime mover can be subjected to. When properly set, the machine will be able to produce maximum speed (flow) or maximum force (pressure), but not both at the same time. Figure 12.18 illustrates the movement of the "corner horsepower" point. The reduced horsepower that results can mean a smaller prime mover, saving machine size and fuel consumption.

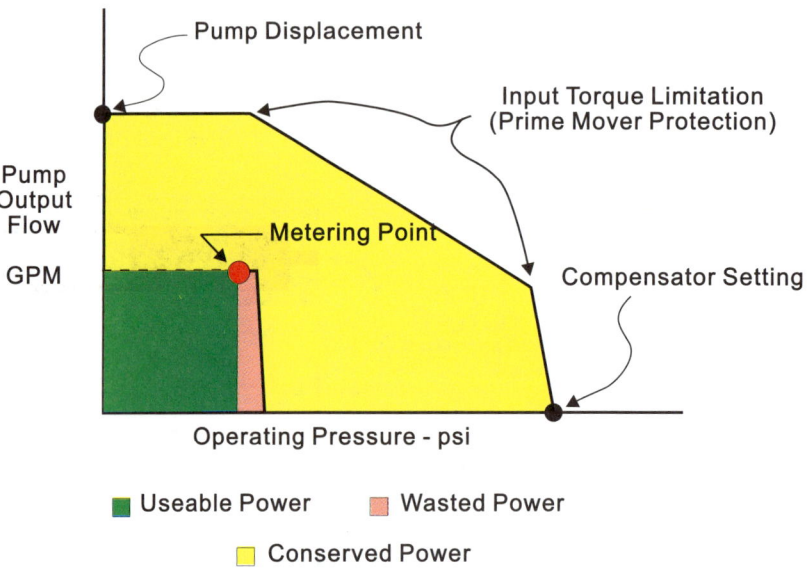

Figure 12.18 Torque limiting reduces the demands for and protects the prime mover

Electrohydraulic Compensator

The electrohydraulic compensator may also be referred to as an electronic displacement control, or EDC. As with all other compensators, its purpose is to control the output of the pump by controlling its displacement. The EDC provides for complete control by a variable voltage signal.

Figure 12.19 is a cross section view of an EDC compensator consisting of a basic control valve (5) and an electronically controlled pilot relief valve (7). The pump rotating group (cylinder block and pistons) discharges fluid to the outlet (1) which is fed constantly to the bias piston (2), the compensator (3) and the system (4). The bias piston is holds the yoke at its maximum angle, such that the pump discharges maximum flow. Any resistance to the flow, due to the valve (4) or the actuator, will create a pressure at the pump outlet (1). This pressure will be transmitted into the control spool (5) area, and attempt to move the spool to the right against the adjustable spring (8). The pressure is also transmitted through the fixed orifice (6) into the spring chamber (8) and to the electronically controlled pilot relief valve (7). The control spool spring is preset to a "standby" pressure, usually in the 200-400 psi range.

Chapter 12 Variable Displacement Pumps

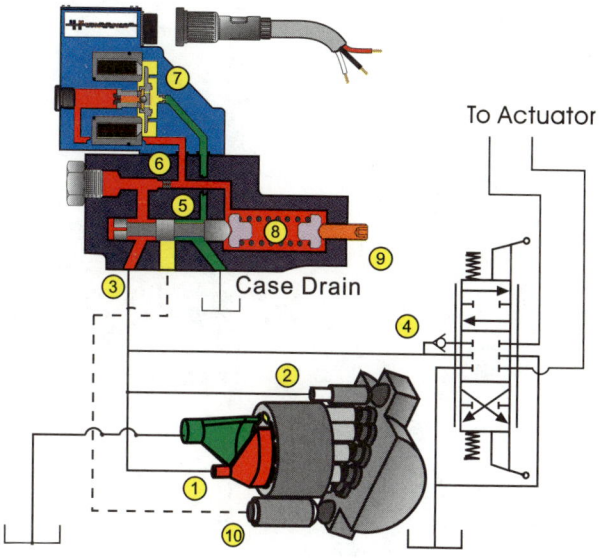

Figure 12.19 Electrohydraulic displacement control for piston pumps

Pressure in the control spool spring chamber (8) is controlled by the pilot relief valve (7). The relief valve will open and meter flow to the case drain (which is directly connected to the system reservoir) in direct proportion to a voltage input. Opening this valve will allow fluid in the control spool spring area (8) to drain out. Flow is also entering the spring chamber through the fixed orifice (6). By controlling the flow out of the spring chamber relative to the flow in, the pressure in the spring chamber can be controlled. This pressure adds to the equivalent pressure of the spring force, the "standby" pressure, and it is the sum of these two pressures that keep the control spool to the left. As output pressure from the rotating group (1) increases higher than the spring chamber pressure, the control spool will move to the right, metering flow to the control piston (10) to stroke the pump back. Pump displacement will be continuously adjusted to maintain output equal to the pressure in the spring chamber, which is controlled by the level of voltage input to the pilot relief valve.

The voltage input signal to the pilot relief valve can be supplied by any number of sources, as described in Chapter 7.

Chapter 12 Variable Displacement Pumps

CHAPTER 13 — Hydrostatic Transmissions

A hydrostatic transmission is simply a pump and motor connected in a circuit. Other components are added to obtain certain operating features. Each component used in a hydrostatic transmission will be discussed in some detail later in the chapter.

Open circuit and closed circuit hydrostatic transmissions will be discussed, as well as open loop and closed loop control functions. While not intended as a design guide, the numerous formulas and the discussion of the variable effects should prove useful in design decisions and aftermarket diagnosis of machine function and operation.

Basic Configurations

The four basic configurations of hydrostatic transmissions are In-line, U-shape, S-shape and Split:

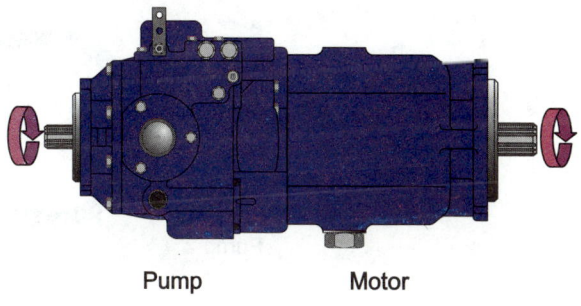

Figure 13.1 In-line configuration

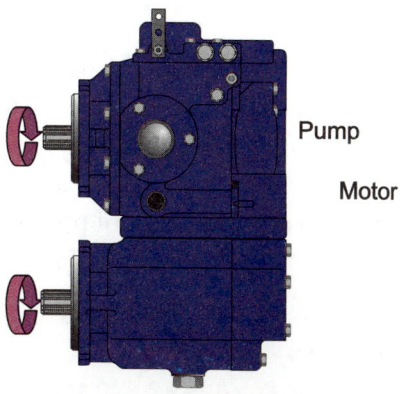

Figure 13.2 U-shape configuration

293

Chapter 13 Hydrostatic Transmissions

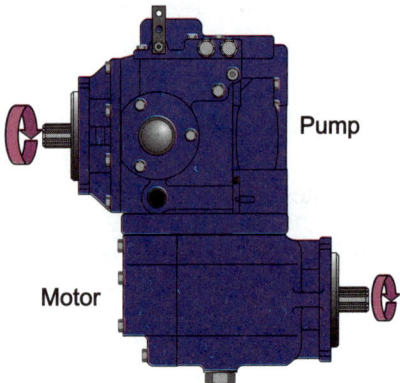

Figure 13.3 S-shape configuration

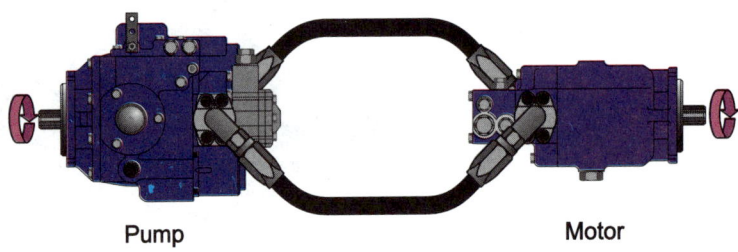

Figure 13.4 Split configuration

Comparison with Mechanical Drives

A vehicle with rear-wheel drive has the components shown in Figure 13.5. To increase speed, the transmission must be shifted to a higher gear ratio. To reverse the direction of travel, the transmission must be shifted to a gear mesh that turns the drive shaft in the opposite direction. Each time the clutch is disengaged to shift the transmission, the flow of power to the drive wheels is interrupted.

Chapter 13 Hydrostatic Transmissions

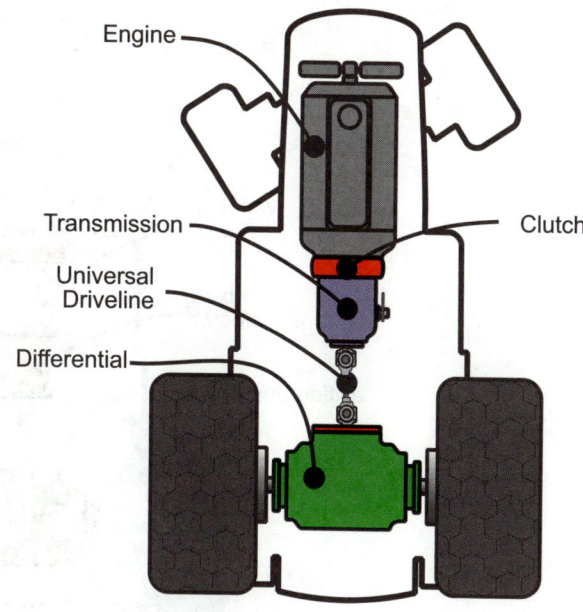

Figure 13.5 Rear wheel drive vehicle with mechanical drive

A differential is needed to allow the outside wheel to rotate faster than the inside wheel when the vehicle turns. If one rear wheel loses traction, the differential will spin that wheel and not transmit power to the other wheel. This disadvantage can be eliminated with a differential locking feature.

The same vehicle, with a hydrostatic transmission, is shown in Figure 13.6. The mechanical transmission has been replaced with a hydrostatic transmission; all other components are the same. To provide a specific example, we specify that the pump is a variable displacement piston pump, and the motor is a fixed displacement piston motor. (Review Chapter 12 for explanation of pump operation and Chapter 5 for explanation of motor operation.) As pump output is increased, motor speed is increased by the increased flow into the motor. The vehicle speed can be increased and reduced by moving the control that positions the pump swash plate. Pump flow can be reversed by moving the swash plate control through the neutral position and displacing it in the opposite direction. The reverse position of the swash plate causes the pump to pump fluid in the opposite direction, which causes the motor to turn in the opposite direction, thus, reversing the vehicle. Vehicle motion can be changed from forward to reverse with a simple hand movement. This maneuverability is often the justification for installing a hydrostatic drive on a vehicle. A good example is a front-end loader, cycling back and forth, loading trucks in a rock quarry. It is tiring and time-consuming for an operator to have to shift a mechanical transmission each time the direction of motion is changed. Vehicle productivity, defined as tons loaded per hour, is increased with a hydrostatic transmission.

Chapter 13 Hydrostatic Transmissions

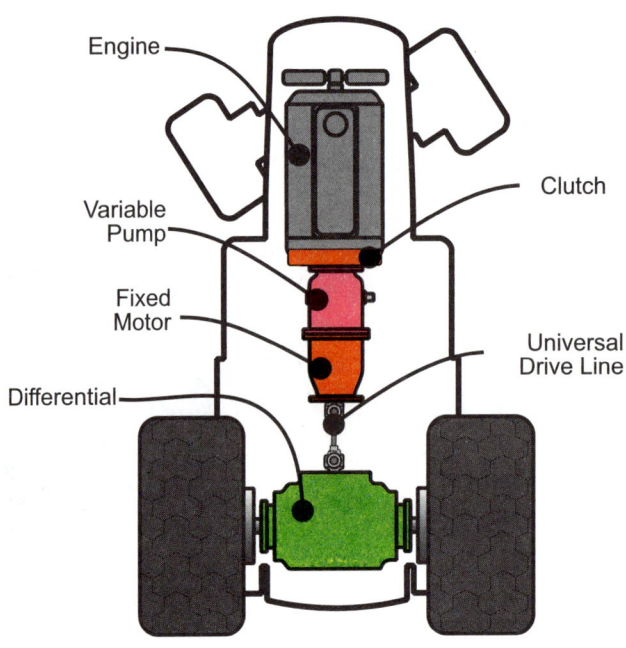

Figure 13.6 Rear wheel drive vehicle with hydrostatic transmission

Advantages of Hydrostatic Transmissions

A hydrostatic transmission, as shown in Figure 13.6, provides improved maneuverability, but at a cost. The efficiency of a hydrostatic transmission is always lower than a mechanical transmission. A mechanical transmission will typically have an efficiency of 92% or greater, meaning that 92% of the input energy is delivered to the load (wheels). A hydrostatic transmission has an efficiency of around 80%. Some well-designed units will have an efficiency slightly above 85%, but none can approach the efficiency of a mechanical transmission. A designer always poses the question: *Does the gain in vehicle productivity offset the loss in efficiency?*

In addition to increased maneuverability, a hydrostatic drive vehicle offers several other advantages:

1. Ability to operate over a wide range of torque/speed ratios. Once a gear ratio is selected with a mechanical transmission, the only speed variation available is that achieved by controlling engine speed. Once the engine speed reaches a maximum, the transmission must be shifted to a lower ratio in order to increase vehicle speed. With a hydrostatic transmission, vehicle speed is continuously variable from a slow creep up to a maximum.

2. Transmit high power per cubic inch of pump/motor displacement with low inertia. When a large mass is rotated at a given speed, it takes an interval of time to change this speed. A hydrostatic transmission adds little inertia to the total rotating mass associated with vehicle operation; consequently, a hydrostatic transmission vehicle tends to change speed quicker (have less inertia) than a mechanical transmission vehicle.

3. a. Provide dynamic braking. A hydrostatic drive vehicle can be stopped by putting it in reverse. Imagine that you are traveling forward and you suddenly move the swash plate control to the reverse position. What will happen? A pressure spike will develop and fluid will flow across the relief valve. The vehicle's mechanical energy will be converted to heat energy and the vehicle quickly slows (probably the wheels will slide). After the pressure drops below the relief valve setting, the wheels will rotate in the reverse direction.

    b. Remains stalled and undamaged under full load. Vehicle hydrostatic transmissions are almost always designed for wheel slip to occur before a relief valve is actuated. The relief valve's role is to clip off pressure peaks and attenuate shocks, as described in part (a). If the vehicle is stuck, the pressure increases until the relief valve opens. Stalling the vehicle in this manner does not damage the transmission. Holding it in a stalled condition, however, causes the fluid temperature to rise, which is undesirable. When a hydrostatic transmission is used for a non-propel application, for example a winch drive, the "remain stalled and undamaged" feature is more important in the overall functionality.

4. No interruption of power to wheels during shifting. Anyone who has watched a driver shift gears while climbing a hill with a heavily loaded truck can appreciate the advantage of continuous power flow over a speed range.

Hydrostatic Transmission with Two-Wheel Motors

Wheel motors, mounted at both rear wheels (Figure 13.7), are a variation of the configuration shown in Figure 13.6. This arrangement eliminates the universal joint drive line, differential, and rear axle, with resultant cost and weight savings. It is understood that the pump mount contains a clutch. With a clutch, the engine can be started under no-load. Because the pump has low inertia, it may be possible to provide enough starting torque to start the engine with a direct-coupled pump. (The swash plate control would have to be set in the neutral position.) A decision to eliminate the clutch must be carefully considered. The remaining configurations, discussed in this section, will show a pump mount, and it is understood that the mount includes a clutch.

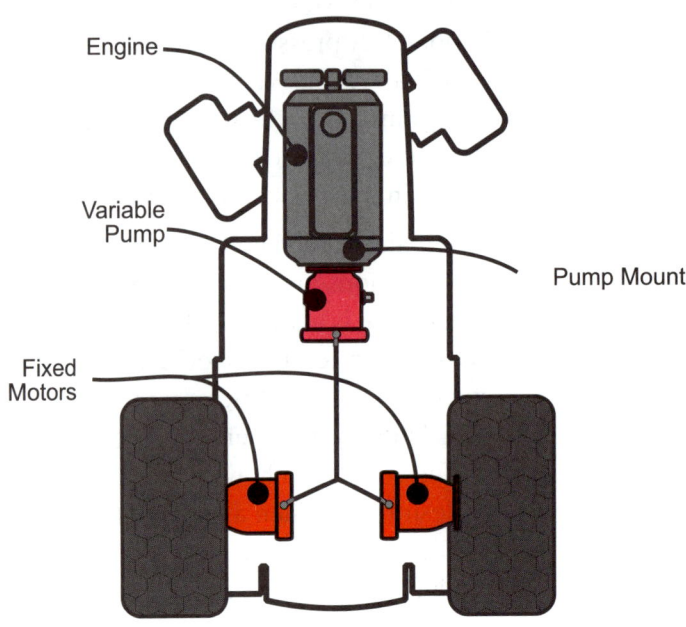

Figure 13.7 Hydrostatic drive with one pump and two wheel motors

In Figure 13.7, the wheels are mounted directly on the shaft of the wheel motors. (Obviously, motors designed for high radial loads must be used.) Hoses or tubing carry the flow from the pump to the motors. A tee is used to divide the flow to the two-wheel motors. When the vehicle turns, more flow goes to the outside wheel as it rotates faster. This action accomplishes the same task as the differential.

With the configuration shown in Figure 13.7, it is important to remember that the speed obtained with a given flow is only one-half of the speed obtained with a single motor. On the other hand, using two wheel motors provides twice the total wheel torque. The configuration shown in Figure 13.7 is useful for a relatively light vehicle that moves slowly but must have high tractive ability. An example would be an agricultural machine used to harvest a vegetable crop.

The wheel motor is quite different from a standard hydraulic motor. It bolts directly to the frame of the vehicle; therefore, the housing has a structural mission relative to the vehicle, in addition to its mission relative to the operation of the motor. A wheel motor is heavy and bulky because of the housing strength required.

As previously mentioned, the wheel is mounted directly on the shaft of the wheel motor. In effect, the wheel motor bearings are the axle bearings for the vehicle. Wheel motors are designed for high radial loads, and care must be taken to insure that the dynamic load does not exceed the radial load rating during vehicle operation.

Chapter 13 Hydrostatic Transmissions

Hydrostatic Transmission with Final Drives

The configuration shown in Figure 13.8 has a two-speed pump mount and the hydraulic motors are mounted in final drives on the rear wheels. Often, it is desirable to provide a road speed for moving the vehicle from one work location to the next. The pump mount is shifted to a higher ratio to drive the pump at higher speed and thus provide the higher flow required for road speed.

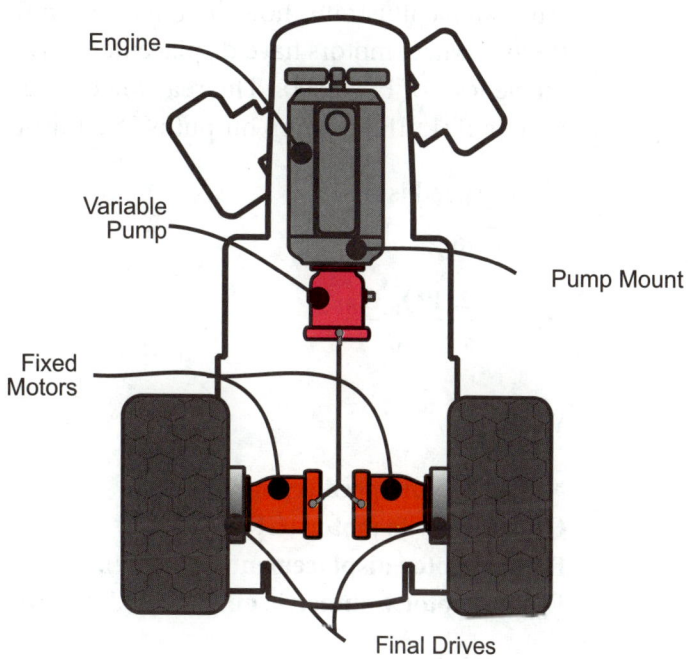

Figure 13.8 Hydrostatic drive with two wheel motors mounted in final drives

An example will illustrate this point. Envision, a transmission where the pump has a displacement $D_p = 1.925$ in.^{3}/rev and is driven with a 1:1 ratio. The pump volumetric efficiency is $E_{vp} = 0.92$. The engine is operated at 2,000 rpm. Total flow from the pump is

Formula 13–1

$$Q_p = \frac{D_p n_p E_{vp}}{231}$$
$$= \frac{1.925\,(2000)\,(0.92)}{231}$$
$$= 15.3 \text{ GPM}$$

If the pump mount is shifted to provide a 1:1.5 ratio, pump speed will be 1.5 x 2000 = 3000 rpm, and pump flow will be

$$Q_p = \frac{1.925\,(2000)\,(0.92)}{231} = 23 \text{ GPM}$$

299

The final drive on the rear wheels is generally some type of planetary gear set. One name given to this gear set is "planetary wheel drive", and other names are "wheel drive", and "power wheel". The housing of the planetary wheel drive mounts to the frame and the wheel mounts to the planet gear carrier, the output side of the final drive. The hydraulic motor bolts to the wheel drive frame and the shaft drives the sun gear, the input side of the final drive. Since the final drive carries the radial load, the hydraulic motor can be a standard high speed motor.

An example illustrates how the final drive influences vehicle performance. Suppose the two-wheel motors have displacements, $D_m = 1.84$ in.^{3}/rev. and volumetric efficiencies of $E_m = 0.90$. The rear wheels are 28 in. diameter and the final drive ratio is 18:1. If the pump output is 15.3 GPM, how fast will the vehicle travel?

Motor speed is

Formula 13-2

$$N_m = \frac{231 \, Q_m E_m}{D_m}$$

Where:

N_m = Motor speed (rpm),
Q_m = Flow to motor (GPM),
D_m = Motor displacement (in.^{3}/rev),
E_m = Motor volumetric efficiency (decimal)

One-half the pump output goes to each wheel motor. For a 15.3 GPM pump output, motor speed is:

$$N_m = \frac{231 \, (15.3/2) \, (0.9)}{1.84}$$

$$= 864 \text{ rpm}$$

Wheel speed is:

Formula 13-3

$$N_w = N_m / G_r$$

Where:

N_w = Wheel speed (rpm),
N_m = Motor speed (rpm),
G_r = Gear ratio
N_w = 864/18 = 48 rpm

Forward speed is:

Formula 13-4

$$v = \frac{2\pi \ (R/12) \ N_w \ (60)}{5{,}280}$$

Where:

v = **Forward speed (mph)**,
R = **Wheel radius (in.)**,
N_w = **Wheel speed (rpm)**

$$v = \frac{2\pi \ (14/12) \ (48) \ (60)}{5{,}280}$$
$$= 4 \text{ rpm}$$

The vehicle can operate at a maximum speed of 4 mph when the pump mount is set in the 1:1 gear ratio. If the pump mount is shifted into the 1:1.5 ratio, then the pump flow will be 23 GPM and the wheel motor speed is

$$N_m = \frac{231 \ (23/2) \ (0.9)}{1.84}$$
$$= 1{,}300 \text{ rpm}$$

Wheel speed is:

N_w = 1,300/18 = 72 rpm

Forward speed is:

$$v = \frac{2\pi \ (14/12) \ (72) \ (60)}{5{,}280}$$
$$= 6 \text{ rpm}$$

Maximum speed for road travel is then 6 mph.

Chapter 13 Hydrostatic Transmissions

Hydrostatic Transmission with Variable Speed Motors

As previously mentioned, it is often necessary to provide a "road speed", so the machine can be moved on the highway between job sites. (Generally, these travel distances are only a few miles, so fast speeds are not necessary.) Shifting a transmission to increase pump speed is one method to increase vehicle speed, and a variable displacement motor is another method. Reducing the motor displacement increases wheel speed. Available wheel torque decreases as displacement decreases, but this is not generally a problem. Torque requirements to propel machines on the highway are relatively low.

A two-speed variable displacement motor has a high displacement position to provide high torque at the work site and a low displacement position for road travel. An infinitely variable displacement motor can also be used, but the cost is higher. In this case, a range of speeds from work speed to road speed is available. Both the two-speed and infinitely variable motors must be sized to prevent driving the wheel at higher than rated speed when they are shifted to the minimum displacement position.

Vehicle with Two Hydrostatic Transmissions

The vehicle shown in Figure 13.9 has a separate in-line hydrostatic transmission for each drive wheel. The engine delivers power via a universal joint drive line to a right angle drive gearbox. Each side of this gearbox powers an in-line hydrostatic transmission. The wheel is connected to the hydraulic motor. A final drive may or may not be used, depending on the vehicle performance criteria.

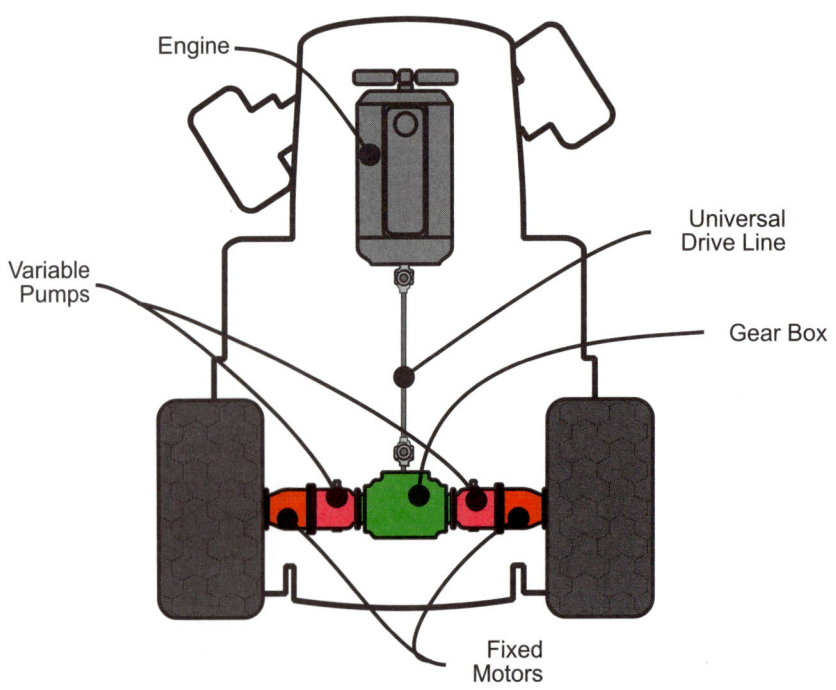

Figure 13.9 Vehicle with separate in-line hydrostatic transmissions for each drive wheel

Steering of this vehicle is an issue. It will not operate like the vehicle shown in Figure 13.8. Suppose, the drive wheels are the rear wheels, and the front wheels are for steering. The swash plate control on both pumps are set for forward travel. When the front wheels turn, both pumps continue to deliver flow for forward travel. There is no differential action. With both rear wheels powering the vehicle forward, they will tend to slide the front wheels sideways and turning will be defeated.

A typical application for this configuration is an agricultural machine called a windrower. This machine cuts hay and rolls it into a continuous pile known as a windrow. The cutting mechanism, or header, is mounted in front of the drive wheels, which are the front (forward) wheels of the machine. The back (rear) wheels are non-steered caster wheels. There is a mechanical linkage from the steering wheel to the swash plate control on both pumps. For straight-ahead travel, the swash plate on both pumps is set at the same position. When the steering wheel is turned, the control on one side is pushed forward, and the control on the other is pushed backward. One pump delivers more flow (the wheel on that side turns faster), and the other pump delivers less flow (the wheel on that side turns more slowly). If the steering wheel is turned far enough, one pump swash plate will be in the full forward position and one will be in the full reverse position. One drive wheel turns forward and the other turns in reverse. In effect, the vehicle "walks" itself around in a tight circle.

The vehicle shown in Figure 13.10 has a separate split hydrostatic transmission for each drive wheel. It does not have a universal joint drive line or a gearbox. Power is transferred via the fluid rather than mechanically.

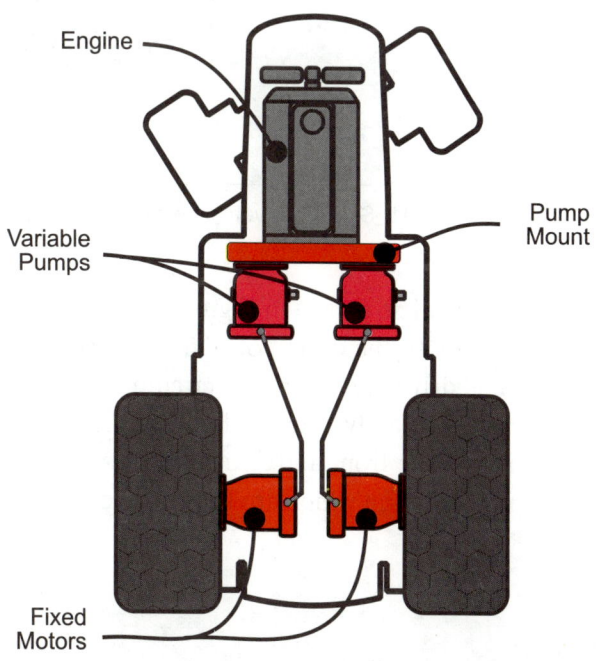

Figure 13.10 Vehicle with separate split hydrostatic transmissions for each wheel

The vehicle shown in Figure 13.11 is a skid-steer machine, and it is a variation of the vehicle in Figures 13.9 and 13.10. Skid-steering can be used for short wheel base machines. Both wheels on each side are connected with a chain drive, consequently both wheels are powered. The machine steering is accomplished as previously described; each swash plate is shifted independently for forward or reverse travel. The wheels slide as the vehicle pivots around. Typically, the swash plates are shifted with a hand lever. The operator can push one lever forward and pull the other back to pivot the machine in a tight circle.

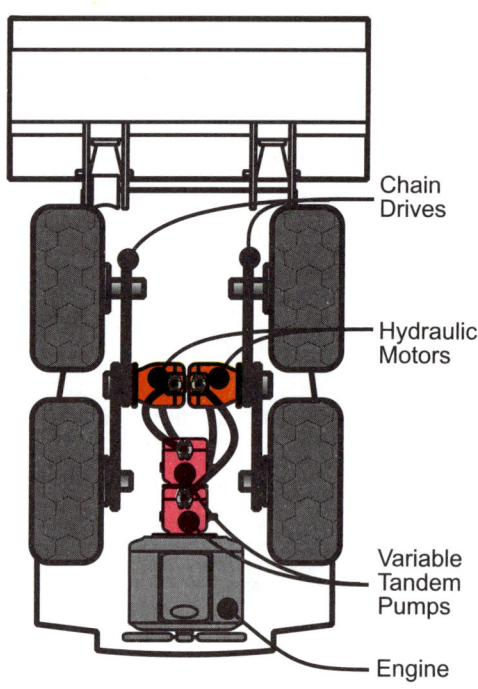

Figure 13.11 Skid steer vehicle with two split hydrostatic transmissions

Hydrostatic Drive for Three-Wheel Vehicle

The vehicle shown in Figure 13.12 has three wheel motors supplied by the same pump. In the design of such a vehicle, care must be taken to size the motors and final drives such that the circumferential speed of the front and rear wheels is approximately equal. The flow does divide in such a manner that the pressure drop across the front and rear motors is approximately equal. There are considerations in the design which are beyond the scope of this discussion.

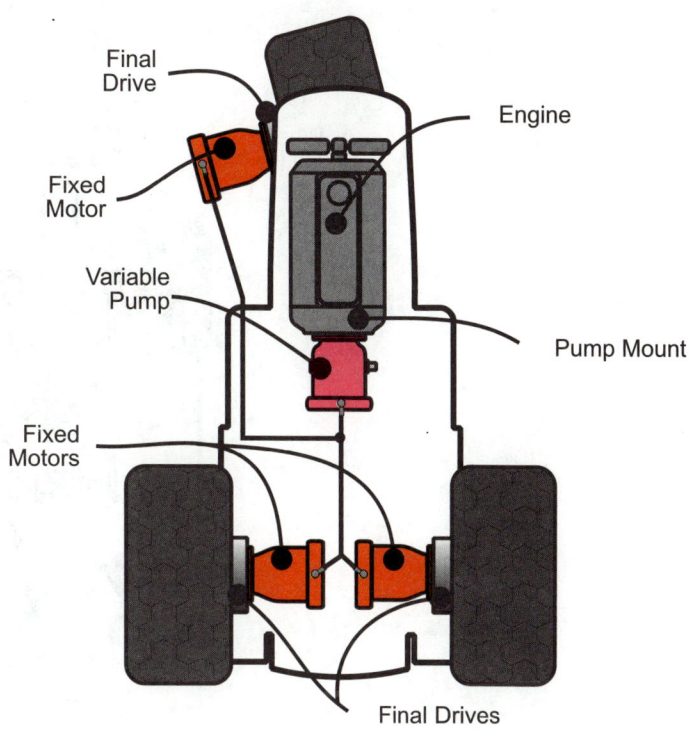

Figure 13.12 Hydrostatic drive with three wheel motors

Hydrostatic Drive for Four-Wheel Vehicle

A configuration in which all four wheels are powered is shown in Figure 13.13. A single pump provides flow to four motors. This machine can be built with two wheels steerable or with all four wheels steerable.

On three-wheel and four-wheel drive vehicles, the front wheels are configured to have a circumferential speed slightly higher (1 or 2%) than the rear wheels. This improves steering and helps to improve tractive effort.

Chapter 13 Hydrostatic Transmissions

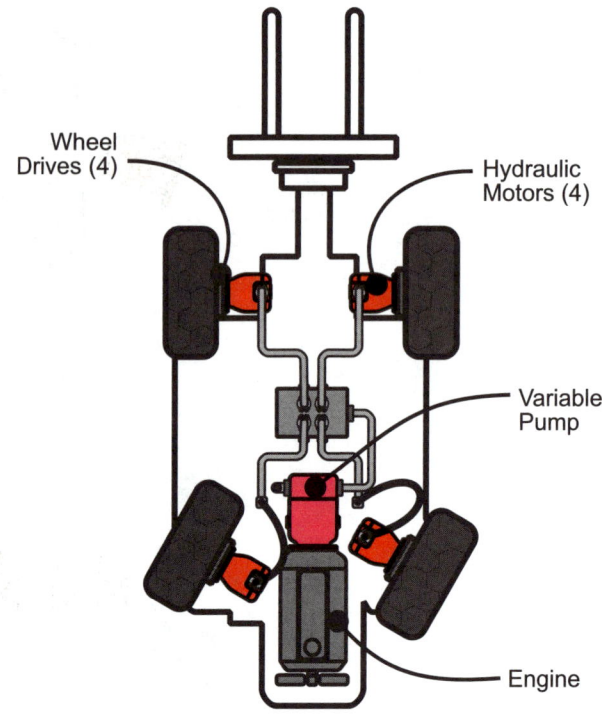

Figure 13.13 Hydrostatic drive with four wheel motors

Summary   The examples described in this section were chosen to give the reader an appreciation of the range of options available to propel a vehicle with a hydrostatic transmission. They are not intended to be a complete list. Also, the examples were somewhat simplified and are not intended as a design guide.

Chapter 13 Hydrostatic Transmissions

Classification of Hydrostatic Transmissions

Hydrostatic transmissions can be classified as shown in Figure 13.14. An open circuit is one in which oil is delivered from the reservoir to the motor by the pump, and flow from the motor returns to the reservoir (Figure 13.15). In a closed circuit hydrostatic transmission, fluid flows from the pump to the motor and back to the pump (Figure 13.16). Provision for insuring that the circuit is always filled with fluid will be discussed in the next section.

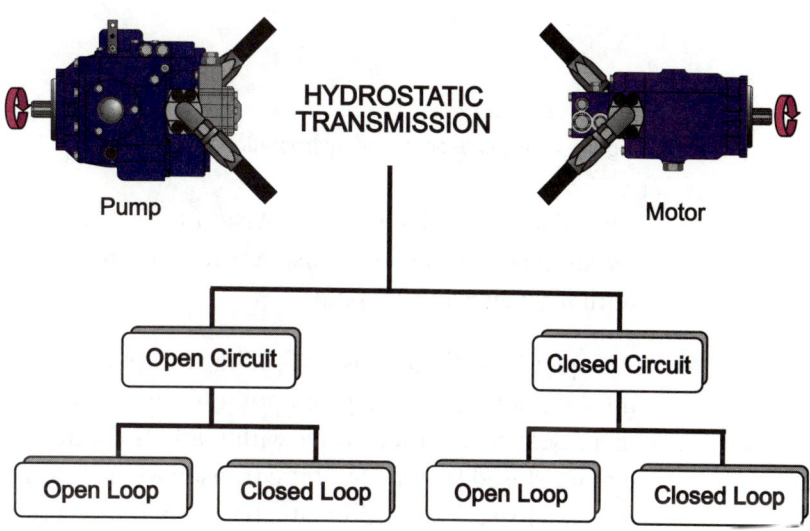

Figure 13.14 Classification of hydrostatic transmissions based on circuit types

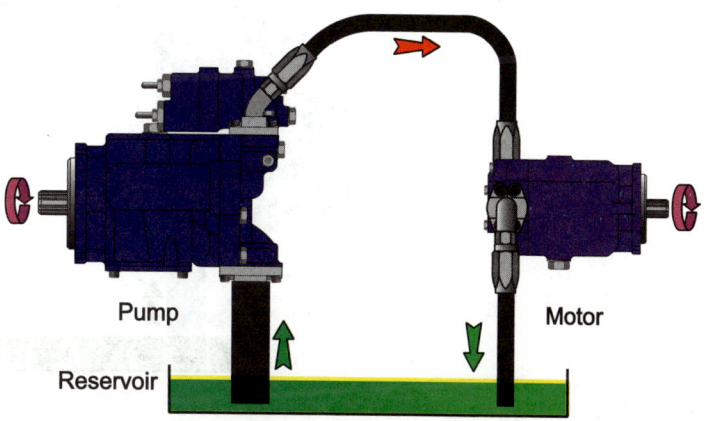

Figure 13.15 Open circuit hydrostatic transmission

307

Chapter 13 Hydrostatic Transmissions

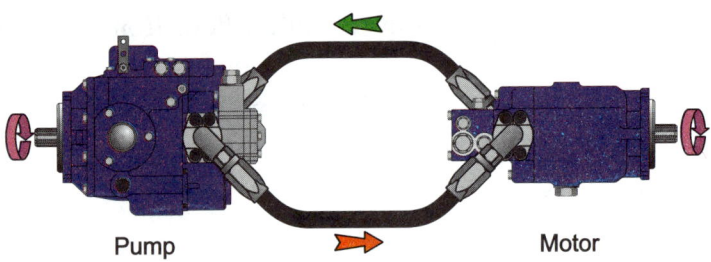

Figure 13.16 Closed circuit hydrostatic transmission

Closed loop circuits have a provision for sensing the motor speed and using this signal to adjust the pump displacement to increase or decrease the motor speed until it reaches the set point.

An open loop circuit has no feedback of the motor speed. When the pump displacement is set at a given point, motor speed decreases as pressure (load) increases. Speed then varies with load. Sometimes, this is undesirable. An open circuit, closed loop hydrostatic transmission is shown in Figure 13.17, and a closed circuit, closed loop hydrostatic transmission is shown in Figure 13.18.

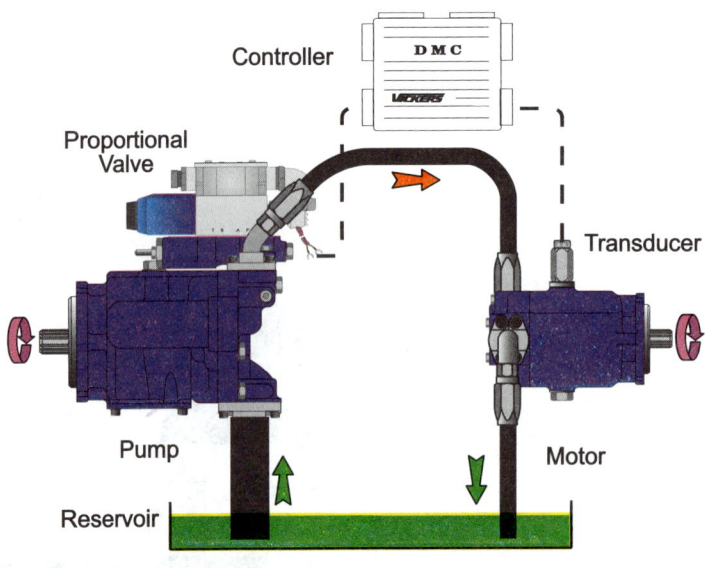

Figure 13.17 Open circuit, closed loop hydrosttic transmission

Chapter 13 Hydrostatic Transmissions

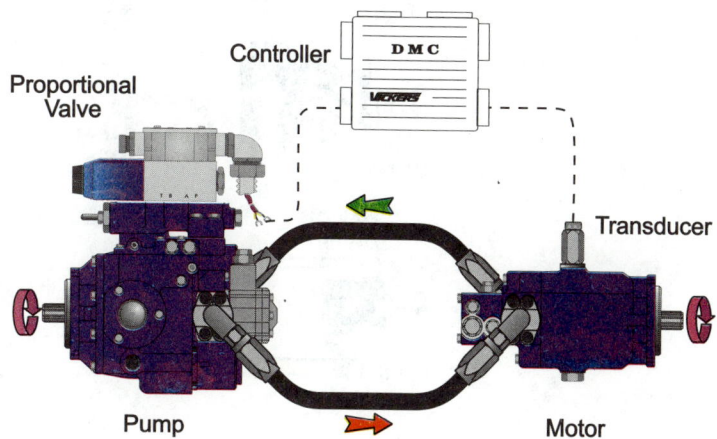

Figure 13.18 Closed circuit, closed loop hydrosttic transmission

The circuit diagrams in Figures 13.15 through 13.18 contain only the key features of the various transmissions. These diagrams are not complete; they are included to illustrate the four classifications.

CLOSED CIRCUIT HYDROSTATIC TRANSMISSIONS

Charge Pumps

Closed circuit hydrostatic transmissions are used for the more sophisticated applications. The technique used to insure that the main circuit is always filled with fluid is illustrated in Figure 13.19. The main pump is an axial piston pump and the motor is an axial piston motor. A charge pump (generally a small fixed displacement pump) is built into the housing with the main pump and operates off the same input shaft as the main pump. The purpose of the charge pump is twofold:

1. It replaces the lubrication and internal leakage fluid that enters the pump housing. This same lubrication and internal leakage flow occurs in the motor. This flow is essential because it provides lubrication to the swash plate, valve plate, shoes, pistons and controls.

2. It provides a flow of cooling fluid through the pump and motor housings. When the high pressure fluid in the main circuit leaks into the housing, mechanical energy is converted into heat energy. In addition, heat results from friction between the moving parts. This heat is removed by the flow of cooling fluid.

Chapter 13 Hydrostatic Transmissions

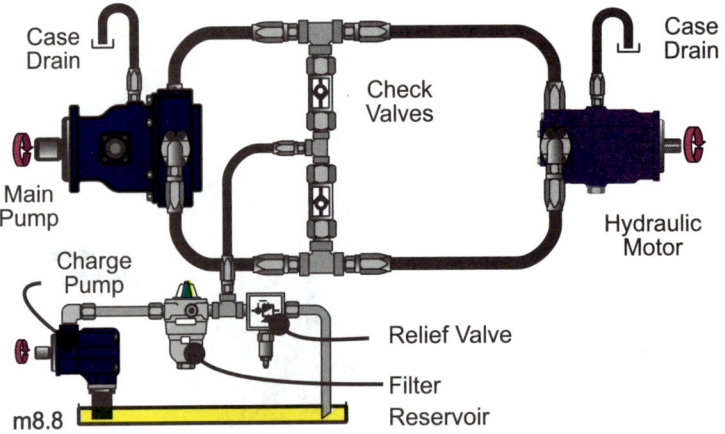

Figure 13.19 Typical charge pump circuit for a closed circuit hydrostatic transmission

Fluid flow in the main circuit is quite simple. The fluid flows to the motor and returns. Flow in the charge pump circuit is as follows.

1. The charge pump receives fluid from the reservoir and delivers it to two check valves, one on each side of the main circuit. Main circuit pressure keeps the check valve seated on the high pressure side. Flow will occur through the check valve on the low pressure side when the main circuit return pressure is less than the charge pump pressure. This pressure will be low if fluid has leaked out of the main circuit. A check valve is needed on both sides because the pump is reversible; either side can be the high pressure side.

2. Fluid that does not flow over a check valve drops across the charge relief valve into the reservoir. In the example shown in Figure 13.19, this fluid flows through a filter, across a relief valve, and into a drain line back to the reservoir.

Because of the fluid that flows past the clearances between the pump parts and into the pump housing, the motor "sees" less fluid than the pump is theoretically pumping. In like manner, the motor has a flow of fluid in the clearances between moving parts. The required charge pump flow is the sum of the fluid lost from the main circuit at both the pump and motor ends. Suppose the volumetric efficiency of the pump is E_{vp}. Volumetric efficiency of the motor is E_{vm}. Required flow of make-up oil is

Formula 13–5

$$Q_l = [1 - (E_{vp} \times E_{vm})] Q_p$$

Where:

Q_p is the theoretical flow from the main pump.

Maximum charge pump flow is required when the main circuit is at maximum operating pressure. Suppose that $E_{vp} = E_{vm} = 0.9$.

$Q_l = 0.19\ Q_p$

The charge pump must be sized to provide at least 19% of the main pump flow.

Shuttle Valve

For larger hydrostatic transmissions, a shuttle valve is incorporated in the motor end (Figure 13.20). Pressure on the high-pressure side shifts the spool of the shuttle valve so that the fluid on the low pressure side has a pathway to a charge relief valve mounted in the end plate of the motor. Fluid drops across this relief valve into the case of the motor where it combines with leakage flow and flows through a drain line to the pump, through the pump housing, and back to the reservoir. The arrangement shown in Figure 13.19 has a charge relief valve only at the pump end. The larger transmission requires more cooling flow at the motor; consequently, it is necessary to have the shuttle valve and second charge relief at the motor end.

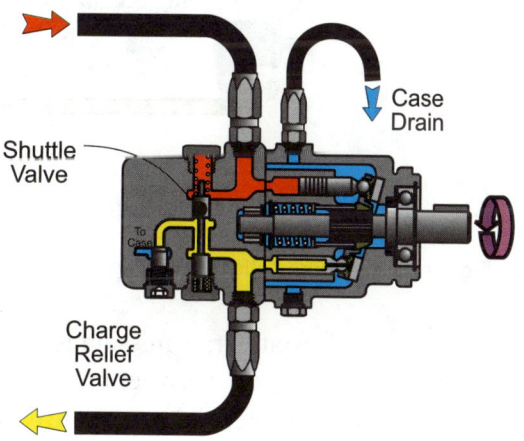

Figure 13.20 Shuttle valve in a hydrostatic transmission motor

The preceding discussion about the second charge relief valve required in the motor end of larger hydrostatic transmissions illustrates a key point about the selection of a hydrostatic transmission. The pump and motor must be engineered to work together as a unit. When the two are supplied as an integral unit, it is understood that a great deal of engineering has gone into the design. Many potential problems are solved for you by the circuitry built into the unit. If the transmission does not perform satisfactorily in a vehicle, the vehicle designer and transmission designer work together to solve the problem.

Chapter 13 Hydrostatic Transmissions

Crossover Relief Valves

A circuit with crossover relief valves is shown in Figure 13.21. The crossover relief valve package has two relief valves, one for each side. Two are needed because either side can be the high pressure side. Relief valves provide for the "remain stalled without damage" feature of a hydrostatic transmission. Once the pressure reaches a certain level, fluid drops across the relief valve to the other side of the main circuit. The motor then stops turning and the transmission is stalled.

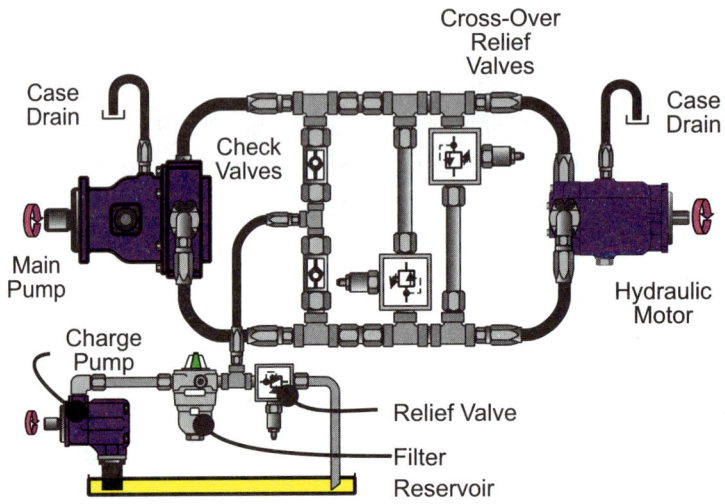

Figure 13.21 Hydrostatic transmission circuit with crossover relief valves

A vehicle hydrostatic transmission is always designed to achieve wheel slip before the relief valve opens. (A brief review of traction mechanics is given in Appendix A.) The slipping wheel is the pressure limiting device. The main purpose of crossover relief valves is to "shave" the pressure peaks resulting from dynamic maneuvers. These valves should pass flow for only short periods.

Mechanical energy is converted to heat energy at the relief valve. (Any time there is a pressure drop in a hydraulic circuit, and no mechanical work is output, heat is generated.) If the crossover relief valve opens, full system horsepower goes into a small hydraulic loop, the loop between the pump and crossover relief valve. The temperature of the fluid rises very quickly. Fluid viscosity decreases as temperature increases. If lubrication is lost, the pump will be destroyed in a very short time.

In applications where wheel slip cannot be depended upon to provide circuit protection, other arrangements must be made. One option is to incorporate a hot-oil replenishing valve in the circuit and increase the size of the charge pump to provide additional cooling flow.

Chapter 13 Hydrostatic Transmissions

Multipurpose Valves

Some manufacturers supply multipurpose valves that incorporate several features in one valve. Typically, two of these valves are installed in the main pump housing, one for each side of the main circuit. One type of multipurpose valve incorporates the following features:

1. High pressure relief [meets the requirement for a crossover relief valve (Figure 13.21)]
2. Check valve [meets the requirement for check valve (Figure 13.19)]
3. Bypass valve [if the vehicle will not start, the bypass valve allows the vehicle to be towed. The transmission will "free wheel".]
4. Pressure limiter [this feature destrokes the pump (reduces flow to near zero) in response to excessive pressure. Typical setting is 6,000 psi.

Summary

Using the broad definition, a hydrostatic transmission is simply a pump and motor connected together. For some simple applications, it is possible to select a pump from one manufacturer and motor (or motors) from another manufacturer, connect them together with hoses or tubing, and get acceptable performance. An example of a simple application would be a conveyor drive on a mobile machine. The load varies over a relatively narrow range and speed control is not critical. Dramatic shock loads are infrequent. Reversing is not required. The pump and motor can be sized to operate at less than 1,500 psi. In this case, a fixed displacement pump and fixed displacement motor can be used satisfactorily with very little engineering design required.

Considerable circuit design is required to get a pump and motor to function satisfactorily as a closed circuit hydrostatic transmission. For high performance applications, such as a vehicle transmission, it is recommended that an integral unit be selected. These units have the following built-in features:

1. Charge pump, charge relief valve, check valves and related circuitry
2. Shuttle valve and motor-end charge relief valve
3. Crossover relief valve
4. Bypass valve

In some cases, a multipurpose valve is used that incorporates several features in one valve.

Chapter 13 Hydrostatic Transmissions

Review of Pump and Motor Operating Characteristics

As explained in Chapter 12, the quantity of fluid delivered by the pump decreases as pressure increases. The actual flow, divided by theoretical flow, is the pumps volumetric efficiency.

Motor speed decreases as pressure increases because of increased flow (lubrication and leakage) in the clearances between the moving parts (Chapter 5). Motor volumetric efficiency is motor output rpm, divided by theoretical motor rpm. Theoretical motor rpm is based on motor displacement and actual flow delivered to the motor.

A simple example illustrates how hydrostatic transmission performance is governed by the characteristics of the pump and motor. Suppose a pump has a displacement of 1.925 in.^{3}/rev and is driven at a speed of 2,400 rpm. Theoretical flow is:

$$\frac{1.925 \text{ in.}^3/\text{rev} \times 2400 \text{ rev/min}}{231 \text{ in.}^3/\text{gal}} = 20 \text{ GPM}$$

The motor used in this hydrostatic transmission also has a displacement of 1.925 in.^{3}/rev. When the pump flow is delivered to the motor, the motor speed is:

$$\frac{20 \text{ gal/min} \times 231 \text{ in.}^3/\text{gal}}{1925 \text{ in.}^3} = 2400 \text{ RPM}$$

Now, consider what happens when operating pressure increases to 1,000 psi. The pump volumetric efficiency at this operating pressure is 0.90, so the actual pump output is:

20 x 0.90 = 18 GPM

Flow to the motor is no longer 20 GPM but 18 GPM. If there is no leakage in the motor, the motor speed will be:

$$\frac{18 \text{ gal/min} \times 231 \text{ in.}^3/\text{gal}}{1925 \text{ in.}^3/\text{rev}} = 2160 \text{ RPM}$$

The 1,000 psi operating pressure also causes leakage in the motor, resulting in a motor volumetric efficiency of 0.90. Actual motor output speed is:

$$\frac{18 \text{ gal/min} \times 231 \text{ in.}^3/\text{gal}}{1925 \text{ in.}^3/\text{rev}} \times 0.90 = 1944 \text{ RPM}$$

Input speed to the pump is 2,400 rpm, and the achieved output speed from the motor is 1,944 rpm. Overall volumetric efficiency of the HST is:

$$\frac{1944}{2400} = 0.81$$

Another way of calculating this efficiency is to multiply the volumetric efficiency of the pump times the volumetric efficiency of the motor.

0.9 x 0.9 = 0.81

Motor speed varies with pressure in a hydrostatic transmission; this important concept should be clearly understood.

To reinforce the concept of speed variation with pressure variation, it is helpful to plot the set of pump and motor performance curves on the same graph (Table 13.1). (Note: the curves in Table 13.1 are for illustration purposes only. Manufacturer's data for a specific pump and motor must be used for a design.) The pump and motor have the same displacement, so the theoretical performance of both is shown as a heavy dark line in the middle of the family of curves. The pump performance curves are shown below the theoretical line, and the motor curves above the line.

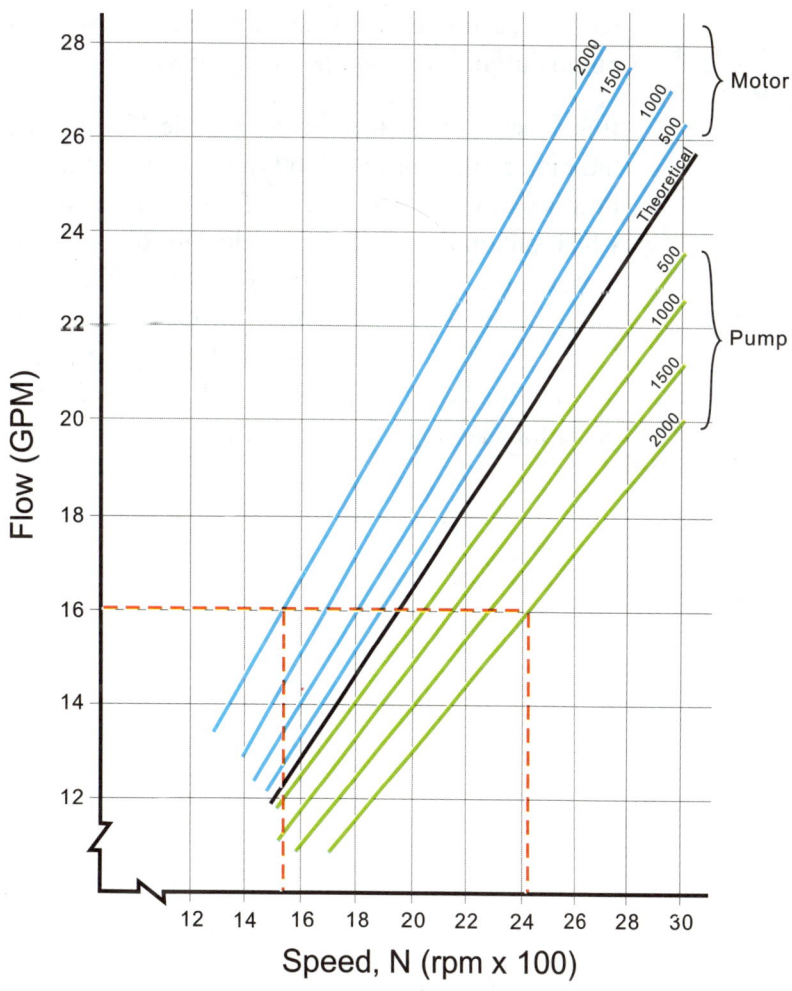

Table 13.1 Typical performance curves for a pump and motor in a hydrostatic transmission.

Chapter 13 Hydrostatic Transmissions

Suppose the pump is driven at 2,400 rpm. Proceeding vertically from 2,400 rpm on the horizontal axis up to the theoretical line and then across to the vertical axis, the theoretical pump flow is 20 GPM. If this flow is delivered to the motor, theoretical speed is 2,400 rpm. (Find 20 GPM on the vertical axis, move horizontally to the theoretical line and down to the horizontal scale to read a motor speed of 2,400 rpm.) With this instruction on how to use the graph, an example can be illustrated.

The pump is driven at 2,400 rpm and the load develops 2,000 psi pressure. What flow is delivered to the motor? Follow the dotted line up to the 2,000 psi curve and then move horizontally to the vertical axis to read a pump flow of 16 GPM. What is the motor speed when this 16 GPM is delivered to the motor? Follow the dotted curve from 16 GPM horizontally to the 2,000 psi motor curve and then down to the horizontal axis, to read a motor speed of 1,530 rpm. If no pressure were developed, the motor speed would have been 2,400 rpm. Actual speed is 1,530 rpm, or 36% less.

With this review of pump and motor operating characteristics, it should be evident how motor speed varies with pressure. If precise control of the motor speed is required, some means must be provided to sense the actual motor speed, and adjust pump output until this speed is achieved.

It is possible to use the curves in Table 13.1 to work a different problem. Suppose the motor must turn at 1,800 rpm when operating pressure is 2,000 psi. What pump flow is required? To avoid confusion, the curves are re-plotted in Table 13.2. Find 1,800 rpm on the horizontal scale and follow the dotted line up to the 2,000 psi motor curve, then across to the vertical axis to read 18.7 GPM. What speed must the pump be driven to deliver 18.7 GPM at 2,000 psi? To answer this question, follow the 18.7 GPM line horizontally to the 2,000 psi pump curve. Then, drop down to the horizontal axis to read 2,800 rpm. The pump must turn at 2,800 rpm in order for the motor to turn at 1,800 rpm.

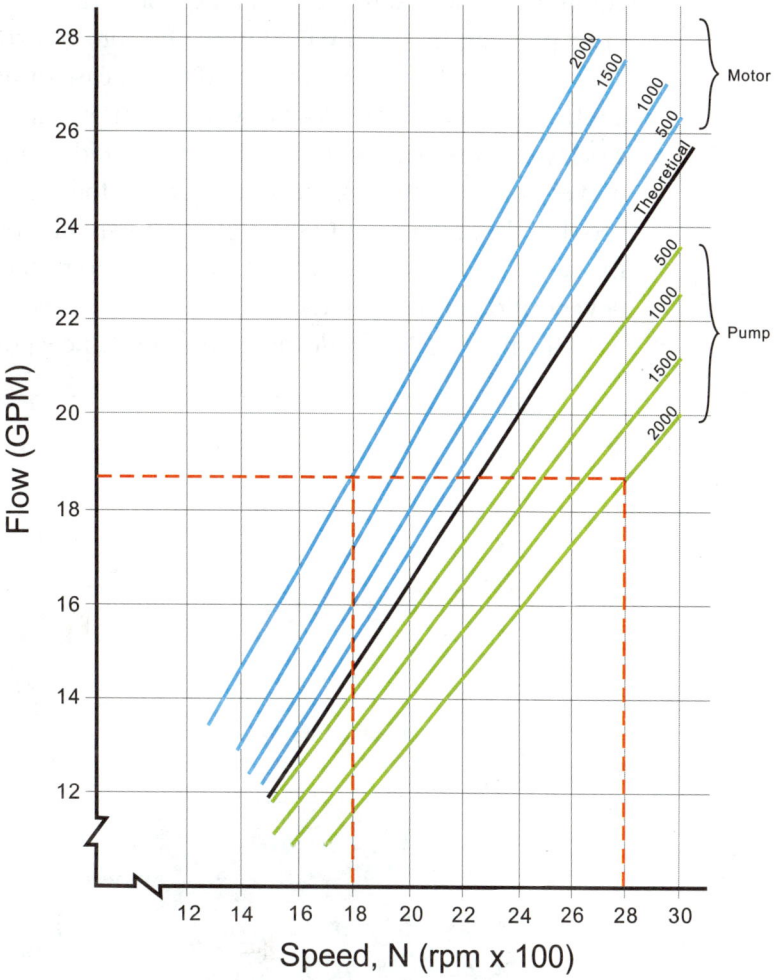

Table 13.2 Typical performance curves for a pump and motor in a hydrostatic transmission, replotted to work example problem.

Servo Controlled Pumps

Before beginning our discussion of closed loop hydrostatic transmissions, it is necessary to first learn how a servo controlled pump operates. A variable displacement axial piston pump will be used as an illustration. This pump can have a control piston mounted in the pump housing. When the control piston extends, it moves the swash plate to increase the amount of fluid pumped by the pistons. (Displacement of the pump is increased.) Flow of fluid to the control piston, and thus its position, is controlled with a servo valve. A pump with these features is called a servo controlled pump.

A servo valve operates like a directional control valve. The spool shifts in one direction to direct pressurized fluid to Port A and in the other direction to direct pressurized fluid to Port B. The spool in a servo valve is precisely machined; consequently, the cost is higher than the cost of a standard directional control valve.

It is helpful to first consider a manually controlled servo pump (Figure 13.22). Suppose the manual control lever is moved to the left (rotated in a counter-clockwise direction). The spool of the servo valve is shifted to the left. High pressure fluid is directed to the bottom control piston causing it to extend. The top control piston is connected to the case drain, thus it retracts as the bottom control piston extends. As the two control pistons move, the swash plate is rotated clockwise, thus increasing the amount of fluid pumped. When the swash plate moves, it pushes the yoke feedback link to the right. The link pivots and pushes the spool of the servo valve to the right. This spool movement closes the two ports of the servo valve, thus locking the control pistons in a position that corresponds to the new position of the manual control lever. This position of the swash plate is held until the manual control lever is moved to a new position.

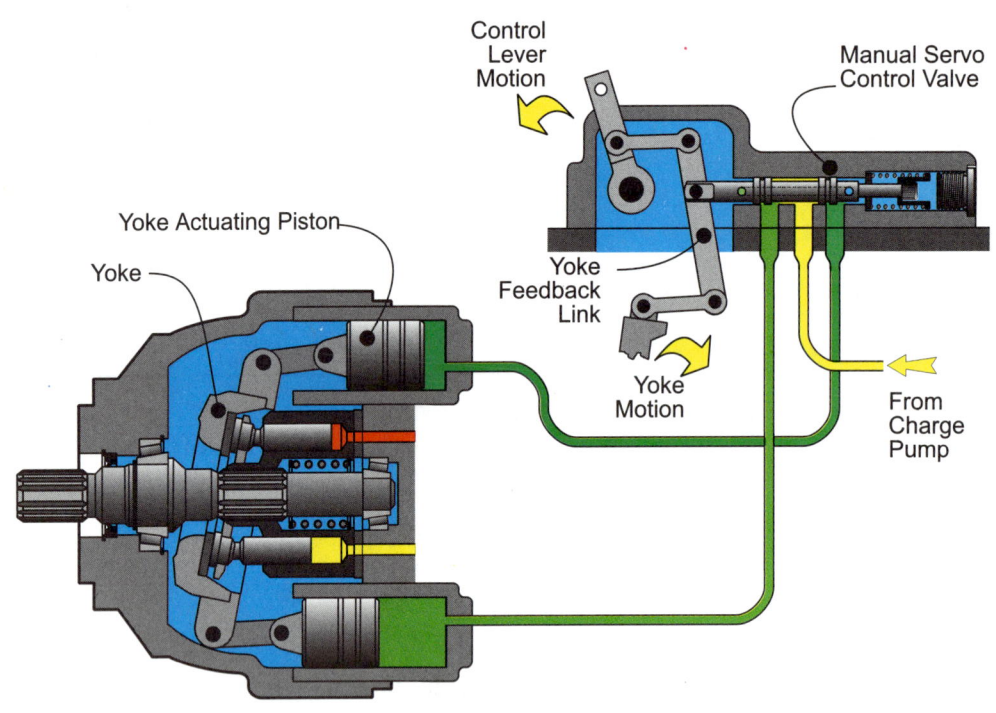

Figure 13.22 Manually operated servo pum,p

The obvious question is, why not connect the manual control lever directly to the swash plate? Then, when the lever is moved, the swash plate is rotated. Smaller hydrostatic transmissions are operated in this manner. With larger transmissions (>25 Hp), the force required to move the swash plate becomes high enough that operator fatigue becomes an issue. The servo valve and control pistons make the swash plate control lever much easier to operate.

The servo pump in Figure 13.23 operates much like the one shown in Figure 13.22, except the control lever is shifted hydraulically, using a pilot hydraulic circuit and a remote valve. The feedback mechanism works in the same way.

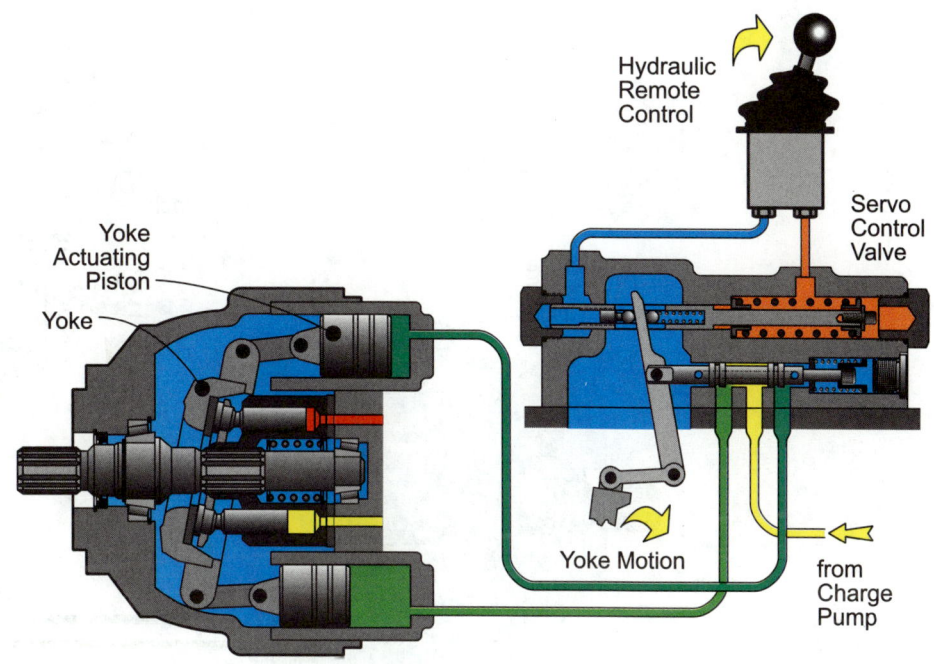

Figure 13.23 Servo pump operated by a pilot circuit

The servo pump shown in Figure 13.24 is like that shown in Figure 13.22, except a torque motor is used to position the spool of the servo valve. A torque motor rotates through several degrees of rotation, when a current is passed through the winding. The torque motor shown has a "flapper" attached to the armature. This flapper is centered in the nozzle, such that the pressure drop on both sides is equal. Pressure on both ends of the servo valve spool is equal. This design is called a "flapper nozzle" servo valve.

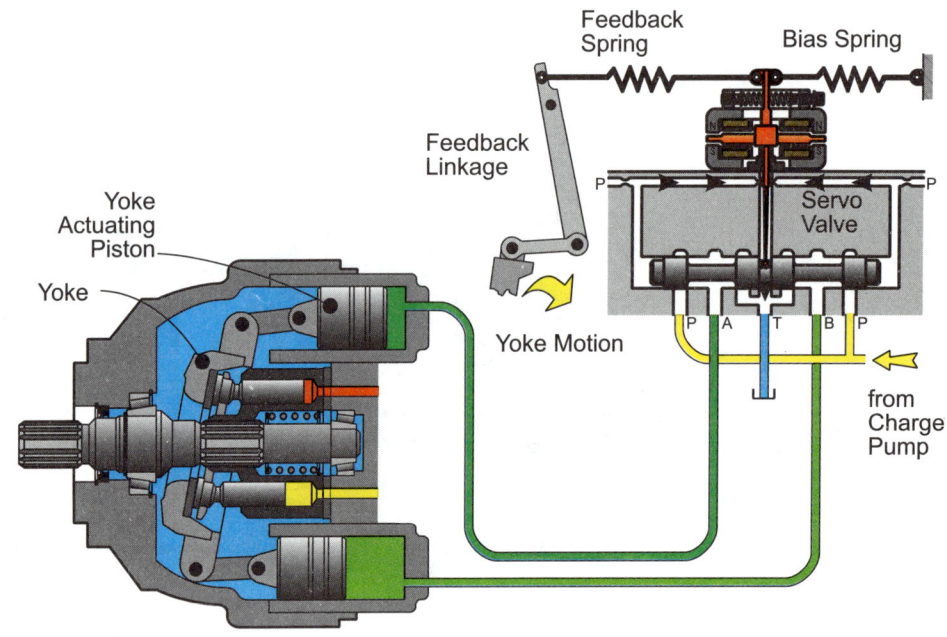

Figure 13.24 Servo pump operated by a servo valve

When the servo valve spool is centered, both ports to the control pistons are blocked, and the control pistons are locked into position. The swash plate is then locked into position.

The following sequence of events occurs when a current is delivered to the torque motor.

1. The armature rotates and moves the flapper to partly block one side of the nozzle.

2. Pressure builds on the side of the partly blocked nozzle and drops on the other nozzle. This pressure shifts the servo valve spool (for example, to the left). Pressurized fluid is directed to port A, and the port B is connected to the case drain.

3. Pressurized fluid from port A causes the top control piston to extend. Fluid from the bottom control piston returns to the reservoir through Port B, and it retracts.

4. The swash plate is moved to a new position.

As the swash plate moves to a new position, some "feedback" is needed to move the servo valve spool back to the center position. Otherwise, the spool will stay shifted, Ports A and B will stay open, and the swash plate will continue to move until it reaches its full displacement in one direction. To accomplish this feedback, a feedback lever is connected to the swash plate. This lever rotates as the swash plate moves. When the lever rotates, it displaces the swash plate feedback spring, which moves the armature until the flapper is re-centered. The nozzle now has equal pressure drop on each side. Equal pressure on both sides causes the servo valve spool to move to the center, Ports A and B are blocked, and the control pistons are locked into position. The swash plate stays in this position until another current is delivered to the torque motor winding.

Servo Valve Circuit

Some explanation of the electrical circuit required for a servo valve is needed in order to understand how the servo valve in the servo pump operates when it is used in a closed loop transmission. The key features of the electrical circuit for a closed circuit, closed loop hydrostatic transmission is shown in Figure 13.25. The motor output shaft drives a transducer. A tachometer generator is a typical transducer used for a closed loop transmission. Voltage output from the tachometer generator is directly proportional to rpm. As rpm increases, voltage increases, and vice versa.

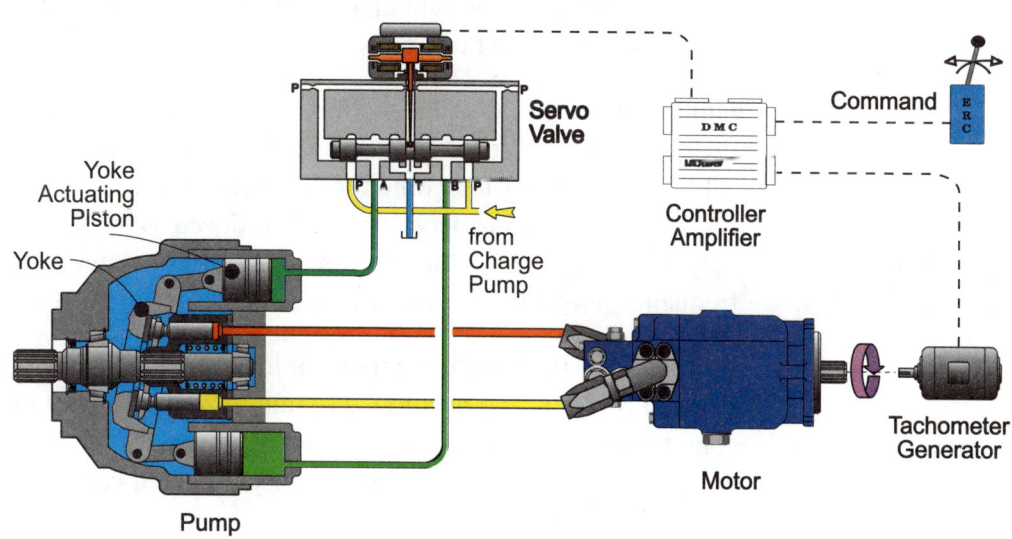

Figure 13.25 Servo speed control

The controller compares the tachometer generator voltage with the command voltage. The difference between the two is the error voltage. If the tachometer generator voltage is equal to the command voltage, meaning that the motor is turning at the desired rpm, then the error voltage is zero.

The error voltage is fed to a servo amplifier. The servo amplifier produces an output current proportional to the input voltage. This current is fed directly to the coil of the torque motor. (Coil is another name for the windings on the armature.) The armature rotates, and this initiates the sequence of events described in the previous section.

Response Time For Closed Loop Circuits

Response time is the time required for motor speed to reach a new set point. Suppose the command signal is changed with a step input; the command signal is 5 volts and it is stepped to 10 volts. The time required to reach the new motor speed is dependent on the natural frequency of the circuit. Neglecting energy dissipation (damping), natural frequency is given by

$$f = \frac{1}{2\pi} \sqrt{\frac{k}{m}}$$

Where

k = elasticity (force required to produce a unit deflection)
m = mass

If k is large, the system is said to be stiff, meaning that little deformation occurs when a large force is applied. A stiff system has a high natural frequency. If m is large, meaning that a heavy load is being moved, then natural frequency will be low.

k is a function of the quantity of fluid under compression. (For this discussion we ignore the swelling of the lines when pressure is applied.) If the lines are long, k will be small, meaning that a small force will cause a deformation. Conversely, if the lines are short, k will be large.

Suppose a heavy load is attached to the motor (m is large), and the lines between the pump and motor are long (k is small). The natural frequency will be low, and it will take a relatively long time for the motor to reach a new speed. To reduce response time, the line length must be reduced as much as possible. Reducing length increases k and thus increases natural frequency. Systems with a high natural frequency have a faster response time.

Fast response time is often important in a closed circuit design. Components must be sized correctly and positioned to minimize the amount of fluid under compression. After doing all that can be done with the hydraulic circuit design, there are methods for reducing response time that can be incorporated in the design of the electronic circuit.

Operation of Closed Circuit, Closed Loop Hydrostatic Transmissions

The hydrostatic transmission, shown in Figure 13.27, is operating under a constant load. Speed is set by the command voltage at 1000 rpm. The load starts to increase, pressure increases, and the pump and motor volumetric efficiencies decrease. The motor begins to slow, and the following sequence of events is initiated:

1. The tachometer generator voltage falls when the motor speed decreases.
2. When the new tachometer generator voltage is compared to the command voltage, a negative error voltage is produced.
3. The negative error voltage causes the servo amplifier to produce a current.
4. Current from the servo amplifier causes the torque motor armature to rotate.
5. Rotation of the armature moves the flapper to create a pressure imbalance on the servo valve spool.
6. The spool shifts to direct high pressure fluid to the bottom control piston, causing it to extend. The top control piston simultaneously retracts.
7. When the control pistons move, the swash plate rotates clockwise to increase pump displacement.
8. More pump flow is delivered to the motor and motor speed increases.
9. Tachometer generator speed increases as motor speed increases.
10. When the tachometer generator voltage equals the command voltage, the error voltage is zero.
11. Zero error voltage produces zero current.
12. Zero current produces zero armature displacement.
13. With the armature centered, the flapper is centered and equal pressure exists on both ends of the servo valve spool.
14. The servo valve spool is centered, Ports A and B are blocked, and the control pistons are locked into position.
15. The swash plate is locked into position and the pump operates at this new displacement.

If the load decreases and the motor speeds up, the reverse of the above sequence of events occurs. The swash plate angle is reduced, the pump displacement is less, and the motor slows until its speed equals the speed set by the command voltage. Each time the load changes, the control compensates to keep speed constant.

The feedback lever, shown in Figure 13.24, is not needed when a servo pump is operated in the closed loop configuration shown in Figure 13.25. When the tachometer generator voltage equals the command voltage, the error voltage is zero. With zero error voltage, current from the servo amplifier drops to zero and the torque motor armature returns to its centered position. The swash plate feedback spring is not needed to move the armature to its centered position.

Chapter 13 Hydrostatic Transmissions

Summary

A hydrostatic transmission is simply a pump and motor connected in a circuit. Other components are included in the circuit design to insure that the functional objective is achieved. The pump and motor can either be included in the same housing or separate components may be connected with hoses or tubing.

Typically, mechanical transmissions have efficiencies of 92% or greater, whereas hydrostatic transmissions have an efficiency of around 80%. Some well-designed units have an efficiency of 85% over a certain operating range.

Hydrostatic transmissions are used to increase vehicle maneuverability. They also provide continuous speed control from a slow creep up to maximum speed. Before a hydrostatic transmission is chosen over a mechanical transmission, a study is done to insure that the advantages yield an increase in vehicle productivity (tons handled per operating hour etc.) to offset the lower efficiency.

If the pump receives fluid from the reservoir, passes it through the motor, and returns it to the reservoir, the design is identified as an "open circuit" transmission. Transmissions which circulate fluid in a continuous loop between the pump and motor are identified as "closed circuit" transmissions.

As the load increases, pressure increases, and leakage inside the pump and motor increases. Motor output speed decreases due to this leakage. If constant speed is important, a transducer is used to sense motor speed, and this signal is used to increase (or decrease) pump displacement by the amount required to maintain motor speed. Sensing motor speed and using this information to control pump displacement is known as "feedback". Feedback can be used for both open and closed circuit transmissions. When feedback is included, the transmission is identified as a "closed loop" transmission.

A variable displacement axial piston pump can have a control piston mounted in the pump housing. This control piston is used to control the position of the swash plate, and thus the displacement of the pump. A servo valve controls the flow which extends (or retracts) the control piston. The servo valve shifts when current is delivered to a torque motor mounted on the valve. A typical closed loop transmission operates as follows. The voltage signal from the transducer that senses motor speed is fed to a servo amplifier to obtain a current. This current causes the torque motor to rotate, which opens the servo valve and ultimately increases (or decreases) the pump displacement. Using a feedback design of this type, transmission output speed can be held constant as loads vary.

For some simple applications, it is possible to select a pump from one manufacturer and motor (or motors) from another manufacturer, connect them together with hoses or tubing, and achieve acceptable performance. In general, however, a hydrostatic transmission is purchased as an integral package with needed valving already installed. Considerable engineering design goes into the circuitry built into the unit. When the transmission is used for a vehicle, the vehicle designer and transmission designer work as a team to insure the desired performance is achieved.

Chapter 13 Hydrostatic Transmissions

Appendix A Basic Concepts in Traction

The ability of a vehicle to develop traction is a function of the weight on the drive wheels and the coefficient of friction between the wheel and surface. Torque at wheel slip is given by:

$$T_s = W_d \mu r$$

Where:

T_s = torque at wheel slip (in-lb_f),
W_d = weight on drive wheel (lb_f),
μ = coefficient of friction (decimal),
r = rolling radius (in.).

The rolling radius is the distance from the center of the axle to the ground. Because the weight of the vehicle causes the tire to deform, the rolling radius is less than the tire radius.

Required wheel torque is the torque that must be supplied to the wheel to move the vehicle (Figure 13.26):

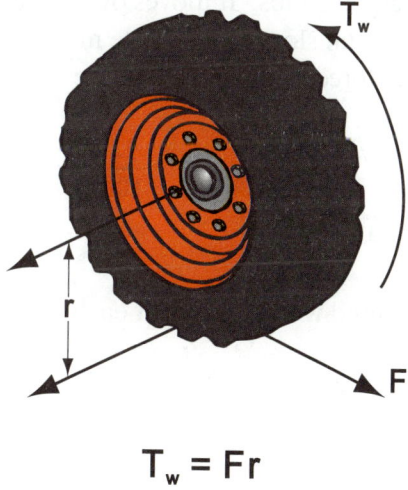

$$T_w = Fr$$

Figure 13.26 Required wheel torque

The total force to move the vehicle is given by

$$F = W_g \frac{R}{1000} + W_g \frac{P_g}{100} + F_d$$

Where:

F = total force (lb_f)
W_g = gross vehicle weight (GVW) (lb_f)
R = rolling resistance (lb_f/1000 lb_f GVW)
P_g = grade (%)
F_d = drawbar pull (lb_f).

Gross vehicle weight is the total weight of the machine, not just the weight on the drive wheels. Rolling resistance is a function of the surface. Typical values are shown below.

Surface	R
Concrete	15
Packed Soil	25
Sandy Soil	37
Mud	37-150

The maximum grade the vehicle must climb is given in percent (rise over run), not degrees.

Drawbar pull is the force the vehicle must develop above the force just to move the vehicle. In the case of a load being towed by the vehicle, drawbar pull is simply the force at the hitch point to move the load forward. Drawbar pull can also be the force required to push a load. An example would be a front-end loader pushing forward to fill the bucket with dirt.

A simple example will reinforce the traction concept. A self-propelled windrower weighs 7785 lbs. It moves over a sod surface with rolling resistance, R = 25. The maximum slope the vehicle must climb is 12%, and the drawbar pull is F_d = 500 lb_f. Total force to move the vehicle is:

$$F = 7785 \left(\frac{25}{1000} + \frac{12}{100} \right) + 500$$

$$= 1630 \ lb_f$$

Required wheel torque at each driving wheel is:

$$T_w = \frac{F_r}{N_w}$$

Where:

n_w = number of driving wheels.

Both front wheels drive and the rolling radius is 23 in. The required torque at each front wheel is:

$$T_w = \frac{1630 \ (23)}{2}$$

$$= 18,745 \ in \text{-} lb_f$$

If the coefficient of friction on a sod surface is $\mu = 0.4$, will the torque at wheel slip be greater than the required torque? Weight on the two drive wheels is 70% of the total vehicle weight.

$$W_a = 0.7 \frac{(7785)}{2} = 2725 \text{ lb}_f$$

$$\begin{aligned} T_s &= W_d\,\mu\,r \\ &= 2{,}725\,(0.4)(23) \\ &= 25{,}070 \text{ in-lb}_f \end{aligned}$$

Since $T_s > T_w$, the vehicle can meet the functional objective.

Peak torque will be developed when the vehicle is operating on a surface with high coefficient of friction. Torque at wheel-slip is a maximum for this condition. The hydrostatic transmission and other drive line components should be designed to deliver this maximum torque.

Chapter 13 Hydrostatic Transmissions

CHAPTER 14 .. Fluid Conductors and Connectors

Fluid conductors are the means to transport fluid from one point to another in a hydraulic circuit. Connectors are the attachments for the conductors, one to another or to a component.

Conductors fall into one of three broad categories; hose, tubing or pipe. There are many sub-categories under each of these, primarily defining the construction, materials and performance criteria.

Connectors are not so easily categorized, as there are numerous types of connectors for each of the categories and sub-categories of conductors. Therefore, connector types will be discussed under each of the conductor headings.

Pipe is occasionally used in hydraulic systems, primarily for its relatively low cost and wide availability. Pipe gained its popularity as a conductor for air, water and steam, and so was a natural choice in the early years of hydraulic development. As the sophistication of hydraulics increased, however, pipe became less attractive to designers of hydraulic systems:

- To adequately seal a tapered pipe thread joint, the connectors must be tightened sufficiently to yield and deform the material that makes up the threads. This frequently causes small particles of the material to dislodge and enter the hydraulic system.

- Even after being securely tightened, a path for ingesting air into, or leaking fluid out of, a system may occur at the root of the threads.

- A sealing compound or tape can be used for a more effective seal, although this introduces another material that frequently finds its way into hydraulic systems.

- When connectors are loosened and then re-tightened, they must be deformed further for the second seal, which aggravates the initial problem of small particles getting into the system, and without the assurance of an adequate seal.

- Mechanical strain, vibration, pressure cycles or pulsations, and thermal expansion/contraction may make even "good" pipe joints leak after a short time in service.

- Pipe is a rather inflexible material, difficult to form and install in close environments. This is particularly important in today's compact machinery.

Pipe is seldom used in mobile hydraulic circuitry today, and is not recommended. Therefore, it is given only superficial coverage in this chapter.

Chapter 14 Fluid Conductors and Connectors

Selection Criteria Hose makes a flexible conductor; tubing makes a rigid conductor. Therefore, the first selection criteria is:

- Do the components being connected move relative to each other? If so, then hose must be used.

- Is there excessive vibration present? Hose can usually accommodate vibration better than tubing, although proper isolation and clamping can make a tube installation successful in many applications.

- Are there frequent pressure pulsations present? Again, hose can usually accommodate pressure pulsations better than tubing, due to the relative movements that take place. Hose will also tend to absorb and dampen pulsations to some degree. As with vibrations, proper isolation and clamping can make for a successful tubing installation in many applications.

The second criteria is the frequency of installation. For a one-time installation, hose is usually more convenient, less costly and quicker. When the installation will be done repeatedly, such as on a production line, tubing is a preferred choice because it can be pre-formed in high quantities and installed quickly.

Third, the space available to accommodate the conductor may influence the selection. For any given inside diameter, the outside diameter of tubing is smaller than the outside diameter of hose.

Fluid Velocity Laminar and turbulent flow are defined in Chapter 2, and illustrated in Figures 2.7 and 2.8. Turbulent flow is an undesirable characteristic, as it causes unnecessary friction within the fluid and against the conductor walls. Excessive fluid velocity is a major contributor to turbulence, which causes unwanted pressure drops between circuit segments. Keeping fluid velocities at reasonable levels is important for maintaining good pump inlet conditions, reducing heat generation and facilitating good component operation.

Fluid velocity (see Chapter 2) is the speed at which fluid moves past a given point, in a conductor, in connectors or in components. The velocity is most commonly stated in feet-per-second (ft/sec) in customary U.S. units, or in centimeters-per-second (cm/sec) or meters-per-minute (m/min) in metric units. However, velocity may be stated in any dimension per unit of time.

Table 14.1 lists the recommended velocities of fluids in hydraulic conductors. Velocities in these ranges result in reasonably low pressure drops through conductors and minimal fluid friction.

Conductor Sizing		
Application	Operating Pressure	Maximum Fluid Velocities
Pump Inlet Line		4 ft/sec
Pressure Line	...0 - 500 psi	15 ft/sec
	...500 - 3000 psi	20 ft/sec
	...over 3000 psi	25 ft/sec
Return Line		15 ft/sec

Table 14.1 Recommended fluid velocities in conductors.

The relationship between fluid velocity and conductor size is:

Flow = Velocity x Area

Care must be taken to assure that units for each of the elements of the above equation are converted correctly to give the desired result. The following formula takes all of the conversion factors into account:

Formula 14-1

$$V_{ft/sec} = \frac{0.3208 \times Q_{GPM}}{A}$$

Where:

v = Fluid velocity in feet-per-second
Q = Fluid flow in gallons-per-minute
A = Inside area of a conductor in square inches

A nomograph such as that shown in Figure 14.1 can also be used. A full size nomograph for fluid velocity, conductor area or inside diameter, and fluid flow is included in the appendix. When a non-standard dimension for inside diameter is determined by either formula 14-1 or the nomograph, the next larger standard figure must be used. The next smaller standard figure would result in velocities higher than recommended.

Chapter 14 Fluid Conductors and Connectors

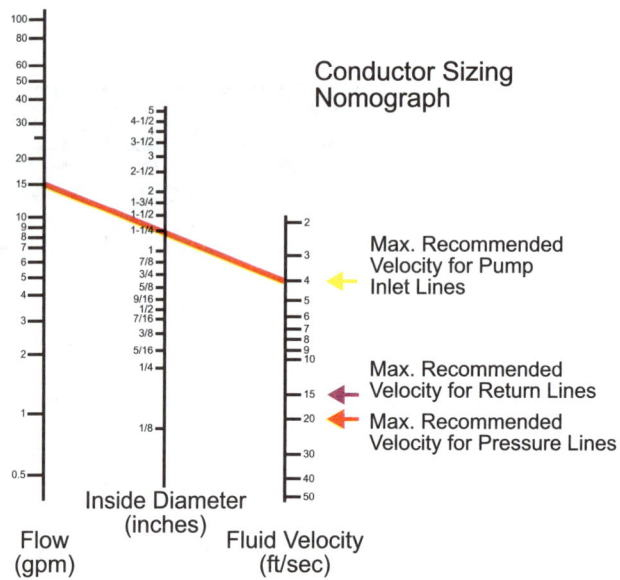

Figure 14.1 Nomograph for determining the inside diameter of a fluid conductor

Fluid velocity varies inversely by the inside <u>area</u> of the conductor, or inversely by the <u>square</u> of the diameter. As Figure 14.2 shows, the velocity of the fluid will increase by 4 if the inside diameter of the conductor is reduced to half its original size.

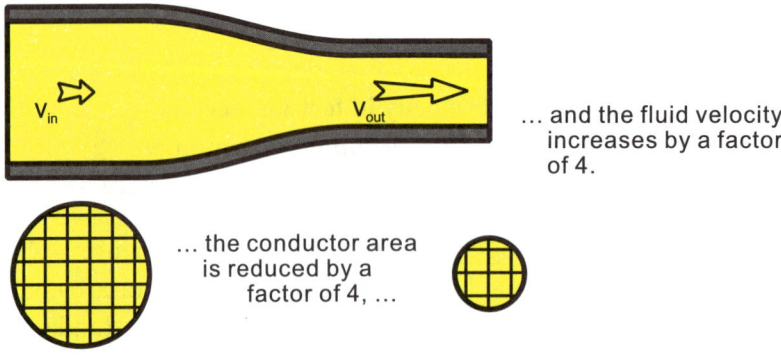

Figure 14.2 Fluid velocity is inversely proportional to the conductor diameter squared

332

Hose

Hose is a flexible conductor that is used when hydraulic lines are subject to movement, flexing or vibration. Figure 14.3 shows the elements that make up hose. A guideline for most hydraulic hose is the SAE J517 Standard. This standard has 100R numbers that define construction, dimension, pressure, fluid compatibility and temperature specifications. A brief description of the 100R numbers is provided in the appendix.

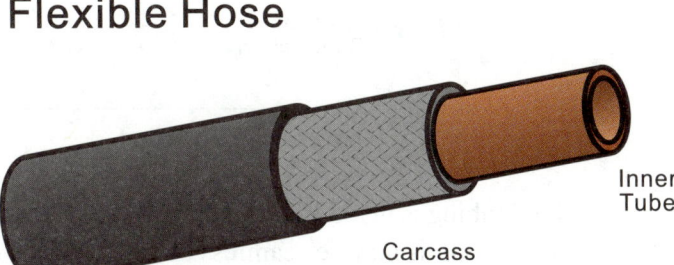

Figure 14.3 Flexible hose is constructed in layers.

The inner tube may be constructed of an oil resistant synthetic rubber material or an oil resistant thermoplastic material. The carcass may consist of a wire braid, multiple wire braids, a braided textile fibre, multiple textile fibre braids, or a combination of both textile and wire braids. The cover consists of an oil and weather resistant synthetic rubber material.

Hose size is designated by a "dash" number, which refers to the nominal inside diameter in sixteenths of an inch. For example, a -6 hose would be 6/16 inches inside diameter, or 3/8 inch ID.

Hose Connectors

Hose connectors, frequently called fittings, may be reusable (screw-together, bolt-together) or non-reusable (crimp or swage). They can be fixed (non-swivel) or swivel type, or split flange type. One end of the hose usually has to have a swivel nut or a split flange so that the connector can be turned for installation. It is recommended that both ends have the swivel or split flange type to insure that the hose does not get twisted during installation.

Figure 14.4 illustrates the types of hose connectors common to hydraulic circuits: reusable, non-reusable, fixed, swivel nut and split flange. Hose connectors are usually designated in size by a "dash" number (such as -6), which references the hose size the connectors are designed for.

333

Chapter 14 Fluid Conductors and Connectors

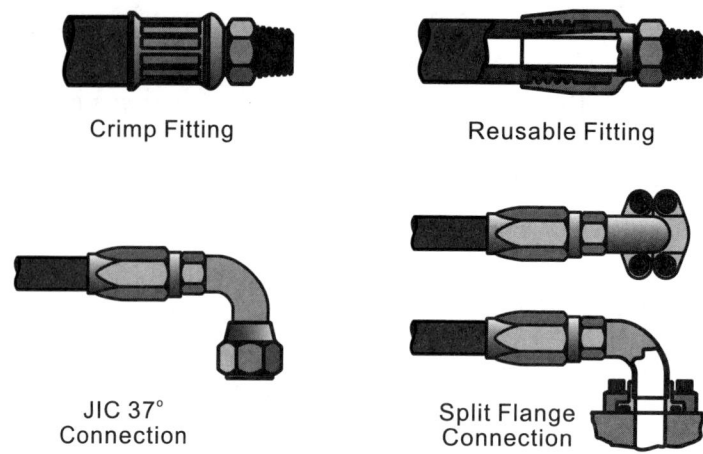

Figure 14.4 Typical hose connectors.

Steel Tubing

Steel tubing is made from a highly ductile, annealed material which is easily bent and flared. It may be seamless or electrically welded. It usually results in a cleaner looking installation and, properly installed, will require less maintenance. It can be made from stainless steel for highly corrosive fluids and high pressure applications.

Tubing size is available in 1/16 inch increments from 1/8 inch up to one inch outside diameter, and in 1/4 inch increments above one inch. Various wall thicknesses are available for each size. Tubing is designated by the outside diameter; the inside diameter is equal to the outside diameter minus two times the wall thickness.

Pressure capability of tubing is based on the diameter and wall thickness. Tubing size charts, as illustrated in Figure 14.5, provide burst pressure and working pressure ratings for each size and wall thickness combination. Burst pressure is some multiple higher than working pressure, usually between three and six times. This provides a working pressure "safety factor".

Tubing Table

Nom. Size	Wall (in.)	Burst psi	Working psi	Tubing Oil Flow Capacities (gpm)					
				2 ft/sec	4 ft/sec	10 ft/sec	15 ft/sec	20 ft/sec	30 ft/sec
1/2	.035	7700	1283	.905	1.81	4.52	6.79	9.05	13.6
	.042	9240	1540	.847	1.63	4.23	6.35	8.47	12.7
	.049	10,780	1797	.791	1.58	3.95	5.93	7.91	11.9
	.058	12,760	2127	.722	1.44	3.61	5.41	7.22	10.8
	.065	14,300	2383	.670	1.34	3.35	5.03	6.70	10.1
	.072	15,840	2640	.620	1.24	3.10	4.65	6.20	9.30
	.083	18,260	3043	.546	1.09	2.73	4.09	5.46	8.18
5/8	.035	6160	1027	1.51	3.01	7.54	11.3	15.1	22.6
	.042	7392	1232	1.43	2.85	7.16	10.7	14.3	21.4
	.049	8624	1437	1.36	2.72	6.80	10.2	13.6	20.4
	.058	10,208	1701	1.27	2.54	6.34	9.51	12.7	19.0
	.065	11,440	1907	1.20	2.40	6.00	9.00	12.0	18.0
	.072	12,672	2112	1.13	2.26	5.66	8.49	11.3	17.0
	.083	14,608	2435	1.03	2.06	5.16	7.73	10.3	15.5
	.095	16,720	2787	.926	1.85	4.63	6.95	9.26	13.9
3/4	.049	7187	1198	2.08	4.17	10.4	15.6	20.8	31.2
	.058	8507	1418	1.97	3.93	9.84	14.8	19.7	29.6
	.065	9533	1589	1.88	3.76	9.41	14.1	18.8	28.2
	.072	10,560	1760	1.75	3.51	8.77	13.2	17.5	26.4
	.083	12,173	2029	1.67	3.34	8.35	12.5	16.7	25.0
					3.07	7.67	11.5	15.3	23.0
							10.4	13.9	

Figure 14.5 Size and capability chart for tubing.

Chapter 14 Fluid Conductors and Connectors

Tubing Connectors

Tubing connectors can be a flared type, compression fitting type, an elastomeric seal type or welded type as shown in Figure 14.6. Tubing connectors do not rely on threads to seal, but use a metal-to-metal seal, a welded seal or an elastomeric "O" ring seal. All of these types are a deterrent to leakage when properly installed and securely tightened. Connectors using elastomeric ("O" ring) sealing are growing in popularity as manufacturers pursue "Zero Leak" performance.

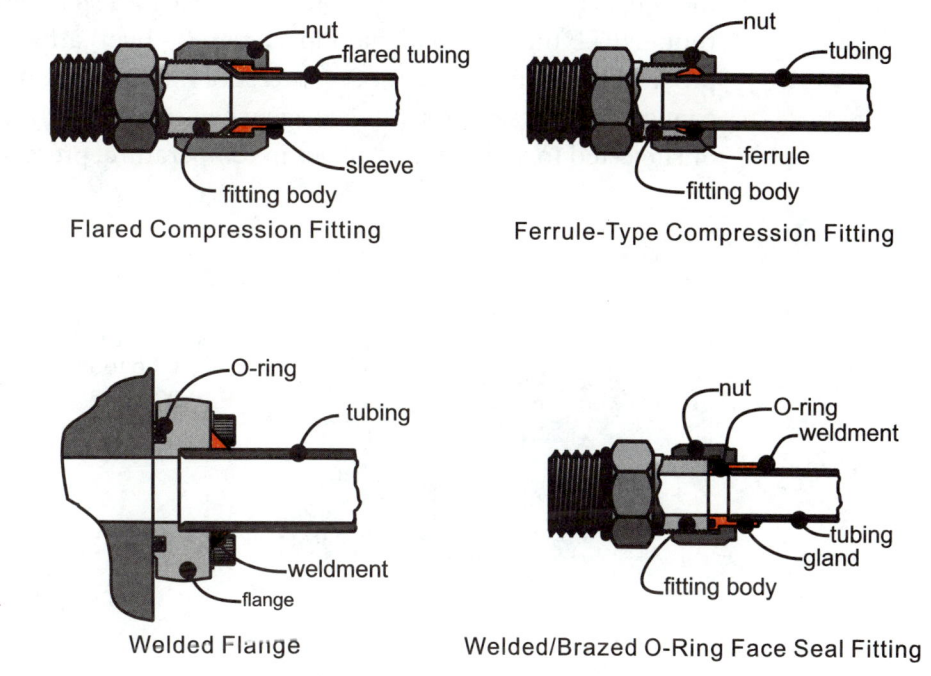

Figure 14.6 Typical tubing connectors.

Probably the most popular tubing connections are the 37° or 45° flared compression fittings. They are metal-to-metal seals created by squeezing the flared tubing against the matching surface of the fitting with a compression sleeve and nut. The sleeve extends well beyond the nut to alleviate the effects of vibration.

Another type of compression fitting uses a "ferrule" to form the metal-to-metal seal around the tube. The tube will be distorted with this type of connector, but does not have to be flared.

The welded connectors shown in Figure 14.6 both rely on an "O" ring to seal, and both are very effective. An important characteristic of these fittings is that there is no relative movement between the "O" ring and the sealing surface during the tightening process.

Pipe

Pipe is not recommended for hydraulic applications, both for its difficulty to install and for the difficulty in maintaining leak-free connections. Pipe connections also have a tendency to introduce foreign material into the hydraulic system.

335

Chapter 14 Fluid Conductors and Connectors

Threads on pipe and connectors are tapered, such that when they are tightened together, an interference fit between the two mating threads creates a seal. The tighter the connection is, the less likely it is to leak. There is never complete assurance that connections are totally leak-free, however, and when a leak is found, it is frequently difficult to correct. Furthermore, air may enter a system through a connection even when fluid may not leak out, making detection very difficult.

NPT threads (National Pipe Tapered) attempt to seal the connection at the sides of the threads (see Figure 14.7), potentially leaving a spiral leakage path at the thread root. NPTF threads (National Pipe Tapered - Fuel) attempt to overcome this problem by sealing at the root and crest of the threads. Although this is a better connection, it is still prone to leakage when subsequently loosened and re-tightened, or subjected to alternating changes in temperature, pressure or external force.

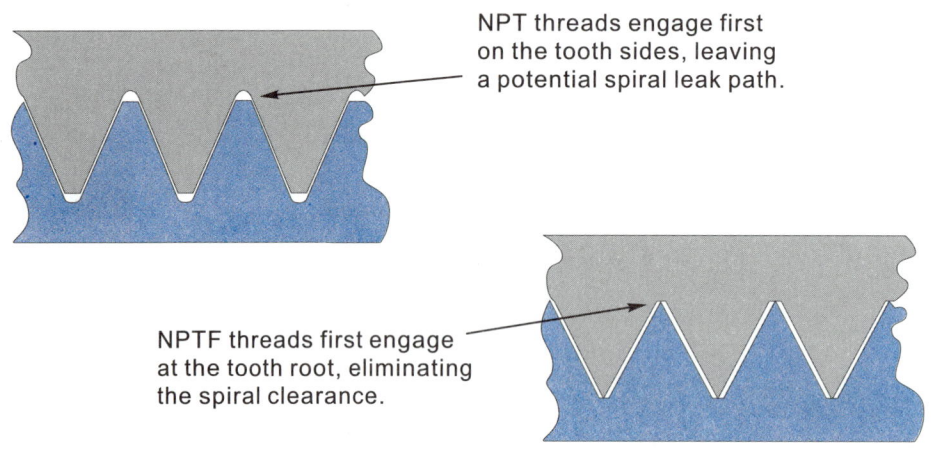

Figure 14.7 Pipe threads are not recommended for hydraulic systems. If used, however, they must be NPTF.

Mating Connectors

Several types of common mating connectors are shown in Figure 14.8. They rely on metal-to-metal sealing, such as the flared or pipe fittings, or on an "O" ring seal. The "O" ring seal is generally considered the most reliable type for assuring a leak-free connection. Of the "O" ring types, the 4-bolt flange and the face seal are considered the most reliable because there is no relative movement between the "O" ring and the mating surface during the tightening process.

Chapter 14 Fluid Conductors and Connectors

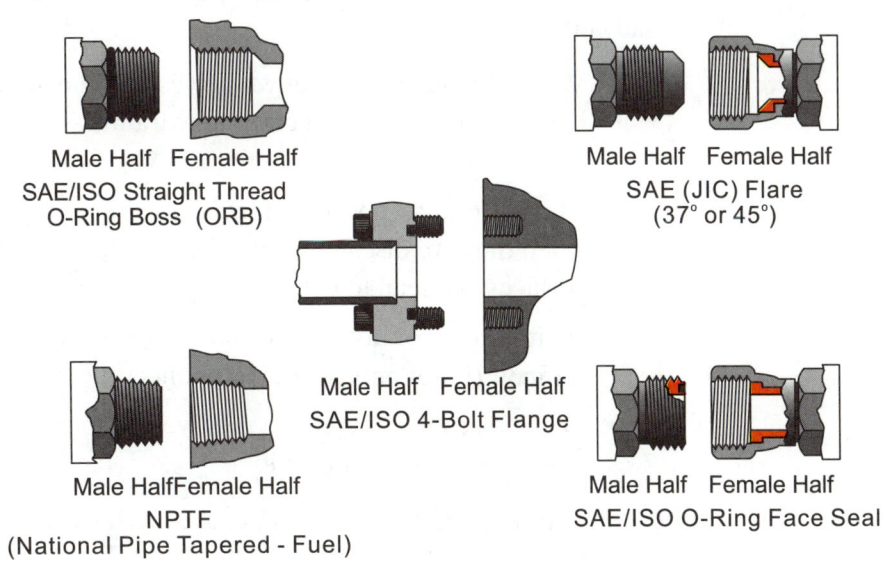

Figure 14.8 Typical mating connectors.

Quick-Disconnect Couplings

Frequently, mobile equipment requires hydraulically operated attachments to be repeatedly connected and disconnected from a primary power source. Examples are hydraulically operated power tools, or "trail behind" farm equipment that is periodically attached and removed. Quick-disconnect couplings are a convenient method of repeatedly making or breaking the hydraulic connection.

Figure 14.9 illustrates the elements of a quick-disconnect coupling. The male and female portions are pushed together to make the connection, and are then locked in place by a spring loaded locking ring. The action of pushing the two halves together forces the check valves open, opening a fluid path within the connection.

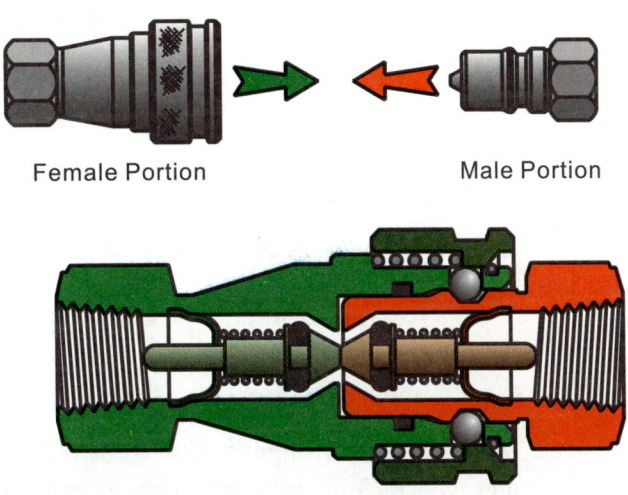

Figure 14.9 Quick-disconnect coupling.

The notable advantage quick-disconnects offer is offset by some serious disadvantages:

- When disconnected, the two halves of the coupling are subject to contamination that will enter a hydraulic system when the two halves are reconnected.
- There are small pockets of hydraulic fluid that drain from the couplings when they are disconnected. This can be unsightly, cause slippery surfaces and lead to environmental contamination.
- If either one, or both, of the couplings are under pressure, it is extremely difficult (if not impossible) to connect the coupling halves.
- Quick disconnects create a relatively high pressure drop, that could cause a significant power loss.

Manufacturers of quick-disconnect couplings have addressed some of these disadvantages:

- A "flat-faced" design coupling is available, as shown in Figure 14.10, that eliminates the pockets that contain fluid during the disconnect process. This can reduce the amount of fluid leakage during disconnect to near zero, and also provide a flat surface that can be wiped free of large contamination particles before they are reconnected.

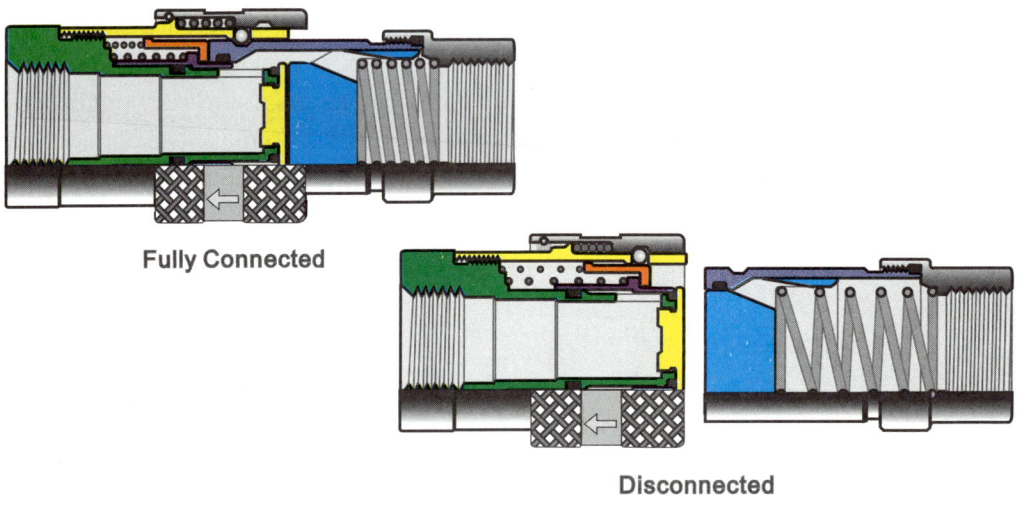

Figure 14.10 Aeroquip DryConnect™ flat-faced quick-disconnect coupling.

- A screw-together type of coupling, where the locking ring threads onto the male connector, can be used if one of the connectors is going to be under pressure.

Hydraulic fluid contamination can still be an issue with quick-disconnect couplings, however, including the ingestion of air into the system. When quick-disconnect couplings are used, properly designed filtration and reservoirs are vital to address the inherent contamination and air problems.

CHAPTER 15 .. Reservoirs

Several containers compete for space on mobile vehicles to store fluids needed to operate: fuel, coolant, lubricating fluid, and hydraulic fluid. These various containers, commonly referred to as "tanks", are more properly called reservoirs.

Hydraulic reservoirs must serve a variety of functions besides being a container for the system fluid. Reservoirs are looked upon to provide sufficient, well conditioned fluid to the hydraulic pump and the hydraulic system, helping to insure predictable operation and long component life.

There are several competing factors relative to installing an adequate reservoir on mobile equipment, not the least of which is space. Some other factors are cost, complexity, service access and capacity. Naturally, some compromises must be made, and these will be discussed in this chapter.

Basic Functions of a Reservoir

Besides being a container for the fluid used in a hydraulic system, reservoirs perform a variety of functions beneficial to systems and components (See Figure 15.1):

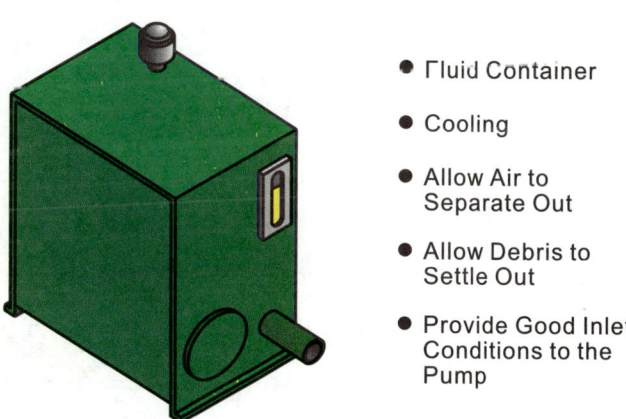

- Fluid Container
- Cooling
- Allow Air to Separate Out
- Allow Debris to Settle Out
- Provide Good Inlet Conditions to the Pump

Figure 15.1 Basic functions of a reservoir

- **Cooling** - Heat generated due to inefficiencies in hydraulic systems is mostly carried to the reservoir, where it must be given off to the atmosphere. If the reservoir is not capable of discharging as much heat as the system is generating, then a cooler must be installed.

- **Air separation** - Air is frequently introduced into hydraulic fluid through stroking cylinders, leaking pump inlet lines, cascading fluid in reservoirs and other sources. It is necessary to remove as much of the air as possible to assure a smooth, well controlled hydraulic system. The reservoir is an excellent place to allow fluid to "rest" briefly, giving the air time to rise to the surface and dissipate.

Chapter 15 Reservoirs

- **Debris separation** - Contaminants that occur in hydraulic systems, either entering from the outside (ingress) or generated from the components that make up the system (internally generated), must be removed to prevent further damage. Larger contaminants, especially metallic ones, can often be removed by allowing them to settle out of the fluid in a reservoir.

- **Good pump inlet conditions** - Reservoirs can help supply fluid to a pump by providing good inlet conditions. Part of that job is done by cooling the fluid and allowing the air and contaminants to settle out. Other things that properly designed reservoirs can do is provide a smooth flow of fluid to the pump, rather than turbulent flow, and to provide fluid at adequate pressure to prevent pump cavitation. Through the proper application of baffles, keeping the fluid level above the pump and/or applying air pressure inside the reservoir, pump inlet conditions can be greatly assisted.

Reservoir Requirements

If space and cost were not a limitation in the design of mobile equipment, reservoirs would contain all of the following elements (See Figure 15.2):

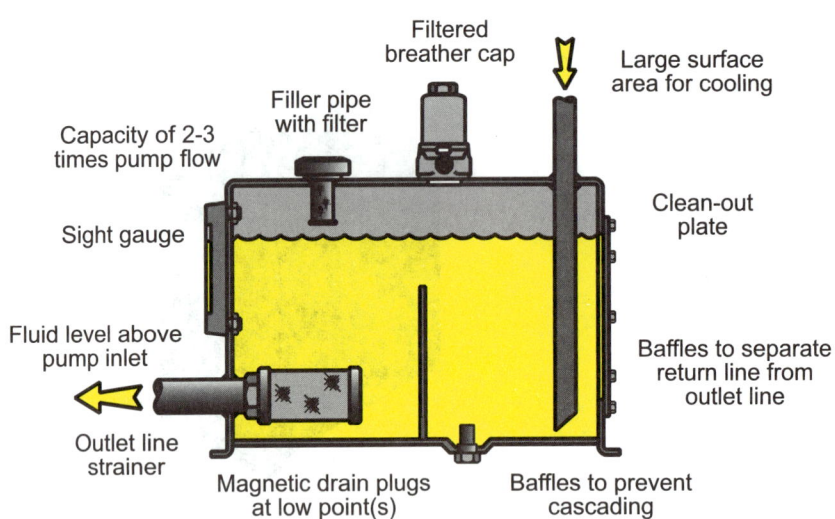

Figure 15.2 Elements of an ideal reservoir.

- Two to three times pump flow as a volume capacity. For example, if the system pump flow is 30 GPM, the reservoir capacity would ideally be 60 to 90 gallons.

- A sealed filler pipe with a filter to clean any fluid being added.
- A filtered breather cap, allowing air to enter and leave the reservoir while being cleaned and de-moisturized on the way.
- A sight gauge to allow visible checking of the fluid level without opening the reservoir to atmosphere. Electronic fluid level indicators can also be used for this purpose.
- Magnetic drain plugs installed at low points in the reservoir to attract and hold metallic particles that may be in the fluid.
- A large clean-out plate that can be easily removed to allow cleaning of the inside of the reservoir on a periodic basis.
- Large, exposed surface areas to enhance cooling through heat dissipation to the atmosphere.
- Strategically placed baffles inside the reservoir that would interrupt any tendency of the fluid coming into the reservoir from "streaming" across to the reservoir outlet.
- Strategically placed baffles that would relieve the "cascading" of fluid (sloshing) inside the reservoir due to vehicle motion, and to prevent any areas of the reservoir from becoming low in fluid level due to cascading or vehicle attitude.
- An elevated installation that would provide a positive pressure, or head, at the pump inlet.

Space and cost are limited, however. Many compromises must be made and alternative provisions must be provided so that as many of the above features as possible can be included.

RESERVOIR LOCATION

Locating the Reservoir

In most mobile applications, the size of a hydraulic reservoir will be dictated, or at least strongly influenced, by the amount of room available to place it. This frequently calls for a great deal of creativity on the designer's part; providing sufficient fluid capacity while still working within the confines of the equipment.

Chapter 15 Reservoirs

Figure 15.3 shows a common reservoir application where the machine structure is used as the reservoir. The vertical upright of the loader frame can hold many gallons of fluid, and in some loaders, both uprights are connected together by a tube, which doubles the capacity.

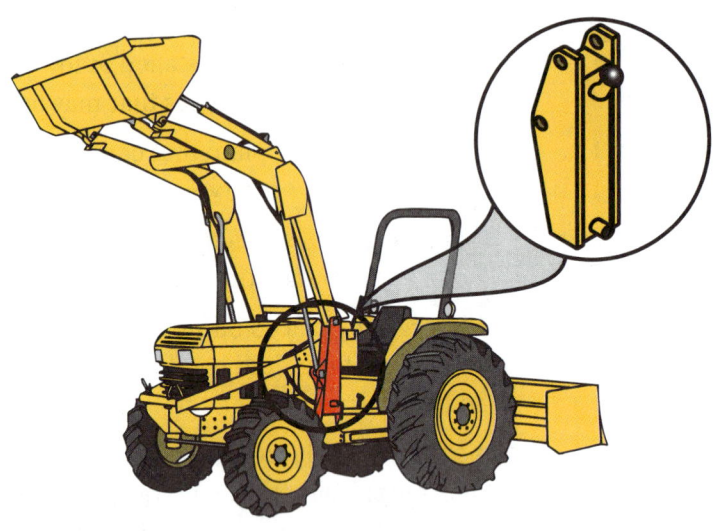

Figure 15.3 Loader uprights being used as a reservoir.

Figure 15.4 shows where the frame member of an industrial lift truck is used for a reservoir. Most sit-down lift trucks of this variety use one side of the frame for the hydraulic fluid, and the other side for the vehicle fuel. Adequate hydraulic fluid in lift trucks is essential. The lift cylinder is usually single acting, and there must be sufficient fluid to fill it when fully extended, while still leaving enough fluid in the reservoir for other hydraulic functions.

Figure 15.4 Lift truck frame being used as a reservoir.

Some machines, such as the turf mower in Figure 15.5, require a specially designed reservoir to fit into a very confined envelope.

Figure 15.5 Unique design reservoir for a turf tractor.

Combining Reservoirs

There are many mobile applications where two or more reservoir requirements have been combined so that the same fluid and reservoir are used for multiple systems. Some examples are:

- Industrial lift trucks (Figure 15.6) typically use a single reservoir, and frequently a single pump, to provide fluid to the operating system (lift, tilt and auxiliary functions), as well as to the power steering system.

Figure 15.6 Lift truck systems usually incorporate power steering using the same reservoir.

- Large farm tractors frequently use the same fluid and reservoir to provide power for their operating system (implement operation) and their transmission / differential lubrication system. Figure 15.7 shows that the transmission and differential housings actually become the reservoir.

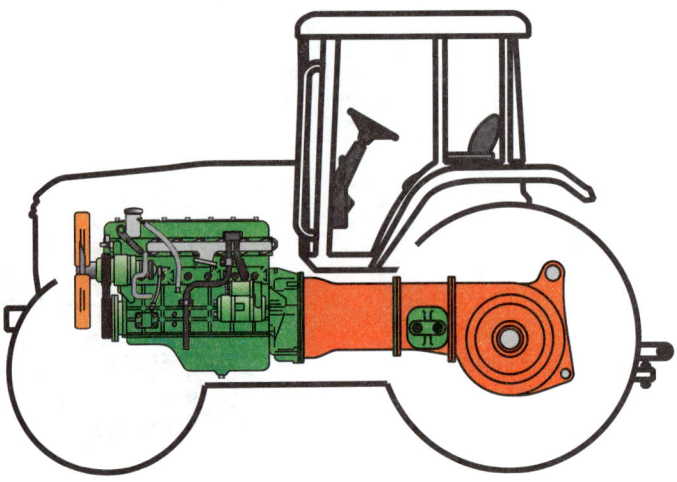

Figure 15.7 Farm tractor transmission and differential housings are frequently used as a hydraulic reservoir.

- Skid-Steer loaders (Figure 15.8) usually use the same reservoir and fluid for their hydrostatic transmission as well as their implement operating systems.

Figure 15.8 Typical skid-steer loader uses the same reservoir for the system and hydrostatic drive.

Chapter 15 Reservoirs

Sizing the Reservoir

One of the difficult compromises a designer must make relates to the volume of the reservoir. Two or three times pump flow would result in a reservoir size that normally precludes any heating or capacity problems. However, more fluid requires more space, and it also adds weight to a vehicle that would increase fuel consumption and reduce the number of hours of operation between refueling. A larger fuel tank may be required, which increases the physical size of the machine, which also adds more weight. Obviously, it is advantageous to keep the size of the hydraulic reservoir as small as possible.

A number of factors can be combined to reduce the reservoir size (Figure 15.9):

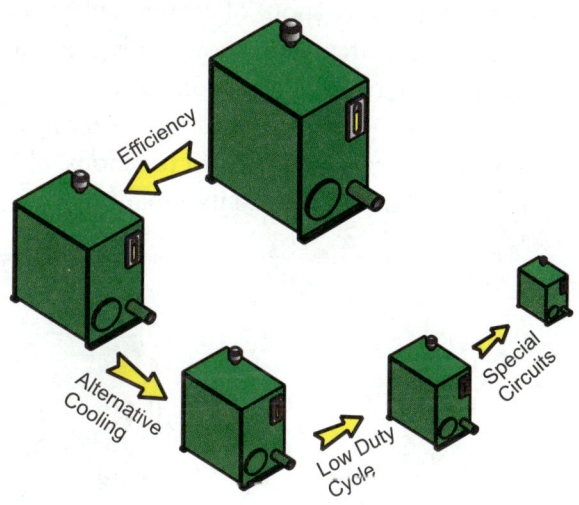

Figure 15.9 Several factors help reduce the reservoir size.

- Hydraulic system efficiency can be kept high by using products intended for that purpose; valves with low pressure drop, variable displacement pumps, load sensing systems, etc. Higher efficiency will result in less heat generated, that would normally have to be dissipated through the reservoir.

- Mobile systems generally have a low "duty cycle", meaning the percentage of time at high loads and pressures is small, with more time spent at low loads. This reduces the heat generation, as well as the stresses on components, that might generate excessive internal contamination.

- Other methods of cooling can be added to systems, such as oil-to-air or oil-to-water coolers. Coolers located in or near the engine cooling radiator are quite common.

- Specially designed hydraulic circuits can be developed which reduce the size of the prime mover. Torque limiting pump controls is an example, where the combination of high pressure and high flow is not allowed to occur simultaneously. A smaller engine results, which will deliver less power to the hydraulic system, and reduce the heat that might otherwise occur.

Other factors must be considered that tend to force a larger reservoir:

- The fluid level must not get too low in the reservoir when many cylinders are extended at once, or when single acting cylinders are extended. Low fluid levels reduce heat rejection, and also open the possibility of air entering the system due to a whirlpool effect at the reservoir outlet.
- A low volume of fluid will mean that the fluid in the reservoir will move through quickly, allowing less time for air and contaminants to settle out. Furthermore, with fluid moving rapidly through a reservoir, turbulence will result and air will be introduced into the fluid.

RESERVOIR COMPONENTS

Filler Cap

Caps on the filler spout of a reservoir usually double as a breather and are referred to as a "filler breather", shown in Figure 15.10. The dual purpose is to provide a mesh screen opening for adding fluid to the reservoir, and providing a filtered path for air to move in and out of the reservoir as the fluid level changes. This is an economical approach to both functions, although the mesh size does not remove other than larger solid contaminants.

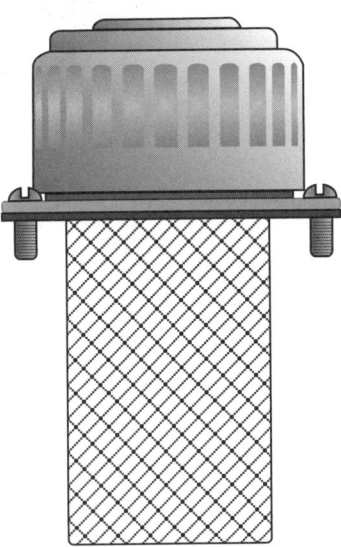

Figure 15.10 Combination filler/breather cap.

A better approach is to address these two functions separately. A filler cap that seals the opening reduces the opportunity for contamination to enter the reservoir, and a more thorough filtering of the incoming air can be done with a separate filter breather.

Breather

As linear actuators extend, the differential area causes the level of fluid in the reservoir to lower. Atmosphere flows into the reservoir as the fluid level drops, and then flows out again as cylinders retract and the fluid level rises. If the air is not properly cleaned, it will carry both moisture and contaminants into the reservoir.

The filler breather, mentioned previously, will remove some of the larger contaminant particles, but moisture and smaller contaminant particles will get through.

A better solution is to seal off the fill spout with a firmly sealed cap, and allow air to enter only through a breather filter. An example of a reservoir breather filter assembly is shown in Figure 15.11. The breather must be sized properly based on the amount of air it has to handle, which is based on the velocity, size and number of cylinders in the system.

Figure 15.11 Reservoir breather filter.

When moisture, or high atmospheric humidity, are a problem, a moisture removing filter is a preferable solution. This type, illustrated by the "H_2O-Gate" filter shown in Figure 15.12, will remove moisture from incoming air and also remove the contaminant particles.

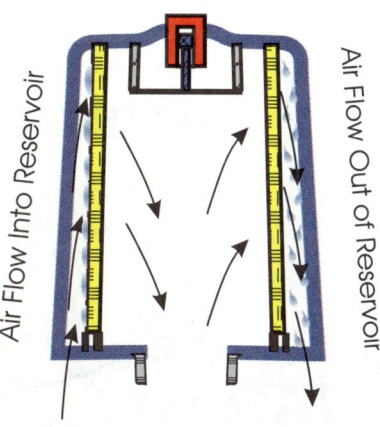

Figure 15.12 H_2O-Gate Breather allows moisture out and traps moisture entering.

Chapter 15 Reservoirs

Fluid Level Gauge The most economical and very common approach to checking fluid level is with a dip-stick assembly. A drawback to this approach is that every time the dip-stick is removed from the reservoir, contaminants are allowed to enter.

Other approaches are more costly, but preclude opening of the reservoir. Sight gauges, mounted on the side of the reservoir, are very effective if they can be placed in an easily viewed area. Barring this, an electric fluid level gauge, similar to a fuel gauge, is an effective solution.

Outlet Line Strainer Outlet line strainers, also called inlet filters or inlet screens (referring to the pump rather than the reservoir), are very common on mobile systems. This may be more traditional than functional. The strainers are usually constructed of 100 mesh screen, having openings of approximately 150 microns. They are intended to keep the larger solid contaminants from entering the hydraulic system.

Outlet strainers must be sized properly based on the hydraulic system pump size. If the strainer is too small, an inlet restriction is created that can lead to pump cavitation. Strainers can be obtained in a variety of sizes, and can be used in tandem if greater flow is required.

A drawback to outlet line strainers is that they are quite inaccessible for service and cleaning. If they become restricted due to excess contamination, they can cause cavitation and damage to system pumps. A more current approach is to assure clean fluid is maintained in the reservoir, precluding the need for an outlet strainer.

Baffles Reservoirs on mobile equipment require baffles, or dividing plates, for two reasons:

- The fluid returning from the system must circulate in the reservoir prior to going back into the system in order to allow air to settle out and heat to dissipate. Therefore, a baffle placed between the return port and the outlet port (Figure 15.13) will force the returning fluid to mix with cooler fluid, and to pass close to the surface to allow air separation. Without baffles, returning fluid will often "stream" directly from the return to the pump inlet without mixing and without discharging air.

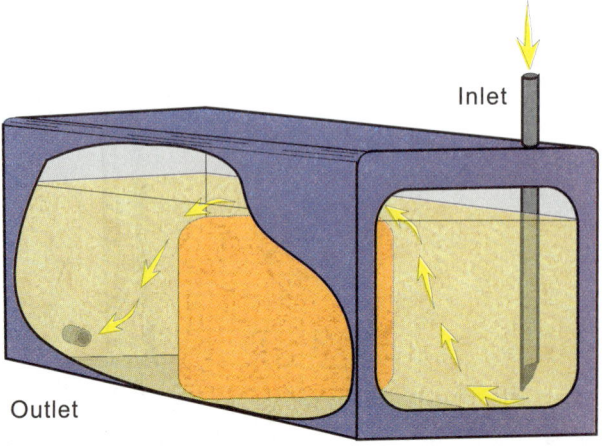

Figure 15.13 A baffle prevents fluid from "streaming" between inlet and outlet.

- Mobile equipment that changes direction frequently, or rapidly, will create violent sloshing in the reservoir. This can cause structural damage to the reservoir mounting or side walls, force fluid to eject through the filler cap or breather assembly, and cause fluid voids to occur near the outlet allowing air to enter the hydraulic system. Baffles that divide the reservoir into smaller sections can prevent these problems.

Properly designed reservoirs can use one or more baffles to satisfy both of the above requirements.

Magnetic Drain Plugs

Reservoirs must have drain plugs so that fluid can be drained for service or for replacement. A common practice is to use a plug with a strong magnet imbedded in it that will collect and hold metallic particles and keep them from re-entering the system.

Pressurized Reservoirs

Reservoirs provide fluid through hoses or tubing to a pump inlet. Atmospheric pressure in the reservoir acts on the surface of the fluid to push it into the pump. Also, if the reservoir is located above the pump, the weight of the fluid provides additional pressure (Figure 15.14). On mobile equipment, however, there are times when these two characteristics are insufficient to supply the pump with enough fluid.

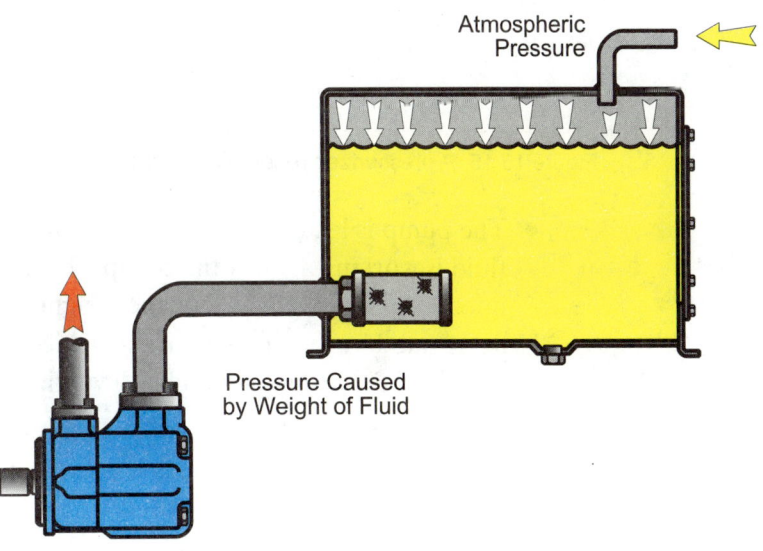

Figure 15.14 Pump inlet pressure is caused by atmospheric pressure and weight of the fluid.

Figure 15.15 shows three conditions as examples of when there may not be sufficient oil entering the pump inlet. When that is the case, cavitation will occur and the pump will be damaged.

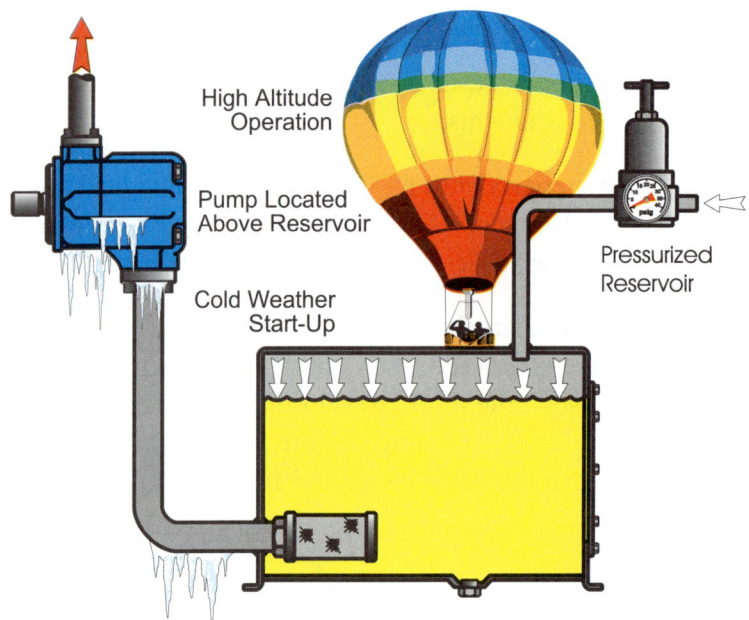

Figure 15.15 A pressurized reservoir can overcome poor pump inlet conditions.

- The pump is located above the reservoir, such that the weight of the fluid is working <u>against</u> the pump. There will be excess vacuum at the inlet of the pump caused by the weight of the fluid. This is often aggravated by the attitude of the vehicle. Facing up hill or down hill can often put the pump at a higher level than the reservoir.

- The equipment is located well above sea level, where the atmospheric pressure is reduced. There is less pressure available to force the fluid into the pump inlet.

- The fluid temperature is cold, so that the viscosity is low and it is more difficult to pump the fluid. It takes greater pressure to push a more viscous fluid.

These situations can be overcome by providing additional pressure in the reservoir to help push the fluid. Care must be taken in doing this so that the strength of the reservoir side wall is not exceeded, or that the seals in the reservoir and in the pump are not damaged by excess pressure.

Reservoirs can be pressurized by removing the breather and replacing it with an air valve, allowing thermal expansion or changing fluid level to create pressure. A small air compressor can also be used.

CHAPTER 16 .. **Hydraulic Fluids**

Technically, a fluid may be defined as any liquid or gas. For the purposes of a mobile hydraulics manual, however, we will consider only liquid fluids. More specifically, we will discuss the liquid fluids used in mobile hydraulic systems.

Hydraulic fluids have numerous requirements in hydraulic systems, as illustrated in Figure 16.1. Foremost, is the efficient transmission of power to linear and rotary actuators (hydraulic cylinders and motors). In addition to the transmission of power, the fluids are also required to: provide cooling of the system and its components via dissipation of heat through the reservoir and/or cooling devices; lubricate sliding or rotating surfaces in components; seal the running clearances in components to minimize internal leakage; and carry away contaminants in the system to filter(s) or to the reservoir where they are allowed to settle out of the fluid.

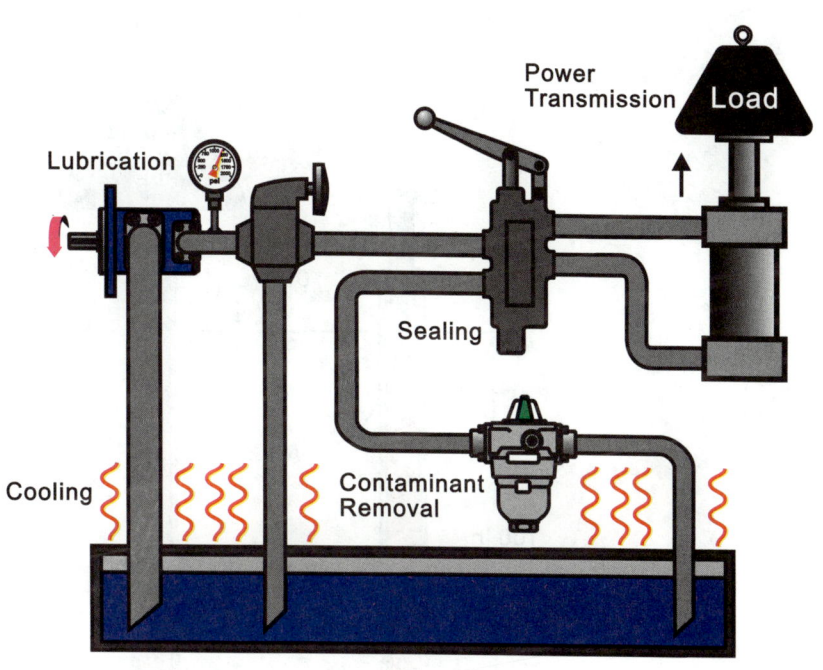

Figure 16.1 Essential properties of fluids in a hydraulic system.

In addition to the above listed attributes, fluids used in hydraulic applications are also required to be: compatible with component materials such as seals and hose interior linings; resistant to foaming; able to expurgate air and water; used over wide operating temperature differentials at acceptable viscosities.

To perform all of these tasks, hydraulic fluids must have the proper characteristics which must be maintained for the life of the fluid in the system. Fluid failure in any of the areas mentioned can lead to varying degrees of component or even total system failure. This chapter concerns itself with the variety of fluids used in hydraulic systems, the characteristics and limitations of these fluids and their application advantages and disadvantages, and some of their basic maintenance requirements. The next chapter, Chapter 17 - Fluid Conditioning, deals with the general care and practices required to provide long and trouble-free operation of hydraulic systems.

Chapter 16 Hydraulic Fluids

PURPOSES OF HYDRAULIC FLUIDS

Power Transmission The fluids used in the hydraulic systems of mobile and stationary machinery must be effective in the transmission of power from the source of power, e.g., an internal combustion engine in the case of mobile equipment; to provide consistent and reliable response, safe operation and optimum efficiency.

To ensure responsiveness of actuation or "stiffness" in a hydraulic circuit, the fluid must experience very little compression, even under high pressures. Typical petroleum-based fluids are said to be virtually incompressible. In fact, even petroleum-based fluids will compress very slightly - 0.4% at 1000 psi and up to 1.1% at 3000 psi operating pressure (Figure 16.2). At a constant operating pressure the oil remains compressed at a given value. However, with the dynamics of loads in mobile machinery, slight decompression or compression can occur and affect actuation slightly. Typically, this is no cause for concern.

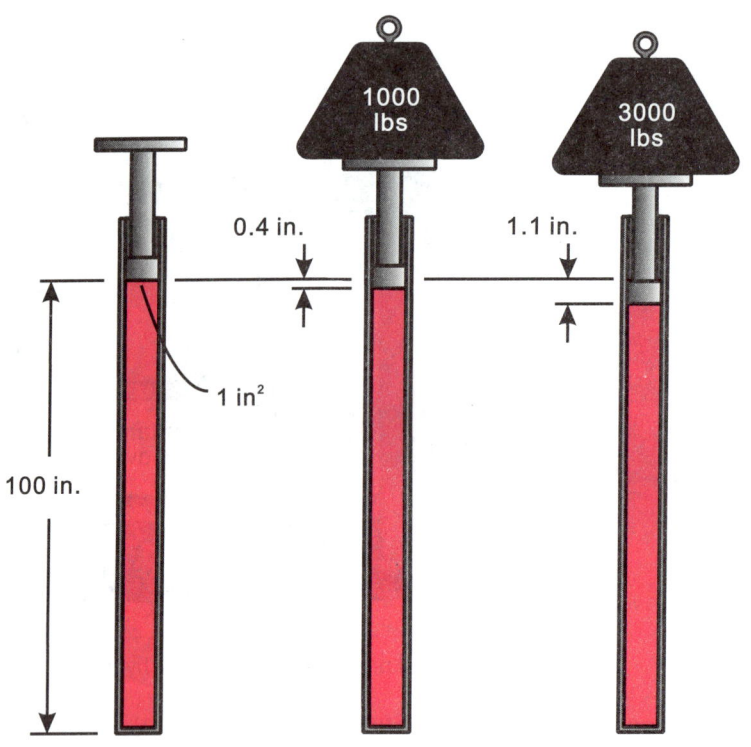

Figure 16.2 Compressibility of fluids.

Because of this property of virtual incompressibility, properly maintained mobile hydraulic systems are extremely responsive and reliable.

Chapter 16 Hydraulic Fluids

Lubrication

Lubricity is defined as the property of a fluid to impart low friction under boundary lubrication conditions. Simply put, we want the fluids we use in our hydraulic systems to prevent excess wear of components and excess production of heat via their properties of lubrication.

All hydraulic systems contain components with moving parts that have the potential to come in contact with each other, particularly under pressure. To minimize wear and reduce heat and excessive damage to fluids and components, all fluids used in hydraulics must have the ability to provide lubrication under a variety of operating conditions.

Ideally, we would like always to have what is called "full-film" lubrication between all moving parts as shown in Figure 16.3. This is a condition in which the metal parts of components are held completely apart by a "full-film" of lubricating fluid. The microscopic "peaks" on the mating surfaces of the parts, called asperities, cannot come in contact with one another causing wear.

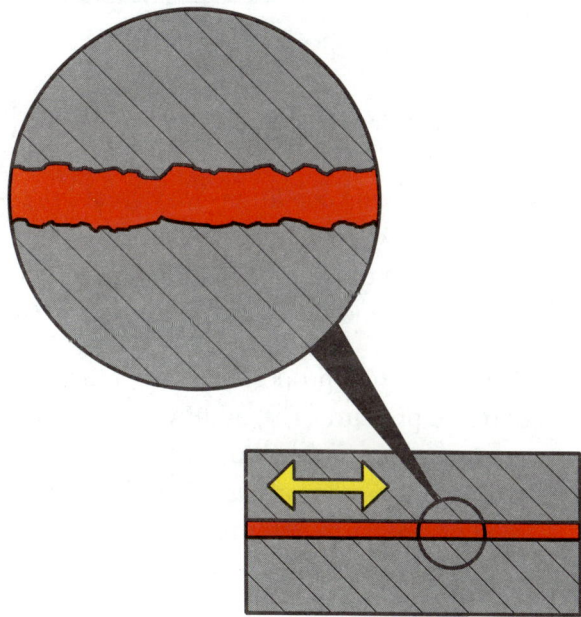

Figure 16.3 Full-film lubrication.

However, with higher pressure demands, the clearances between moving surfaces have been reduced by manufacturers to reduce leakage at higher pressure drops across clearances. This creates a condition called "boundary lubrication". As shown in Figure 16.4, with boundary lubrication the asperities on the surfaces of mated parts can now contact one another. This can cause galling of the component surfaces, which creates work-hardened metal wear contamination, and severe heat, which will increase the oxidation rate of fluids, reducing their useful life.

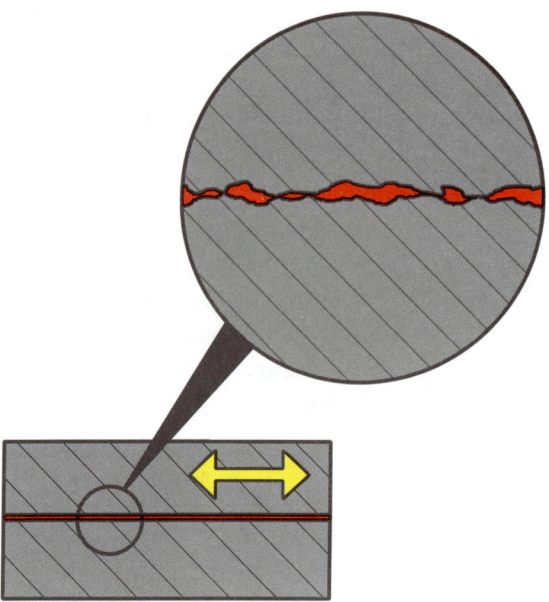

Figure 16.4 Boundary lubrication.

Severe working conditions may create pressure "spikes" or transient pressure conditions within components that force a reduction of running clearances. This may cause a break in the lubrication film between running clearances resulting in component and fluid destruction via excess wear and heat.

The lubrication properties of fluids are enhanced by the addition of anti-wear and extreme pressure (EP) additives. When using petroleum-based fluids, these compounds are added to the base stock, which also has specific properties. Because additives typically have a life shorter than that of the base stock, they may be replenished when determined necessary by competent spectrographic fluid analysis, as long as the base stock of the oil is still viable.

These types of additives are also present in synthetic and biodegradable fluids.

Chapter 16 Hydraulic Fluids

Sealing

Clearances inside hydraulic components cause internal leakage that affects the efficiency of systems as well as the potential to create excess heat. As shown in Figure 16.5, we rely on the fluid in the system to minimize leakage across these clearances to improve efficiencies and reduce the production of heat. The physical size of clearances, pressure drop across the clearances and the operating viscosity of the fluid determine the leakage rate.

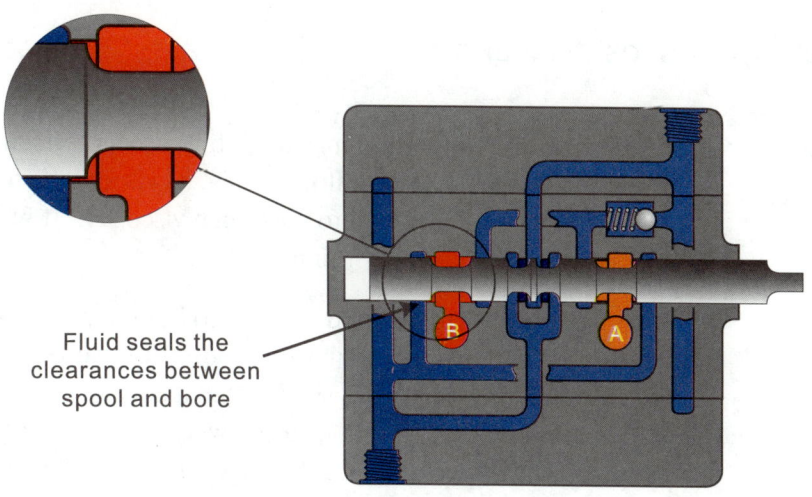

Figure 16.5 Fluid helps to seal between components.

Cooling

Any fluid used in mobile machinery absorbs and carries heat away from heat generating components such as cylinders and pumps. As discussed in Chapter 15, the fluid then must be allowed to circulate as much as possible against the heat dissipating sides of a reservoir before it is allowed to reenter the pump.

Some system designs may not allow sufficient transfer of fluid to the reservoir, particularly with long lines from the rod end of cylinders. This can cause a buildup of heat and oxidized fluid in an isolated segment of a circuit and result in destruction of the fluid and components. Provision should be made in machine design for the ability to "flush" these segments regularly to prevent cumulative damage to components and the fluid.

Contaminant Removal

While contamination control will be examined in depth in Chapter 17, the transport, filtration and settling out of contamination are important functions of the hydraulic fluid. All types of contamination, solid, fluid or gas, must be dealt with via the hydraulic fluid. It is the primary carrier of contamination through filters, a baffled reservoir, coolers, strainers and other fluid conditioning devices designed to remove it from the system.

Demulsifiers, anti-foaming agents and rust and corrosion inhibitors are some of the additives put in hydraulic fluids to aid in the removal or prevention of contaminants.

Chapter 16 Hydraulic Fluids

FLUID PROPERTIES To adequately carry out the previously stated purposes, the fluids used in hydraulic systems must possess, to varying degrees, specific desirable characteristics. Not all fluids will have all of the necessary attributes in equal strength. Consequently, when selecting fluids it is sometimes necessary to compromise some properties in favor of others that may be more important for a specific application requirement. In general, these properties include: viscosity and viscosity index, pour point, lubricating ability, oxidation resistance, compatibility with system elements, rust and corrosion protection and demulsibility.

VISCOSITY AND VISCOSITY INDEX (VI)

Viscosity The viscosity of a fluid may be defined as its resistance to flow at a given temperature. If a fluid flows easily, its viscosity is low. A fluid that flows with difficulty would be said to have high viscosity. It has more resistance to flow than the lower viscosity or "thinner" fluid.

Viscosity affects the fluid's ability to be pumped, transmitted through the system, carry a load and maintain separation (lubrication) between moving surfaces. The selection of proper viscosity is often a compromise in order to optimize system performance. Too high, or too low a viscosity for a given system could present problems of performance, leakage, energy usage, etc.

Viscosity too high (fluid is too thick):

- High resistance to flow
- Increased energy consumption due to increased friction, increased input torque requirement at the pump
- High temperatures created by power loss to friction
- Increased pressure drops (ΔP) due to increased resistance to flow
- Slow or sluggish operation/actuation
- Inefficient separation of air from the oil in the reservoir
- Pump cavitation

Viscosity too low (fluid is too thin):

- Increased internal leakage
- Excess wear. Seizure, particularly of pumps, could occur under heavy load because of a breakdown in lubrication film between clearances of moving parts
- Decreased pump efficiency (volumetric) due to increased leakage and possible cylinder blow-by. This could cause increased cycle times or slower machine operation.
- Internal leakage causing an increase in operating temperatures

Chapter 16 Hydraulic Fluids

There are numerous methods for defining viscosity. Among these are: absolute or dynamic viscosity (centipoise/cP), kinematic viscosity (centistoke/cSt) and relative viscosity as measured in Saybolt Universal Seconds (SUS or SSU). Most hydraulic systems run with oil in the range of 150 to 300 SUS with typical ISO viscosity grade (ISOVG) range of ISOVG-22 to -68.

Viscosity Index

The viscosity of virtually all fluids used in hydraulic systems is affected by variations in temperature. Viscosity Index or VI is an arbitrary number that characterizes the variation of viscosity of a fluid with variations of temperature. Simply put, the higher a fluid's VI number, the less change there is in the fluid's viscosity over a given range of temperature as shown in Figure 16.6.

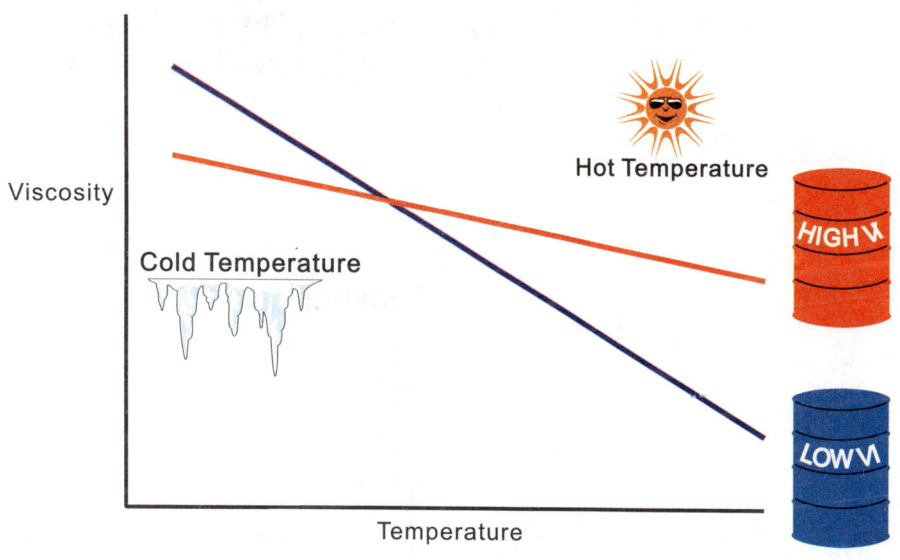

Figure 16.6 The effect of improved viscosity index.

Multiple viscosity oils typically include additives (polymers) to improve viscosity index. These oils can exhibit temporary and permanent decrease of viscosity due to oil shear during hydraulic machine operation. Consequently, multiple viscosity fluids with high shear stability are desirable.

Viscosity and Temperature

The viscosity of hydraulic fluids, particularly petroleum-based fluids and vegetable oils, is directly and sometimes adversely affected by changes in temperature. For this reason it is essential that the startup and operating temperatures of mobile machinery be monitored. Machinery should not be put into high speed or heavily loaded operation until the system fluid is warmed up to operating temperatures to provide adequate lubrication.

Chapter 16 Hydraulic Fluids

Arctic conditions will cause an increase in viscosity that could result in pump cavitation and severe damage to the hydraulic system at startup. Many hydraulic components for mobile applications are designed with cold starting conditions taken into consideration and offer some protection against cavitation until the fluid is warmed to operating temperatures. Unless otherwise indicated by the manufacturer, mobile machinery is typically run at fast idle until satisfactory operating temperature is attained.

Excessive heat in a system can cause a loss of viscosity. This will result in severe wear due to a loss of lubricity as well as the destruction of the fluid through oxidation. Severe oxidation, in turn, may cause the fluid to thicken by causing varnish, sludge, corrosive acids, and the destruction of additives.

SAE Viscosity Numbers

The Society of Automotive Engineers (SAE) has established numbers to specify ranges of viscosity for engine oils at specified test temperatures, illustrated in Figure 16.7. Winter numbers (0W, 5W, 10W, 15W, etc.) are specified viscosity ranges at cold temperatures. Summer oil viscosity numbers (20, 30, 40, etc.) are measured at 212°F.

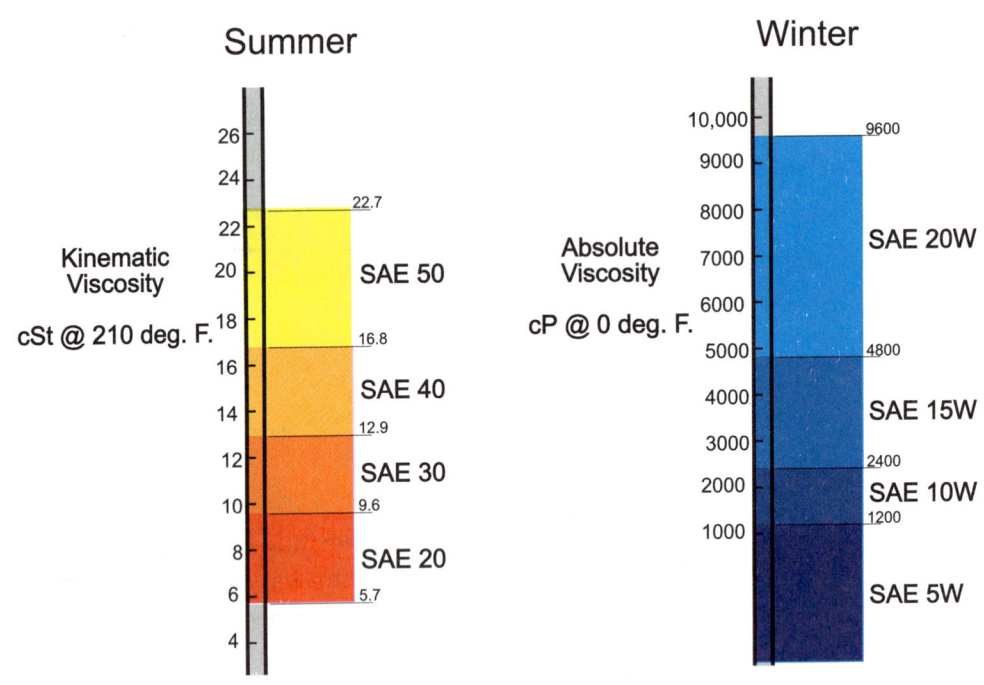

Figure 16.7 SAE viscosity numbers.

Chapter 16 Hydraulic Fluids

ISO Viscosity Grades

International Standards Organization (ISO) viscosity grades are shown in Figure 16.8. The ISO viscosity grade number represents a range of kinematic viscosity values at 40°C (104°F), and is the midpoint number of the range. The ISO designation is becoming more popular, and many fluid suppliers are incorporating the designation in their brand description.

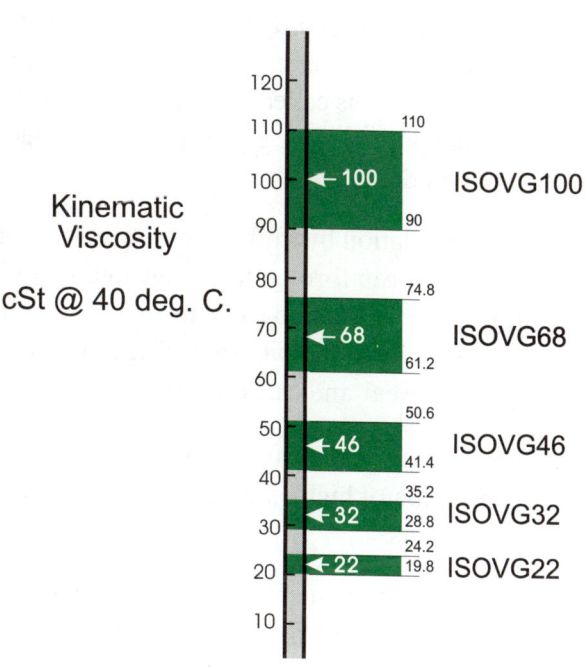

Figure 16.8 ISO viscosity grades.

Viscosity Analysis

It is a useful tool, particularly with mobile equipment exposed to radical changes in temperature, to obtain occasional verifications of a fluid's adherence to its specified viscosity. Of course, this specified viscosity should be defined within the operating parameters of the application. When viscosity analysis is used as a fluid monitoring tool, you must realize that the actual viscosities of new fluids can vary somewhat from their specifications. Consequently, *changes* in viscosity over a period of time are usually considered more important than an actual reading. Changes of more than 10% from the original reading are considered an indication of problems.

Lubricity

As stated under Fluid Properties and illustrated by Figure 16-3, it is desirable for hydraulic component moving parts to have sufficient clearance to run on a substantial film of fluid. This condition is called *full film lubrication* and does not allow contact between the moving parts. However, in certain types of applications and in higher performance equipment, increased speeds and pressure along with smaller clearances, cause the fluid to be squeezed very thin as in Figure 16-4. This creates the previously described condition called *boundary lubrication*. To prevent excess wear of potential contact between moving parts, extreme pressure (EP) and antiwear additives are required.

Chapter 16 Hydraulic Fluids

Pour Point

Pour point is the lowest temperature at which a fluid will flow. It is an important specification if the hydraulic system will be exposed to extremely low temperatures. Typically, the pour point of a selected fluid should be 20°F (-11°C) below the lowest temperature to be encountered. Pump manufacturers provide information about the maximum viscosity acceptable for a specific pump or other components.

Oxidation Resistance

Oxidation, or chemical combination of the oil with oxygen, can seriously reduce the service life of fluids. Both petroleum and vegetable oils are susceptible to oxidation, particularly in hydraulic applications. In petroleum, oxygen readily combines with the carbon and hydrogen present. Low temperature oxidation of vegetable oils is called autoxidation and should be prevented by low temperature storage of these oils, generally with nitrogen introduced in the storage vessel to deter oxidation.

Most oxidation byproducts are soluble in oil causing additional reactions to take place that can form gum, sludge and varnish. The first stage products which stay in the oil are acidic and can cause corrosion throughout the system and may increase the viscosity of the oil. Insoluble gums, sludge and varnish can plug orifices, increase wear and cause valves to stick.

Operating temperature of petroleum and vegetable-based oils is important for the prevention of high oxidation rates and the production of destructive oxidation byproducts.

Catalysts in Oxidation

There may be numerous oxidation catalysts present in poorly maintained hydraulic systems, i.e., conditions or elements that increase the potential for oxidation. Heat, pressure, contaminants, water, metal surfaces, some metals, and agitation all accelerate oxidation. Temperature is particularly important. Below 140°F (60°C), petroleum oil oxidizes very slowly. However, the rate of oxidation approximately doubles for every 18°F (10°C) increase in operating temperature above 140°F.

Rust and Corrosion Prevention

Rust is the chemical union of ferrous metal (iron or steel) with oxygen. Corrosion is a chemical reaction between a metal and a chemical - typically an acid. Acids result from the union of water with some elements.

Because it is extremely difficult to keep air and airborne moisture out of hydraulic systems, there is always the possibility that rust and corrosion may occur. With corrosion, particles of metal are dissolved and washed away. Both rust and corrosion contaminate the system and increase component wear. They may also increase internal leakage past the affected parts causing high temperatures, and may also cause components to seize through heat and closure of running clearances with debris.

Particular care must be taken when operating and cleaning mobile equipment to prevent the contamination of the hydraulic system with water or cleaning solvents - either from the environment or when washing down equipment.

Chapter 16 Hydraulic Fluids

Demulsibility

Demulsibility is the ability of a fluid to separate out or reject water. The negative affects of water contamination will be discussed in Chapter 17. However, it is desirable when using petroleum or vegetable-based oils that water content be kept to an absolute minimum and that free water not be allowed into the hydraulic system.

Fire Resistance and Flash Point

In some applications, environmental conditions may dictate that the fluid used in hydraulic systems must have fire resistant properties.

The *flash point* or temporary ignition point of typical hydrocarbons (petroleum oils) may be as low as 200°F or be higher than 500°F. It is important to analyze the working environment of your specific application to determine fire hazards. Governmental regulations, local or Federal, could also impact your choices. *Fire point* is the temperature the fluid must attain for continuous burning. Typical petroleum products may continue to burn even after the point of ignition is removed.

Fire resistance in fluids generally means that, while most fluids can be ignited under the right conditions, a fire resistant fluid will not sustain combustion when an ignition source is removed. Resistant fluids will not allow flame to flash back to the ignition source.

ADDITIVES

Virtually all fluids used in hydraulic systems contain a variety of additives to improve or augment the performance of the fluid under various conditions. It is important that your fluid supplier understand the nature of your application: the environment, the types of components and their manufacturer's specifications relative to fluids, duty cycles, loads (pressure), storage ability, temperature extremes, and any unusual or special considerations in the operation of your mobile machinery that could affect the life of the fluid or its performance.

Although additives often enhance the natural abilities of the particular fluid being used, they must be monitored and kept at specified levels to prevent failure of the fluid and eventually the entire system.

Rust and Corrosion Inhibitors

Because virtually every environment contains some moisture, the addition of rust inhibitors to the hydraulic fluid will prevent the formation of iron oxides (rust) in components. Rust inhibitors typically coat metal parts so natural air and moisture do not interact with the metal to form oxide compounds.

Corrosive elements are often created through oxidation, but may be introduced to the system via poor maintenance practices. Care must be exercised whenever the hydraulic system is exposed to atmosphere to minimize the introduction of incompatible elements that may react with the fluid chemistry. Acids formed by the interaction of water in hydraulic fluid can corrode metal components. Corrosion inhibitors either form a protective coating on surfaces or neutralize acids as they form.

Some component materials such as alloys containing magnesium, lead and zinc are very susceptible to corrosion and oxidation and should be avoided in hydraulic systems even if anticorrosion compounds are added.

Chapter 16 Hydraulic Fluids

Antioxidants Because oxidation is accelerated by the presence of air and excess heat, preventive measures should be taken to complement the antioxidant additives put in the fluid by refiners. Periodic testing of the fluid can indicate problems with fluids that, if detected early, can be corrected and remedied thus extending the life of the fluid.

The presence of some metals, such as copper, will act as oxidation catalysts and should be prevented from induction into hydraulic systems.

Demulsifiers These are additives that aid the fluid in the rejection of water. Proper maintenance dictates that rejected water contained in reservoirs be removed periodically to prevent re-emulsification and/or reaction with the fluid chemistry. Water in the bottom of a reservoir could freeze in cold weather conditions and cause a serious potential for cavitation of the pump on startup.

Anti-wear Three types of anti-wear additives are used to provide increased wear resistance and enhanced lubrication qualities in hydraulic systems. These additives are classified as: *anti-wear (AW), wear resistant (WR) and extreme pressure (EP)*.

Most anti-wear additives form a protective film on metal surfaces when exposed to low frictional heat. EP types of additive come out of solution when exposed to high frictional heat and either prevent the surfaces from coming in contact or prevent the surfaces from welding to each other. In general, EP additives should be present when operating above 3000 psig.

Currently, the most common anti-wear additive is zinc dithiophosphate. However, others, such as sulfurized olefins are also used. Fluid suppliers should be contacted to determine the nature of anti-wear additives in your fluid.

Anti-foaming Agents Aeration, the introduction of outside air into the hydraulic system is extremely destructive to pumps. Aeration can also create severe safety hazards when air is present in cylinders creating dangerous load dynamics, or being conducted through lines where fluid velocity changes can allow the decompression of the air causing erratic response of actuators and loads. Air in thermoplastic lines, e.g., on utility trucks for high-voltage line work, can compromise the dielectric properties of the hydraulic lines creating a potential for conductivity.

The presence of excess air in fluids also promotes more rapid oxidation and the destruction of the fluid. Anti-foaming agents help prevent foaming and the retention of air in the reservoir. However, close attention must be paid to the integrity of the hydraulic system to prevent the introduction of air in the first place.

Foam and entrained air can be caused by:

- Air leaks, typically at pump inlet (low oil level, leaky fittings) and cylinder rod seals.
- High velocities through orifices and servovalve spools
- Rapid discharge of accumulators

Additives typically promote the combination of small bubbles into larger ones that are then more easily forced out of the fluid in the reservoir.

VI (viscosity index) Improvers

VI improvers are usually long-chain polymers which help prevent waxes from forming crystalline structures at very low temperatures. Specific attention to viscosity index must be paid when operating machinery at very low temperatures or over a wide range of temperature change.

FLUID SELECTION

The general categories of available hydraulic fluids are: petroleum oils (hydrocarbon based), fire resistant fluids (high water based - HWBF; water glycol, phosphate esters and polyol esters) and specialty or biodegradable fluids which include vegetable oils and polyol esters.

Because no single fluid is exceptional in all requirements, some compromise may be necessary when selecting a fluid for a specific application. Since mobile applications frequently subject fluids to extremes of temperature and shock, properties of lubricity and a high viscosity index (the ability of a fluid to maintain acceptable viscosity over a wide range of temperature variation) are often important considerations.

Increasing concern about environmental issues has caused heightened interest in biodegradable fluids for use in hydraulic systems, particularly mobile machinery that operates in environmentally sensitive areas such as woodlands, watersheds and farmland. The use of organic vegetable-based fluids such as rapeseed or soy oil has expanded dramatically and using these types of "environmentally friendly" fluids will be discussed in this chapter.

Another concern in fluid selection relative to the working environment of machinery is flammability of the fluid. This may be of more concern with stationary machinery such as that present in a steel foundry where molten metals are present, however, some mobile applications may present fire or flammability issues that must be addressed in the selection of hydraulic fluids. The use of synthetic or water-based fluids for their fire-resistant properties may present serious design problems for mobile machinery.

Synthetic and water-based fluids are typically heavier than petroleum fluids (specific gravity of these fluids may exceed 1.15) and may require flooded inlet conditions to prevent cavitation of pumps. In the case of water-based fluids, the constant monitoring of water levels to maintain the effectiveness and additive/water ratio of the fluids is a serious drawback. These requirements can be very difficult to accommodate in mobile applications.

Chapter 16 Hydraulic Fluids

FLUID TYPES

Hydraulic fluids for use in hydraulic machinery can be grouped in three general categories: petroleum oils, bio-degradable fluids and fire resistant fluids. Figure 16.9 shows the various types of fluids that fall into these categories.

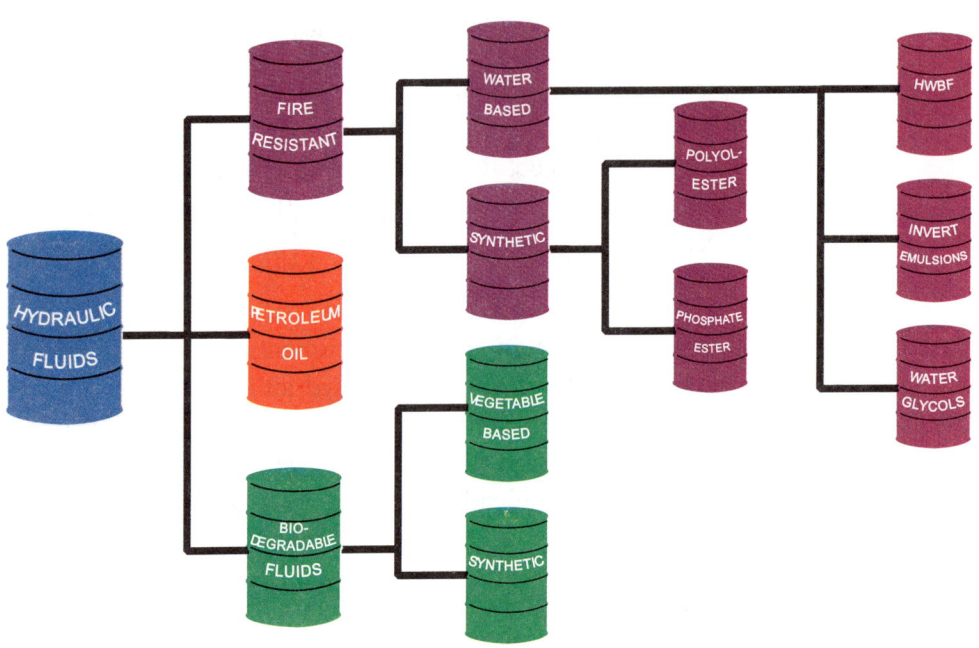

Figure 16.9 Fluid types typically used in hydraulic systems.

Hydrocarbon Based – Petroleum Oil

Petroleum or hydrocarbon-based fluids are the most common used in hydraulic systems. Straight oils (crankcase or automotive) may be used with the proper additives present for use in hydraulic applications. The major difference between hydraulic fluids and these straight oils is their additive package.

The advantages of petroleum-based fluids are: low cost, good lubricity, ready availability, and relatively low toxicity.

Petroleum oils specifically blended as hydraulic oils, or antiwear hydraulic oils, are the fluids of choice for mobile hydraulic systems. All major oil suppliers produce one or more varieties of this fluid, which have very good VI characteristics as well as additives for antiwear, antioxidation and antirust. The VI improvers used are exceptionally stable and resist shear at the higher temperatures typical of mobile systems. Oil suppliers conduct the Vickers "Pump Test Procedure for Evaluation of Antiwear Hydraulic Fluids for Mobile Systems" as described in publication M-2952-S, and will provide these details upon request.

ATF (automatic transmission fluid) is frequently used in hydraulic systems. ATF provides excellent low temperature viscosity and typically has a very high viscosity index, accomplished by the elevated addition of VI improvers. With the exception of demulsifiers, ATFs contain many other hydraulic fluid additives. The presence of high quantities of additives, particularly VI improvers, places the fluid at risk of breakdown (temporary and permanent shear of the VI additive molecular structure) during high temperature operation. Fluid temperatures above 165°F should be avoided for any more than very short periods of time. Because hydraulic fluid temperatures can frequently exceed 180°F continuously during warm weather operation, the use of ATF should be carefully evaluated.

While ATF is readily available, it is more costly than hydraulic oils. Some OEMs combine hydraulic and transmission reservoirs into single units. Any time ATF is used, the reservoir must be clearly marked to avoid mixing incompatible fluids whose reactions to each other could cause severe damage to the fluid and components.

ATF is colored red as are military hydraulic fluids such as MIL-H-5606 and MIL-H-83282. In fact, military fluids are often referred to as *red oil*. Before using MIL-H-5606, you should consult your pump manufacturer as this oil has a very low viscosity and lubricity for compatibility with very low temperature operation.

The use of crankcase oils is common in mobile hydraulic applications. Oils used should meet performance classifications SC, SD or SE of SAE J183. The following chart shows oil viscosity recommendations for use with Vickers equipment in mobile hydraulic systems:

Hydraulic system operating temperature range (Min. ambient startup to Maximum)*	SAE Viscosity Designation
** -10°F to 130°F (-23°C to 54°C°)	5W, 5W-20, 5W-30
0°F to 180°F (-18°C to 83°C)	10W
0°F to 210°F (-18°C to 99°C)	10W-30 (ensure VI stability)
50°F to 210°F (10°C to 99°C)	20 – 20W

- * The temperatures shown in the above table are cold startup to maximum operating temperatures. Proper startup procedures must be followed to ensure adequate lubrication during warm-up.

- ** Arctic conditions represent a specialized field with widely practiced heating of equipment before starting. Oils especially developed for use in arctic conditions such as synthetic hydrocarbons, esters or mixtures of the two may be used. Dilution of SAE 10W oil with a maximum of 20% kerosene or low temperature diesel fuel may be permissible. However, dilution of special oils should only be done with the knowledge and approval of the fluid supplier.

During cold startup, avoid high speed operation of hydraulic systems until the system fluid is at operating temperatures to ensure proper lubrication. Exceeding 130°F (54°C) with diluted oils should be avoided and closely monitored.

Chapter 16 Hydraulic Fluids

High Water-Based Fluids (HWBF) These fluids are used because of their fire-resistant characteristics. They include *water-in-oil emulsions, oil-in-water emulsions* and *water-glycols*. Because of the increased specific gravity of these fluid, flooded inlet conditions are recommended with a lower allowable pump inlet vacuum than with petroleum.

Although water glycol mixtures have comparable viscosities and additives with hydrocarbon oils, they are more expensive and present considerable maintenance requirements in mobile applications.

Phosphate Esters These fluids are somewhat fire resistant and allow higher temperature operation. They generally have lubrication attributes similar to hydrocarbon-based fluids.

Phosphate esters attack elastomers, some paints and plastics and require special seals such as ethylene propylene rubber (EPR), butyl rubber and occasionally fluorocarbon (Viton).

Special material requirements, as well as higher cost, do not enhance the use of these fluids in mobile applications.

Synthetics Most synthetic fluids and blends, typically phosphate esters and chlorinated hydrocarbons or blends come with the same cautions associated with phosphate esters previously mentioned. For these reasons, they are not normally recommended for mobile applications even though many are readily biodegradable.

Vegetable-Oil Based Fluids Ongoing research into the viability of vegetable oils as lubricants and power transmitting fluids for industry has created an important option for users of mobile equipment. Problems of heat and oxidation as well as stability are being addressed to enhance the performance of these oils in hydraulic applications.

Vegetable oils achieve superior biodegradability and essentially zero toxicity relative to the environment. These oils are typically more expensive than petroleum-based fluids, but still only about half the cost of synthetics with similar environmental advantages.

Soy, rapeseed and other vegetable-based oils typically have high viscosity indexes (VI) and possess good thermal stability, low volatility and high flashpoint as well as additive compatibility.

On the negative side relative to petroleum-based fluids, vegetable oils possess poorer hydrolytic stability and poorer low-temperature characteristics. Their oxidation stability and response to pour point depressants (cold weather performance additives) are also poor compared to petroleum oils.

Regardless of some of the shortcomings of these vegetable-based fluids, they can provide very good performance in many mobile applications where environmental considerations are important. They are also readily available and renewable from a supply standpoint.

Desirable Characteristics for Hydraulic Fluids*

Property	Characteristics
Suitable viscosity	The viscosity characteristics of selected fluids must be suitable for all aspects of the application, including pour point, flashpoint, etc.
Viscosity index	The ability of the fluid to maintain an acceptable viscosity over the operating temperature range of the application is vital for lubrication, efficiency and system longevity.
Chemical/environmental stability	Minimal chemical changes whether in storage or in application. Also, low volatility to minimize evaporation and bubble formation to reduce loss of fluid and potential for cavitation/aeration damage. Fluids, in some cases, may have to meet safety and environmental regulatory standards for toxicity and biodegradability.
Lubricity	Proper maintenance of adequate lubrication film over the pressure and temperature operating ranges of the application.
System compatibility	Zero or minimum reaction with other system materials such as seals, hose inner tubes and covers, platings and metal components.
Heat transfer	Good thermal conductivity and high specific heat for dissipation and cooling.
Minimum compressibility	High bulk modulus for efficient transmission of energy. Fluid should resist foaming as entrained air increases compressibility.

* Many of these attributes are augmented by the additives put in the fluid by manufacturers. It is essential that a fluid supplier be aware of all operating parameters of hydraulic applications before a fluid is selected.

The selection of fluids used in hydraulic systems is of considerable importance. When discussing fluid selection with a supplier, it may be useful to keep in mind the following general observations in relation to your application and final fluid selection:

- Temperatures affect viscosity
- Viscosity doubles for every 5000 psi increase in pressure.
- Oxidation rates for petroleum-based oils double for each 18°F rise in temperature higher than 140°F. This effectively reduces the life of the fluid by half for each 18°F rise.
- Petroleum oil at atmospheric pressure (14.7 psia) contains 8-10% air in solution. This air cannot be observed and is normal.
- Cold starts at less than 20°F above a fluid's pour point can damage components and the system.

Chapter 16 Hydraulic Fluids

- Do not put additives in your fluid without the advice of a professional.
- Fluids and additives break down and wear out. While laboratory testing and analysis are extremely useful, test results should be analyzed by knowledgeable individuals who can then recommend proper action. Fluid conditioning will be discussed at length in Chapter 17.

Hydraulic Fluid Comparisons

Fluid Type (Rel. Cost)	Specific Gravity	Viscosity Index	Lubricity	Oxidation Stability	Fire Resist.	Corrosion Protection	Heat Transfer	Temp. Range °F	Effect on Conv. Elastomers	Effect on Conv. Paints
Anti-wear petroleum (1)	0.85-0.89	Good	Excellent	Excellent	Poor	Excellent	Good	20-150	Minimal	None
Soybean-Based Oil (2-3)	0.92	Fair	Excellent	Fair	Excellent	Excellent	Good	-13 - 175	Minimal	Minimal
Synthetic Blends (4)	0.8-1.0	Good	Good	Good	Fair	Good	Good	20-150	Severe	Severe
Phosphate Esters (5)	1.15	Poor	Excellent	Good	Good	Good	Good	20-150	Severe	Severe
Water Glycol (2)	1.1	Excellent	Good	Good	Excellent	Fair	Excellent	0-120	Minimal	Minimal to Moderate

Water-in-oil (invert) emulsions, oil-in-water emulsions (soluble oil) and water additive fluids are not included because of the problems associated with mobile applications. Your fluid supplier should be consulted regardless of the fluid selected.

Leakage Practices

Proper assembly practices and usage can prevent the existence of any external leakage in modern hydraulic systems. It is estimated by industry analysts that up to 7 million barrels of hydraulic fluid are lost annually via external leaks and line breaks.

From 70 to 85% of hydraulic problems are associated with contamination of the fluid (see Chapter 17) or the improper choice and handling of hydraulic fluids. Proper maintenance practices and the sensible use of mobile hydraulic equipment within manufacturers operating specifications are essential for safety and to prolong the useful operating life of systems and components

Trained, informed operators and repair crews, as well as purchasing and design personnel are necessary to the effective operation of hydraulic equipment, particularly in the demanding area of mobile hydraulics. Elsewhere in this book, there is found all of the information necessary to optimize the capabilities of personnel and equipment for the effective application of hydraulics as an efficient and cost-effective means of power and motion control.

CHAPTER 17 .. Fluid Conditioning

As illustrated in the previous chapter, *Hydraulic Fluids,* the selection of the fluids used in mobile hydraulic applications is extremely important to the proper function of fluid power machinery. However, this selection is only the beginning of the "care and feeding" of the fluids once the machinery is put to work.

Fluid conditioning includes: proper filtration to ensure adequate cleanliness of the fluid, temperature monitoring and moderation, elimination or control of shock in the system, proper maintenance practices and procedures, and a basic understanding of options on mobile equipment to properly optimize the life and utility of the fluid being used.

This chapter will deal with systemic contamination control, i.e., proper filtration and maintenance practices to reduce contamination; design elements geared to reduce shock and heat damage to the fluid; proper maintenance procedures used to minimize or eliminate fluid contamination and degradation in demanding mobile hydraulics applications.

DEFINING CONTAMINATION

More than anything, contamination of hydraulic fluids is the major contributor to failures of hydraulic components and systems. Upward of 80% of hydraulic failures are believed to be caused by contamination of the fluid and thus the system.

Contamination can be defined as anything in a hydraulic system that is not supposed to be there. Although contamination is often defined as solid particles, serious contamination damage may be caused by air or other gasses, incompatible fluids such as cleaning solvents or hydraulic fluids, and even failed additives that have come out of solution and are now considered contamination as they no longer perform a useful function.

Types of Contamination

Solid, or particle contamination. This category of contamination may consist, literally, of dirt: sand, silica, loam, etc. It may also be composed of microscopic metal particles, seal materials, Teflon sealing tape, cloth or wood fibers or virtually any other solid material.

The size of particles in hydraulic fluids is measured in micrometers, commonly referred to as microns or units of one millionth of a meter. One micrometer or micron (symbol µ) is equivalent to 0.0000394 inch. There are 25,400 microns in one inch.

Chapter 17 Fluid Conditioning

The table in Figure 17.1 illustrates the relatives sizes of common objects. The smallest particle observable without assistance with the human eye is about 40μ. This is considerably larger than the largest running clearances in most hydraulic components.

Component	Micron
100 mesh screen	159
Grain of table salt	100
Smog particle	90
Machining particle	80+
200 mesh screen	74
Human hair	70
Pollen	60
Fog droplet	50
Particle visible to unaided eye	40
White blood cell	24
Grinding particles	10+
Red blood cell	6
Honing particles	2+
Yeast cells	1
Bacteria	0.2+
Viruses	0.1+

Figure 17.1 Micron sizes of common objects.

As seen in Figure 17.2, different sized particles affect clearances in different ways. If the particle is larger than a clearance or orifice, it may not enter the clearance and cause interference, but it could block the opening. This could present serious problems in the case of small damping or control orifices in components. Particles that are very close to the size of the clearance may get caught between clearances and cause abrasive damage or jamming of the component. Very small particles, typically smaller than 5μ, are called silt. These particles usually pass through most clearances, but due to high fluid velocity, turbulence or impingement on surfaces with flow direction changes, they may erode metal surfaces or metering control edges. This can reduce efficiency and create heat by increasing internal leakage. It can also affect response, control and reliability. *This type of erosive damage generates additional wear particles that further aggravate contamination damage and failures.*

Chapter 17 Fluid Conditionings

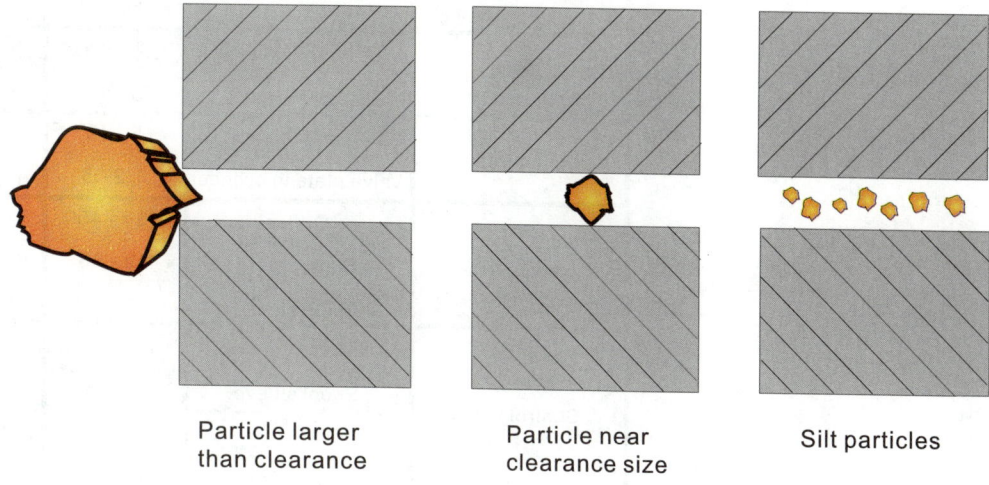

Particle larger than clearance   Particle near clearance size   Silt particles

Figure 17.2 Particle size differences.

Figure 17.3 illustrates typical clearances in hydraulic components in micrometers and inches. From these dimensions, it is readily observable that very small solid contamination, i.e., between 2µ and 15µ, will present serious problems in hydraulic applications. Micron ratings of filters and cleanliness requirements of various components will be illustrated and discussed under Principles of Systemic Contamination Control, often called *proactive* or *predictive* maintenance, later in this chapter.

Chapter 17 Fluid Conditioning

COMPONENT	CLEARANCE LOCATION	MICRONS	INCHES
Gear Pump	Gear to side plate	1/2 - 5	0.00002-0.0002
Gear Pump	Gear tip to case	1/2 - 5	0.00002-0.0002
Vane Pump	Tip of vane	1/2 - 1	0.00002-0.00004
Vane Pump	Sides of vane	5 - 13	0.0002-0.0005
Piston Pump	Piston to bore	5 - 40	0.0002-0.0015
Piston Pump	Valve plate to cylinder	1/2 - 5	0.00002-0.0002
Servo Valve	Orifice	130 - 450	0.005-0.018
Servo Valve	Flapper wall	18 - 63	0.0007-0.0025
Servo Valve	Spool sleeve	1 - 4	0.00005-0.00015
Control Valve	Orifice	130 - 10,000	0.005-0.4
Control Valve	Spool sleeve	1 - 23	0.00005-0.0009
Control Valve	Disk type	1/2 - 1	0.00002-0.00004
Control Valve	Poppet type	13 - 40	0.0005-0.0015
Actuators		50 - 250	0.002-0.01
Hydrostatic Bearings		0 - 25	0.00005-0.001
Antifriction Bearings		1/2	0.00002-
Slide Bearings		1/2	0.00002-

Figure 17.3 Typical component clearances.

Heat or thermal contamination. Hydraulic fluids are typically not considered "high temperature" fluids. Their useful life, as well as maintenance of viscosity, chemistry and the other attributes discussed in Chapter 16, is based upon continuous operation below a critical temperature. This critical temperature is 140° F (60° C). Every 18° F increment (10° C) higher than 140° F effectively doubles the oxidation rate of the hydraulic fluid (petroleum based) thus cutting its useful life in half, e.g., running a system at a consistent 176° F (80° C) would reduce the useful life of the fluid by 75%.

Most fluid manufacturers specify optimum ranges of temperature for their products -typically from 90° F to 120° F (32° C to 49° C) or 100° F to 130° F (38° C to 54° C) even though many fluids are operated in excess of these temperature ranges. The effects of temperature on the fluid are the same, regardless of the source of heat.

Air and gas contamination. Air and gas contamination is classified as dissolved, entrained or free. All hydraulic fluids contain *dissolved* air. This air is *in solution*, that is, the air molecules occupy the spaces (interstices) between the molecules of the fluid. The amount of air present is measured as a percent of volume. At atmospheric pressure (14.7 psia, or pounds per square inch, absolute) a typical petroleum-based hydraulic fluid may contain as much as 10% air by volume. This means that 1 gallon of fluid may contain about 0.1 gallon of air.

The actual amount of air dissolved in the fluid depends on the fluid itself and the pressure to which it is subjected. The percentage of air in the fluid at any given pressure is called the *saturation level* of the fluid. As seen in Figure 17.4, different types of fluid have different saturation levels. In all cases, however, saturation level follows increases or decreases in pressure.

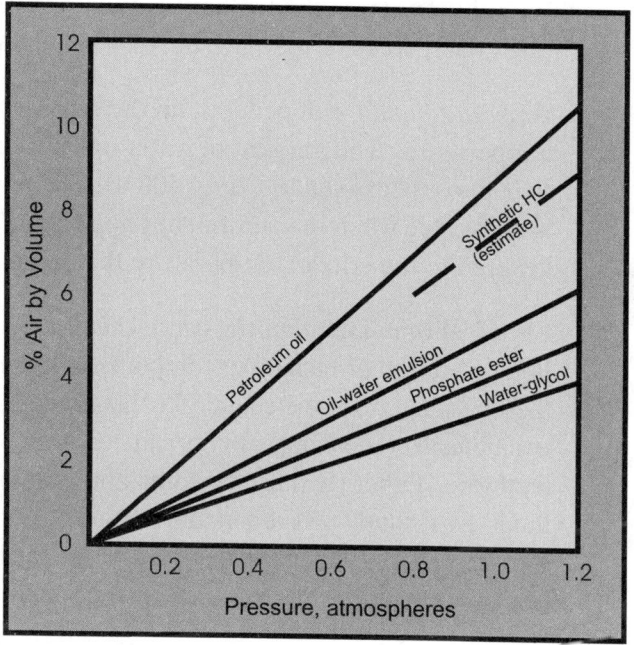

Figure 17.4 All hydraulic fluids contain some air.

Entrained air is any air present in excess of the saturation level of the fluid at a given pressure. Entrained air bubbles may range from a few microns in diameter to bubbles that are visible.

Free air is found in large bubbles or pockets of air within the system. This air is often trapped in high points in the system piping or in hydraulic cylinders. Pockets of air trapped in pump cases can cause severe damage and catastrophic pump failure by causing a loss of lubrication and increased heat from friction and leakage.

Water contamination. In petroleum oils, water can degrade the fluid and the system in numerous ways, even though water may be present in small quantities. In general, water content up to 700 ppm (parts per million) may be tolerated. However, some tests indicate that levels as low as 140 ppm can significantly reduce pump life under certain circumstances.

Water typically exists in three forms in hydraulic oils - free water, emulsified water, and water in solution.

Free water is water which can readily separate from the oil. This water will usually settle to the bottom of the reservoir and should be removed by periodic draining.

Emulsified water is the result of the mixing action of the pump on water contaminated fluid or from high turbulence or violent agitation at points in components or fluid conveyors. These actions cause the water droplets to reduce in size so that they will not readily separate by settling or centrifuging. Some additives, including detergents, polar rust inhibitors and EP (extreme pressure) additives, can increase emulsification. Emulsified water in fluid imparts a milky appearance to the fluid. Severe emulsification (high water content) may give the oil a gelatinous texture.

Water in solution is dependent on the type of oil, base stock, additives, and temperature. The amount of water in solution may range from 60 to 220 ppm (0.006 to 0.022%) at 100° F to 500 to 1000 ppm (0.05 to 0.1%) at 180° F. This water is not visible when in solution, but appears as a cloud in the oil as temperature is lowered to the critical temperature that begins to force the water out of solution.

Chemical contamination can cause chemical changes or breakdown of the hydraulic fluid, corrosion or etching of components, seal failure, and the damage associated with these results. Airborne chemical contaminants may enter the system via air breathers or maintenance. Cleaning solvents can be left in overhauled components or as residue on hardware. Other chemical contamination can be caused by mixing different hydraulic fluids, particularly petroleum and non-petroleum based fluids. Even mixing different formulations of the same type of fluids can lead to fluid degradation such as viscosity and viscosity index changes, additive precipitation and such.

Radiation contamination. Nuclear radiation, as might be present in mining or environmental clean up operations, can be a major contaminant to exposed systems due to chemical changes caused in the fluid and many elastomeric sealing materials. The seals and elastomeric materials will be more likely to degrade at lower radiation levels than the fluid.

Microbial growth can be a problem in all water-containing fluids as well as petroleum oils that have become contaminated with water. These growths are primarily bacteria, fungi and mold. These contaminants can degrade fluid, produce acids that corrode metal surfaces, and clog filtration devices. They may be toxic to humans or produce harmful chemicals.

Chapter 17 Fluid Conditionings

SOURCES OF CONTAMINATION

Contamination can be introduced into systems from both external and internal sources. Equipment designers, component manufacturers, and users must understand these sources in order to prevent or minimize contamination. It is much easier, through proper practices, to exclude and prevent contamination than it is to remove it or deal with the consequences.

Built-in Contamination

Much of the solid contamination in fluid power systems results from the manufacture and assembly of the individual components and the system itself. Manufacturers must be aware of this problem and ensure that contaminants are removed from delivered components and systems.

Every step in the manufacture of a component generates a significant number of solid particles. Valve bodies, and pump and motor housings are typically cast, so there is the potential for casting material to remain in the component. Machining operations - drilling, tapping, milling, grinding and so forth - produce hard metal particles that must be removed. Burrs left by machining operations that are not removed can break loose and cause severe damage during operation.

Possibly the worst source of built-in contaminants is the *fluid reservoir* (Figure 17.5). Because of the relative simplicity of construction, reservoirs may be overlooked by manufacturers relative to contamination. Attention must be paid to the surface condition and cleanliness of the metal used. Weld beads, grinding and buffing may leave debris in the reservoir. It is not uncommon to hear of rags, broken drill bits, weld slag, chips and even expired animals being found in hydraulic reservoirs. With debris of this size often found, it is easy to see why even careful procedures can leave particles that will cause severe damage even though they may be microscopic.

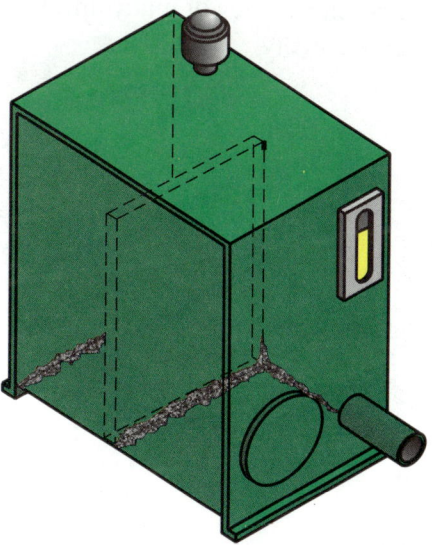

Figure 17.5 Built in reservoir contamination.

Chapter 17 Fluid Conditioning

Another major source of built-in contamination is fluid conveyors - hose, pipe and tubing - used to connect the system components. All of these are likely to contain particles generated by their manufacture or assembly. In addition, contamination may enter during storage, shipping and handling.

New hoses and hose assemblies can be highly contaminated from manufacture and assembly processes. Material generated by cutting hose to length or skiving must be removed from the final assembly before use.

The assembly and disassembly of threaded fittings and connectors generate particles that can enter the system. Fittings themselves may be dirty from storage and handling. The use of Teflon® tape, pipe dope and thread lubricants can introduce contamination into a system. Machine failures resulting from jammed valves and plugged orifices are commonly caused by small pieces of tape that have worked loose into the system.

Steel pipe, while not common in most mobile hydraulic applications because of weight, is a source of serious contamination from rust and scale if not stored and treated properly before use. Most in the hydraulics industry recommend the discontinuation of tapered thread, steel pipe in fluid power applications because of external leakage problems.

External Contamination

External contamination is contamination that enters the system from the environment. It includes airborne dirt and moisture, but may also include metal particles, chemical pollutants, microbial life and the media that support their growth, industrial pollutants, molds, pollens and so forth.

A 1973 study reported that the ingression rate of contaminants larger than 10 microns into a fluid power system is 100 <u>million</u> particles per minute. This number is based on a survey of many different types of equipment conducted at Oklahoma State University.

A comparison of various types of equipment can be made from Figure 17.6.

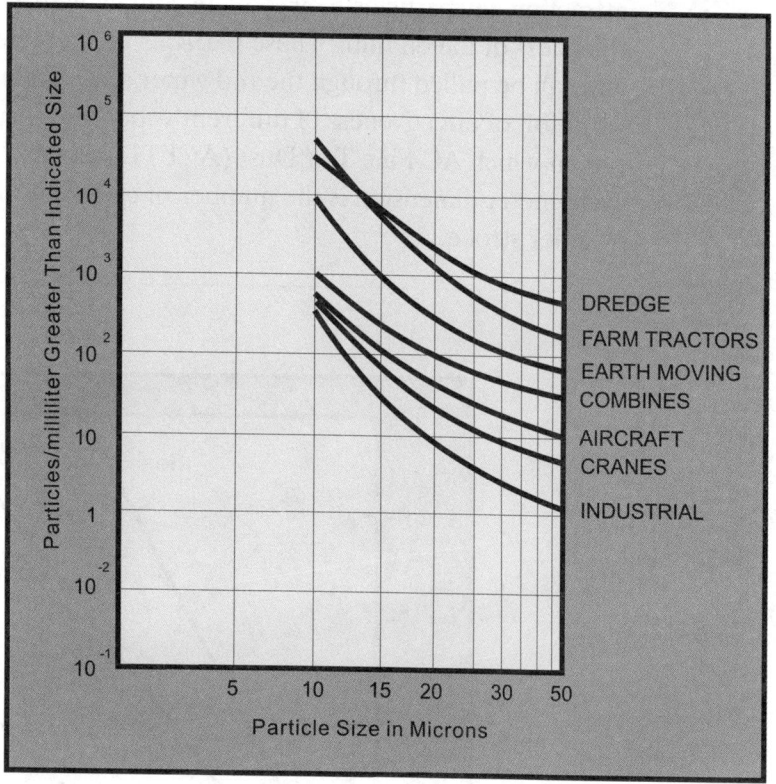

Figure 17.6 Contamination ingression rates by application.

The two most common points for the ingression of external contamination are the reservoir breather and cylinder rod seals. In some pump designs, a worn shaft seal can allow ingression as can loose fittings on the suction side of the pump.

A reservoir vented to atmosphere will "breathe" as the fluid level rises and falls during machine operation. Unless a proper breather vent filter is present, this breathing action will allow significant amounts of airborne particles, water, chemicals and so forth, to enter the reservoir and contaminate the fluid and system. Breather vent media should remove particles 3 microns and larger in size. In some environments, moisture resistant media may also be desirable. Reservoir covers and access panels must be properly sealed and tightened or large amounts of contamination could enter the system.

Cylinder rods are particularly good at contaminant ingression. They move in and out of the system, regularly being exposed to the outside environment. Because they are coated with a thin film of oil from the system, they become "dirt magnets", attracting any particles they come in contact with, some as large as ten times the thickness of the oil film. These particles are held by the surface tension of the fluid and can be pulled through the rod wiper seal. Figure 17.7 illustrates the wide variation of effectiveness of different wiper seals or scrapers. The graph shows the rate at which AC Fine Test Dust (ACFTD) passes through the wiper seal. The cycle-meter dimension is the number of cycles multiplied by the length of the cylinder stroke.

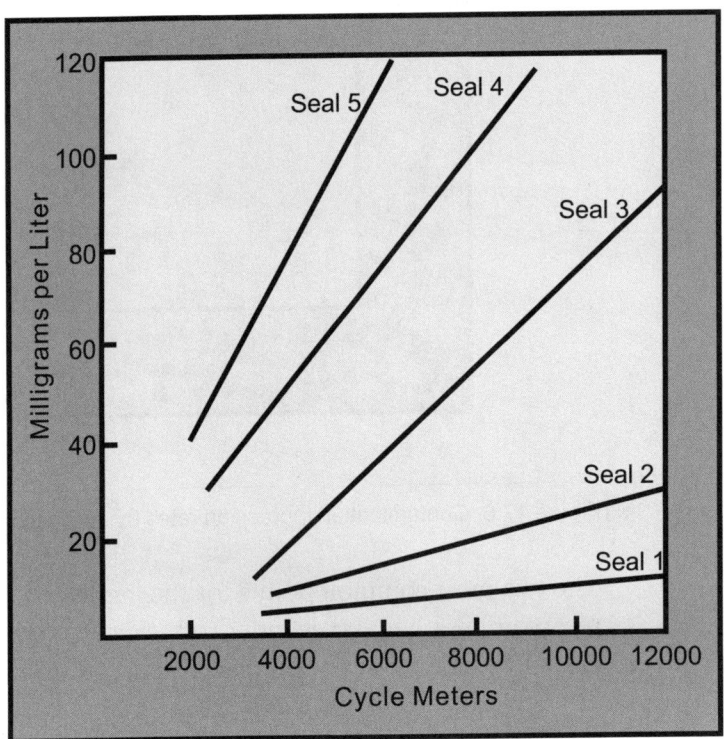

Figure 17.7 Wiper seal effectiveness by type.

Side-loading, shock and poor maintenance practices can seriously increase the ingression rate of contaminants at the rod wiper seal. Any indication of excess wear should be acted upon promptly with the seals being replaced and the cause of side loading or shock corrected.

Because of the dynamic nature of loads on mobile equipment, worn rod seals or loose port fittings may also allow air to enter the cylinder. This air could pose serious operational and safety problems because of its compressibility.

Internally Generated Contamination

As fluid power systems operate, wear particles are generated because of the interaction of surfaces within the components. Figure 17.8 illustrates a typical contamination curve for hydraulic systems. Called a "bathtub" curve because of its shape, it is a typical wear or failure curve for mechanical components.

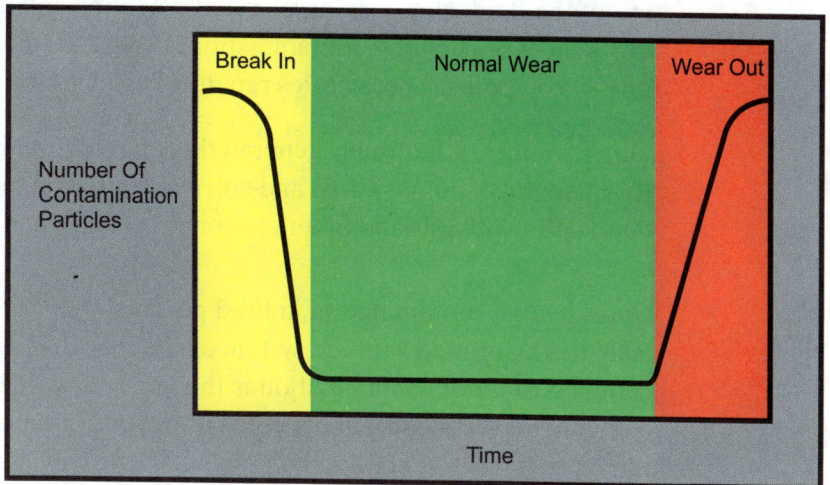

Figure 17.8 Typical wear curve for mechanical components.

Relatively high numbers of large particles are present in the break-in portion of the curve. This is the result of built-in contamination and the initial mating of moving surfaces in the components. With proper filtration to remove this initial contamination and maintain an acceptable cleanliness level, the curve reaches the normal wear zone in which the number and size of particles reaches an equilibrium. As a component begins to wear out, the number and size of particles begins to increase. This increase indicates a failure mode.

Although the curve in Figure 17.8 is typical, the actual number of particles and their size is dependent on numerous factors, including: fluid characteristics (viscosity, anti-wear additives, etc.), condition of the fluid, system filtration, operating pressures, shock, and the rate of ingression and nature of external contaminants.

The more contaminated a system is, the more contamination will be generated by wear mechanisms in the system. High pressure and/or shock further aggravate these conditions. The wear rate is increased by increased pressure; at a given pressure, the wear rate is increased by increased contamination levels.

An important factor in the wear rate of components is the fluid viscosity. The ability of the fluid to keep relative moving surfaces separated is called "load carrying capability". This load carrying capability is strongly dependent on the fluid's viscosity. Changes in the fluid's viscosity, caused by heat, shock, abuse or chemicals, or the use of incorrect fluid will allow moving metal parts to contact each other without a lubricating film between them. This can cause severe wear, heat and eventual failure.

Chapter 17 Fluid Conditioning

Anti-wear additives such as ZDP (zinc dialkyldithiophosphate), play an important role in reducing wear generated contamination and extending the life of the system. These additives are based on compounds that contain sulfur and/or phosphorous. These compounds deposit polysulfide or phosphate films on the metal surfaces of components. The low shear films created by these additives prevent adhesion of the surfaces as one moves past another. These additives are depleted through normal use of the fluid. However, accelerated depletion will occur in the presence of water, excess heat or chemical contamination. As these additives are depleted, the wear rate of components accelerates rapidly.

Other sources of internally generated contamination include cavitation and aeration (particularly of pumps), rust and corrosion, erosion, metal fatigue, seal wear, degradation of seals and other elastomeric components.

Maintenance Generated Contamination

Considerable contamination in fluid power systems is caused by maintenance activities associated with the system or the machine on which it is installed. A common cause of contamination is the unnecessary removal and replacement of components dictated by an arbitrary "preventive maintenance" schedule. Many of these programs require that components be removed and replaced after a specific number of operating hours or calendar time. This can be harmful in that removing the components opens the system to external contamination; the manipulation of fittings and conveyors generates and introduces contamination into the system.

If replacement components have not been properly stored or prepared, they will likely contain significant levels of contamination to be introduced into the system. Air will also likely be added to the system. The wear debris from breaking in the new components will cause a rise in overall contamination levels that increase wear damage.

On many mobile applications, filters are not equipped with bypass indicators and thus are changed on an arbitrary schedule. If this is necessary, the machine and fluid should be monitored and the fluid tested at frequent intervals until a schedule is devised that minimizes incursions into the system with filter element changes and maximizes the useful life of the filter elements. Maintenance personnel should be well trained in the requirements and practices to ensure minimum contamination generation when changing out components and filter elements.

Another cause of maintenance generated contamination is poor troubleshooting procedures. "Hit and miss" troubleshooting is that in which the maintenance personnel do not have sufficient knowledge or experience to allow them to properly evaluate the cause of failure, so they replace components until they find one that has failed or that corrects the system failure. This opens the system to huge amounts of contamination each time a component is replaced.

The ability to read circuit diagrams and interpret measurable specifications in the system and relate them to component function is essential for targeted troubleshooting. Users must be aware of the importance of machine circuit diagrams that include all power and motion control components and their connections to each other. They should insist that OEMs provide current prints for all new machines. They should also ensure proper hydraulics technical training for machine repair personnel and their supervisors.

Fluid servicing can be a major contributor to high levels of contamination. One possible problem is the addition of an incompatible fluid to the system. This is more probable when different types of fluids are used in different machinery. Service personnel must remember that brands of fluids and different fluids from the same manufacturer are not necessarily compatible just because they may be petroleum-based.

A major problem with fluid replenishment is introducing contaminated oil into the system. New oil is typically not clean enough for introduction into a hydraulic system without pre-filtration. Figure 17.9 shows the contamination levels of $\geq 5\mu$ and $\geq 10\mu$ particles in new oils taken directly from their containers. Although these types of studies have led suppliers to take actions to improve the cleanliness of the new fluids, most new oils must be pumped through a high efficiency filter before being introduced into the system.

Sample number	Particle Count / Milliliter	
	$\geq$ 5 micron	$\geq$10 micron
1	180	20
2	21401	4087
3	1703	342
4	11230	3209
5	309	63
6	2303	406
7	1933	313
8	2437	183
9	3342	419
10	963	155
Average of all Samples	4580	920

Figure 17.9 Particle counts for new hydraulic fluids.

Chapter 17 Fluid Conditioning

The already unacceptable contamination levels of most new fluids can be severely compounded by poor storage and handling procedures. Open storage drums or cans, the use of hand pumps in drums, the use of screwdrivers or other contaminated tools to open cans, carrying fluids in open buckets, allowing servicing hoses to drag on the ground or lie unprotected in dirty truck beds, and removing the filler screen at the reservoir are all commonly observed occurrences that significantly add to contamination problems and failures.

These potential problems with new oil and fluid handling procedures may be aggravated by the common practice of some users to periodically drain, flush and refill their systems. Rather than being conscientious about contamination control, this practice is often a useless, if not harmful, exercise that increases operating costs, may increase contamination levels, and may negatively impact the environment and lead to fines if the used oil is not disposed of properly.

A proactive approach is much more effective. This approach requires proper system design, periodic fluid sampling and analysis, and operating and maintenance practices that address and remedy the fluid's condition. Using this approach, the life of the oil can be increased almost indefinitely. Changing the oil in a fluid power system should be a very rare requirement.

Particle Contamination and Clearances

Particles of solid contamination in hydraulic fluids are expressed in micrometers or microns - one-millionth of a meter. A human hair is about 70 microns (70μ) in diameter. The smallest particle seen by the unaided human eye is approximately 40μ in size. The table in Figure 17.3 illustrates typical clearances in hydraulic components. These clearances and the effects of particle contamination on them determine component sensitivity to contamination. Particle analysis and sizing is discussed later in this chapter.

RESULTS OF CONTAMINATION

The results of contamination may take many different forms. System performance may degrade slowly over a long period of time before it becomes unacceptable. In some cases intermittent failures may occur, in others, sudden, catastrophic failures may cause total machine failure and create safety concerns. In any case, the results of these failures may be very costly. These costs of failure are not only the expense of replacement components, but the cost of unplanned downtime and the resultant loss of productivity.

Wear

Hydraulic system failures typically fall into three general categories: termed degradation, transitory or intermittent, and catastrophic.

Long term degradation due to internal wear is the most common. This process is the wearing away of the internal surfaces and clearances of components until internal leakage is such that the component can no longer properly perform its function. This wear may be two-body or three-body abrasion, erosion, or the presence of air or gas in the fluid. All of these processes can be intensified by contamination in the fluid.

Chapter 17 Fluid Conditionings

Transitory or intermittent failures may occur only once or intermittently, usually due to large particles that can cause temporary blockage or the jamming of a component. These failures can often be cleared by cycling or shutting down and starting the equipment.

Catastrophic failures may be the result of long term wear, metal fatigue, lubrication failure, bearing failure or any other event that causes the system to fail suddenly and completely.

Proactive maintenance procedures can prevent the majority of failures in these categories and hence up to 80% of all hydraulic failures said to be caused by contamination.

Two-body and Three-body Wear

Two body wear is commonly called abrasion, or the result of two moving surfaces coming into intimate contact with each other. The result is slight distortion of the surfaces and the galling of material from one or both surfaces. The removed material may be transferred from one surface to another, but more likely it will be broken loose to become a contaminant particle.

Figure 17.10 illustrates the microscopic peaks, or asperities, present on even finely finished metal surfaces. The lubrication properties of hydraulic fluids are meant to prevent the interaction, or abrasion, of these asperities by providing a film of lubrication between the two surfaces. The type of lubrication film present will determine the success of the fluid to prevent wear and is dependent on a number of variables.

Figure 17.10 Asperities on a finished surface.

Types of lubrication conditions

There are three generally accepted types of lubrication conditions: Full film or hydrodynamic lubrication, boundary lubrication and mixed film lubrication. Figure 17.11 illustrates these types of film thicknesses.

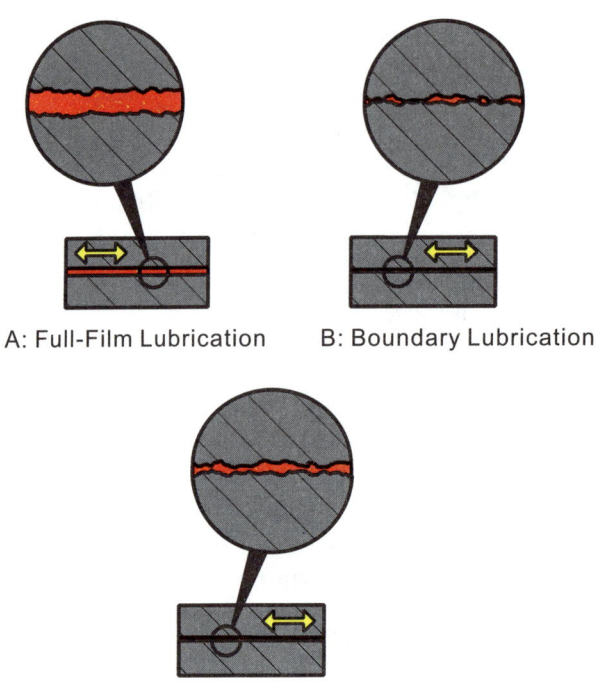

Figure 17.11 Types of lubrication conditions.

Full film lubrication, Figure 17.11A, indicates that the metal surfaces theoretically never contact one another. Oil films of this type are typically more than 2μ in thickness.

Boundary lubrication, Figure 17.11B, illustrates that total separation of the two surfaces is not effected and in fact the effectiveness of this type of lubrication is somewhat dependent on the physical contact of the surfaces. As the asperities are broken off, creating heat from friction, a chemical lubricant in the fluid plates out and covers the damaged area. This chemical lubricant, or extreme pressure (EP) additive, forms a protective, lubricant coating that prevents further wear at that spot. Antiwear additives such as ZDP are also used to provide protection in boundary lubrication situations. Boundary oil films are typically less than 0.05μ in thickness.

Mixed film lubrication, Figure 17.11C, is also called elastohydrodynamic lubrication. In this case, some separation of the surfaces is present, but some asperity contact is expected due to the increased speeds and pressures, and small clearances in high performance fluid power equipment. Extreme pressure and antiwear additives are essential to reduce wear with this type of lubrication. Mixed film lubrication thicknesses range from 0.05 to 2μ.

Another form of two-body wear is erosion that can occur when small particles, carried by the fluid, impinge on surfaces at high velocities. This may result in further wear of these surfaces and a loss of efficiency or control. Eroded metering edges of lands in valves or bores in valve bodies may result in loss of control, increased internal leakage and heat generation, and can affect safety when trying to control dynamic loads with mobile equipment.

Three-body wear is particularly common to cylinders, where softer seal materials hold particles which then damage the cylinder barrel and rod surface as the cylinder reciprocates. Figures 17.12 and 17.13 illustrate two types of three-body wear.

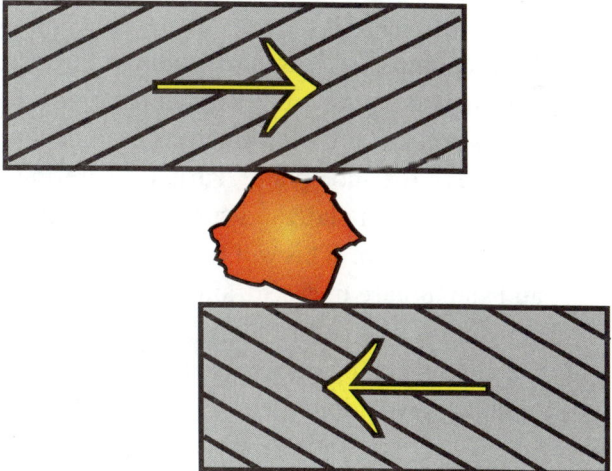

Figure 17.12 Three body wear between two moving surfaces.

Chapter 17 Fluid Conditioning

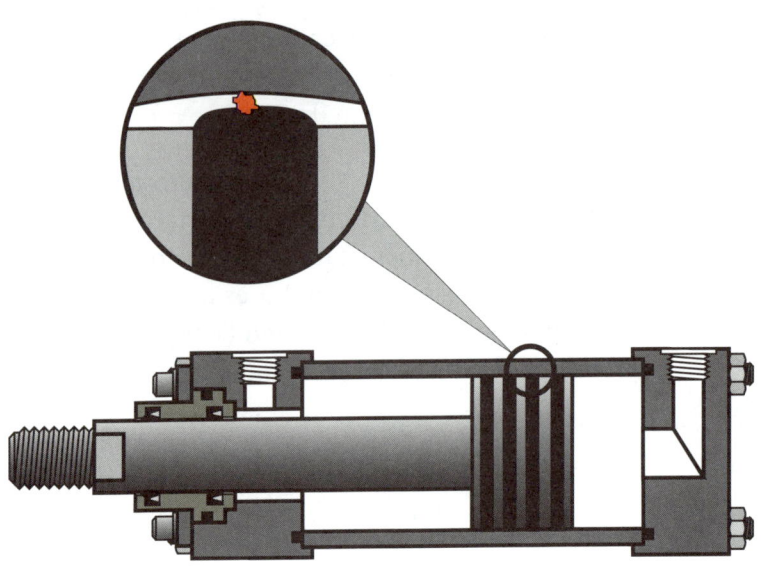

Figure 17.13 Three body cylinder wear.

Piston pumps are very susceptible to three-body wear between the piston shoes and swash plate on the outlet side of rotation. The lubrication film under high pressure in this area is only a few microns in thickness and is held in place by the machined surface of the shoe plate. Particles entering this area can be trapped between the shoe and swash plate and cut grooves in the two surfaces that allow for loss of lubricating film. This will lead to increased wear particles, loss of pump output flow and an increase in case drain flow from the pump - all of which increase the generation of heat by the pump. As this pattern progresses, increasing metal-to-metal contact between the shoes and the swash plate increases wear dramatically and may lead to the catastrophic failure of the pump.

PRINCIPLES OF SYSTEMIC CONTAMINATION CONTROL

There are essentially three basic approaches to maintenance - reactive, preventive and proactive. Reactive maintenance is a crisis management philosophy of "breakdown" maintenance - don't fix it unless it has failed, or, more familiar, "if it ain't broke, don't fix it". Of course, this philosophy practically guarantees the ultimate breakdown of equipment, usually at the most inopportune time.

Preventive maintenance, "change it whether you need to or not" is typically associated with time-scheduled changes of pumps, filters, and other components, regardless of their condition, in the hope of avoiding an unplanned failure. In most cases, the change is probably either too early - the component still has useful life, or too late - the component may have already contributed to other system damage or loss of productivity because it has exceeded its useful life.

Proactive maintenance, also called predictive maintenance, unlike reactive or preventive maintenance, combines the elements of sensible circuit design, fluid analysis and condition monitoring to track the "health" of the fluid and, hence, the entire system. However, the nature of mobile hydraulics can severely limit the opportunities for ongoing diagnostics and other proactive means of measuring fluid properties and conditions that allow optimum results. Consequently, a compromise between sensible preventive maintenance and fully proactive maintenance must be made for the reality of mobile hydraulic applications and fleet management programs.

Concepts of Proactive Maintenance

The goal of any properly constructed maintenance program is to save money while minimizing unplanned downtime due to machine failure. First, the system fluid must be maintained to optimize the attributes of the fluid that contribute to proper machine operation: lubrication, sealing, etc. Dirty fluid is compromised fluid that can contribute to machine failure. In order to develop a sensible contamination control program, an understanding of what cleanliness means and how it is measured is essential.

Fluid Cleanliness Standards - ISO Particle Range Coding

It was not until the mid-1970s that a method and accepted standard of assessing solid contamination in hydraulic fluid was established. ISO code 4406 established a range code of particle contamination concentration, independent of particle size, and a classification to indicate the concentration of particles larger than 5 microns and larger than 15 microns. Figure 17.14 illustrates the use of a standard range number to represent the number of particles (of any size) in a specific volume of sampled fluid. The range number for the upper limit of the range is used. For example, if the number of particles of a given size being measured, say 5μ and larger, is 7800, the range is from 5000 to 10,000, so the range number would be 20. Indicating an ISO range code rating of 20 for particles 5μ and larger in one milliliter of fluid. Thus a fluid subject to particle count that was determined to have a cleanliness level of 18/16 would be determined to have a range of 18 for 5μ and larger particles/ml, and a range of 16 for 15μ and larger particles per ml.

Chapter 17 Fluid Conditioning

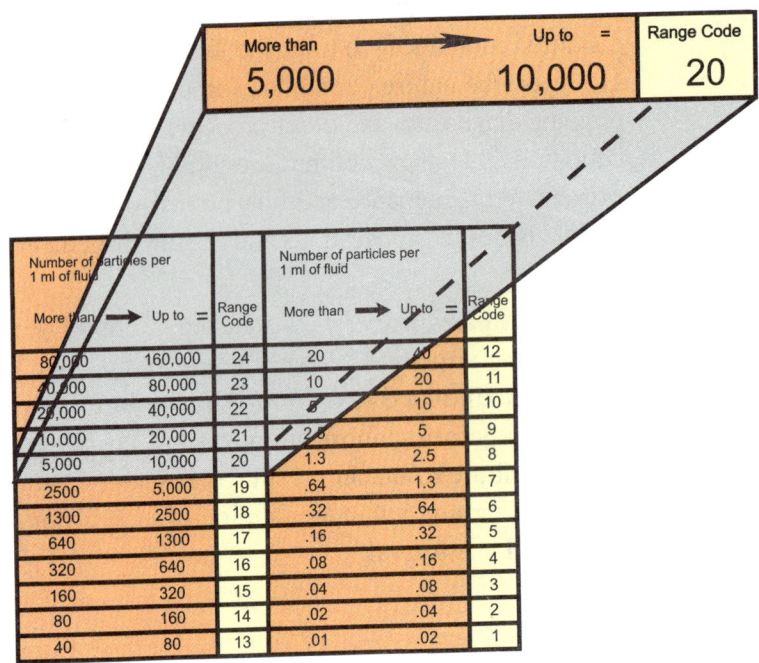

Figure 17.14 ISO range number chart.

The current ISO standard is currently being considered for revision to include particles as small as 2μ (silt), a practice long adhered to by Vickers. Because the accumulation of particles finer than 5μ can cause serious damage to components and degradation of machine efficiencies and operation, it is good practice to require particle analysis to include these finer particles. In the example above, a sample could then be determined as having a range code of 20/18/16 to include the more numerous particles of a size 2μ and larger. Because of the design characteristics of mobile equipment that may limit space for filters, the use of more finely rated filters (1 and 3μ) may not be practical with larger pump flows. In this instance, machines may be put on off-line filtration during planned downtime to clean up the system oil with finer filters than on-board design allows.

Figure 17.15 illustrates a cleanliness sample chart used to determine the cleanliness level range code for 2µ, 5µ and 15µ particles in a specific sample of oil. While laser light extinguishing technology allows a track of particles from less than 2µ to greater than 50µ, the numbers extracted for the cleanliness level reflect the 2, 5 and 15µ particles.

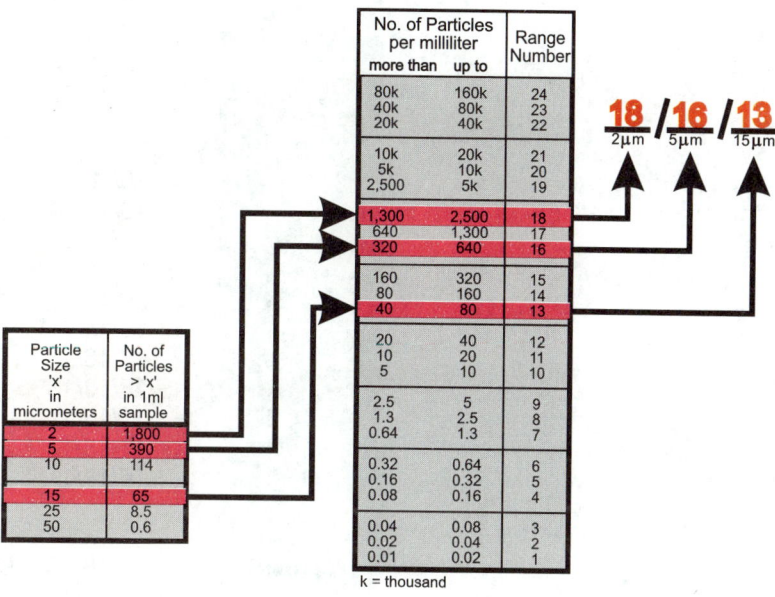

Figure 17.15 Sample three-code cleanliness code chart.

Determining System Cleanliness Requirements

Major fluid power component and systems manufacturers provide the cleanliness level requirements of their products. This cleanliness level, or tolerance for contamination, reflects the materials used in the components, as well as internal clearances, pressure requirements of the system the components are used in, and the effects of contamination on internal passages, orifices, and mating parts that must be separated and lubricated by the fluid. Figure 17.16 provides general component category information for cleanliness requirements for Vickers components at typical pressures. It should be noted that higher pressures require cleaner fluid, and more complex components, particularly with smaller tolerances, typically require cleaner fluid.

Chapter 17 Fluid Conditioning

	1000 PSI	2000 PSI	3000 PSI+I	4000 PSI
PUMPS				
Fixed Gear	20/18/15	19/17/15		
Fixed Vane	20/18/15	19/17/14	18/16/13	
Fixed Piston	19/17/15	18/16/14	17/15/13	
Variable Vane	18/16/14	17/15/13		
Variable Piston	18/16/14	17/15/13	16/14/12	
VALVES				
Directional (solenoid)		20/18/15	19/17/14	
Pressure Control (modulating)		19/17/14	19/17/14	
Flow Controls (standard)		19/17/14	19/17/14	
Proportional Directional (throttle valves)		17/15/12	15/13/11*	
Check Valves		20/18/15	20/18/15	
Servo Valves		16/14/11*	15/13/10*	
Cartridge Valves		18/16/13	17/15/12	
H.R.C.		18/16/13	17/15/12	
Proportional Pressure Controls	16/14/12*	15/13/11*		
Flow controls (pressure compensating)	17/15/13	17/15/13		
Proportional Cartridge Valves	17/15/12	16/14/11*		
ACTUATORS				
Cylinders	20/18/15	20/18/15	20/18/15	
Vane Motors	20/18/15	19/17/14	18/16/13	
Axial Piston Motors	19/17/14	18/16/13	17/15/12	
Gear Motors	21/19/17	20/18/15	19/17/14	
Radial Piston Motors	20/18/14	19/17/13	18/16/13	
Cam Wave Motors	18/16/14	17/15/13	16/14/12*	
HYDROSTATIC TRANSMISSIONS				
Hydrostatic Transmissions (in loop fluid)		17/15/13	16/14/12*	16/14/11*
BEARINGS				
Ball Bearing Systems		15/13/11*		
Roller Bearing Systems		16/14/12*		
Journal Bearings (high speed)		17/15/13		
Journal Bearings low speed)		18/16/14		
General Industrial Gearboxes		17/15/13		

*Requires precise sampling practices to verify cleanliness levels.

Figure 17.16 Vickers component cleanliness requirements.

Chapter 17 Fluid Conditionings

Figure 17.17 is an example of a systemic contamination control worksheet that may be used to determine the oil cleanliness level requirements of a specific system. Note that reductions are made in initial component-determined cleanliness levels for such variables as fluid type, system pressure requirements, temperature and safety.

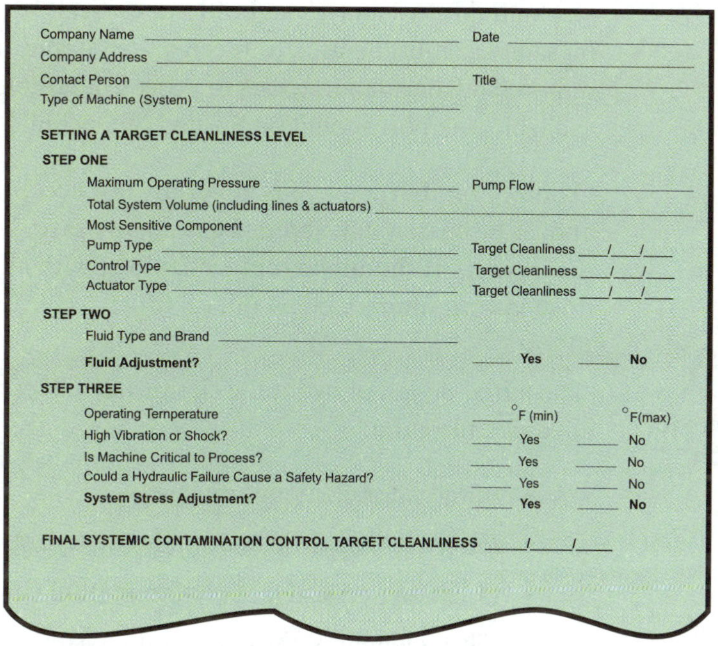

```
Company Name _____       Date _____
Company Address _____
Contact Person _____       Title _____
Type of Machine (System) _____

SETTING A TARGET CLEANLINESS LEVEL
STEP ONE
    Maximum Operating Pressure _____       Pump Flow _____
    Total System Volume (including lines & actuators) _____
    Most Sensitive Component
    Pump Type _____        Target Cleanliness ___/___/___
    Control Type _____        Target Cleanliness ___/___/___
    Actuator Type _____        Target Cleanliness ___/___/___
STEP TWO
    Fluid Type and Brand _____
    Fluid Adjustment?                        ___ Yes    ___ No
STEP THREE
    Operating Temperature                    ___°F (min)   ___°F (max)
    High Vibration or Shock?                 ___ Yes    ___ No
    Is Machine Critical to Process?          ___ Yes    ___ No
    Could a Hydraulic Failure Cause a Safety Hazard?  ___ Yes    ___ No
    System Stress Adjustment?                ___ Yes    ___ No

FINAL SYSTEMIC CONTAMINATION CONTROL TARGET CLEANLINESS ___/___/___
```

Figure 17.17 Systemic Contamination Control Worksheet

Achieving Target Cleanliness

In order to achieve the target cleanliness levels determined for the system, two major areas must be considered. First, the ingression of contamination must be minimized. This encompasses initial start up and commissioning as well as the pre-filtering of initial fill or replenishment fluid, proper maintenance techniques in changing filters, and the proper placement and ongoing maintenance of filters.

Because many, if not the majority, of filtration devices used on mobile applications do not include bypass indicators, time in service and calendar schedules are often used to change the filter elements. While this may not be ideal, it is a real world fact that must be addressed. In determining the scheduled change of filter elements, an initial effort should be made to determine the amount of time under typical duty cycles that filter elements will maintain acceptable cleanliness levels in systems. Once this is determined through frequent sampling and analysis of particle build up over time or duty cycles, a sensible schedule of element maintenance can be determined.

Chapter 17 Fluid Conditioning

Sampling Techniques

Obtaining a proper sample is the first and one of the most important elements of fluid analysis. Because of the limitations of mobile systems, the lack of sampling ports being the most limiting factor, reservoir samples are perhaps the most common. Before taking a sample, the system fluid should be brought to operating temperature and the machine cycled to suspend active contamination in the fluid.

Once operating temperature is reached, a sample should be taken from the center of the reservoir with a properly applied vacuum pump that has been flushed with the system fluid. A super-clean bottle, specifically designed and cleaned for receiving oil samples, must be used to receive the sample. A minimum of 60 ml is typically required by most laboratories to process a sample for analysis. If sampling ports are available, the port should be flushed of any build-up before a sample is taken.

Oxidized or "burned" oil is generally not acceptable for a particle count sample and, as discussed in Chapter 16 is, in any case, not acceptable for continued use in the system. It should be replaced, along with all filter elements, and the system flushed with filtered, clean oil.

Design Considerations for Contamination Control

The initial design of hydraulic circuits should consider the following to achieve adequate filtration:

- Initial filter element efficiency
- Filter element efficiency under system stress
- Location and sizing of contamination controls devices
- Filter element service life in the system

These design elements are critical to the life, durability and safe function of mobile machinery.

Figure 17.18 illustrates proper filter placements for a variety of system variables. This may be used with the cleanliness worksheet (Figure 17.17) to determine if a machine has adequate filtration.

Chapter 17 Fluid Conditionings

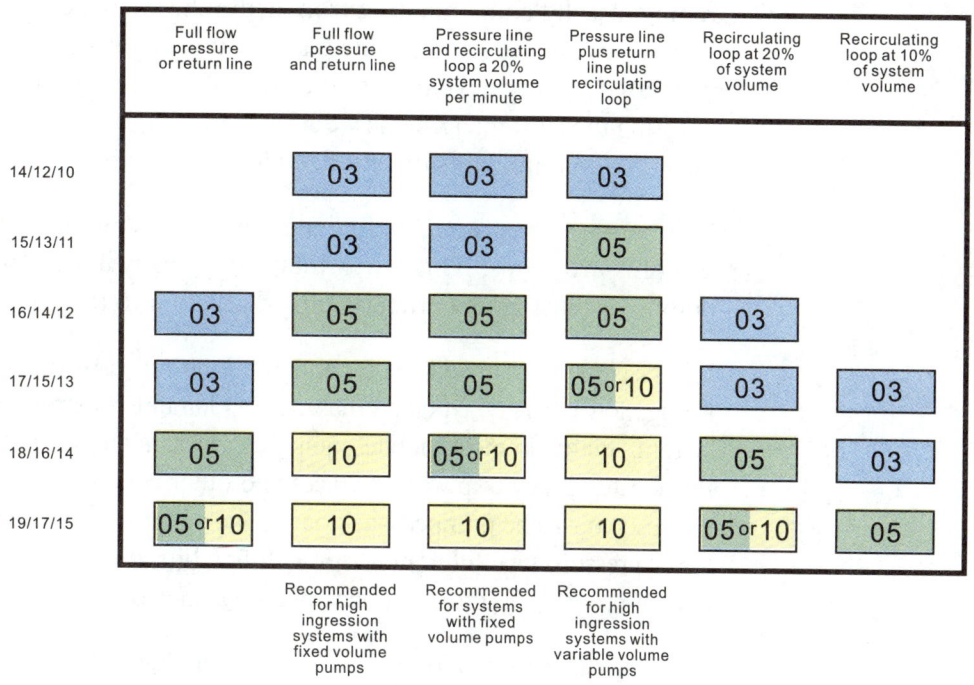

Figure 17.18 Filter placements for achieving desired cleanliness levels.

Determining Sampling Frequency

Once target cleanliness levels are achieved, charts should be developed to indicate to maintenance personnel how often each vehicle or piece of equipment should be sampled. This can be an effective tool, particularly for fleet management, to ensure the ongoing effectiveness of equipment and to reduce unplanned downtime and critical failures.

While it may seem excessive, sampling a new or overhauled or modified vehicle on a weekly basis can be very informative. The effectiveness of the filters can be readily determined. Problems with ingression will be obvious. This frequency will also provide evidence of unusually high wear in components - possibly due to excess contamination. Steps can then be taken early to correct problems and prevent serious damage or catastrophic failures.

Once the target level has been achieved and equilibrium reached in the system, sampling frequency can be relaxed gradually until a sensible schedule of filter changes and fluid analysis can be established.

Chapter 17 Fluid Conditioning

Filters and Strainers **The difference between a filter and a strainer**

The hydraulic symbols for filters and strainers are identical. Both do similar jobs in removing solid contamination from hydraulic systems. The currently accepted definitions of filters and strainers are: A *strainer* is a coarse filter - that is it removes very large particles, typically at the inlet of a pump, to prevent coarse materials from entering the system; a *filter* is a device whose primary function is the retention of insoluble contaminants in a fluid, typically via a porous medium.

Strainers are constructed of fine mesh wire screen that allows freer flow of fluid at low pressure drops. This is why they are typically used in pump inlets as first defense against large particles being ingested into the pump.

Filters, come in many different configurations and construction. Filters differ in micron rating, flow capacity, type and material of element construction and location(s) in a system. Filters in most mobile equipment circuits are located in the tank return line. In this location, they trap wear particles and other contaminants before the oil returns to the reservoir to be pumped back out to the system. This location also allows use of a lower pressure type filter, however with fine filtration, pressure drop may be 25 psi (1.7 bar) or more. This creates a back pressure on actuators.

Filter media, particularly in high-shock mobile applications, must be adequately supported to prevent collapse of pleats or separation of the media. Load induced or operation created pressure transients (shock) that damage the media could create "bursts" of trapped contamination being released into the system. Figure 17.19 illustrates the breakdown of filter media that could cause the destruction of the media layer creating large flow paths and particle generation. Figure 17.20 shows the advantages of properly supported filter media vs. unsupported media.

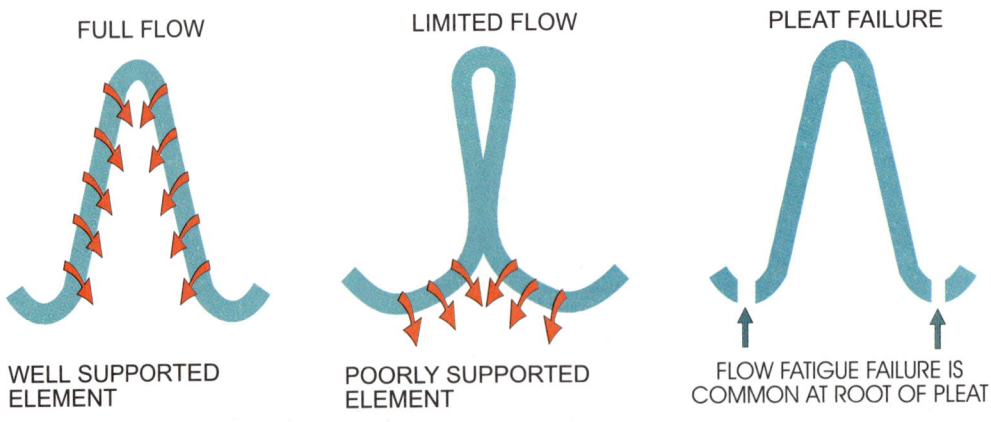

Figure 17.19 Unsupported filter media failures.

Chapter 17 Fluid Conditionings

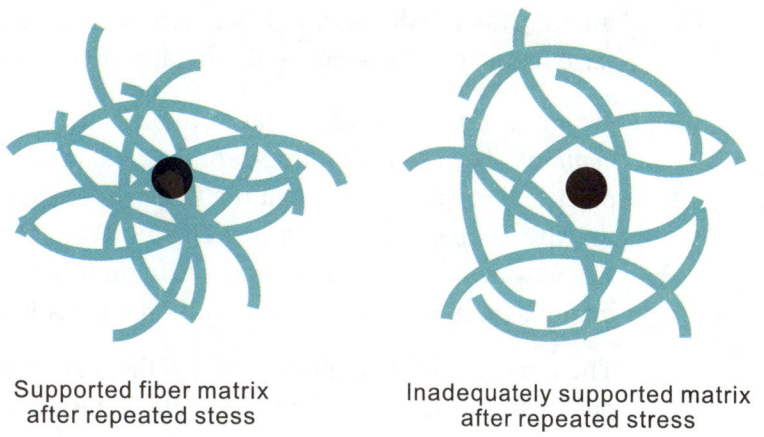

Figure 17.19 Unsupported filter media failures.

Paper filters are not generally compatible with high pressure hydraulic systems. Figure 17.21 shows the construction of Vickers V-Pak media with proper media support as well as diffuser layers to increase the life and effectiveness of the filter.

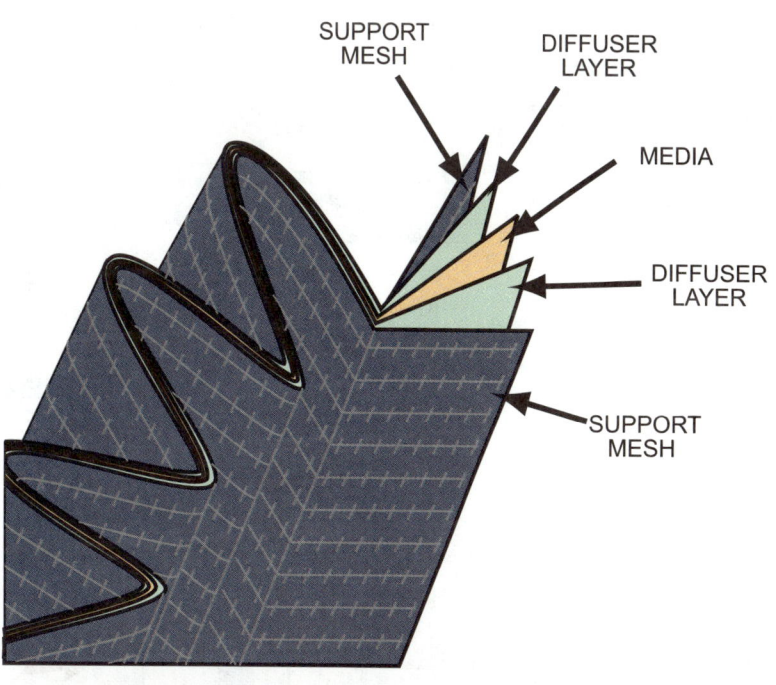

Figure 17.21 Vickers V-Pack Construction.

Chapter 17 Fluid Conditioning

Filters Ratings

The filter rating is used to define the ability of the filter to remove particles from the fluid stream. This is often expressed simply as a particle size; for example a "10 micron" filter. Although widely accepted, though not necessarily understood, this term has no official meaning that relates to any industry standard.

Two frequently used terms that were originated by the U.S. Navy are *absolute* and *nominal*. The absolute rating defines the diameter of the largest hard spherical particle that will pass through the filter under specified test conditions. It indicates the largest opening in the filter element. While this rating presumably defines the size at which the filter is 100% efficient, in fact it can mean whatever the filter manufacturer decides it means. It is an unreliable and often misleading term.

The nominal rating originally defined the diameter of the largest hard spherical particles at which capture was 98% efficient. This has become an arbitrary micron value defined by the filter manufacturer. It is also unreliable and not reproducible and is not endorsed by standards groups.

The international standard for the rating of efficiency of hydraulic and lubrication filters is the Multipass Filter Performance Beta Test (ISO 4572). In this test, fluid that is continuously injected with a standard contaminant (AC Fine Test Dust) is circulated through the test filter using a test stand such as is illustrated in Figure 17.22. Fluid samples are taken upstream and downstream of the element. An automatic particle counter determines the size and distribution of particles. The Beta rating of the filter is the ratio of the number of particles of a given size or larger in the upstream sample to the number of particles in the same size range in the downstream sample. Thus, a filter rating of $\beta_{10} = 100$ indicates that the ratio of the number of particles upstream to downstream that are 10µ and larger is 100. That is, for every 100 particles 10 micron and larger that entered the filter, only one passed through it, hence 99 particles were captured making the filter 99% efficient.

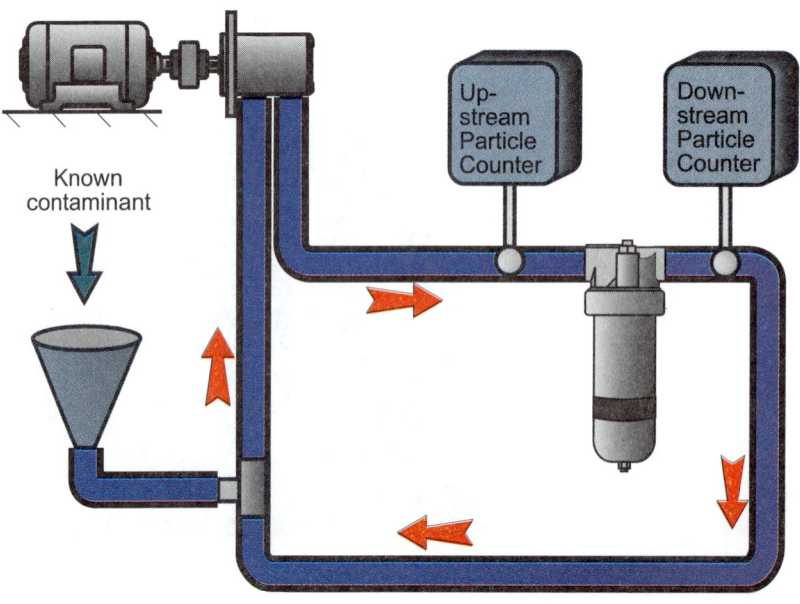

Figure 17.22 Multipass filter performance test.

The Beta rating can be expressed to capture efficiency by the following formula:

$$\text{eff} = \frac{\beta - 1}{\beta}$$

While Beta rating is the only established rating system for filters, it is necessary to realize that this rating must be taken into account with micron rating, filter element construction and placement to effectively control contamination. While a filter element may have a Beta rating of 500 indicating 99.8% efficiency, if this is its rating for particles 20 microns and larger, it may not be suitable for most hydraulic applications because its efficiency at smaller particles is not specified. Because the largest particles specified in fluid sampling specifications and ISO range codes measure 15µ, a filter rated for the removal of larger particles is contraindicated. Figure 17.23 shows a comparison of Beta ratios and their efficiencies. It is generally accepted that filters with less than a 75 Beta rating are not acceptable in hydraulic systems. Most reputable filter element manufacturers rate their filters at Beta 100 or better.

Beta Ratio	Efficiency
1	0%
2	50.00%
5	80.00%
10	90.00%
20	95.00%
75	98.70%
100	99.00%
200	99.50%
1000	99.90%
5000	99.98%

Figure 17.23 Beta ratios and corresponding efficiencies.

There are some limitations to the multipass test. The first is that ACFTD is not the same as the "real" dirt present in hydraulic systems. Another limitation is that the flow is held constant throughout the test, with no attempt to duplicate the flow surges, pressure spikes, and other disturbances experienced in the operation of actual hydraulic machinery. Maintenance of constant temperature and the use of an antistatic agent at low relative humidities further reduce the reality of the test. In spite of these limitations, the test provides a repeatable, reproducible method for comparing the abilities of filters to remove particles. While efficiencies are unlikely to be duplicated in the field, a good result on the test stand will typically translate into a good result in the field.

Chapter 17 Fluid Conditioning

Water Removal

Water can be removed from hydraulic fluids by several methods. Water intrusion in mobile hydraulic systems can be a constant problem because of an outdoor environment, the need to power wash vehicles, and severe weather duty. The following paragraphs indicate methods of water removal. Practicality for a specific application is a concern for design and maintenance personnel.

Settling and draining

Because water is heavier than petroleum oils, free water in oil will tend to settle to low places in the system. Properly constructed reservoirs should allow the draining of water from a low area of the reservoir supplied with a drain plug.

Although settling can remove gross amounts of free water, it is not effective for removing emulsified water and suspended microscopic droplets. Settling will occur effectively only in quiescent fluid, typically with the equipment shut down. Allow 15 minutes of settling time for each foot of reservoir depth. Of course, fluid that remains in components and piping will not benefit from settling in the reservoir.

Centrifuging

The separation of water from oil can be effected with centrifugal force that mechanically separates the heavier water from the oil. While fairly effective for removing free water, it is less effective at removing emulsified water. Because centrifuging requires considerable time it lends itself to batch processing of large amounts of water contaminated fluid. Large fleet maintenance programs for machines prone to water contamination might prove more cost effective than with smaller fleets or single vehicles.

Water removal elements

Water removing filters are typically of two types - coalescers and chemical. Both types can be used with some effectiveness to remove water from petroleum based fluids in hydraulic applications.

Coalescing filters involve passing the oil through a dense inorganic filter mat. Water droplets impinge on the fibers and adhere to them when the oil film is broken and the oil passes through the fibers with the water droplets being retained by the fibers. As more droplets are retained, they coalesce to form larger droplets which then flow down the fibers to a collector. The collector may be manually drained periodically, or it may drain automatically.

Chemical water removing filters are impregnated with chemicals that have a strong affinity for water. They may be adsorptive (the water remains external to the chemical structure) or absorptive (where the water becomes part of the chemical structure). Absorptive filters are preferable because of their permanent retention characteristics. Adsorptive filters have a tendency to "unload" some of the captured water under the right circumstances.

Evaporation

Evaporation of water from hydraulic fluid requires considerable effort and special equipment. It may be done by air drying, heating and vacuum dehydration. These processes are typically applied to stationary equipment systems, ring main systems or are done by fluid reconditioners.

It must be recognized that the long-term presence of water in oil can accelerate damage to the base chemistry of the oil that cannot be remedied only by the removal of the water. Reliable spectrometric analysis of the oil may be necessary to determine if removal by any means is practical and cost-effective, or if the fluid must be replaced.

Filter Locations

As previously mentioned, typical filter placement on mobile equipment is in the system return line. Separate filtration may be designed to protect specific parts of the system or separate tools or auxiliary devices run off the same system. Filter elements may not always be easily found because of space limitations and mounting requirements. Current hydraulic circuit prints of mobile machines should be available along with personnel trained to read them to determine the functions of all components, the controls used, safety interlocks, and the element requirements and placement of filters.

System Design Considerations

The hydraulic circuit design and the selection of certain components is important in the sense that good design and proper component selection with compatible functions will contribute to extended fluid life and more efficient machine operation.

Center bypass and *closed center* systems and tools are both used on mobile equipment. Typically center bypass control valves and tools are used with fixed displacement pumps to reduce energy waste and heat generation. Closed center systems are used with variable displacement, pressure compensated pumps for exactly the same reasons. However, using closed center tools with center bypass control valves and fixed displacement pumps will damage the machine and may cause serious problems. Conversely, using open center tools with a closed center system will not allow the pressure compensator to reduce pump output relative to the compensator setting. Both of these scenarios may see heat, shock and destruction of the fluid and components as a result.

Air removal - by design

The presence of air in hydraulic lines and actuators can present serious safety problems. Original equipment manufacturers often include air bleed valves in cylinders and other portions of mobile hydraulic systems to allow the removal of air from lines and components.

Chapter 17 Fluid Conditioning

Because of the compressibility and other properties of air, load control dynamics, dielectric strength of hose, and the reliability of consistent machine operation may be seriously compromised by the presence of air in hydraulic systems. Maintenance personnel must be acutely aware of the potential hazards of free air in mobile hydraulic systems and ensure that trapped air is bled from the system before it interferes with safety or machine operation.

Design modifications

Machine modifications should be done only after consultation with the machine manufacturer or a reliable fluid power expert - this includes modification of filtration devices and their placement. Ignorance of basic hydraulic principles as well as the requirements of size, pressure drop, power requirements and flow dynamics can defeat the best intentions of "making it work better". Considerations of load safety, component/circuit design compatibility, physical strength potential of machine load-bearing members and many other aspects of machine design must be analyzed before modifications of original machine design are undertaken.

As with "hit and miss" troubleshooting, "hit and miss" machine modifications more often create more problems than they solve. Only personnel with the requisite expertise and experience should attempt redesign of hydraulic machinery.

Cooling

While it would be ideal to moderate the temperature of the oil and hydraulic circuit in mobile machinery, cooling is, at best, a difficult proposition in many mobile applications. As related in Chapter 16, heat creates serious problems in hydraulic oils, machines and operations. Air and water coolers are available for use in hydraulic circuits.

Unfortunately, the nature of mobile machinery is such that the most efficient coolers, water coolers, are impractical on mobile machines - can you a imagine a garden hose several miles long to supply water for cooling to an off-highway piece of equipment out on Interstate highway construction! Air coolers are a possibility, however space and ambient temperature can be serious constraints. On vehicles with internal combustion engines, it may be practical to use a radiator bottom tank for cooling hydraulic oil.

CHAPTER 18 .. Circuits

Mobile hydraulic systems have some basic similarities; one or more pumps are connected to a directional control valve which leads to linear or rotary actuators. The actuators perform a work function on command from an operator. Other valves are needed in the circuit to protect the components or to condition the fluid before it enters the actuator.

Nearly all similarity ends there, however. Practically every machine requires some unique hydraulic arrangement to perform its task in an efficient and productive way. Considerations of safety, economy, energy conservation, pollution control, performance specification, machine size, operator convenience, etc., become reasons for developing a distinctive combination of valves and connectors for particular models, series or types of machines.

This chapter will describe several types of conventional applications, and will define a special configuration for each circuit that has been created to perform a specific function. In each instance, the requirement of the circuit and the operation of each special configuration is explained.

SYSTEM FLOW CONTROL

Requirement Designate a maximum flow for a circuit, regardless of the input pump speed.

Application 1 Vehicle power steering circuits will operate at best performance with rated flow, and efficient performance depends on flow that does not exceed this value by more than a small percentage. For example, power steering on a hypothetical farm tractor may perform best with 12 GPM input flow to the steering valve. However, any flow above this amount will waste power by either providing performance that is unnecessary (possibly even destructive), or by diverting excess flow and heat to the reservoir.

Application 2 A refuse hauling truck may require 20 GPM to compact at an acceptable performance level. Preferably this would be done at a low engine speed in order to keep noise levels low in residential areas. The pump must then be disengaged to prevent damage by over-speeding when the engine speed is increased during road travel. Furthermore, high flows at high engine speeds creates excessive heat, and also prevents packing the refuse during travel because the packing mechanism may be damaged.

Chapter 18 Circuits

Resolution

A pressure compensated pump with a pressure limiting, load sensing compensator is used as shown in Figure 18.1. The pump is sized to provide the desired flow at the desired operating speed. For example, the tractor power steering pump might be a 3.75 CIR displacement pump, giving about 12 GPM output at 800 RPM, a low engine operating speed. The refuse compactor may utilize a 4 CIR displacement pump that will provide about 20 GPM at 1200 rpm, a relatively low operating speed that will give quieter operation during the packing cycle.

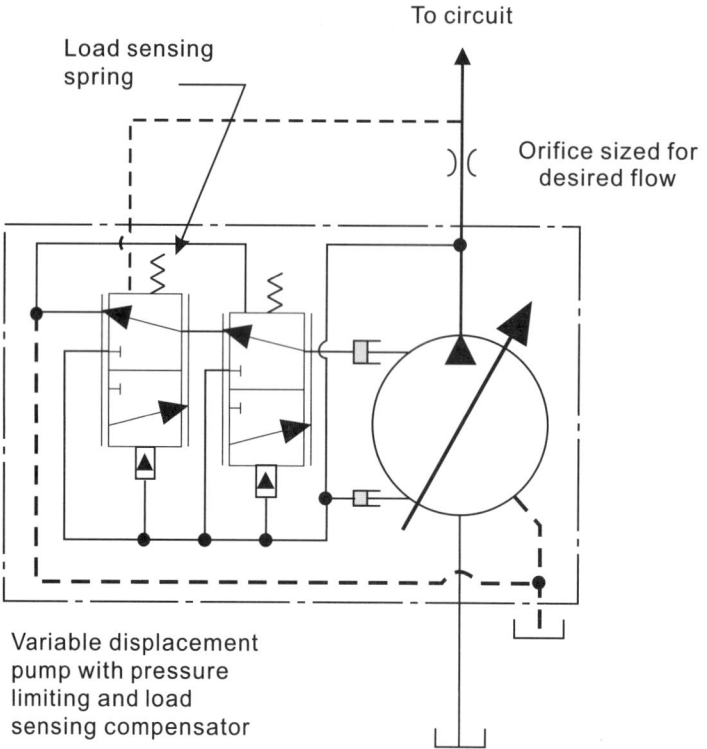

Figure 18.1 Flow control circuit.

An orifice is placed at the outlet of the pump, sized for the desired flow in each situation (12 GPM for the tractor and 20 GPM for the refuse hauler) at a pressure drop equal to the spring setting of the load sensing compensator on the pump. A typical load sensing spring setting is 300 psi, which would result in a 0.18 inch orifice for the 12 GPM flow, and a 0.24 inch orifice for the 20 GPM flow. The load sensing port is connected to the down-stream side of the orifice, between the orifice and the system control valves.

The pump will continuously adjust its displacement to maintain the 300 psi pressure drop across the orifice, maintaining the fixed flow rate regardless of the pump speed or working pressure.

Chapter 18 Circuits

ALTERNATE PILOT PRESSURE SOURCE

Requirement Provide an emergency power source for hydraulic remote control valves.

Application Larger vehicles that utilize pilot operated proportional control valves, such as the Vickers CMX valve, require a source of pressure and flow to the remote hydraulic pilot valves. When there is a "power-down" situation, such as when the engine inadvertently stops, there is no hydraulic power to the HRC valves. This may occur while a loader bucket or a crane boom is in the raised position. Without power to the HRC valves, the bucket or boom cannot be easily or safely lowered.

Resolution While the equipment (boom or bucket) is in the air, there is a source of fluid under pressure in the base end of the lift cylinders. The circuit in Figure 18.2 utilizes this source to operate the hydraulic remote control valves to lower the equipment.

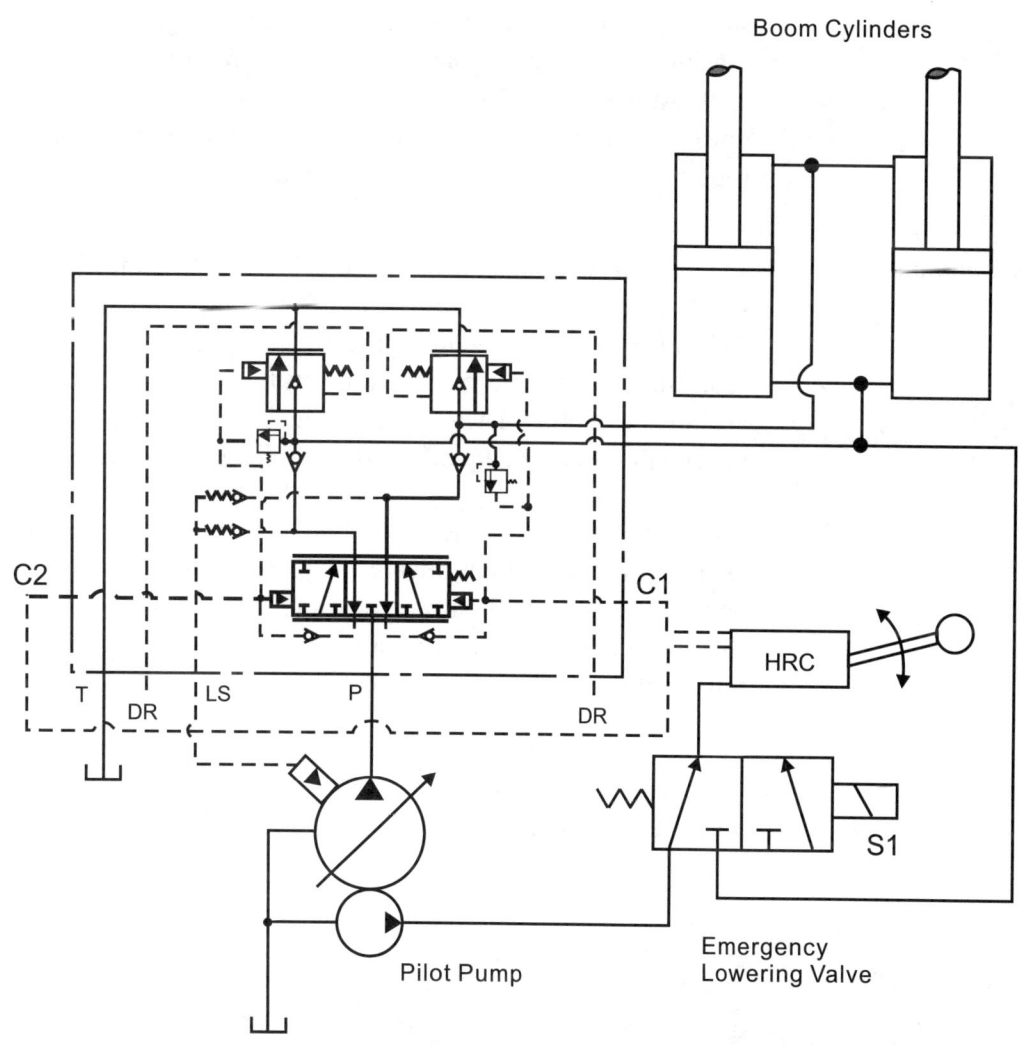

Figure 18.2 Emergency pilot pressure source.

Chapter 18 Circuits

A line is connected from the base end of the lift cylinders to the emergency lowering valve, a 2-way, 2-position solenoid operated spring offset control valve. During normal operation, the pilot pump is directed to the HRC valve while the emergency lowering valve is in the spring offset position. When a power-down situation occurs, the solenoid on the emergency lowering valve may be activated, and the HRC now receives its flow from the base of the lift cylinders.

When the equipment has been lowered and the lift cylinders are collapsed, there is no longer any flow and pressure available to the HRC. Presumably, at this point, a pilot source is no longer needed until power is restored.

PRIME MOVER PROTECTION

Requirement   Prevent system flow and pressure from exceeding the prime mover (engine) capability.

Application   A mud auger requires sufficient speed for optimum operation, and also requires sufficient torque to overcome exceptionally tough conditions. However, when both maximums occur at the same time, power requirements are such that the engine will stall. At the point of highest torque, maximum speed is not required. The need is to have a method that automatically adjusts system flow at the higher pressure requirements to reduce horsepower and protect the prime mover.

Resolution   The system consists of a variable displacement pump with a load sensing, pressure limiting compensator, a Valvistor® flow control valve and an adjustable pressure reducing valve as shown in Figure 18.3. The flow control valve and the pressure reducing valve are placed in parallel at the pump outlet and feed directly to the auger drive motor. The pressure reducing valve is set to a pressure below which the auger will provide adequate torque at higher speed.

As system pressure, caused by torque requirement at the auger, rises above the setting of the pressure reducing valve, the valve will close and cause all pump flow to be directed through the flow control valve. The load sensing control of the pump will now maintain a fixed pressure drop across the flow control, equal to the spring setting of the load sensing spool, by reducing pump displacement as required.

The system pressure can now rise to the setting of the pressure limiting compensator, providing maximum torque, at a reduced flow so that horsepower will not exceed the engine capability.

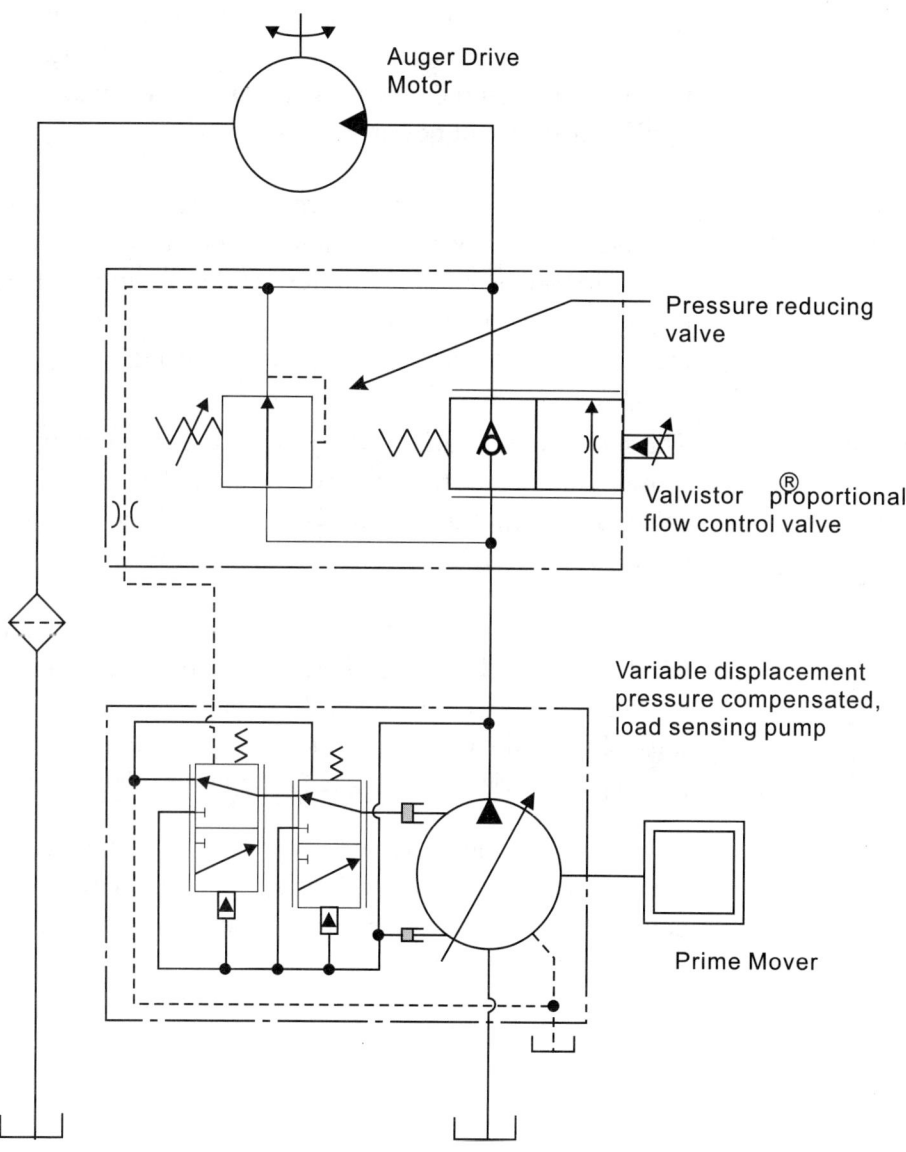

Figure 18.3 Prime mover protection on auger drive.

Chapter 18 Circuits

PROPORTIONAL COOLING FAN DRIVE

Requirement Regulate an IC engine cooling fan to run only to the extent needed to maintain engine temperature.

Application The cooling fan of an internal combustion (IC) engine absorbs power based on its size and operating speed. When driven directly from the engine crankshaft, its speed is a function of engine speed, and the horsepower required to drive it increases by N^3, that is, the cube of the operating speed. Controlling fan speed to that which is required only for maintaining engine coolant temperature is a very significant source of power savings.

Engines for mobile equipment require a varying degree of cooling due to the variable nature of working conditions and air temperature. There are times when very little cooling air needs to flow, and other times when maximum air flow is required. A desirable system would detect engine operating temperature and adjust the running speed of the cooling fan accordingly.

Resolution The system uses a variable displacement piston pump with a pressure limiting, load sensing compensator, a fan drive motor and a four-way, two position proportional valve. The proportional valve is controlled by an amplifier that receives its input signal from an engine coolant temperature sensor.

The proportional valve regulates the pressure in the load sensing spring chamber, which regulates the system pressure that drives the fan motor. When minimum cooling is required, the valve is fully shifted and all load sense flow is diverted to tank. The pump will operate at "standby" pressure, which is usually in the 150-300 psi range. This is enough to keep the fan revolving slowly with very little air flow.

When working conditions are severe, and maximum cooling is required, the proportional valve closes and the system will operate at maximum pressure, governed by the pressure limiting compensator setting. A full range of pressures, and therefore fan speed, will be available between minimum and maximum.

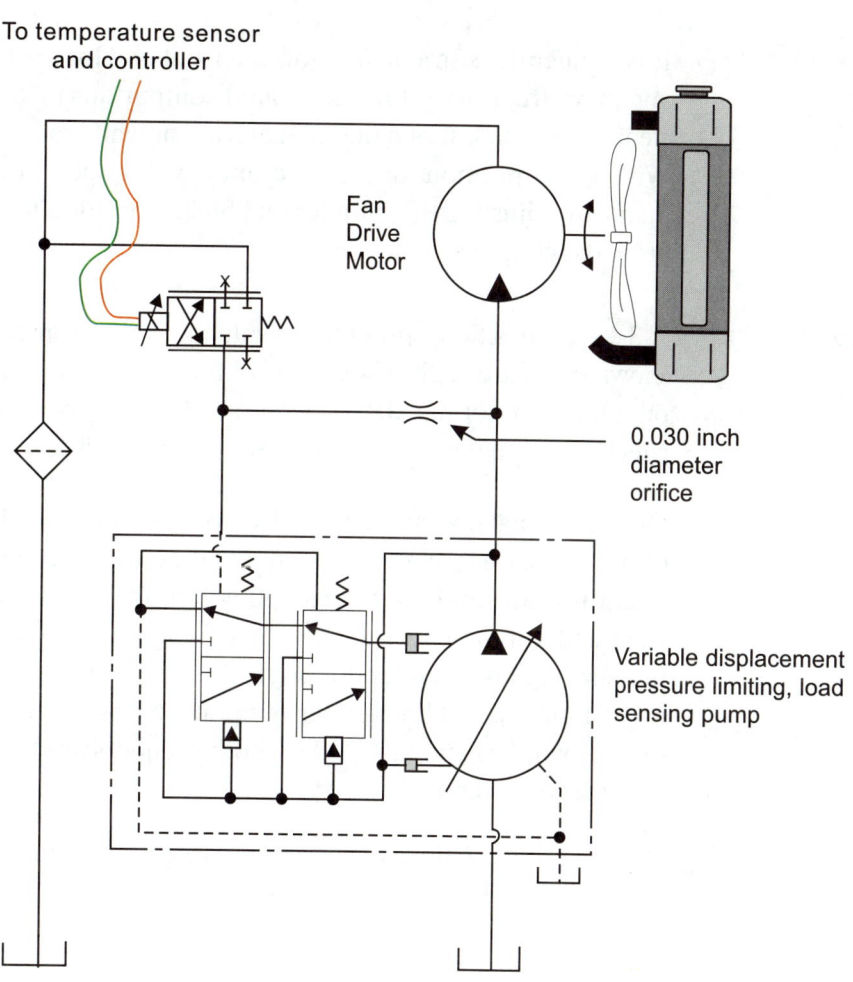

Figure 18.4 Energy conserving engine cooling fan drive system.

Chapter 18 Circuits

FLOAT POSITION USING CMX VALVE

Requirement Provide a solenoid activated float position for the bucket of a front end loader when using the CMX type of valve.

Application The CMX type of valve uses a meter-in, meter-out control for the bucket lift cylinders on a front end loader. This provides finite control of the lift cylinders while raising and lowering the bucket. This configuration of the valve also precludes a float position.

It is frequently desirable to allow the front end loader bucket to set on the ground and move freely to follow the ground contour during travel. This is useful while cleaning up the last of a pile of material, or while ramming the bottom of the pile. With a float position for the lift cylinders, this operation can be done without manually adjusting the cylinder and bucket position to compensate for uneven ground conditions.

Resolution A 3-way, 2-position, spring offset solenoid valve is installed in the pilot supply line as shown in Figure 18.5. The A port of this valve connects to the HRC valve to supply pilot fluid for normal HRC operation. The B port connects through two shuttle check valves to the HRC control lines to the bucket lift cylinders.

During normal operation, the HRC controls the pilot flow to one side or the other of the CMX valve, controlling lift and lowering in a conventional way. When the float valve solenoid is activated, pilot pump flow is directed away from the HRC valve and, through the shuttle valves, to both pilot ports of the CMX valve. By pressurizing both CMX pilot ports, the meter-in spool is not shifted (equal pressure on both sides), and both meter-out poppets are raised to allow free flow between the lift cylinders and tank. Thus, both cylinders are free to float, exchanging fluid with the reservoir as they move.

De-activating the float valve solenoid returns the pilot flow to the HRC valve, and reinstates conventional control.

Chapter 18 Circuits

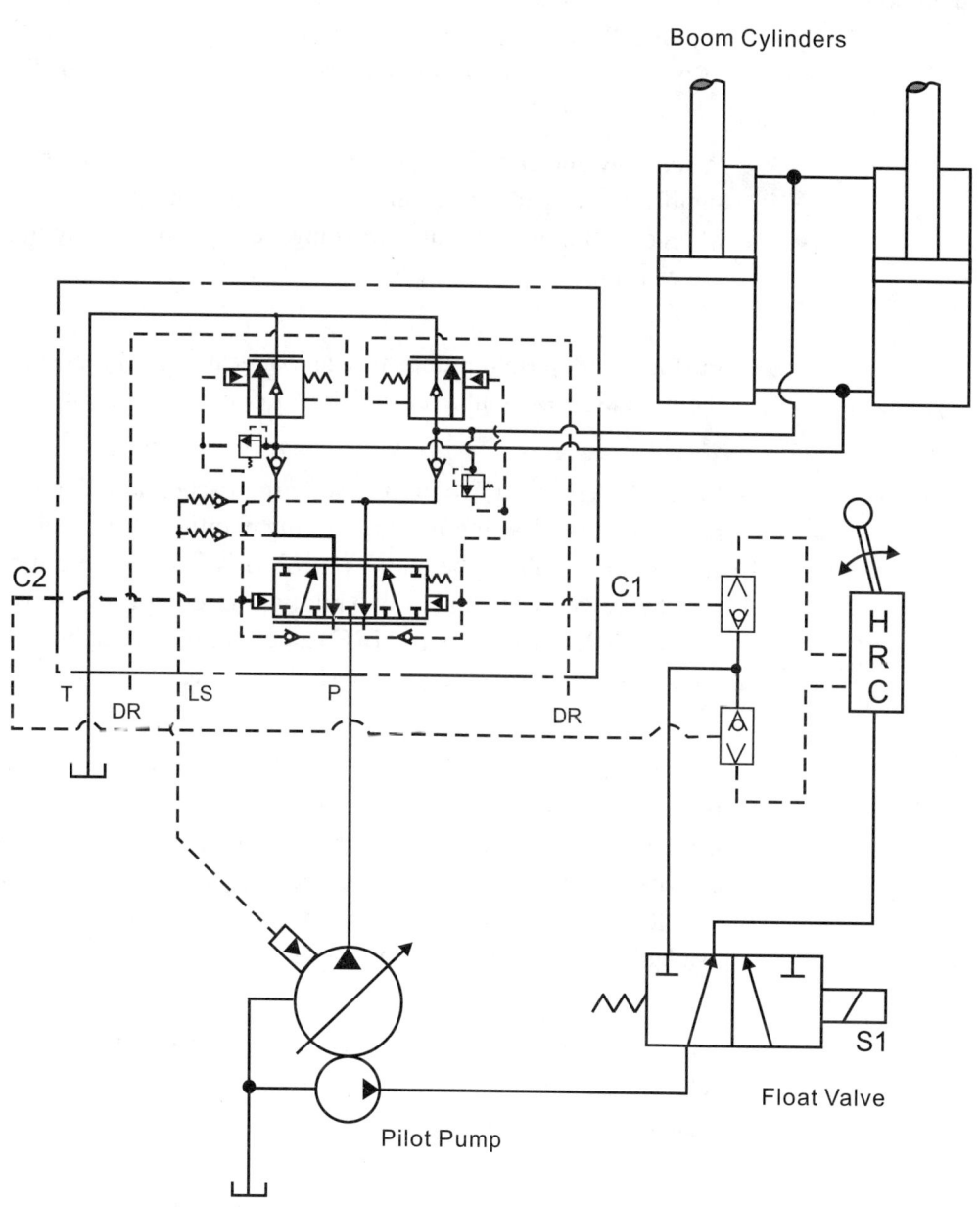

Figure 18.5 CMX loader boom circuit with float position.

Chapter 18 Circuits

TWO PUMP COORDINATION SYSTEM

Requirement Obtain the maximum benefit of a two pump system by using two pumps for a single circuit operation and individual pumps for dual circuit operation.

Application Many larger machines utilize a two pump system, where each pump serves its own circuit. When both circuits are operating, each circuit has all the flow available from its own pump and is not affected by operation of the other circuit.

When only one circuit is operating, however, the other circuit pump is idle. A significant benefit can be derived by having both pumps operating in either circuit when only that one circuit is working, yet operating individually when both circuits are working.

Furthermore, if both pumps utilize load sensing compensators, they must receive their load sense signal from whichever circuit they are working in.

Resolution Figure 18.6 shows a two pump, two circuit schematic that will satisfy the application requirements of both pumps operating in either circuit together, or in their own circuit individually. The main control for selecting these functions is the circuit selector valve, consisting of two pilot operated, 4-way, 2-position, spring offset valves and a shuttle valve. If both HRC valves are operated, both 4-way valves are piloted and there is no flow through the shuttle valve.

If only one HRC is operated, one of the 4-way valves is piloted, and flow is directed through the shuttle valve to the circuit combining valve. The circuit combining valve is a 1:1.2 slip-in cartridge valve that is held normally closed by either of the two main circuits until the 4-way, 2-position valve controlling it is activated by the circuit selector valve. The circuit combining valve is then opened and both pump flows are available to either main circuit 1 or main circuit 2, whichever one is being operated.

Flow from the circuit selector valve shuttle is also directed to the load sensing selector valve, piloting the 4-way, 2-position valve that will then connect both load sensing compensators to either load sensing signal.

Thus, operation of either HRC by itself will connect both system pumps to the same circuit, and the load sensing compensators will react to the pressure signals from that single circuit. When both HRCs are operated, there is no flow through the circuit selector shuttle, the circuit combining valve is closed and the load sensing selector valve switches to separate the two compensators. The pumps will then operate in each individual circuit independently.

Chapter 18 Circuits

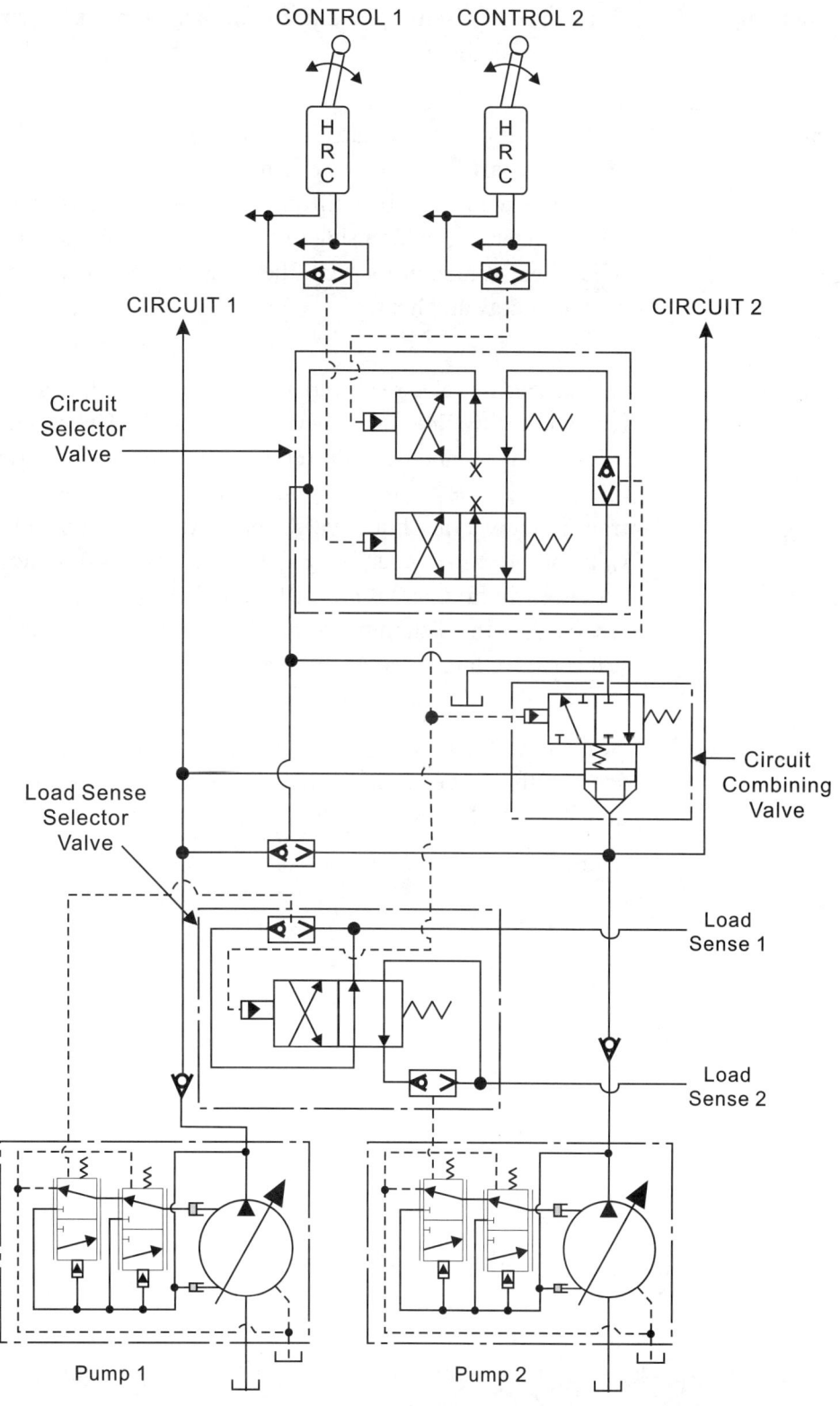

Figure 18.6 Two pump individual and combined control circuit.

FIXED PUMP LOAD SENSING SYSTEM

Requirement   To provide load sensing to a fixed pump system for improved operating economy and less heat generation.

Application   Metering flow to an actuator in a fixed pump system divides the pump flow into two parts: one part flows to the actuator, and the second part flows over the relief valve to the reservoir. The flow through the relief valve wastes energy and creates heat in the reservoir. A load sensing system would allow the unused fluid to flow over the relief valve at load pressure, rather than maximum system pressure, reducing the wasted energy and heat.

Resolution   A load sensing valve manifold assembly is connected across the pump outlet and reservoir return lines as shown in Figure 18.7. The manifold consists of a relief valve cartridge, and two differential pressure sensing cartridge valves. The first differential valve maintains system pressure at the pressure required by the actuator, transmitted via the shuttle valve and load sensing line, plus the differential pressure value of the valve, which is set at 160 psi. Excess fluid not being used by the actuator will flow across this differential valve to the return line, and to the reservoir, at the system pressure. At low pressure requirements of the actuator, this will conserve large amounts of energy.

As actuator load pressure, and therefore system pressure, continues to rise to the value of the relief valve, the second differential valve will open at 80 psi above this and all pump flow will flow to the return line. This valve acts as the system relief.

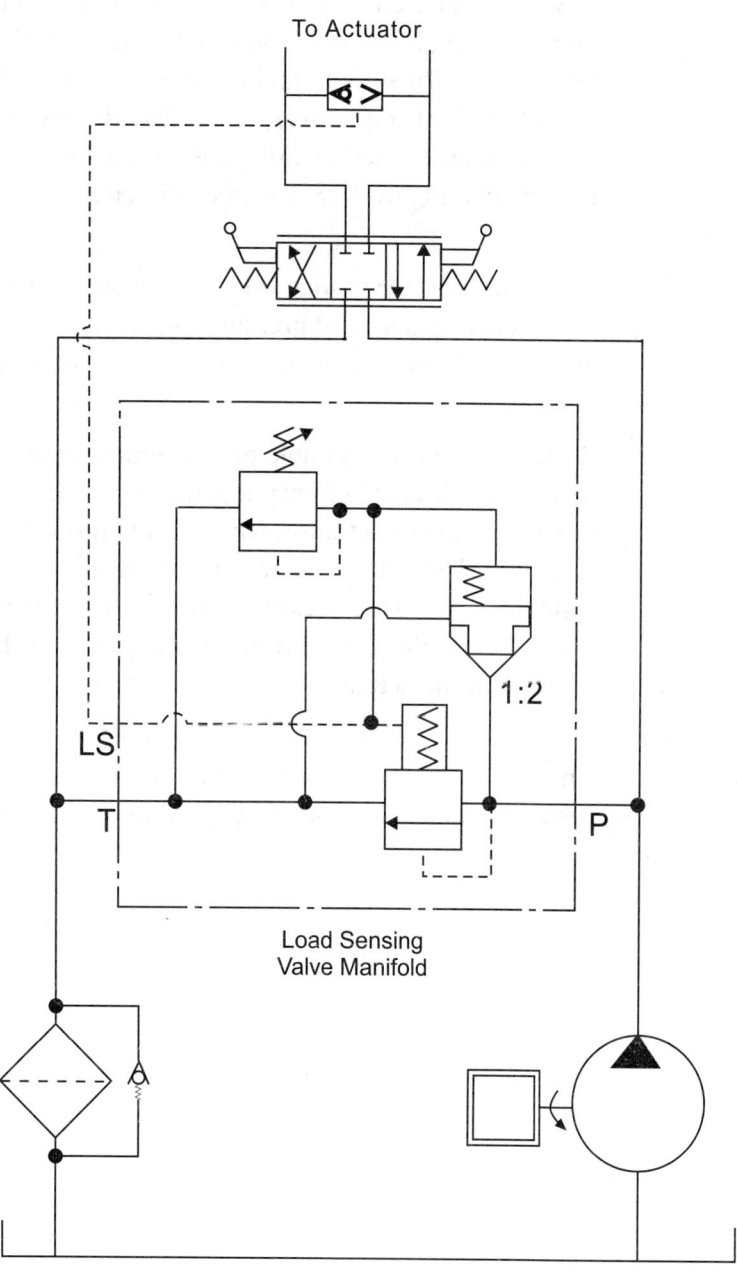

Figure 18.7 Fixed pump load sensing system.

METER-OUT USING COUNTERBALANCE VALVES

Requirement Provide meter-in and meter-out control of actuators using cartridge valves and electric remote controls.

Application Lower pressure and lower flow applications that cannot utilize the capabilities of high pressure, high flow valves such as the CMX may still have a meter-in/meter-out requirement for smooth and reliable operation. Meter-in can be accomplished with Valvistor® cartridge valves, controlled electrically by ERCs. Meter-out can be effectively controlled by using counterbalance valves and a dedicated reservoir return line. Figure 18.8 describes this circuit.

Resolution Four Valvistor® cartridge valves are grouped in a manifold, connected in parallel to a variable displacement pressure limiting, load sensing pump. Each Valvistor® acts as a proportional meter-in valve to each inlet port of the actuators.

Both actuators have double pilot operated counterbalance valves connected across their ports. Meter-in flow pilots the counterbalance valve to meter flow out of the return side of the actuator, which flows through a dedicated return line back to the reservoir. The counterbalance valve prevents overrunning loads under all inlet flow conditions, up to the pressure setting of the internal pilot, which then acts as a brake valve. Under these conditions, makeup flow to the inlet side of the actuator will come from the actuator outlet side or from the reservoir.

Check valves installed downstream of the Valvistor® flow controls act as a shuttle valve providing a load sensing signal to the pump compensator.

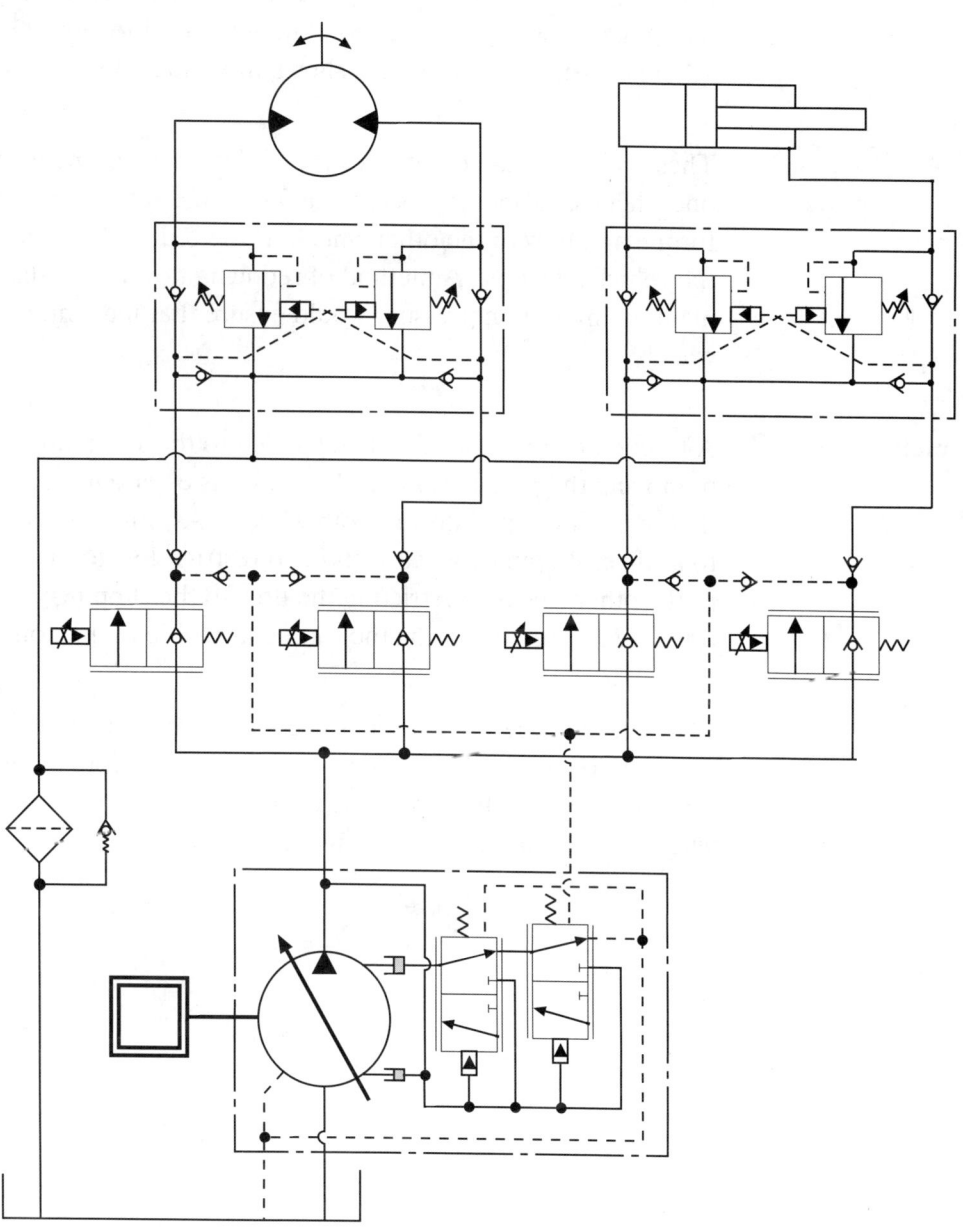

Figure 18.8 Meter-in/meter-out circuit using Valvistor® and counterbalance valves.

Chapter 18 Circuits

HYDROSTATIC PROPEL TRACTION CONTROL

Requirement Provide traction control for a hydrostatic single pump/split motor transmission system.

Application Many machines utilize a single hydrostatic transmission pump and two motors for propel, such as lift trucks, rough terrain lift trucks, self propelled agriculture equipment (mowers, mower/conditioners, windrowers, etc.) and sweepers. A typical schematic for such a vehicle is shown in Figure 18.9.

These vehicles can potentially become immobile if one wheel loses traction. When one wheel slips and begins turning faster, the system pressure drops, reducing torque and flow to the other wheel. If the reduced torque is insufficient to propel the vehicle, it stops. A method of reducing flow to the slipping wheel and maintaining system pressure would ensure that maximum torque is available at both wheels.

Resolution A traction control valve has been added to the schematic in Figure 18.9, between the pump and the propel motors. This valve is essentially a flow divider/combiner valve that maintains equal flow to both wheels. As flow begins to become unbalanced due to a wheel slipping, the valve shifts to restrict flow to that wheel and maintain flow to the other wheel. Restricting the flow to the slipping wheel will cause the system pressure to rise, increasing the torque available at the non-slipping wheel.

The difference between a traction control valve and a flow divider/combiner valve is the wider band of unequal flow that is allowed before adjustment occurs. This prevents the valve from reacting during vehicle turns (during turns, the outside wheel revolves faster than the inside wheel).

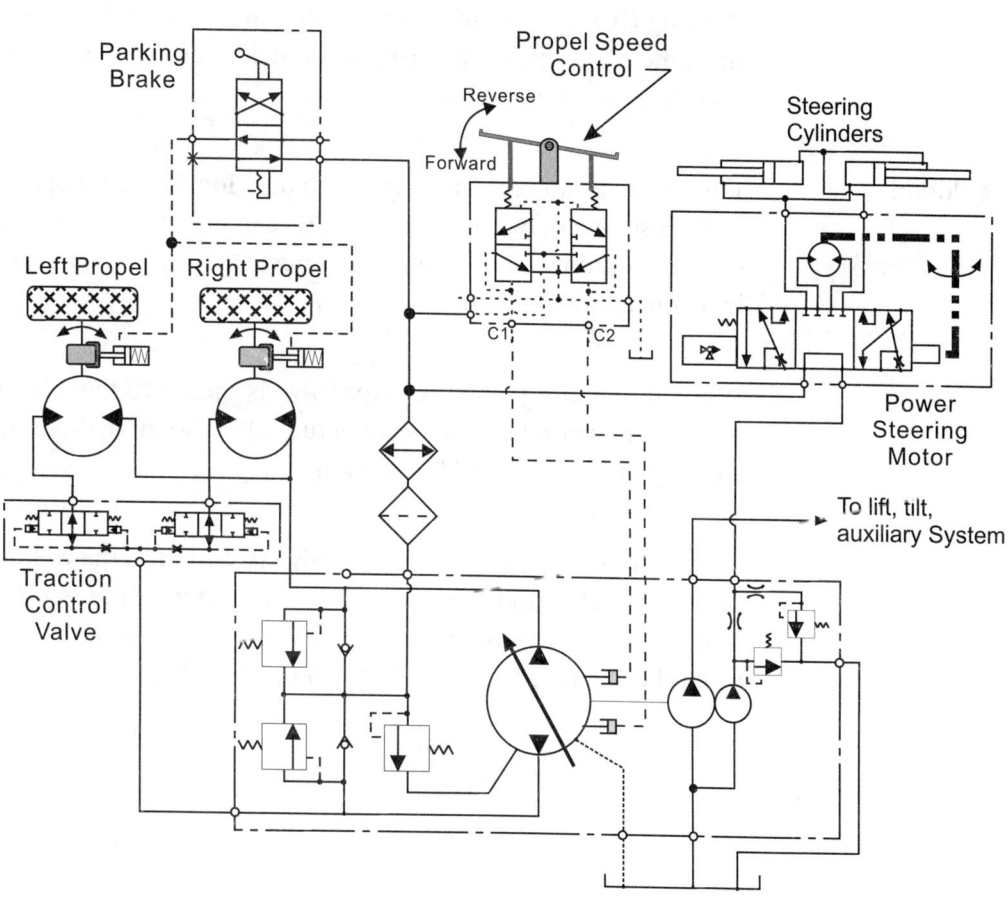

Figure 18.9 Hydrostatic single pump, split motor transmission circuit.

Chapter 18 Circuits

STEERING MODE SELECTION

Requirement To quickly select a four wheel steering mode between conventional, crab and articulated.

Application Large four wheel drive vehicles, such as pneumatic front end loaders or farm tractors, are difficult to maneuver with conventional two wheel steering. Therefore, four wheel steering is provided that will allow conventional two wheel steering for travel, four wheel articulated steering for turning (front wheels and rear wheels turning in opposite directions for minimum turning radius), and four wheel crab steering (front and rear wheels turning in the same direction to provide lateral movement). A means of easily and quickly selecting between the three steering modes is required.

Resolution Figure 18.10 shows a four way, three position, solenoid operated directional control valve installed between the front and rear steering cylinders. With the valve in the center position as shown, the rear wheel steering is locked and the front wheels steer in a conventional manner.

When the steering mode selector valve is shifted to the left, crab steering is provided as all four steering cylinders are actuated in the same direction when the power steering motor is turned left or right.

When the steering mode selector valve is shifted to the right, articulated steering is provided. When the steering motor is activated to the left, the front wheels will steer to the left and the rear wheels will steer to the right. The opposite will occur when the steering motor is activated to the right.

Chapter 18 Circuits

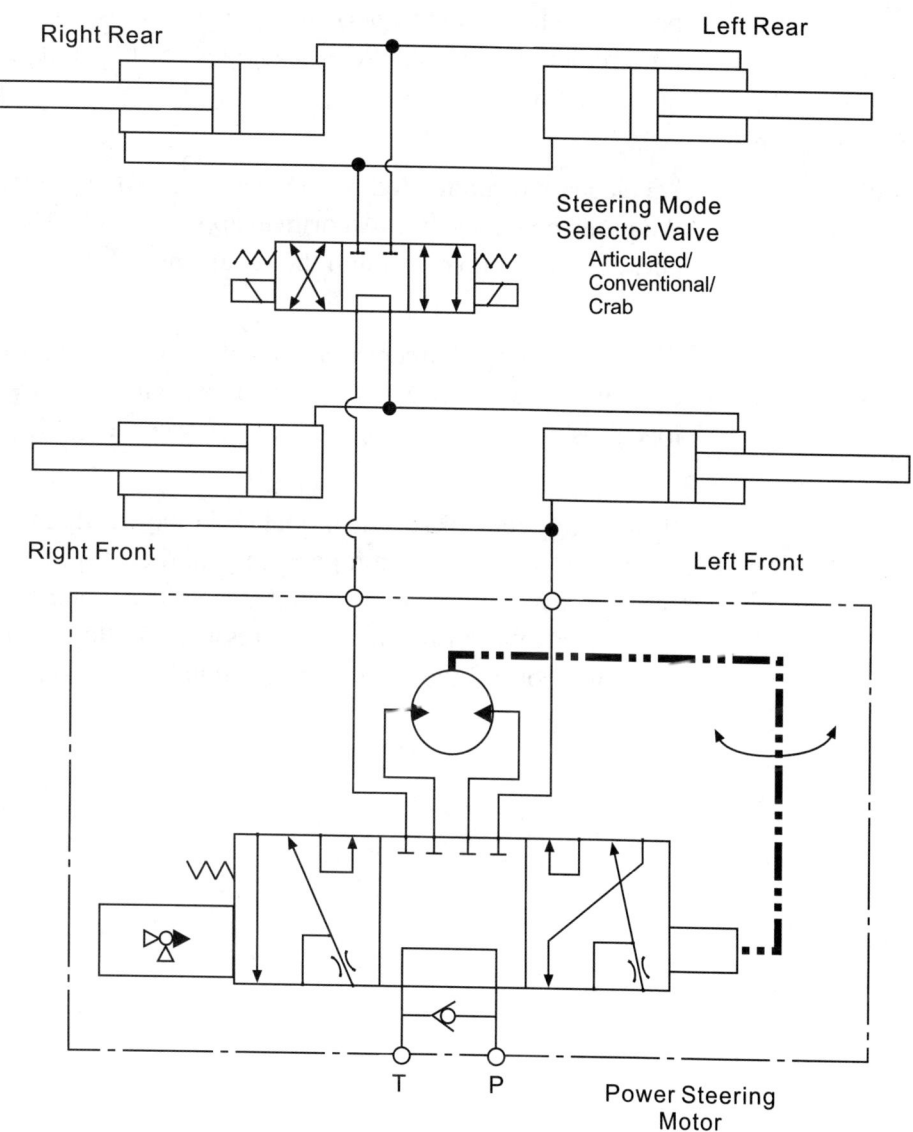

Figure 18.10 Conventional/articulated/crab four wheel steering circuit.

WATER PUMP FLOW CONTROL

Requirement Provide an adjustable water discharge control on a hydraulically driven water pump.

Application Water pumps are typically an impeller pump that discharges liquid based on the speed and power of the input driver. In this case, the driver is a hydraulic motor driven by a variable displacement pressure limiting, load sensing pump. The water pump may be used for emergency fire apparatus, crop spraying, etc.. The requirement is to be able to easily adjust the flow output of the water pump for different circumstances.

Resolution An electrically modulated relief valve is placed in the load sensing line leading to the pump compensator. An orifice, approximately 0.020 inches diameter, is placed between the pump outlet and the relief valve. Figure 18.11 illustrates this circuit.

The load sensing compensator will adjust the pump output flow to maintain a constant pressure drop across the orifice. This pressure drop will be the value of the load sensing standby spring, typically between 150-300 psi.

By adjusting the electrically modulated relief valve, the pressure in the load sensing line will be varied. The compensator will then adjust the pump flow to attain a system pressure equal to the relief valve setting, plus the value of the standby spring. This effectively controls the inlet pressure and flow to the water pump drive motor, which will control the pump speed and liquid flow output.

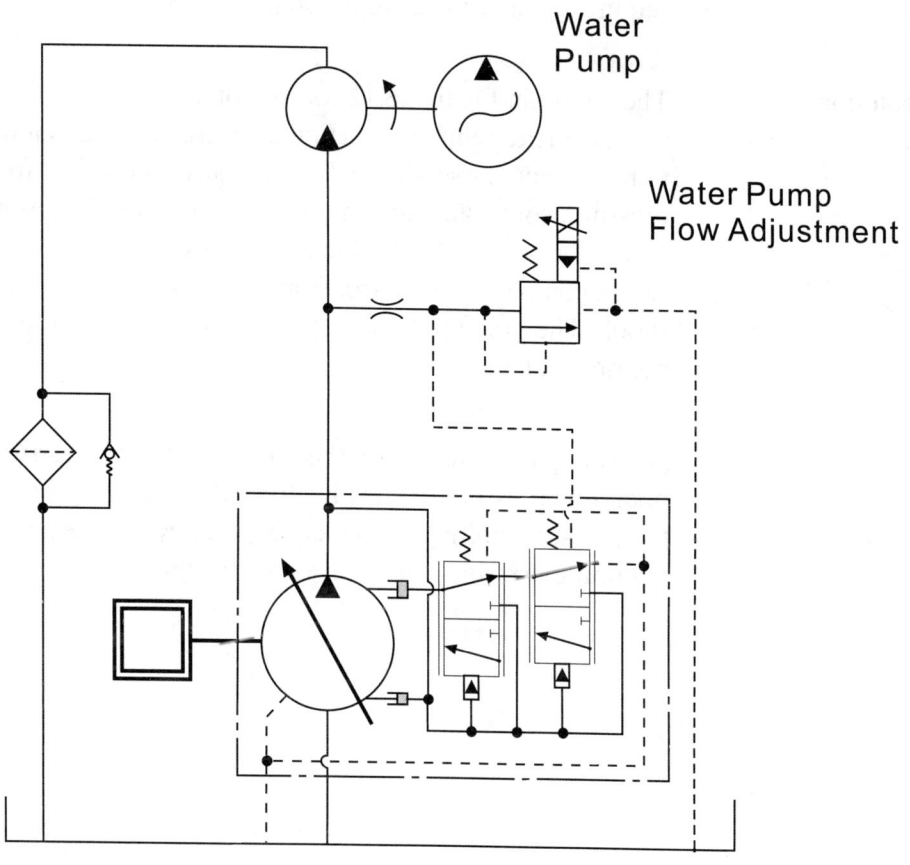

Figure 18.11 Water pump flow control circuit.

INTERMITTENT HIGH PRESSURE OPERATION

Requirement Provide an intermittent higher pressure capability on demand to address high load circumstances.

Application Frequent situations with mobile machinery require a marginally higher load capacity for a short duration. Examples are: winch drives where maximum pulling capacity is needed only occasionally; propel drives where higher drive capability is needed only in unusual circumstances; crane applications where overloading occurs and the capability is needed to remove the load; off-road equipment where off road propel requires greater pressure capability than on-highway operation; etc.

Resolution The circuit in Figure 18.12 consists of a pressure limiting, load sensing variable displacement pump, a directional control valve and an unspecified actuator. Added to the circuit is a small relief valve that would be set to normal system operating pressure, controlling maximum pressure in the load sensing line. A 2-way, 2-position spring offset solenoid valve is also in the load sensing line, which will switch compensator control from the load sensing spool to the pressure limiting spool. The pressure limiting portion of the compensator is set to the higher pressure value.

Activating the solenoid on the momentary high pressure switch will remove the operating pressure relief valve from the circuit, and the higher pressure setting of the pressure limiting compensator will govern. Deactivate the solenoid, and the operating pressure relief valve governs the system pressure.

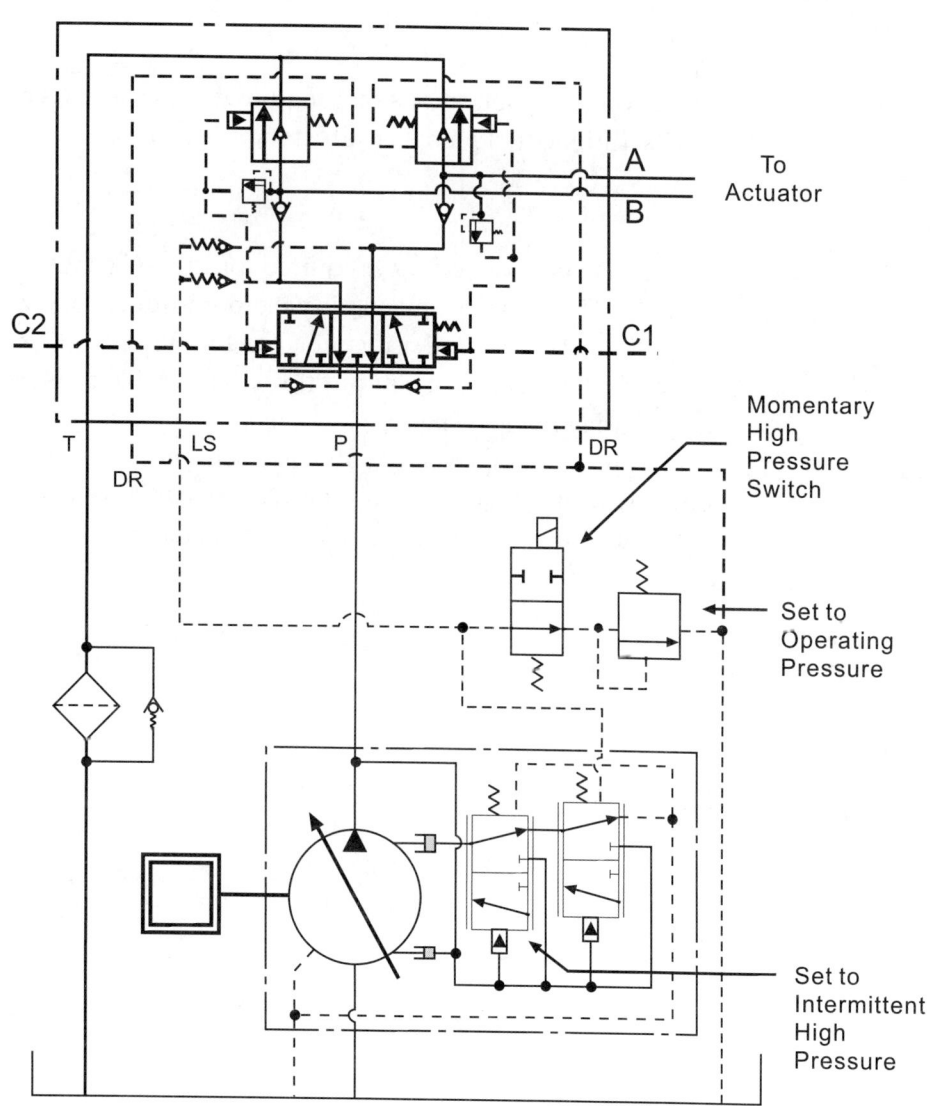

Figure 18.12 Circuit with intermittent high pressure capability.

Chapter 18 Circuits

CLOSED LOOP VELOCITY CONTROL

Requirement   To accurately control rotary or linear velocity.

Application   Figure 18.13 illustrates an application where rotary velocity must be very accurate. The rotary speed of an AC generator must be held very close because the voltage frequency is dictated by the RPM. Frequency must be reliable to prevent damage to motors and appliances.

Other velocity applications that require various degrees of accuracy are header reel drives, turbine fuel pumps, hydrostatic transmission anti-stall controls and mill saw traverse controls.

Resolution   The high degree of accuracy required for the AC generator application depicted in Figure 18.13 requires a magnetic pickup transducer and a closed loop velocity controller. The controller is connected to a Valvistor® proportional flow control valve which regulates the output flow of a variable displacement pressure limiting, load sensing pump. The load sensing line connects from downstream of the flow control to the compensator. The load sensing compensator maintains a constant pressure drop across the Valvistor®, allowing the velocity controller to regulate a predictable flow regardless of the load created by the generator.

Chapter 18 Circuits

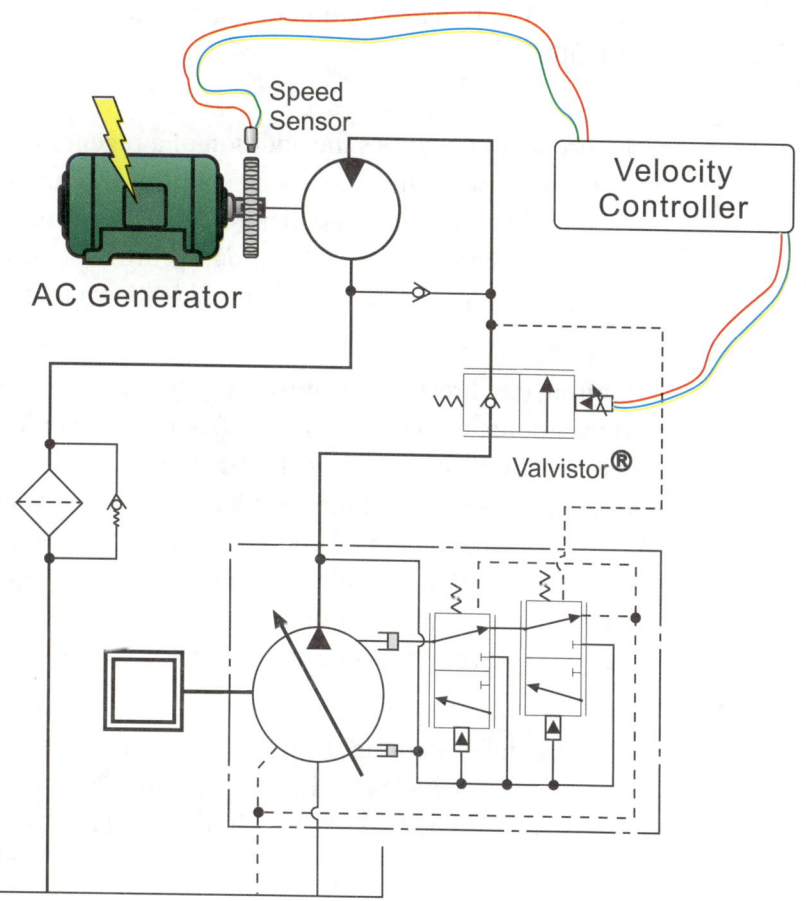

Figure 18.13 Speed control.

Chapter 18 Circuits

REGENERATION SELECTOR CIRCUIT

Requirement Select and de-select regeneration mode.

Application Applications such as refuse compactors are more productive if they move the compacting ram quickly, at low force, to begin the compaction cycle, and then slowly at high force to complete compaction. This will conserve a considerable amount of power as there is no need for coincident high force and high speed. Regeneration can serve this requirement well if there is a convenient method to switch the circuit from conventional actuation to regenerative actuation and back to conventional actuation again.

Resolution Figure 18.14 illustrates the addition of a regeneration selector valve manifold installed between the directional control valve and the actuators (two actuators are used in the example). The manifold consists of two 2-way, 2-position, spring offset solenoid valves, a 4-way, 2-position, spring offset solenoid valve and a 1:2 ratio slip-in cartridge valve. The manifold is shown in its conventional mode.

Shifting the directional control valve to the left will extend the cylinders in a conventional mode. Return flow from the rod end of the cylinders will pass through the directional control valve to the reservoir. Shifting the directional control valve to the right will retract the cylinders. The 4-way valve is also shifted at this time to provide a second path for fluid returning from the cap end of the cylinders. Because of the differential area of the cylinders, and the large rod diameter of typical compaction cylinders, this second path keeps the directional control valve from being very large or becoming a restrictor.

Shifting the directional control valve to the left while also shifting the two 2-way valves will provide a regeneration mode while extending the cylinders. Return flow from the rod end of the cylinders will bypass the directional control valve and go directly to the cap end of the cylinders. Thus, a high speed, low force mode is achieved.

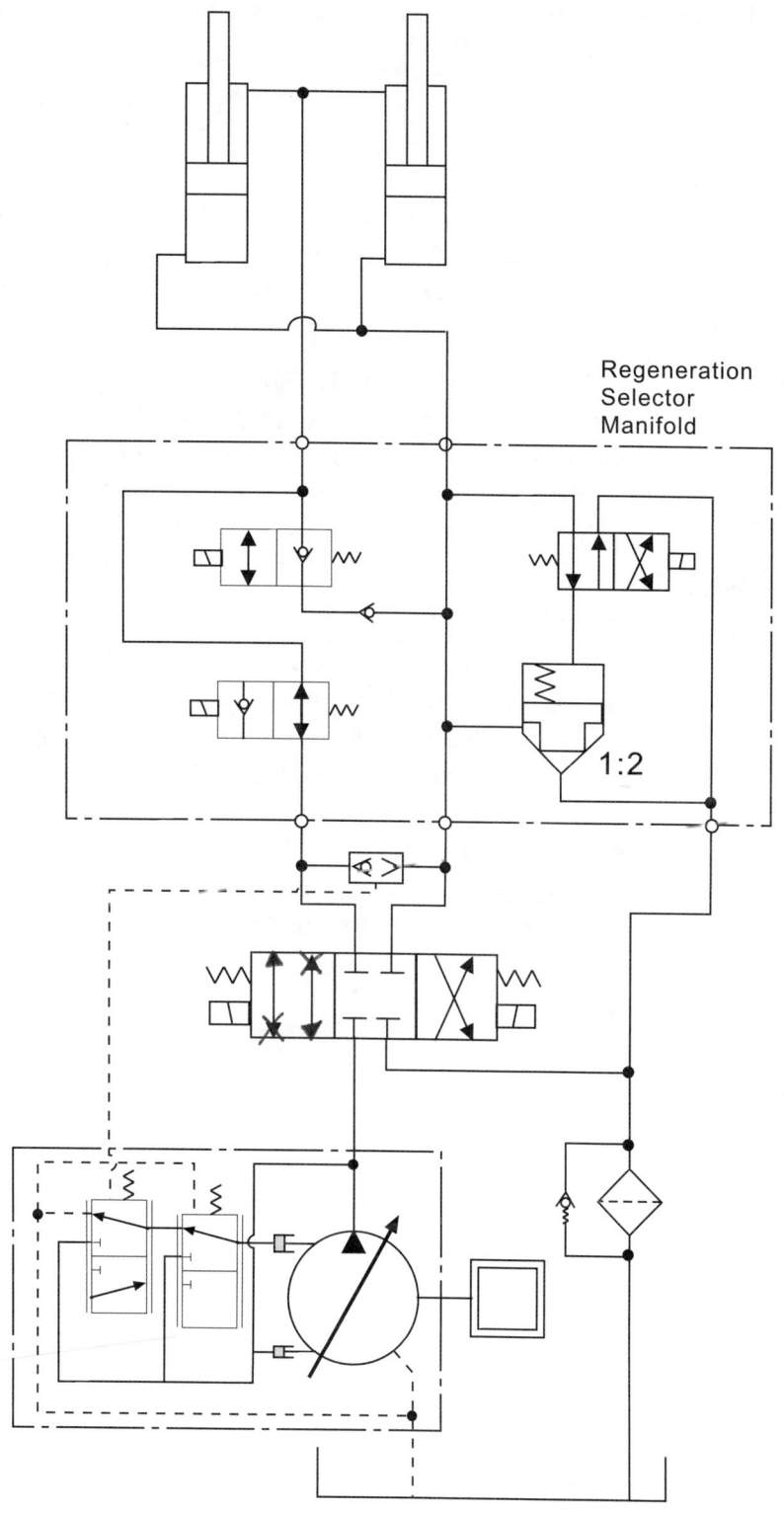

Figure 18.14 Conventional/Regeneration selector circuit.

Chapter 18 Circuits

PROPEL CREEP CONTROL

Requirement Provide a means of maintaining a slow (creep) travel speed separate from the normal propel circuit.

Application Some hydraulically propelled machines require a continuous slow speed, or creep, while in a working mode. It would be convenient to place the propel control into an adjustable creep mode and rely on the selected speed to be maintained without continuous monitoring. Asphalt or concrete paving machines, ditchers, cable plows, etc. are the type of machines where this method of control can be advantageous.

Resolution The circuit in Figure 18.15 shows the addition of two Valvistor® flow control valves connected in parallel to the CMX propel valve and to the propel motor. A double pilot operated counterbalance valve is added to the propel motor, and a dedicated return line is provided to the reservoir. A shuttle valve is also added across the Valvistor® flow control valves and connected to the load sensing spool of the pump compensator.

When a creep speed is desired, the CMX valve is maintained in neutral, and the appropriate Valvistor® flow control (forward or reverse) is activated. It is proportionally adjustable; once the correct creep speed has been selected, the pressure drop, and therefore the flow, across the flow control will remain constant by virtue of the load sensing control of the pump. Return flow from the propel motor will pass through the counterbalance valve, which will act as a meter-out control, and flow directly to the reservoir. The creep speed will remain constant, regardless of the load on the propel motor, until the Valvistor® flow control is altered.

Chapter 18 Circuits

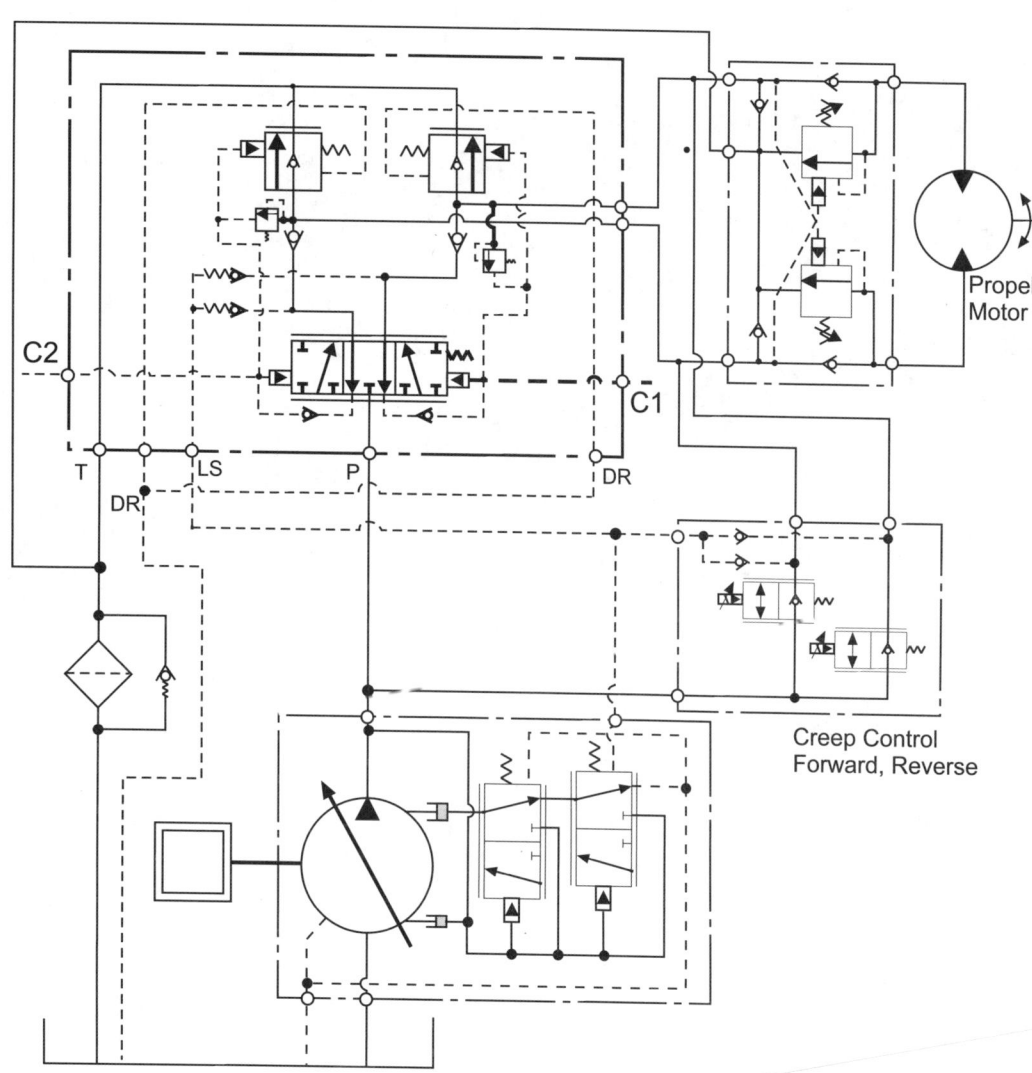

Figure 18.15 Propel circuit with the addition of a creep control

Chapter 18 Circuits

..... Appendix

A

ABSOLUTE PRESSURE
The indicated value of the weight of the earth's atmosphere. At sea level this value is approximately 14.65 psi (pounds per square inch).

ACCUMULATOR
A vessel, normally cylindrical, which is used to store fluid and gas for future release of the energy in the compressed fluid and gas. Normally contains a diaphragm or piston between the fluid (liquid) and gas chambers. Fluid is normally introduced at one end and the gas at the opposite end.

ACCURACY
The ability of the servo system to achieve the desired output.

ACTUATOR
A device for converting hydraulic energy into mechanical energy, i.e., a motor or cylinder.

ADAPTER
A mechanical device used to align the shaft of an electric motor (or other rotary device) with the shaft of a hydraulic pump to maintain radial and parallel shaft alignment.

AERATION
Air trapped in the hydraulic fluid. Excessive aeration causes the fluid to appear milky and components to operate erratically.

AIR BLEEDER AUTOMATIC
A valve that is fit into a hydraulic pipe line to facilitate automatic release of air trapped in the pipeline. See also AIR BLEEDER MANUAL.

AIR BLEEDER MANUAL
A valve that is fit into a hydraulic pipe line to facilitate manually initiated release of air trapped in the pipeline. See also AIR BLEEDER AUTOMATIC.

AIR BREATHER
A mechanical device which contains a fine mesh filter element. Normally attached to the top of a reservoir or tank to allow air to pass in and out of the reservoir or tank.

AMBIENT
The current condition of temperature, humidity and atmospheric pressure.

AMBIENT NOISE LEVEL (BACKGROUND NOISE)
The noise level in the area surrounding the machine or component to be tested with machine being tested not operating.

AMPLIFIER
An electronic device that receives an input voltage or current signal and modifies the signal into a driving voltage or current at a different level.

ANALOG
An infinitely variable device.

ANALOG DEVICE
An electronic device that requires or produces an infinitely variable signal, usually voltage or current, in response to a state change within the device.

ANNULAR AREA
A ring shaped area - often refers to the net effective area of the rod slide of a cylinder piston, i.e. the piston area of the rod.

ANSI FLANGE
A mechanical device that is used to connect two pieces of pipe together to form a pressure tight joint. ANSI flanges are round, use through bolts and/or nuts to attach two matched flanges together or to a valve or other mechanical device. See ANSI standards for pressure and temperature ratings.

ATMOSPHERE (ONE)
A measure of pressure equal to about 14.7 psi.

ATMOSPHERIC PRESSURE
See absolute pressure.

ATTENUATION
Opposite of gain (see gain).

B

BACK PRESSURE
The level of pressure on the return or downstream side of a device or system.

BACKUP BOTTLE
A vessel, normally cylindrical, which is used to store gas for future release of the energy in the compressed gas to an accumulator.

BACK-UP RING
A fabric or plastic device that is used with an o-ring or other gasket to present extrusion of the o-ring or gasket into an adjacent space or crevice.

BAFFLE
A separator found in a reservoir, tank or other chamber to divert fluid flow in specific direction(s) for de-aeration of moving fluid.

BALL VALVE
A valve that may be used to divert the flow of fluid in a passage. Most normally configured in a two way pattern which is either open or closed.

BAR
The measure of pressure in the metric system. One (1) bar = 14.5 psig.

BETA RATIO ( )
The amount, expressed as a ratio, of particles in a fluid stream upstream of a filter, divided by the amount of particles downstream, for a particular size particle. See PARTICLE COUNT.

BLADDER
A separator or diaphragm, usually found in a chamber to facilitate separation of two (2) fluids of gases. See ACCUMULATOR.

BLEED-OFF
To divert a specific, controllable portion of pump delivery directly to reservoir.

BOLT KIT
A set of bolts or screws that are selected to suit a particular application, i.e. pre-selected length, threads and strength to match the mounted component.

BREATHER
A device which permits air to move in and out of a container or component to maintain atmospheric pressure.

Definition of Technical Terms

BURST PRESSURE
The level of pressure at which a component, pipe, tube, hose or other fluid passage will burst during application of internal pressure. Normally 2.5 – 4.0 times working pressure. See also WORKING PRESSURE and PROOF PRESSURE.

BYPASS
A secondary passage for fluid flow.

C

CASE DRAIN LINE
The line or passage from the internal cavity of a pump or other component that will carry fluid leakage from the device to a low pressure reservoir or tank.

CAVITATION
A localized gaseous condition within a liquid stream which occurs where the pressure is reduced to the vapor pressure.

CHARGING ASSEMBLY
A system of valves and passages that allow addition or deletion of gas to the gas chamber of an accumulator without discharging any existing gas.

CHECK VALVE
A valve that allows fluid flow in one direction, yet stops flow in the opposite direction.

CIRCUIT
A combination of passages, components and devices that form a working set of logic for a particular application.

CLEAN OUT
A hole in a reservoir or tank that is normally covered with a plate that may be removed to allow cleaning of the interior of the reservoir or tank.

CLOSED CENTER VALVE
A condition where pump output is not unloaded to pump when the valve is in its center or neutral operating position.

CLOSED CIRCUIT
A piping arrangement in which pump output, after passing through other hydraulic components, returns directly to pump inlet.

CLOSED LOOP
In a control system, a type of control that has an input signal and a feedback of the result of the input signal which is used to modulate the input signal automatically. See OPEN LOOP.

COMMAND SIGNAL
An external signal to which the servo must respond.

COMPENSATOR CONTROL
A displacement control for variable pumps and motors which alters displacement in response to pressure changes in the system as related to its adjusted pressure setting.

COMPOUND GAUGE
A visual indicator of pressure that is set for 'zero' psi at atmospheric pressure and includes a dial which will continue to indicate the level of pressure above or below atmospheric pressure.

COMPRESSIBILITY
The change in volume of a unit of fluid when it is subjected to a unit change in pressure (in^2/lb).

COMPRESSION
The name used to describe the change in pressure in a hydraulic system from low pressure to an elevated pressure. Normally the change in pressure is made in a controlled amount of time to cause an even application of energy into the system. See DECOMPRESSION.

CONNECTOR
A mechanical device used to attach two pieces of tubing together or to attach a piece of tubing to a component.

CONTROLLABILITY
The finest adjustable increment of a system.

COOLER
A mechanical device used to transfer heat from a fluid to air or another fluid. Normally constructed of finned tubes with one fluid on the inside and the other fluid or air on the outside of the tubes. See HEAT EXCHANGER.

COUNTERBALANCE VALVE
A valve used to balance the weight of a machine or dead load by causing a back pressure in the system cylinders of sufficient magnitude to support the weight. Normally closed, opened by internal pressure in the counterbalance valve or from a separate source of fluid, that is connected to the opposite end of the balanced cylinder.

COUPLING
A mechanical device used to attached the shaft of an electric motor or other motive power device to a hydraulic pump.

CRACKING PRESSURE
The pressure at which a pressure operated valve begins to pass fluid.

CUSHION
A mechanical device fitted into a hydraulic cylinder that closes off the flow path of fluid to effect a smooth deceleration and stop of the cylinder at the end of the stroke.

CYCLE
The time of activation of a device or system that is one complete movement from the start position to an extreme position and back to the original position.

CYCLING
A rhythmic change of the factor under control.

CYLINDER, DOUBLE ACTING
A hydro–mechanical device, usually a cylindrical chamber with one closed end and a movable shaft at the other end. When fluid flow is applied to a port in the closed end, the shaft extends until the collar or piston reaches the shat end. When fluid is applied to the shaft end port, the shaft will retract into the chamber until the piston or collar reaches the closed end. The cylinder will produce an output force at the shaft end in proportion to its internal area multiplied times the pressure potential of the fluid power system.

Definition of Technical Terms

CYLINDER, SINGLE ACTING
A hydro-mechanical device, usually a cylindrical chamber with one closed end and a movable shaft at the other end. When fluid flow is applied to a port in the closed end, the shaft extends until the collar or piston reaches the shaft end. When an external weight or load is placed on the shaft, the shaft will retract into the chamber until the piston or collar reaches the closed end. The cylinder will produce an output force at the shaft end in proportion to its internal closed end area multiplied times the pressure potential of the fluid power system.

CYLINDER, TELESCOPING SINGLE ACTING
A hydro-mechanical device, usually a cylindrical chamber with one closed end and a telescoping movable shaft at the other end. When fluid flow is applied to a port in the closed end, the telescoping shaft extends until all collars or pistons reach their limit. When an external weight is applied to the shaft end, the shaft will retract into the chamber until all the pistons or collars reach their closed end limits. The cylinder will produce an output force at the shaft end in proportion to its internal closed end area multiplied times the pressure potential of the fluid power system.

CYLINDER, DOUBLE ROD
A hydro-mechanical device, usually a cylindrical chamber with movable shafts at the both ends. When fluid flow is applied to a port in either end, the shaft extends until the collar or piston reaches the shaft end. When fluid is applied to the other port, the shaft will retract into the chamber until the piston or collar reaches the opposite end. The cylinder will produce an output force at the shaft end in proportion to its internal area multiplied times the pressure potential of the fluid power system.

CYLINDER, TELESCOPING, DOUBLE ACTING
A hydro-mechanical device, usually a cylindrical chamber with one closed end and a telescoping movable shaft at the other end. When fluid flow is applied to a port in the closed end, the telescoping shaft extends until all collars or pistons reach their limit. When fluid is applied to the shaft end port, the shaft will retract into the chamber until all the pistons or collars reach their closed end limits. The cylinder will produce an output force at the shaft end in proportion to its internal area multiplied times the pressure potential of the fluid power system.

D

DEADBAND
The region or band of no response where an error signal will not cause a corresponding actuation of the controlled variable.

DEADTIME
Any definite delay between two related actions. Measured in units of time.

DECIBEL (dB)
A non-dimensional number used to express sound pressure and sound power. It is logarithmic expression of the ratio of a measure quantity to a reference quantity.

dB (A) & (C)
A sound level reading in decibels made on the A- & C- weighted network, respectively of a sound level meter.

DECOMPRESSION
The name used to describe the change in pressure in a hydraulic system from elevated pressure to a lower pressure. Normally the change in pressure is made in a controlled amount of time to cause an even release of energy in the system. See COMPRESSION.

DELIVERY
the volume of fluid discharged by a pump in a given time, usually expressed in gallons per minute (gpm).

DE-VENT
To close the vent connection of a pressure control valve, permitting the valve to function at its adjusted pressure setting.

DEVICE
A combination of individual components that are arranged to form a unit with a specific set of operating parameters.

DIAGRAM
A formal drawing showing the arrangement of components or devices.

DIFFERENTIAL CURRENT
The algebraic summation of the current in the torque motor; measures in MA (milliamperes).

DIFFERENTIAL CYLINDER
Any cylinder I which the two opposed pistons are not equal.

DIFFERENTIAL PRESSURE
The value or magnitude of pressure measured as the asoluteb difference of the inlet pressure and outlet pressure.

DIGITAL
The production of a discrete signal based on a change in state. See ANALOG.

DIGITAL DEVICE
A device or component that responds to or produces a discrete function based on a change in state. See ANALOG.

DIRECTIONAL VALVE
A valve whose primary function is to direct or prevent flow through selected channels.

DIRT CAPACITY
The measure of volume (or weight) of particles that a filter or strainer will hold at the limit of operation.

DISPLACEMENT
The volume for one revolution or stroke or for one radiant when so stated.

DITHER
A cyclic application of voltage across a solenoid or coil. Most often used to assure that the device driven by the coil or solenoid remains in a state of constant motion, thus reducing breakaway friction.

DOWNSTREAM
The passage beyond a device, normally at the outlet of direction of flow.

DRIFT
The measure of movement of a device after a preset condition is applied. Normally drift is measured with varying temperature, although drift may be plotted against any variable, such as humidity, etc.

DUROMETER
The measure of hardness of a rubber or other synthetic compound.

DYNAMIC BEHAVIOR
Describes how a control system or an individual unit reacts with time when subjected to an input signal.

Definition of Technical Terms

DYNAMIC ERROR
The error that results during the transient state, that is, the state when the system is moving from one steady state condition to another.

E

EFFICIENCY
The ratio of output to input. Volumetric efficiency of a pump is the actual output, in gpm, divided by the theoretical or design output. The overall efficiency of a hydraulic system is the output power divided by the input power. Efficiency is usually expressed as a percent.

ELECTRIC MOTOR
An electro-mechanical device that converts electrical power into rotary motion. The resultant power output is measured in horsepower.

ELECTROHYDRAULIC SERVO-VALVE
A servo-valve which is capable of continuously controlling hydraulic output as a function of an electrical input.

ENERGY
See JOULE.

ELEMENT
See FILTER ELEMENT.

EROSION
Degradation of a surface which is the result of mixtures of fluid and air or fluid and dirt particles passing over the surface at the same time as a change in pressure occurs.

ERROR (SIGNAL)
The signal which is the algebraic summation of an input signal and feedback signal.

EXHAUST LINE
A passage that is open to atmosphere. Normally used in systems using pressurized air or gas which may be dispersed into the atmosphere.

F

FITTING
A mechanical device used to attach two pieces of tubing/piping together or to attach a piece of tubing/pipe to a component.

FEEDBACK
Part of a closed loop system which monitors back information about the condition under control for comparison.

FEEDBACK LOOP
Any closed circuit consisting of one or more forward elements and one or more feedback elements.

FILLER CAP
A mechanical device which provides an access for filling a reservoir or tank. Normally equipped with a fine screen to strain out dirt particles.

FILTER
A mechanical device used to house a filter element. See FILTER ELEMENT.

FILTER ELEMENT
A series of wire or fabric meshes which are bonded together by caps or perforated cylinders and are fitted into hydraulic system passages to strain fine particles and silt from fluid passed through the passage.

FITTING
A mechanical device used to attach two pieces of tubing/piping together or to attach a piece of tubing/pipe to a component.

FLOW CONTROL VALVE, PRESSURE COMPENSATED
A valve used to cause a variable pressure drop in a fluid passage, thus reducing the amount of fluid that may pass through the passage regardless of the pressure level at the inlet of the valve. Often fitted with a check valve that permits free flow of fluid in the opposite direction.

FLOW CONTROL VALVE, NON-PRESSURE COMPENSATED
A valve used to cause a variable pressure drop in a fluid passage, thus potentially reducing the amount of fluid that may pass through the passage regardless of the pressure level at the inlet of the valve. Varying pressures at the inlet of the valve will change the flow capacity. Often fitted with a check valve that permits free flow of fluid in the opposite direction.

FLOW DIVIDER
A mechanical device used to divide the fluid in a passage into two or more separate fluid streams.

FLOW RATE
The volume mass, or weight of a fluid passing through any conductor per unit of time.

FLOW SWITCH
A digital device that opens or closes a contact when a preset flow passes over the sensing element. Normally mounted in a fluid flow passage with a paddle or want perpendicular to the fluid stream.

FLOWMETER
An analog device which indicates the volume of fluid passing through its interior passage. The output signal may be a visual one or a low level electrical signal.

FLUID
A media used in a fluid power system for transfer of energy (work). See FLUID POWER SYSTEM.

FLUID FRICTION
The measure of the resistance of flow of fluid in a passage, measured in psi (pounds per square inch) or other measures of pressure. Fluid friction results in increased fluid temperature and loss of work potential in the fluid power system.

FLUID MOTOR
A mechanical device that transforms the flow of pressurized fluid into rotary motion.

FLUID POWER SYSTEM
The term used to describe a system of components that use a pressurized fluid to transfer energy (do work).

FORCE
The measure of the result of pressurized fluid acting upon a chamber in a fluid power system. Normally the measure is in pounds and is most often used to state the force in pounds that will be available at the rod of a cylinder when acted upon by pressurized fluid from a fluid power system. The system of units normally used are square inches, pounds per square inch, and pounds.

FOUR WAY
A term used to describe a valve that has four ports, normally a pressure (inlet) port, a return (tank) port, an 'A' ('1') work port and a 'B' ('2') work port. Used to change direction of a cylinder or other output device.

Definition of Technical Terms

FOUR WAY VALVE, MANUALLY & DIRECT OPERATED
A valve having a four way functional capability that may be manually activated to directly control the operating spool. Movement of the spool from extreme end to extreme end reverses the flow paths of the ports. See FOUR WAY.

FOUR WAY VALVE, PROPORTIONAL CONTROL & DIRECT OPERATED
A valve having a four way functional capability that may be proportionately actuated by a solenoid to control the operating spool in infinite resolution. Movement of the spool from extreme end to extreme end completely reverses the flow paths of the ports. See FOUR WAY.

FOUR WAY VALVE, SOLENOID & DIRECT OPERATED
A valve having a four way functional capability that may be solenoid activated to directly control the operating spool. Movement of the spool from extreme end to extreme end reverses the flow paths of the ports. See FOUR WAY.

FOUR WAY VALVE, SOLENOID & PILOT OPERATED
A valve having a four way functional capability that may be solenoid activated to directly control the operating spool which then controls a secondary, larger spool. Movement of the secondary spool from extreme end to extreme end reverses the flow paths of the ports. See FOUR WAY.

FREQUENCY BANDS
A division of the audible range of frequencies into sub-groups for detailed analysis of sound.

FREQUENCY RESPONSE ANALYSIS
A control system analysis which by introducing a varying rhythmic change (like alternating current) into a process or control unit observes what effect these changes have on the output. Since the information determines how a system or control unit will react, it is possible to use this method of analysis to predict what the addition of new equipment will mean to an operation.

FULL FLOW
A filter in which all the fluid must pass through the filter element or medium.

G

GAIN
Ratio of increase in a signal (or measurement) as it passes through a control system or a specific control element. If a signal gets smaller, it is said to be attenuated.

GAS BOTTLE
See BACK-UP BOTTLE.

GASKET
A seal, made from rubber or other synthetic material in the shape of a circle and of polygonal cross-section. See O-RING.

GATE VALVE
A two-way valve that may be opened or closed to block the flow of fluid in a passage. Normally manually operated, but may be automated, especially for larger sizes. Normally designed so that when open, the opening of the passage is not restricted, but there will be some small pressure loss. See GLOBE VALVE and NEEDLE VALVE.

GAUGE PRESSURE
A term used to state that any pressure stated is corrected for atmospheric pressure. Normally abbreviated psig (pounds per square inch gauge)

GLAND
A mechanical device that is used to contain a seal, o-ring or gasket in a specified space to result in a leak-proof connection between two or more mechanical components.

GLOBE VALVE
A two-way valve that may be opened or closed to block the flow of fluid in a passage. Normally manually operated, but may be automated, especially for larger sizes. Normally designed so that the flow of fluid must make a non-straight turn inside the valve body which results in a loss of pressure across the valve when open, which is greater than the loss across a gate valve. See GATE VALVE and NEEDLE VALVE.

H

HEAD
The measure of pressure at the base or other reference point of a column of fluid. Normally measured in feet of water.

HEAT
The form of energy that has the capacity to create warmth or to increase the temperature of a substance. Any energy that is wasted or used to overcome friction is converted to heat. Heat is measured in calories or British Thermal Units (BTU's). Once BTU is the amount of heat required to raise the temperature of one pound of water one degree Fahrenheit. In the metric system one calorie is the amount of heat required to raise the temperature of one gram of water from 3.5 C to 4.5 C (called a small calorie). If the temperature change is from 14.5 C to 15.5 C, the unit is the normal calorie.

HEATER
An electro-mechanical device that converts electricity into heat, normally for use in raising the temperature of fluid stored in a reservoir or tank.

HEAT EXCHANGER
See COOLER.

HORSEPOWER
The measure of energy used in description of the normal power level in a system. 1 horsepower = 550 lb.-ft./min. of work.

HOSE
A passage used to transport fluid between components in a fluid power system. Normally constructed from multiple layers of rubber or other synthetic materials interlaced and bonded with wire mesh to form a flexible passage. Normally fitted with metal end connections to permit connection to pipe threads or other joints.

HUNTING
Tendency for a system to oscillate continuously.

HYDRAULIC BALANCE
A condition of equal opposed hydraulic forces acting on a part in a hydraulic component.

HYDRAULIC CONTROL
A control which is actuated by hydraulically induced forces.

HYDRAULIC POWER
See FLUID POWER.

HYDRAULIC MOTOR
See FLUID MOTOR.

HYDRODYNAMICS
Engineering science pertaining to the energy of liquid flow and pressure.

Definition of Technical Terms

HYDROKINETICS
Engineering science pertaining to the energy of liquids in motion.

HYDROPNEUMATICS
Pertaining to the combination of hydraulic and pneumatic fluid power.

HYDROSTATICS
Engineering science pertaining to the energy of liquids at rest.

HYSTERESIS
The difference between the response of a unit r system to an increasing signal and the response to a decreasing signal.

HZ (HERTZ)
A measure of the number of cycles that occur in a specific period of time. Usually the time base is the second, but the time base may be any acceptable measure of time. Synonymous term for "cycles per second".

I

INDICATOR
A mechanical device with points to a scale to provide a visual perspective of the state of a component. See NEEDLE.

INPUT
Incoming signal to a control unit or system.

INTAKE LINE
A passage at the inlet port of a component, normally at the inlet port of a pump.

J

JOULE
A unit of work, energy, or heat.
1 J (joule) = 1 Nm (Newton meter).

K

KINETIC ENERGY
Energy that a substance or body has by virtue of its mass (weight) and velocity.

L

LAG
Preferred engineering term for delay in response (usually in degrees).

LAMINAR FLOW
A condition of flow in a passage that is typified by slow movement of fluid in a relatively straight path along the centerline of a passage. See TURBULENT FLOW.

LEVEL TRANSMITTER
An electro-mechanical device which senses the level of fluid in a chamber and produces an analog signal that corresponds with the change of state in the chamber. See LEVEL SWITCH.

LEVEL SWITCH
An electro-mechanical device which senses the level of fluid in a chamber and opens or closes a digital switch to indicate a change of state. See LEVEL TRANSMITTER.

LEVERAGE
A gain in output force over input force by sacrificing the distance moved. Mechanical advantage or force multiplication.

LIFT
The measure of the capability of a pump to raise fluid from a lower to higher level at its inlet port without damage to the pump. Normally expressed in feet of water.

LINE
A connection between components, a passage for fluid or gas transfer. See PIPE, and TUBE and HOSE.

LINEAR ACTUATOR
A device for converting hydraulic energy into linear motion, i.e. a cylinder or ram.

LINEAR VARIABLE TRANSFORMER (LVT)
An electro-mechanical linear device that produces an analog signal in proportion to the difference in velocity between a magnet and a separate fixed coil.

LINEAR VARIABLE DIFFERENTIAL TRANSFORMER (LVDT)
An electro-mechanical linear device that produces an analog signal in proportion to the difference in distance between a magnet and separate fixed coil.

LINEARITY (SERVOVALVE)
The degree of straightness of the hysteresis plot.

LIQUID LEVEL GAUGE
Gauge to visually indicate the fluid level in a reservoir or tank.

LITER
A metric measure of volume. One (1) liter = 0.2642 gallons.

LUBRICATOR
A mechanical device which is used to inject drops or mist of oil into an air line for lubrication purposes.

M

MANIFOLD
A fabricated system of passages to which various components are attached to form a working assembly or sub-assembly.

MANUAL CONTROL
A control actuated by the operator.

MANUAL OVERRIDE
A means of manually actuating an automatically-controlled device.

MECHANICAL CONTROL
A control actuated by linkages, gears, screws, cams or other mechanical elements.

METER
39.37 inches. The measure of distances in the metric system.

METER
To regulate the amount or rate of fluid flow.

METER-IN
To regulate the amount of fluid flow into an actuator or system.

METER-OUT
To regulate the flow of the discharge fluid out of an actuator or system.

MICRON
1/1000th of a millimeter or 0.00003937 inches. The measure used to determine the particle size of contaminants in a fluid system.

MICRON RATING
The size, in microns, of the particles a filter will remove.

MUFFLER
A mechanical device which provides a complex path for exhaust of air from a pressurized chamber, thus reducing the noise level of the exhausting air.

N

NEEDLE
See INDICATOR.

NEEDLE VALVE
A two-way valve that may be opened or closed to block the flow of fluid in a passage. Normally manually operated, but may be automated, especially for larger sizes. Normally designed so that the flow of fluid must make an non-straight turn inside the valve body which results in a desired loss of pressure across the valve when open which is greater than the loss across a gate valve. See GATE VALVE and GLOBE VALVE.

NEWTON
A unit of force based on the unit of mass, Kg (kilogram), multiplied by the acceleration, m/s^2 (meters per second per second) which produces Kgm/s^2, called the Newton. $1 N = 1 Kgm/s^2 = 0.1225$ lbs. (F) – (pounds force).

NIPPLE
A short length of pipe. May be threaded or plain end.

NITROGEN
An inert gas used to serve as an energy source for accumulators or to be used as a cleaning agent when pure, non-explosive gases are required.

NULL
The position of a device that is its normal or otherwise preset 'zero' condition.

O

O-RING
A seal, made from rubber or other synthetic material in the shape of a circle and of circular or other polygonal cross-section. See GASKET.

OPEN LOOP
In a control system, a type of control that has an input signal, but no feedback of the result of the input signal. See CLOSED LOOP.

OPERATING PRESSURE
The level of pressure at which a component, pipe, tube, hose or other fluid passage will experience during application of maximum expected fluid pressure. See also BURST PRESSURE and PROOF PRESSURE.

ORIFICE
A narrowing of the passage size. Normally constructed in a connector or fitting of a sharp edged metallic component.

OUTPUT STAGE
A spool or other device that is controlled by a smaller spool or torque motor.

OVERLAP
The condition of a spool and body in a servo valve or other spool valve wherein the spool must move a specified amount (the overlap) before exposing two adjacent cavities.

OVERSHOOT
Occurs when the process exceeds the target value as operating conditions change.

OXIDATION
The absorption of oxygen into fluid and the subsequent plating of the oxygen/fluid mixture onto metal surfaces.

P

PACKING
A seal or gasket. See SEAL, O-RING and GASKET

PARTICLE
A piece of debris (sand, dirt, metal, fabric, etc.) Found in a fluid.

PARTICLE COUNT
The visual or electronic summation of the quantity of particles, grouped by size, in a fluid sample of specified size.

PASSAGE
A hole through which fluid is passed in a fluid power system. See TUBE, PIPE, HOSE and MANIFOLD.

PETROLEUM FLUID
A hydraulic oil (fluid) that is made from a petroleum base. Normally will support combustion if heated to a specific temperature.

PH (PHASE)
A term used to describe the quantity of cyclic electrical power sources in a high voltage system. Most commonly 1-phase or 3-phase.

PHASE SHIFT
A time difference between the input and output signal of a control unit or system, usually measures in degrees.

PHOSPHATE ESTER FLUID
A hydraulic oil (fluid) that is made from an ester base. A synthetic fluid, manufactured to specific characteristics. Normally will not support combustion if heated to a specific temperature.

PILOT LINE
A passage in a fluid power system that is used to transport a fluid at a pressure lower than the normal operating pressure to facilitate controlled shifting of spool valves.

PILOT-OPERATED CHECK VALVE
A special check valve that may be opened against a check load by applying pilot pressure from a secondary source to open the check to free reverse flow.

PILOT PRESSURE
The pressure in the pilot circuit.

PILOT VALVE
A valve applied to operate another valve or control. The controlling stage of a 2-stage valve.

PIPE
A passage in a fluid power system that is constructed of metal and conforms dimensionally to standard established by the ANSI. May be acquired by size and schedule, where increase in wall thickness does not increase the outside diameter. See TUBE.

PISTON, CYLINDER
A cylinder in which the movable element has a greater cross-sectional area than the piston rod.

PISTON RING
A metal ring that is used to seal high pressure fluid inside a passage to prevent (limit) leakage across the passage. Normally found in cylinders.

PLUNGER, CYLINDER
A cylinder in which the movable element has the same cross-sectional area as the piston rod.

POPPET
That part of certain valves which blocks flow when it closes against a seat.

PORT
An internal or external terminus of a passage in a component.

Definition of Technical Terms

POSITIVE DISPLACEMENT
A characteristic of a pump or motor when a constant volume is delivered for each revolution or stroke.

POTENTIOMETER
An electrical device that changes its internal resistance when moved to a specified point. Most commonly found in electronic control panels. Used to change the voltage in a control system for required control changes (position, speed, pressure, etc.)

POUNDS PER SQ. INCH, GAUGE (PSIG) & ABSOLUTE (PSIA)
The measure of pressure, corrected for atmospheric pressure, that is 'zero' psig= 14.65 psia. 'Zero' psia = absolute zero vacuum.

POWER SUPPLY
Term used to describe a fluid power source. A hydraulic power unit.

PRE-FILL VALVE
A valve that is arranged so its inlet port is connected to a reservoir or tank and so that fluid will flow from the inlet of the valve into a cylinder or ram when opened. When closed, the valve must close off the ram or cylinder from the reservoir or tank to permit application of high pressure from another source on the cylinder side of the valve. Most commonly used to fill large rams on presses to take up non-operating stroke.

PRECHARGE PRESSURE
The pressure of compressed gas in an accumulator prior to the admission of liquid.

PRESSURE COMPENSATOR
A hydro-mechanical device fitted to a pump or other flow producing/controlling device that reduces flow when pressure rises and increases flow as pressure decreases, to preset limits.

PRESSURE DIFFERENTIAL
The difference in pressure between any two points in a system or a component.

PRESSURE DIFFERENTIAL SWITCH
A digital device that opens or closes a switch when the internal pressure differential changes state. Most commonly used to sense clogging of filter elements.

PRESSURE DROP
See Pressure, Differential.

PRESSURE GAUGE
A visual indicator of pressure that is set for 'zero' psi at atmospheric pressure and includes a dial which will continue to indicate the level of pressure above atmospheric pressure. See VACUUM GAUGE and COMPOUND GAUGE.

PRESSURE LINE
A passage that carries fluid from the source of flow to various operating elements of a fluid power system. Rated for operating pressure at the maximum expected pressure of the system.

PRESSURE OVERRIDE
The measure of pressure increase over the nominal setting of a device when additional fluid flow is passed over the device after it initially opens.

PRESSURE REDUCING VALVE
A pressure control valve whose primary function is to limit outlet pressure.

PRESSURE SWITCH
A digital device that opens or closes a switch when the internal pressure changes state.

PRESSURE TRANSDUCER
An analog device that produces a change in voltage or current when the internal pressure changes state. Normally a fast response device for use in servo control systems. See PRESSURE TRANSMITTER.

PRESSURE TRANSMITTER
An analog device that produces a change in voltage or current when the internal pressure changes state. Normally a slow acting device for use in display systems where update time is not crucial. See PRESSURE REDUCER.

PROOF PRESSURE
The level of pressure at which a component, pipe, tube, hose or other fluid passage will not yield during application of internal pressure. Normally 1.5 times working pressure. See WORKING PRESSURE and BURST PRESSURE.

PROPORTIONAL FLOW
In a filter, the condition where part of the flow passes through the filter element in proportional to pressure drop.

PUMP, AIR-OIL
A mechanical device containing two sets of isolated pistons and control valving that are used to intensify fluid pressure by use of a multiplication effect across the two sets of pistons. The air piston being larger than the fluid piston.

PUMP, FIXED DISPLACEMENT
A mechanical device that creates a flow of fluid when its shaft is rotated in the proper direction and when its inlet is connected to a chamber filled with fluid (a reservoir or tank). The outlet port may be connected to a passage leading to a fluid power system or exhausted into another chamber that is at a higher pressure. The higher pressure chamber must be equipped with a pressure limiting device. The output flow rate is fixed by the pump displacement per revolution.

PUMP, VACUUM
A mechanical device that creates a pressure that is lower than atmospheric at its inlet when the shaft is rotated. The outlet port is normally connected to a higher pressure chamber or atmosphere.

PUMP, VARIABLE DISPLACEMENT
A mechanical device that creates a flow of fluid when its shaft is rotated in the proper direction and when its inlet is connected to a chamber filled with fluid (a reservoir or tank). The outlet port may be connected to a passage leading to a fluid power system or exhausted into another chamber that is at a higher pressure. The higher pressure chamber must be equipped with a pressure limiting device. The output flow rate is fixed by the pump displacement per revolution but variable by the operator in a manual or servo controlled system, depending on the design.

Q

QUICK DISCONNECT
A mechanical device that may be engaged or dis-engaged to attach two fluid passages. Typically, dis-engagement is possible by manual means.

R

RECIPROCATION
Back-and-forth straight line motion or oscillation.

RAM
A cylinder that has an extend port only. Usually accompanied by auxiliary cylinders that are mechanically linked to the ram to facilitate retraction action.

RAMP
The rate of change of a specific output, such as the ramp of a pressure compensator.

RAMP MODULE
An electronic device that controls the rate of rise of a servo or proportional valve by using capacitors to limit the rate of voltage or current change to the servo or proportional valve.

RATED FLOW
The maximum flow that a manufacturer assigns to a specific component as the maximum desirable flow at which the device will function properly. Also the flow that a designer assigns to a system as the nominal maximum flow. See WORKING PRESSURE.

RATED PRESSURE
The maximum pressure that a manufacturer assigns to a specific component as the maximum desirable pressure at which the device will function properly. See WORKING PRESSURE.

REDUCING VALVE
A valve that decreases the downstream pressure (at the valve outlet) in order to control the flow and therefore the outlet pressure to some preset level. Normally accomplished by balancing the outlet pressure against a precision spring.

REGENERATIVE CIRCUIT
A piping arrangement for a differential type cylinder in which discharge fluid from the rod end combines with pump delivery to be directed into the head end.

REGULATOR
A term used to describe a valve or device that limits the pressure in a passage.

RELIEF VALVE
A valve that limits the pressure at its inlet port by exhausting flow present at its inlet port to another chamber of lower pressure potential through its outlet port.

REPLENISH
To add fluid to maintain a full hydraulic system.

RESERVOIR
A chamber used to store fluid.

RESPONSE TIME
The elapsed time that occurs after the beginning of a function until its completion. For example, the time elapsed between application of electrical power to a solenoid and its full excursion or stroke.

RESTRICTION
A reduced cross-sectional area in a line of passage producing a pressure drop.

RESTRICTOR
See ORIFICE.

RETURN LINE
A passage that is used to route fluid to a reservoir or tank after use in some function. Normally limited to low pressures of 0-150 psig, but may be higher in special applications if so designed.

REVERSING VALVE
A four-way directional valve used to reverse a double-action cylinder or reversible motor.

ROTARY VARIABLE DIFFERENTIAL TRANSFORMER (RVDT)
An electro-mechanical rotary device that produces an analog signal in proportion to the difference in distance between a magnet and a separate fixed coil.

ROTARY VARIABLE TRANSFORMER (RVT)
An electro-mechanical rotary device that produces an analog signal in proportion to the difference in velocity between a magnet and a separate fixed coil.

ROTARY ACTUATOR
A hydro-mechanical device that converts fluid flow into incremental rotary motion as compared to a fluid motor which produces infinite numbers of turns. See FLUID MOTOR.

ROTARY JOINT
A connector or fitting that is equipped with seals or o-rings that allow it to rotate while passing one or more fluid paths through sealed internal passages.

S

SAE 4 BOLT PORT, CODE 61
A system for flange and surface mounting configurations that are used to attach pipes, tubes or hoses to a component or manifold. Normally rated at 3000 psig. See SAE 4 BOLT PORT, CODE 62.

SAE 4 BOLT PORT, CODE 62
A system of flange and surface mounting configurations that are used to attach pipes, tubes or hoses to a component or manifold. Nominally rated at 6000 psig, although larger sizes are only rated for 500 psig. See SAE 4 BOLT PORT, CODE 61.

SAE PORT
A threaded hole and stud system that may be used to attach fittings to a component or manifold. Sealed with an o-ring or gasket.

SAFETY FACTOR
The ratio of burst pressure to rated pressure under specific static pressure and temperature conditions. See BURST PRESSURE.

SCRAPER RING
A metal or synthetic ring that is fitted to the shaft of a cylinder to remove particles from the shaft so to prevent them from entering the cylinder seal chamber.

SEAL
See O-RING and GASKET.

SENSITIVITY
The minimum input signal required to produce a specified output signal.

SEQUENCE VALVE
A valve that is normally closed or normally open and changes to the opposite state when pilot pressure is applied to its spring chamber at a preset pressure level. Normally used to initiate a secondary set of operations in a system, based on application of the pilot signal.

Definition of Technical Terms

SERVO CONTROL
A term used to describe the type of electronic system used for finite, analog control of a function. See SOLENOID CONTROL.

SERVO VALVE
A valve that uses a torque motor type coil to control a small stream of fluid. Direction of the fluid stream is used to position a large spool. Therefore a low level power signal may provide precise spool position. Normally, the spool had mechanical feedback of spool position to the torque motor, creating a closed loop spool position system.

SHUTTLE VALVE
A valve that has three ports and a common ball or spool check valve. When flow is applied at either of the two inlet ports, the third or output port receives flow from the higher pressure inlet port.

SILENCER
See MUFFLER.

SILT
Fine particles of debris. Normally found in chambers with little or no circulation, such as at the bottom of a reservoir or tank. See SLUDGE.

SLIP
Internal leakage of hydraulic fluid.

SLUDGE
Partially hardened silt. See SILT.

SOLENOID
A coil of metallic wire, usually copper, wound around a bobbin. Used to magnetize the bobbin and produce linear motion of a companion spool when electricity is applied.

SPOOL
A term loosely applied to almost any moving cylindrically shaped part of a hydraulic component which moves to direct flow through the component.

STABILITY
Ability of a system to maintain control when subject to severe outside disturbances.

STATIC BEHAVIOR
Describes how a control system, or an individual unit, carries on under fixed conditions (As contrasted to dynamic behavior which refers to behavior under changing conditions).

STEP CHANGE
The change from one value to another in a single step.

STATIC HEAD
A measurement of pressure that is present when no fluid flow exists in a passage. The static head is normally expressed in feet of water.

STRAINER
A series of wire or fabric meshes which are bonded together by caps or perforated cylinders and are fitted into hydraulic system passages to strain particles from fluid passed through the passage.

SUBPLATE
A metal base to which a specific valve may be attached using a specified bolt kit.

SUCTION LINE
A passage that leads from a reservoir or tank to the inlet port of a pump.

SUPERCHARGE
To replenish a hydraulic system above atmospheric pressure.

SURGE
An increase in pressure that occurs for a specified short period of time over the normal expected working pressure.

SWASH PLATE
A stationary canted plate in an axial type-piston pump which causes the pistons to reciprocate as the cylinder barrel rotates.

SWITCH
A digital device which closes or opens a discrete set of contacts at a pre-set condition.

SWIVEL JOINT
A connector or fitting that is equipped with seals or o-rings that allow it to partially rotate while passing a fluid path through a sealed internal passage.

SYNCHRO
A rotary electromagnetic device generally used as an AC feedback signal generator which indicates position. It can also be used as a reference signal generator.

SYNTHETIC FLUID
A hydraulic oil (fluid) that is made from a synthetic base. A fluid, manufactured to specified characteristics. Normally will not support combustion if heated to a specific temperature.

SYSTEM PRESSURE
See OPERATING PRESSURE.

T

TACHOMETER
A digital or analog device that produces a pulse train of electrical signals that is proportional to its rotational speed.

TANK
See RESERVOIR.

TEMPERATURE SWITCH
A digital device that opens or closes a switch when the internal temperature changes state to a preset temperature limit.

THERMOCOUPLE
A precision resistive element that changes resistance in proportion to the temperature of the element. May be used, therefore, with proper DC electrical voltage to indicate temperature on a voltmeter style indicator.

THREE WAY
A term used to describe a valve that has three ports, normally a pressure (inlet) port, a normally closed (n.c.) port and a normally open (n.o.) port. Used to block or open a common flow passage.

THREE WAY VALVE, PROPORTIONAL CONTROL AND DIRECT OPERATED
A valve having a three way functional capability that may be proportionately actuated by a solenoid to control the operating spool in infinite resolution. Movement of the spool from extreme end to extreme rod completely reverses the flow paths of the ports. See THREE WAY.

Definition of Technical Terms

THREE WAY VALVE, MANUALLY AND DIRECT OPERATED
A valve having a three way functional capability that may be manually activated to directly control the operating spool. Movement of the spool from extreme end to extreme end reverses the flow paths of the ports. See THREE WAY.

THREE WAY VALVE, SOLENOID AND PILOT OPERATED
A valve having a three way functional capability that may be solenoid activated to directly control the operating spool which then controls a secondary, larger spool. Movement of the secondary spool from extreme end to extreme end reverses the flow paths of ports. See THREE WAY.

THREE WAY VALVE, SOLENOID AND DIRECT OPERATED
A valve having a three way functional capability that may be solenoid activated to directly control the operating spool. Movement of the spool from extreme end to extreme end reverses the flow paths of the ports. See THREE WAY.

THROTTLE
To permit passing of a restricted flow. May control flow rate or create a deliberate pressure drop.

TIE ROD
A metal rod that is used to prevent two or more components from separating. Normally used to restrain the end plates of cylinders against the cylinder tube.

TORQUE
The measure of force applied to a lever arm. Normally expressed in lb.-ft. (pound-feet) or lb.-in. (pound-inch).

TORQUE CONVERTER
A rotary fluid coupling that is capable of multiplying torque.

TORQUE MOTOR
A coil of wire and bobbin assembly used in a servo valve that causes the internal mechanism of the servo valve to be offset when current passes through the coil.

TRANSDUCER
An analog device which produces a change in signal level during state changes. Normally used for high speed control systems.

TRANSFER FUNCTION
A mathematical expression of the relationship between the outgoing and incoming signals of a process or control element.

TRANSMITTER
An analog device which produces a change in signal level during state changes. Normally used for indication systems.

TUBE
A term used to describe a passage for fluid in a hydraulic system. Normally specified by outside diameter, wall thickness, material type and material strength.

TURBINE
A rotary device that is actuated by the impact of a moving fluid against blades or vanes.

TURBULENT FLOW
A condition of flow in a passage that is typified by rapid movement of fluid in a passage, where the fluid is churning and bouncing off the passage walls. See LAMINAR FLOW.

TWO WAY
A term used to describe a valve that has two ports, normally a pressure (inlet) port and an outlet port. Used to open or close a flow passage. May be configured as normally closed (n.c.) Or normally open (n.o.).

TWO WAY VALVE, SOLENOID AND PILOT OPERATED
A valve having a two way functional capability that may be solenoid activated to directly control the operating spool which then controls a secondary, larger spool. Movement of the secondary spool from extreme end to extreme end opens or closes the flow paths of the ports. See TWO WAY.

TWO WAY VALVE, SOLENOID AND DIRECT OPERATED
A valve having a two way functional capability that may be solenoid activated to directly control the operating spool. Movement of the spool from extreme end to extreme end opens or closed the flow paths of the ports. See TWO WAY.

TWO WAY VALVE, MANUALLY AND DIRECT OPERATED
A valve having two way functional capability that may be manually activated to directly control the operating spool. Movement of the spool from extreme end to extreme end opens or closed the flow paths of the ports. See TWO WAY.

U

UNDERLAP
The condition of a spool and body in a servo valve or other spool valve wherein the spool is displaced a specified amount (the underlap) to expose two adjacent cavities to each other.

UNLOAD
To release flow (usually directly to the reservoir), to prevent pressure being imposed on the system or portion of the system.

UNLOADING VALVE
A valve that is normally closed and opens from a separate fluid source on rising pressure that is balanced against a precision spring. Re-set point is normally fixed.

UPSTREAM
The passage ahead of a device, normally at the inlet of direction of flow.

V

VACUUM
Pressure less than atmospheric pressure. It is usually expressed in inches of mercury (Hg) as referred to the existing atmospheric pressure.

VACUUM GAUGE
A visual indicator of pressure that is set for 'zero' psi at atmospheric pressure and includes a dial which will continue to indicate the level of pressure below atmospheric pressure.

VALVE
A mechanical device that is used in a fluid power system, which is used to provide some change of state of the fluid.

VAPOR PRESSURE
The measure of pressure at which a specific fluid will change to a gas.

Definition of Technical Terms

VARIABLE
A factor or condition which can be measured, altered or controlled, i.e., temperature, pressure, flow, liquid level, humidity, weight, chemical composition, color, etc.

VELOCITY
The speed of fluid flow through a hydraulic line. Expressed in feet per second (fps), inches per second (ips), or meters per second (mps). Also, the speed or a rotating component measure in resolutions per minute (rpm).

VENT VALVE
A valve that may be manually opened to allow air or fluid or a combination of both to be exhausted into a lower pressure chamber or to the atmosphere.

VISCOSITY
The measure of resistance to flow of a fluid against an established standard. See SUS and SSU.

VISCOSITY INDEX
A measure of the viscosity-temperature characteristics of a fluid as referred to that of two arbitrary reference fluids (ASTM Designation D2270-64).

VOLUME
The size of a space or chamber in cubic units. Loosely applied to the output of a pump in gallons per minute.

W

WAFER VALVE
A two way valve that may be opened or closed to block the flow of fluid in a passage. Normally manually operated, but may be automated, especially for larger sizes. Normally designed so that when open, the opening of the passage is only restricted by the thickness of the wafer. There will be some pressure loss. See GATE VALVE, GLOBE VALVE and NEEDLE VALVE.

WATER GLYCOL FLUID
A hydraulic fluid that is comprised of a mix of distilled or other pure water and glycol to form a fluid that has enough lubricity to function as a fluid power fluid, but is relatively fire-resistant, i.e., will not support combustion.

WIPER RING
A rubber or other synthetic seal that is fitted around a moving shaft to form a low pressure seal. Normally used to prevent fluid from entering the sealed volume.

WORK
The transfer of power from one state to another. The movement of weight over a specified distance.

Common Abbreviations

AAR	Association of American Railroads
abs	absolute
acfm	actual cubic feet per minute
AISI	American Iron and Steel Institute
alt	altitude
amb	ambient
amp	ampere
AN	Army Navy Aeronautical Standards
AND	Air Force–Navy–Aeronautical Design Standards
ANSI	American National Standards Institute
API	American Petroleum Institute
ARP	Aeronautical Recommended Practice
ASA	American Standards Association
ASAE	American Society of Agriculture Engineers
ASHRAE	American Society of Heating, Refrigerating and Air Conditioning
ASME	American Society of Mechanical Engineers
ASTM	American Society of Testing Materials
ASTME	American Society of Tool and Manufacturing Engineers
atm	atmosphere
bhp	brake horsepower
BTU	British thermal unit
C	circumference
C or cent.	centigrade
cSt	centistokes
cal	calorie
cc	cubic centimeter
cm	centimeter
cfm	cubic feet per minute
cu	cubic
cu ft	cubic feet
cu in	cubic inch
cyl	cylinder
db	decibel
deg	degrees
D or dia	diameter
F or fahr	Fahrenheit
fpm	feet per minute
fps	feet per second
FPS	Fluid Power Society
ft	foot
flgd	flanged
FM	approved by Factory Mutual Laboratories
gal	gallons
gpm	gallons per minute
gm	gram
GN_2	gaseous nitrate
hp or HP	horsepower
"Hg	inches mercury
hr	hour
Hz	Hertz (cycles per second)
ICC	Interstate Commerce Commission
in	inch
ID	inside diameter
Imp gal	Imperial gallon
IPS	iron pipe size
ISO	International Organization for Standardization
IEC	International Electrotechnical Committee
JIC	Joint Industry Conference
kip	thousand pounds
kg	kilograms
km	kilometer
kva	thousand volt amperes
kw	kilowatt
kwhr	kilowatt hour
kHz	thousand Hertz (cycles per second)
lbs	pounds
liq	liquid
log	logarithm (common)
LOX	liquid oxygen
lb per cu ft	pounds per cubic foot
MS	military standard
MIL	military
M	Meter
mm	millimeter
m	micron (micro–meter)
max	maximum
mech	mechanical
min	minute
min	minimum
MSS	Manufacturers Standardization Society of the Valve and Fittings Industry
NAS	National Aerospace Standard
NASA	National Aeronautics and Space Administration
NEMA	National Electrical Manufacturers Association
NFPA	National Fluid Power Association
NMTBA	National Machine Tool Builders Association
NEC	National Electrical Code
NPT	National Standard Pipe – Tapered
NPTF	National Standard Pipe, Tapered – Fuel (dryseal)
NPSC	National Standard Pipe, Straight–Couplings
NPSH	National Standard Pipe, Straight–Hose Couplings and Nipples

Common Abbreviations

NPSI	National Standard Pipe, Straight–Internal (dryseal)
NPSL	National Standard Pipe, Straight–Lock Nuts
NPSM	National Standard Pipe, Straight–Mechanical
NPTR	National Standard Pipe, Tapered–Railroad
OD	outside diameter
oz	ounces
psi	pounds per square inch
psia	pounds per square inch absolute
psig	pounds per square inch gauge
π	(Pi) 3.1416
QPL	Military Qualified Products List
r or rad	radius
rpm	revolutions per minute
red	reducing
RMS	root mean square
SAE	Society of Automotive Engineers
scfm	standard cubic feet per minute
sec	second
SSU	Saybolt Universal seconds (or SUS)
SSF	Saybolt Furol seconds (or SFS)
shp	shaft horsepower
sol	solenoid
sp gr	specific gravity
sq ft	square foot
std	standard
sch	schedule
temp	temperature
UL	Underwriters Laboratory
USASI	USA Standards Institute
va	volt amperes
vel	velocity
VI	viscosity index

Prefixes

Prefix	U.S. Term
deka	Ten
hecto	Hundred
kilo	Thousand
mega	Million
giga	Billion
tera	Trillion
deci	Tenth
centi	Hundredth
milli	Thousandth
micro	Millionth
nano	Billionth
pico	Trillionth
femto	Quadrillionth
atto	Quintillionth

Conversions

To convert	into	Multiply by	To convert	into	Multiply by
Acre	Rods	160.0	Btu	Gram-calories	252.0
Acre	Hectare or		Btu	Horsepower-hrs.	3.931×10^{-4}
	Sq. hectometer	0.4047	Btu	Joules	1,054.8
Acres	Sq. feet	43,560.0	Btu	Kilogram-calories	0.2520
Acres	Sq. meters	4,047.0	Btu	Kilogram-meters	107.5
Acres	Sq. miles	1.562×10^{-3}	Btu	Kilowatt-hrs.	2.928×10^{-4}
Acres	Sq. yards	4,840.0	Btu/hr	Foot-pounds/sec.	0.2162
Acre-feet	Cu. feet	43,560.0	Btu/hr	Gram-cal./sec.	0.0700
Acre-feet	Gallons	3.259×10^{-5}	Btu/hr	Horsepower-hrs.	3.929×10^{-4}
Amperes/sq.cm	Amps/sq.in.	6,452	Btu/hr	Watts	0.2931
Amperes/sq.cm	Amps/sq.meter	10^4	Btu/min	Foot-lbs./sec.	12.96
Amperes/sq.in	Amps/sq.cm.	0.1550	Btu/min	Horsepower	0.02356
Amperes/sq.in	Amps/sq.meter	1,550.0	Btu/min	Kilowatts	0.01757
Amperes/sq.meter	Amps/sq.cm.	10^{-4}	Btu/min	Watts	17.57
Amperes/sq.meter	Amps/sq.in.	6.452×10^{-4}	Btu/sq.ft./min	Watts/sq.in	0.1221
Ampere-hours	Coulombs	3,600.0	Bushels	Cu.ft.	1.2445
Ampere-hours	Faradays	0.03731	Bushels	Cu. In.	2,150.4
Ampere-turns	Gilberts	1.257	Bushels	Cu. Meters	0.03524
Ampere-turns/cm	Amp-turns/in.	2.540	Bushels	Liters	35.24
Ampere-turns/cm	Amp-turns/meter	100.0	Bushels	Pecks	4.0
Ampere-turns/cm	Gilberts/cm	1.257	Bushels	Pints (dry)	64.0
Ampere-turns/in	Amp-turns/cm	0.3937	Bushels	Quarts (dry)	32.0
Ampere-turns/in	Amp-turns/meter	39.37	Calories, gram (mean)	Btu (mean)	3.9685×10^3
Ampere-turns/in	Gilberts/cm	0.4950	Centares (centiares)	Sq. Meters	1.0
Ampere-turns/meter	Amp/turns/cm.	0.01	Centigrade	Fahrenheit	(C x9/5) +32
Ampere-turns/meter	Amp-turns/in.	0.0254	Centigrade	Kelvin	C +273
Ampere-turns/meter	Gilberts/cm.	0.01257	Centegrams	Grams	0.01
Angstrom unit	Inch	3937.0×10^{-9}	Centiliters	Liters	0.01
Angstrom-unit	Meter	1.0×10^{-10}	Centimiters	Feet	3.281×10^{-2}
Angstrom unit	Micron or (Mu)	1.0×10^{-4}	Centimeters	Inches	0.3937
Astronomical unit	Kilometers	1.495×10^{-8}	Centimeters	Kilometers	10^{-5}
Atmospheres	Bar	1.013	Centimeters	Meters	0.01
Atmospheres	Cms. of mercury	76.0	Centimeters	Miles	6.214×10^{-6}
Atmospheres	Mm of mercury	760.0	Centimeters	Millimeters	10.0
Atmospheres	Ft. of water (at 4 C)	33.90	Centimeters	Mils	393.7
Atmospheres	In. of mercury		Centimeters	Yards	1.024×10^{-2}
	(at 0 C)	29.92	Centimeter-dynes	Cm.-grams	1.020×10^{-3}
Atmospheres	Kgs./sq. cm.	1.0333	Centimeter-dynes	Meter/kgs.	1.020×10^{-8}
Atmospheres	Kgs./sq.meter	10,332.0	Centimeter-dynes	Pound-feet	7.376×10^{-8}
Atmospheres	Pounds/sq. in.	14.70	Centimeter-grams	Cm.-dynes	980.7
Barrels (U.S. dry)	Cu. inches	7056.0	Centimeter-grams	Meter-kgs.	10^{-5}
Barrels (U.S. dry)	Quarts (dry)	105.0	Centimeter-grams	Pound-feet	7.233×10^{-5}
Barrels (U.S. liquid)	Gallons	31.5	Centimeters of mercury	Atmospheres	0.01316
Barrels (Oil)	Gallons (oil)	42.0	Centimeters of mercury	Feet of water	0.4461
Bars	Atmospheres	0.9869	Centimeters of mercury	Kgs./sq. Meter	136.0
Bars	At. (Tech.)	1.0197	Centimeters of mercury	Pounds/sq. ft.	27.85
Bars	Dynes/sq. cm.	10^{-4}	Centimeters of mercury	Pounds/sq. in.	0.1934
Bars	Kgs./sq. meter	1.020×10^{-4}	Centimeters/sec.	Feet/min.	1.1969
Bars	Kilopascal	100.0	Centimeters/sec.	Feet/sec.	0.03281
Bars	Newtons/Sq. meter	10^{-5}	Centimeters/sec.	Kilometers/hr.	0.036
Bars	Pounds/sq. ft	2,089.0	Centimeters/sec.	Knots	0.1943
Bars	Psi	14.504	Centimeters/sec.	Meters/min.	0.6
Baryl	Dyne/sq. Cm	1.000	Centimeters/sec.	Miles/hr.	0.02237
Btu	Liter-Atmosphere	10.409	Centimeters/sec.	Miles/min.	3.728×10^{-4}
Btu	Ergs	1.0550×10^{-10}	Centimeters/sec./sec.	Feet/sec./sec.	0.03281
Btu	Foot-lbs.	778.3	Centimeters/sec./sec.	Kms./hr./sec.	0.036

Conversions

To convert	into	Multiply by	To convert	into	Multiply by
Centimeters/sec./sec..	Meters/sec./sec.	0.01	Cubic yards	Cu. cms	7.646×10^5
Centimeters/sec./sec..	Miles/hr./sec.	0.02237	Cubic yards	Cu. feet	27.0
Centipoise	Gram/cm.sec.	0.01	Cubic yards	Cu. inches	46,656.0
Centipoise	Pound mass/ft. sec.	0.000672	Cubic yards	Cu. meters	0.7646
Centistokes	Sq. feet/sec.	1.076×10^{-6}	Cubic yards	Gallons (U.S. liq.)	202.0
Circular mils	Sq. cms	5.067×10^{-6}	Cubic yards	Liters	764.6
Circular mils	Sq. mils	0.7854	Cubic yards	Pints (U.S. liq.)	1,615.9
Circumference	Radians	6.283	Cubic yards	Quarts (U.S. liq.)	807.9
Circular mils	Sq. inches	7.854×10^{-7}	Cubic yards/min.	Cubic ft./sec.	0.45
Coulomb	Statcoulombs	2.998×10^{-9}	Cubic yards/min.	Gallonssec.	3.367
Coulombs	Faradays	1.036×10^{-5}	Cubic yards/min.	Liters/sec.	12.74
Coulombs/sq. cm.	Coulombs/sq.in	64.52	Days	Seconds	86,400.0
Coulombs/sq. cm.	Coulombs/sq.meter	10^{-4}	Decigrams	Grams	0.1
Coulombs/sq. in.	Coulombs/sq.cm	0.1550	Deciliters	Liters	0.1
Coulombs/sq. in.	Coulombs/sq.meter	1,550.0	Decimeters	Meters	0.1
Coulombs/sq. meter.	Coulombs/sq.cm	10^{-4}	Degrees (angle)	Minutes	60.0
Coulombs/sq. meter.	Coulombs/sq.in	6.452×10^{-5}	Degrees (angle)	Quadrants	0.01111
Cubic centimeters	Cu. feet	3.531×10^{-5}	Degrees (angle)	Radians	0.01745
Cubic centimeters	Cu. inches	0.06102	Degrees (angle)	Seconds	3,600.0
Cubic centimeters	Cu. meters	10^{-6}	Degrees/sec.	Radians/sec.	0.01745
Cubic centimeters	Cu. yards	1.308×10^{-6}	Degrees/sec.	Revolutions/min.	0.1667
Cubic centimeters	Gallons (U.S. liq.)	2.642×10^{-4}	Degrees/sec.	Revolutions/sec	2.778×10^{-3}
Cubic centimeters	Liters	0.001	Dekagrams	Grams	10.0
Cubic centimeters	Pints (U.S. liq.)	2.113×10^{-3}	Dekaliters	Liters	10.0
Cubic centimeters	Quarts (U.S. liq.)	1.057×10^{-3}	Dekameters	Meters	10.0
Cubic feet	Cu. cms.	28,320.0	Drams	Grams	1.7718
Cubic feet	Cu. inches	1728.0	Drams	Grains	27.3437
Cubic feet	Cu. meters	0.02832	Drams	Ounces	0.0625
Cubic feet	Cu. yards	0.03704	Dyne/sq.cm	Atmospheres	9.869×10^{-7}
Cubic feet	Gallons (U.S. liq.)	7.48052	Dyne/sq.cm	In. of mercury at 0 C	2.953×10^{-5}
Cubic feet	Liters	28.32	Dyne/sq.cm	In. of water at 4 C	4.015×10^{-4}
Cubic feet	Pints (U.S. liq.)	59.84	Dynes	Grams	1.020×10^{-3}
Cubic feet	Quarts (U.S. liq.)	29.92	Dynes	Joules/cm.	10^{-7}
Cubic feet/min.	Cu. cms./sec.	472.0	Dynes	Joules/meter(newtons)	10^{-5}
Cubic feet/min.	Gallons/sec.	0.1247	Dynes	Kilograms	1.020×10^{-6}
Cubic feet/min.	Liters/sec.	0.4720	Dynes	Poundals	7.233×10^{-5}
Cubic feet/min.	Pounds of water/min.	62.43	Dynes	Pounds	2.248×10^{-6}
Cubic feet/sec	Million gal./day	0.646317	Dynes/sq.cm	Bars	10^{-6}
Cubic feet/sec	Gallons/min.	448.831	Erg/sec.	Dyne-cm./sec.	1.000
Cubic inches	Cu. cms.	16.39	Ergs.	Btu	9.480×10^{-11}
Cubic inches	Cu. feet	5.787×10^{-4}	Ergs.	Dyne-centimeters	1.0
Cubic inches	Cu. meters	1.639×10^{-5}	Ergs.	Foot-pounds	7.367×10^{-8}
Cubic inches	Cu. yards	2.143×10^{-5}	Ergs.	Gram-calories	0.2389×10^{-7}
Cubic inches	Gallons	4.329×10^{-3}	Ergs.	Gram-cms	1.020×10^{-3}
Cubic inches	Liters	0.01639	Ergs.	Horsepower-hrs.	3.7250×10^{-14}
Cubic inches	Mil.-feet	1.061×10^5	Ergs.	Joules	10^{-7}
Cubic inches	Pints (U.S. liq.)	0.03463	Ergs.	Kg.-calories	2.389×10^{-11}
Cubic inches	Quarts (U.S. liq.)	0.01732	Ergs.	Kg.-meters	1.020×10^{-8}
Cubic meters	Cu. cms	10^6	Ergs.	Kilowatt-hrs.	0.2778×10^{-13}
Cubic meters	Cu. feet	35.31	Ergs.	Watt-hours	0.2778×10^{-10}
Cubic meters	Cu. inches	61,023.0	Ergs/sec.	Btu/min.	$5,688.0 \times 10^{-9}$
Cubic meters	Cu. yards	1.308	Ergs/sec.	Ft.-lbs./min.	4.427×10^{-6}
Cubic meters	Gallons (U.S. liq.)	264.2	Ergs/sec.	Ft.-lbs./sec.	7.3756×10^{-8}
Cubic meters	Liters	1,000.0	Ergs/sec.	Horsepower	1.341×10^{-10}
Cubic meters	Pints (U.S. liq.)	2,113.0	Ergs/sec.	Kg.-calories/min.	1.433×10^{-9}
Cubic meters	Quarts (U.S. liq.)	1,057.0	Ergs/sec.	Kilowatts	10^{-10}

Conversions

To convert	into	Multiply by
Fahreheit	Centigrade	5/9 (F − 32)
Fahrenheit	Rankine	F + 460
Farads	Microfarads	10^6
Faraday/sec.	Ampere (absolute)	9.6500×10^4
Faradays	Ampere-hours	26.80
Faradays	Coulombs	9.649×10^4
Fathom	Meter	1.828804
Fathoms	Feet	6.0
Feet	Centimeters	30.48
Feet	Inches	12.0
Feet	Kilometers	3.048×10^{-4}
Feet	Meters	0.3048
Feet	Miles (naut.)	1.645×10^{-4}
Feet	Miles (stat.)	1.894×10^{-4}
Feet	millimeters	304.8
Feet	Yards	1/3
Feet of water	Atmospheres	0.02950
Feet of water	In. of mercury	0.8826
Feet of water	Kgs./sq. cm.	0.03048
Feet of water	Kgs./sq. meter	304.8
Feet of water	Pounds/sq. ft.	62.43
Feet of water	Pounds/sq. in.	0.4335
Feet/min.	Cms./sec.	0.5080
Feet/min.	Feet/sec.	0.01667
Feet/min.	Kms./Hr.	0.01829
Feet/min.	Meters/min.	0.3048
Feet/min.	Miles/hr.	0.01136
Feet/sec.	Cms./sec.	30.48
Feet/sec.	Kms./hr.	1.097
Feet/sec.	Knots	0.5921
Feet/sec.	Meters/min.	18.29
Feet/sec.	Miles/hrs.	0.6818
Feet/sec.	Miles/min.	0.01136
Feet/sec./sec.	Cms./sec./sec.	30.48
Feet/sec./sec	Kms./hr/sec.	1.097
Feet/sec./sec	Meters/sec./sec	0.3048
Feet/sec./sec	Miles/hrs./sec	0.6818
Feet/100 feet	Per cent grade	1.0
Foot-pounds	Btu	1.286×10^{-3}
Foot-pounds	Ergs	1.356×10^{-7}
Foot-pounds	Gram-calories	0.3238
Foot-pounds	Hp-hrs.	5.050×10^{-7}
Foot-pounds	Joules	1.356
Foot-pounds	Kg.-calories	3.24×10^{-4}
Foot-pounds	Kg.-meters	0.1383
Foot-pounds	Kilowatt-hrs.	3.766×10^{-7}
Foot-pounds	Newton-meters	1.356
Foot-pounds/min.	Btu/min.	1.286×10^{-3}
Foot-pounds/min.	Foot-pounds/sec.	0.01667
Foot-pounds/min.	Horsepower	3.030×10^{-5}
Foot-pounds/min.	Kg.-calories/min.	3.24×10^{-4}
Foot-pounds/min.	Kilowatts	2.260×10^{-5}
Foot-pounds/sec.	Btu/sec.	4.6263
Foot-pounds/sec.	Btu/min.	0.07717
Foot-pounds/sec.	Horsepower	1.818×10^{-3}
Foot-pounds/sec.	Kg.-calories/min.	0.01945
Foot-pounds/sec.	Kilowatts	1.356×10^{-3}
Foot-pounds/sec.	Newton-meters/Sec	1.356
Furlongs	Miles (U.S.)	0.125
Furlongs	Rods	40.0
Furlongs	Feet	660.0
Gallons	Cu. cms.	3,785.0
Gallons	Cu. feet	0.1337
Gallons	Cu. inches	231.0
Gallons	Cu. meters	3.785×10^{-3}
Gallons	Cu. yards	4.951×10^{-3}
Gallons	Liters	3.785
Gallons	Pints (liq.)	8.0
Gallons	Quarts (liq.)	4.0
Gallons (liq. Br.Imp.)	Gallons (U.S. liq.)	1.20095
Gallons (U.S.)	Gallons (Imp.)	0.83267
Gallons/min.	Cu. ft./sec.	2.228×10^{-3}
Gallons/min.	Liters/sec.	0.06308
Gallons/min.	Cu. ft./hr.	8.0208
Gallons/min.	Pounds of water/hrs.	500.0
Gausses	Lines/sq.in.	6.452
Gausses	Webers/sq.cm.	10^{-8}
Gausses	Webers/sq.in.	6.452×10^{-8}
Gauesses	Webers/sq.meter	10^{-4}
Grade	Radian	0.01571
Grains (troy)	Grains (avdp.)	1.0
Grains (troy)	Grams	0.06480
Grains (troy)	Ounces (avdp.)	2.0833×10^{-3}
Grains /U.S. gal.	Parts/million	17 118
Grains/U.S gal.	Pounds/million gal.	142.86
Grains/Imp. gal.	Parts/million	14.286
Grams	Dynes	980.7
Grams	Grains	15.43
Grams	Joules/cm.	$9.807/10^{-5}$
Grams	Joules/meter (newtons)	$9.807/10^{-3}$
Grams	Kilograms	0.001
Grams	Milligrams	1,000.0
Grams	Ounces (avdp.)	0.03527
Grams	Ounces (troy)	0.03215
Grams	Poundals	0.07093
Grams	Pounds	2.205×10^{-3}
Grams/cm	Pounds/inch	5.600×10^{-3}
Grams/cu. cm.	pounds/cu.ft.	62.43
Grams/cu.cm.	Pounds/cu.in.	0.03613
Grams/cu. cm.	Pounds/mil.-foot	3.405×10^{-7}
Grams/liter	Grains/gal.	58.417
Grams/liter	Pounds/1,000 gal.	8.345
Grams/liter	Pouns/cu.ft.	0.062427
Grams/liter	Parts/million	1,000.0
Grams/sq. cm.	Pounds/sq. ft.	2.0481
Gram-calories	Btu	3.9683×10^{-3}
Gram-calories	Foot-pounds	3.0880
Gram-calories	Horsepower-hrs.	1.5596×10^{-6}
Gram-calories	Kilowatt-hrs.	1.1630×10^{-6}
Gram-calories	Watt-hrs.	1.1630×10^{-6}
Gram-calories/sec.	Btu/hr.	14.286

Conversions

To convert	into	Multiply by	To convert	into	Multiply by
Gram-centimers	Btu	9.297×10^{-8}	Joules	Watt-hrs	2.778×10^{-4}
Gram-centimers	Joules	9.807×10^{-5}	Joules/cm.	Grams	1.020×10^{4}
Gram-centimers	Kg.-cal.	2.343×10^{-8}	Joules/cm.	Dynes	10^{7}
Gram-centimers	Kg.-meters	10^{-5}	Joules/cm.	Joules/meter (newtons)	100.0
Hectares	Acres	2.471	Joules/cm.	Poundals	723.3
Hectares	Sq. feet	1.076×10^{5}	Joules/cm.	Pounds.	22.48
Hectograms	Grams	100.0	Kilograms	Dynes	980,665.0
Hectoliters	Liters	100.0	Kilograms	Grams	1,000.0
Hectometers	Meters	100.0	Kilograms	Joules/cm.	0.09807
Hectowatts	Watts	100.0	Kilograms	Joules/meter(newtons)	9.807
Henries	Millihenries	1,000.0	Kilograms	Poundals	70.93
Horsepower	Btu/min.	42.44	Kilograms	Pounds	2.205
Horsepower	Foot-lbs./min.	33,000.0	Kilograms	Tons (long)	9.842×10^{-4}
Horsepower	Foot-lbs./sec.	550.0	Kilograms	Tons (short)	1.102×10^{-3}
Horsepower (metric) (542.5 ft. lb./sec.)	Horsepower (550.ft. lb./sec.)	0.9863	Kilograms/cu. meter	Grams/cu. cm.	0.001
			Kilograms/cu. meter	Pounds/cu.ft.	0.06243
Horsepower (550.ft. lb./sec.)	Horsepower (metric) (542.5 ft. lb./sec.)	1.014	Kilograms/cu. meter	Pounds/cu.in	3.613×10^{-5}
			Kilograms/cu. meter	Poundsmil.-foot	3.405×10^{-10}
Horsepower	Kg.-calories/min.	10.68	Kilograms/meter	Pounds/ft.	0.6720
Horsepower	Kilowatts	0.7457	Kilograms/sq. cm.	Bar	0.981
Horsepower	Watts	745.7	Kilograms/sq. cm.	Dynes	980,665.0
Horsepower-hrs.	Btu	2,547.0	Kilograms/sq. cm.	Atmospheres	0.9678
Horsepower-hrs.	Foot-lbs.	1.98×10^{6}	Kilograms/sq. cm.	Feet of water	32.81
Horsepower-hrs.	Gram-calories	641,190.0	Kilograms/sq. cm.	Inches of mercury	28.96
Horsepower-hrs.	Joules	2.684×10^{6}	Kilograms/sq. cm.	Pounds/sq. ft..	0.2048
Horsepower-hrs.	Kg.-calories	641.1	Kilograms/sq. cm.	Pounds/sq. in.	14.228
Horsepower-hrs.	Kg.-meters	2.737×10^{5}	Kilograms/sq. meter	Atmospheres	$9.678.10^{-5}$
Horsepower-hrs.	Kilowatt-hrs.	0.7457	Kilograms/sq. meter	Bars	98.07×10^{-6}
Hours	Days	$4.167.10^{-2}$	Kilograms/sq. meter	Feet of water	3.281×10^{-3}
Hours	Weeks	5.952×10^{-3}	Kilograms/sq. meter.	Inches of mercury	2896×10^{-3}
Inches	Centimeters	2.540	Kilograms/sq. meter	Pounds/sq. ft.	0.2048
Inches	Meters	2.540×10^{-2}	Kilograms/sq. meter	Pounds/sq. in	1.422×10^{-3}
Inches	Miles	1.578×10^{-5}	Kilograms/sq. mm	Kgs./sq. meter	10^{4}
Inches	Millimeters	25.40	Kilogram-calories	Btu	3.968
Inches	Mils	1,000.0	Kilogram-calories	Foot-pounds	3,088.0
Inches	Yards	2.778×10^{-2}	Kilogram-calories	Hp.-hrs.	$1.560-10^{-3}$
Inches of mercury	Atmospheres	0.03342	Kilogram-calories	Joules	4,186.0
Inches of mercury	Feet of water	1.133	Kilogram-calories	Kg.-meters	426.9
Inches of mercury	Kgs./sq. cm.	0.03453	Kilogram-calories	Kilojoules	4.186
Inches of mercury	Kgs./sq.meter	345.3	Kilogram-calories	Kilowatt-hrs.	1.163×10^{-3}
Inches of mercury	Pounds/sq. ft.	70.73	Kilogram-meters	Btu	9.294×10^{-3}
Inches of mercury	Pounds/sq. in.	0.4912	Kilogram-meters	Ergs.	9.804×10^{7}
Inches of water (at 4 C)	Atmospheres	2.458×10^{-3}	Kilogram-meters	Foot-pounds	7.233
Inches of water (at 4 C)	Inches of mercury	0.07355	Kilogram-meters	Joules	9.804
Inches of water (at 4 C)	Kgs./sq. cm.	$2,540 \times 10^{-3}$	Kilogram-meters	Kg.-calories	2.342×10^{-3}
Inches of water (at 4 C)	Ounces/sq. in.	0.5781	Kilogram-meters	Kilowatt-hrs.	2.723×10^{-6}
Inches of water (at 4 C)	Pounds/sq. ft.	5.204	Kiloliters	Liters	1,000.0
Inches of water (at 4 C)	Pounds/sq. in.	0.03613	Kilometers	Centimeters	10^{5}
International Ampere	Ampere (absolute)	0.9998	Kilometers	Feet	3,281.0
International Volt	Volts (absolute)	1.0003	Kilometers	Inches	3.937×10^{4}
International Volt	Joules	9.654×10^{4}	Kilometers	Meters	1,000.0
Joules	Btu	9.480×10^{-4}	Kilometers	Miles	0.6214
Joules	Ergs	10^{7}	Kilometers	Millimeters	10^{4}
Joules	Foot-pounds	0.7376	Kilometers	Yards	1,094.0
Joules	Kg.-calories	2.389×10^{-4}	Kilometers/hr.	Cms./sec.	27.78
Joules	Kg.-meters	0.1020	Kilometers/hr.	Feet/min.	54.68

Conversions

To convert	into	Multiply by	To convert	into	Multiply by
Kilometers/hr.	Feet/sec.	0.9113	Meters/min.	Knots	0.03238
Kilometers/hr.	Knots	0.5396	Meters/min.	Miles/hrs.	0.03728
Kilometers/hr.	Meters/min.	16.67	Meters/sec.	Feet/min.	196.8
Kilometers/hr.	Miles/hr.	0.6214	Meters/sec.	Feet/sec.	3.281
Kilometers/hr./sec.	Cms./sec./sec.	27.78	Meters/sec.	Kilometers/hr.	3.6
Kilometers/hr./sec.	Ft./sec./sec.	0.9113	Meters/sec.	Kilometers/min.	0.06
Kilometers/hr./sec.	Meters/sec./sec.	0.2778	Meters/sec.	Miles/hr.	2.237
Kilometers/hr./sec.	Miles/hr./sec.	0.6214	Meters/sec.	Miles/min.	0.03728
Kilowatts	Btu/min.	56.92	Meters/sec./sec.	Cms./sec./sec.	100.0
Kilowatts	Foot-lbs./min.	4.426×10^4	Meters/sec./sec.	Ft./sec./sec.	3.281
Kilowatts	Foot-lbs./sec.	737.6	Meters/sec./sec.	Miles/hr./sec.	2.237
Kilowatts	Horsepower	1.341	Meters-kilograms	Cm-dynes	9.807×10^7
Kilowatts	Kg.-calories/min.	14.34	Meters-kilograms	Cm-grams	10^5
Kilowatts	Watts	1,000.0	Meters-kilograms	Pound-feet	7.233
Kilowatts-hrs.	Btu	3,413.0	Micrograms	Grams	10^{-6}
Kilowatt-hrs.	Foot-lbs.	2.655×10^6	Microhms	Megohms	10^{-12}
Kilowatt-hrs.	Gram-calories	859,850.0	Microhms	Ohms	10^{-6}
Kilowatt-hrs.	Horsepower-hrs.	1.341	Microliters	Liters	10^{-6}
Kilowatt-hrs.	Joules	3.6×10^4	Microns	Inches	39×10^{-6}
Kilowatt-hrs.	Kg.-calories	860.5	Microns	Meters	1×10^{-6}
Kilowatt-hrs.	Kg.-meters	3.671×10^5	Miles (naut.)	Feet	6,080.27
Knots	Feet/hr.	6,080.0	Miles (naut.)	Kilometers	1.853
Knots	Kilometers/hr.	1.8532	Miles (naut.)	Meters	1,853.0
Knots	nautical miles/hr.	1.0	Miles (naut.)	Miles (statute)	1.1516
Knots	Statue miles/hr.	1.151	Miles (naut.)	Yards	2,027.0
Knots	Yards/hr.	2,027.0	Miles (statute)	Centimeters	1.609×10^5
Knots	Feet/sec.	1.689	Miles (statute)	Feet	5,280.0
League	Miles (approx.)	3.0	Miles (statute)	Inches	6.336×10^4
Light year	Miles	5.9×10^{12}	Miles (statute)	Kilometers	1.609
Light year	Kilometers	9.46091×10^{12}	Miles (statute)	Meters	1,609.0
Liters	Bushels (U.S. dry)	0.02838	Miles (statute)	Miles (naut.)	0.8684
Liters	Cu. cm.	1,000.0	Miles (statute)	Yards	1,760.0
Liters	Cu. decimeters	1.0	Miles/hr.	Cms./sec.	44.70
Liters	Cu. feet	0.03531	Miles/hr.	Feet/min.	88.0
Liters	Cu. inches	61.02	Miles/hr.	Feet/sec.	1.467
Liters	Cu. meters	0.001	Miles/hr.	Kms./hr.	1.609
Liters	Cu. yards	1.308×10^{-3}	Miles/hr.	Kms./min.	0.02682
Liters	Gallons (U.S. liq.)	0.2642	Miles/hr.	Knots	0.8684
Liters	Pints (U.S. liq.)	2.113	Miles/hr.	Meters/min.	26.82
Liters	Quarts (U.S. liq.)	1.057	Miles/hr.	Miles/min.	0.1667
Liters/min.	Cu. Ft./sec.	5.886×10^{-4}	Miles/hr./sec.	Cms/sec./sec.	44.70
Liters/min.	Gals./sec.	4.403×10^{-3}	Miles/hr./sec.	Feet/sec./sec.	1,467
Megohms	Microhms	10^{12}	Miles/hr./sec.	Kms./hr./sec.	1.609
Megohms	Ohms	10^6	Miles/hr./sec.	Meters/sec./sec.	0.4470
Meters	Centimeters	100.0	Miles/min.	Cms./sec.	2,682.0
Meters	Feet	3.281	Miles/min.	Feet/sec.	88.0
Meters	Inches	39.37	Miles/min.	Kms./min.	1.609
Meters	Kilometers	0.001	Miles/min.	Knots/min.	0.8684
Meters	Miles (naut.)	5.396×10^4	Miles/min.	Miles/hr.	60.0
Meters	Miles (stat.)	6.214×10^4	Milliers	Kilograms	1,000.0
Meters	Millimeters	1,000.0	Millimicrons	Meters	1×10^{-9}
Meters	Yards	1.094	Milligrams	Grains	0.01543236
Meters/min.	Cms./sec.	1.667	Milligrams	Grams	0.001
Meters/min.	Feet/min.	3.281	Milligrams/liter	Parts/million	1.0
Meters/min.	Feet/sec.	0.05468	Millihenries	Henries	0.001
Meters/min.	Kms./hr.	0.06	Milliliters	Liters	0.001

Conversions

To convert	into	Multiply by	To convert	into	Multiply by
Millimeters	Centimeters	0.1	Poundals	Grams	14.10
Millimeters	Feet	3.281×10^{-3}	Poundals	Joules/cm.	1.383×10^{-3}
Millimeters	Inches	0.03937	Poundals	Joules/meter (newtons)	0.1383
Millimeters	Kilometers	10^{-6}	Poundals	Kilograms	0.01410
Millimeters	Meters	0.001	Poundals	Pounds	0.03108
Millimeters	Miles	6.214×10^{-7}	Pounds	Drams	256.0
Millimeters	Mils	39.37	Pounds	Dynes	44.4823×10^{4}
Millimeters	Yards	1.094×10^{-3}	Pounds	Grams	453.5924
Millimeters of mercury	Psi	0.0194	Pounds	Joules/cm.	0.04448
Million gals./day	Cu. ft./sec.	1.54723	Pounds	Joules/meter(newtons)	4.448
Mils	Centimeters	2.540×10^{-3}	Pounds	Kilograms	0.4536
Newton	Dynes	$1 \times 10^{5.0}$	Pounds	Newtons (N)	4.44
Newton	Kilograms	0.1020	Pounds	Ounces	16.0
Newton	Pounds	8.85	Pounds	Ounces (troy)	14.5833
Newton/sq. meter	Pascal	1.0	Pounds	Poundals	32.17
Newton-meter	Foot-pounds	0.7375	Pounds	Pounds (troy)	1.21528
Newton-meter	Joule	1.0	Pounds	Tons (shorts)	0.0005
Newton-meter/sec.	Foot-pounds/sec.	0.7375	Pounds (troy)	Ounces (avdp.)	13.1657
Newton-meter/sec.	Watts	1.0	Pounds (troy)	Ounces (troy)	12.0
OHM (International)	OHM (absolute)	1.0005	Pounds (troy)	Pounds (avdp.)	0.822857
Ohms	Megohms	10^{-6}	Pounds (troy)	Tons (long)	3.6735×10^{-4}
Ohms	Microhms	10^{6}	Pounds (troy)	Tons (metric)	3.7324×10^{-4}
Ounces	Drams	16.0	Pounds (troy)	Tons (short)	4.1143×10^{-4}
Ounces	Grams	28.349527	Pounds of water	Cu. feet	0.01602
Ounces	Pounds	0.0625	Pounds of water	Cu. inches	27.68
Ounces	Ounces(troy)	0.9115	Pounds of water	Gallons	0.1198
Ounces	Tons (long)	2.790×10^{-5}	Pounds of water/min.	Cu. ft./sec.	2.670×10^{-4}
Ounces	Tons (metric)	2.835×10^{-5}	Pounds/inch	newton-meters	0.113
Ounces (fluid)	Cu. inches	1.805	Pound-feet	Cm.-dynes	1.356×10^{7}
Ounces (fluid)	Liters	0.02957	Pound-feet	Cm.-grams	13,825.0
Ounces (troy)	Ounces avdp.)	1.09714	Pound-feet	Meter-kgs.	0.1383
Ounces (troy)	Pounds (troy)	0.08333	Pounds/foot	Newton-meters	1.356
Ounce/sq. inch	Dynes/sq. cm.	4309.0	Pounds/cu.ft	Grams/cu.cm.	0.01602
Ounces/sq.in.	Pounds/sq. in.	0.0625	Pounds/cu. ft.	Kgs./cu.meter	16.02
Parsec.	Miles	$19. \times 10^{12}$	Pounds/cu. ft.	Pounds/cu.in.	5.787×10^{-4}
Parsec	Kilometers	3.084×10^{13}	Pounds/cu. ft.	Pounds/mil–foot	5.456×10^{-9}
Parts/million	Grains/U.S. gal.	0.0584	Pounds/cu. in.	Gm./cu. cm.	27.68
Parts/million	Grains/Imp.gal.	0.07016	Pounds/cu. in	Kgs./cu. meter	2.768×10^{4}
Parts/million	Pounds/million gal.	8.345	Pounds/cu. in.	Pounds cu. ft.	1,728.0
Pecks (British)	Cubic inches	554.6	Pounds cu. in.	Pounds/mil–foot	9.425×10^{-6}
Pecks (British)	Liters	9.091901	Pounds/ft.	Kgs./meter	1.488
Pecks (U.S.)	Bushels	0.25	Pounds/in.	Gms./cm.	178.6
Pecks (U.S.)	Cubic Inches	537.605	Pounds/sq.ft.	Atmospheres	4.725×10^{4}
Pecks (U.S.)	Liters	8.809582	Pounds/sq.ft.	Feet of water	0.01602
Pecks (U.S.)	Quarts (dry)	8	Pounds/sq.ft.	Inches of mercury	0.01414
Pints (dry)	Cu. inches	33.60	Pounds/sq.ft.	Kgs./sq. meter	4.882
Pints (liq.)	Cu.cms.	473.2	Pounds/sq.ft.	Pounds/sq. in.	6.944×10^{-3}
Pints (liq.)	Cu. feet	0.01671	Pounds/sq.in.	Atmospheres	0.06804
Pints (liq.)	Cu. inches	28.87	Pounds/sq.in.	Bar	0.0690
Pints (liq.)	Cu. meters	4.732×10^{-4}	Pounds/sq.in.	Feet of water	2.307
Pints (liq.)	Cu. yards	6.189×10^{-4}	Pounds/sq.in.	Inches of mercury	2.036
Pints (liq.)	Gallons	0.125	Pounds/sq.in.	Inches of water	27.7
Pints (liq.)	Liters	0.4732	Pounds/sq.in.	Kgs./sq. meter	703.1
Pints (liq.)	Quarts (liq.)	0.5	Pounds/sq.in.	Kilopascal	6.895
Pounds (avoirdupois)	Ounces (troy)	14.5833	Pounds/sq.in.	Pounds/sq.ft	144.0
Poundals	Dynes	13,826.0	Pounds/hr.	Kilograms/hr.	0.454

Conversions

To convert	into	Multiply by	To convert	into	Multiply by
Pounds/sec.	Kilograms/hr.	1,633.0	Square Centimeters	Sq. feet	1.076×10^{-3}
Pounds-sec./sq.ft.	Pound mass/ft. sec.	32.2	Square Centimeters	Sq. inches	0.1550
Quadrants (angle)	Degrees	90.0	Square Centimeters	Sq. meters	0.0001
Quadrants (angle)	Minutes	5,400.0	Square Centimeters	Sq. miles	3.861×10^{-11}
Quadrants (angle)	Radians	1.571	Square Centimeters	Sq. millimeters	100.0
Quadrants (angle)	Seconds	3.24×10^5	Square Centimeters	Sq. yards	1.196×10^{-4}
Quarts (liq.)	Cu. cms.	946.4	Square Feet	Acres	2.296×10^{-5}
Quarts (liq.)	Cu. feet	0.03342	Square Feet	Circular mils.	1.833×10^8
Quarts (liq.)	Cu. inches	57.75	Square Feet	Sq. cms.	929.0
Quarts (liq.)	Cu. meters	9.464×10^{-4}	Square Feet	Sq. inches	144.0
Quarts (liq.)	Cu. yards	1.238×10^{-3}	Square Feet	Sq. meters	0.09290
Quarts (liq.)	Gallons	0.25	Square Feet	Sq. miles	3.587×10^{-8}
Quarts (liq.)	Liters	0.9463	Square Feet	Sq. millimeters	9.290×10^4
Radians	Degrees	57.30	Square Feet	Sq. yards	0.1111
Radians	Minutes	3,438.0	Square Feet/sec.	Centistokes	92,903.0
Radians	Quadrants	0.6366	Square Inches	Circular mils.	1.273×10^6
Radians	Seconds	2.063×10^5	Square inches	Sq. cms.	6.452
Radians/sec.	Degrees/sec.	57.30	Square inches	Sq. feet	6.944×10^{-3}
Radians/sec.	Revolutions/min.	9.549	Square inches	Sq. millimeters	645.2
Radians/sec.	Revolutions/sec.	0.1592	Square inches	Sq. mils	10^6
Radians/sec./sec.	Revs./min./min.	573.0	Square inches	Sq. yard	7.716×10^{-4}
Radians/sec./sec.	Revs./min./sec.	9.549	Square kilometers	Acres	247.1
Radians/sec./sec.	Revs./sec./sec.	0.1592	Square kilometers	Sq. cms	10^{10}
Revolutions	Degrees	360.0	Square kilometers	Sq. ft.	10.76×10^4
Revolutions	Quadrants	4.0	Square kilometers	Sq. inches	1.550×10^9
Mils	Feet	8.333×10^{-5}	Square kilometers	Sq. meters	10^6
Mils	Inches	0.001	Square kilometers	Sq. miles	0.3861
Mils	Kilometers	2.540×10^{-8}	Square kilometers	Sq. yards	1.196×10^4
Mils	Yards	2.778×10^{-5}	Square meters	Acres	2.471×10^{-4}
Minutes (angles)	Degrees	0.01667	Square meters	Sq. cms.	10^4
Minutes (angles)	Quadrants	1.852×10^{-4}	Square meters	Sq. feet	10.76
Minutes (angles)	Radians	2.909×10^{-4}	Square meters	Sq. inches	1,550.0
Minutes (angles)	Seconds	60.0	Square meters	Sq. miles	3.861×10^{-7}
Myriagrams	Kilograms	10.0	Square meters	Sq. millimeters	10^6
Myriameters	Kilmeters	10.0	Square meters	Sq. yards	1.196
Myriawatts	Kilowatts	10.0	Square miles	Acres	640.0
Revolutions	Radians	6.283	Square miles	Sq. feet	27.88×10^6
Revolutions/min.	Degrees/sec.	6.0	Square miles	Sq. kms.	2.590
Revolutions/min.	Radians/sec.	0.1047	Square miles	Sq. meters	2.590×10^6
Revolutions/min.	Revs./sec.	0.01667	Square miles	Sq. yards	3.098×10^6
Revolutions/min./min.	Radians/sec./sec.	1.745×10^{-3}	Square millimeters	Circular mils	1,753.0
Revolutions/min./min.	Revs./min./sec.	0.01667	Square millimeters	Sq. cms	0.01
Revolutions/min./min.	Revs./sec./sec.	2.778×10^{-4}	Square miles	Sq. feet	1.076×10^{-5}
Revolutions/sec.	Degrees/sec.	360.0	Square miles	Sq. inches	1.550×10^{-3}
Revolutions/sec.	Radians/sec.	6.283	Square mils	Circular mils.	1.273
Revolutions/sec.	Revs./min.	60.0	Square mils	Sq. cms	6.452×10^{-6}
Revolutions/sec./sec.	Radians/sec./sec.	6.283	Square mils	Sq. inches	10^6
Revolutions/sec./sec.	Revs./min./min.	3,600.0	Square yards	Acres	2.066×10^{-4}
Revolutions/sec./sec.	revs./min./sec.	60.0	Square yards	Sq. cms	8,361.0
Seconds (angle)	Degrees	2.778×10^{-4}	Square yards	Sq. feet	9.0
Seconds (angle)	Minutes	0.01667	Square yards	Sq. inches	1,296.0
Seconds (angle)	Quadrants	3.087×10^{-6}	Square yards	Sq. meters	0.8361
Seconds (angle)	Radians	4.848×10^{-6}	Square yards	Square millimeters	8.361×10^5
Slug	Kilograms	14.59	Square yards	Sq. miles	3.228×10^{-7}
Slug	Pounds	32.17	Temperature (C) +273	Absolute Temp. (C)	1.0
Square Centimeters	Circular mils.	1.973×10^5	Temperature (C) +17.78	Temperature (F)	1.8

Conversions

To convert	into	Multiply by	To convert	into	Multiply by
Temperature (F) +460	Absolute Temp. (F)	1.0	Watts(Abs.)	Joules/sec.	1.0
Temperature (F) −32	Temperature (C)	5/9	Watts/hours	Btu	3.413
Tons (long)	Kilograms	1,016.0	Watts/hours	Foot–pounds	2,656.0
Tons (long)	Pounds	2,240.0	Watts/hours	Gram-calories	859.85
Tons (long	Tons (short)	1.120	Watts/hours	Horsepower-hrs.	1.341×10^{-3}
Tons (metric)	Kilograms	1,000.0	Watts/hours	Kilograms-calories	0.8605
Tons (metric	Pounds	2,205.0	Watts/hours	Kilograms-meters	367.2
Tons (short)	Kilograms	907.1848	Watts/hours	Kilowatt-hrs.	0.001
Tons (short)	Ounces	32,000.0	Watt (International)	Watt (absolute)	1.0002
Tons (short)	Ounces (troy)	29,166.66	Yards	Centimeters	91.44
Tons (short)	Pounds	2,000.0	Yards	Kilometers	9.144×10^{-4}
Tons (short)	Pounds (troy)	2,430.56	Yards	Meters	0.9144
Tons (short)	Tons (long)	0.89287	Yards	Miles (naut.)	4.934×10^{-4}
Tons (short)	Tons (metric)	0.9078	Yards	Miles (stat.)	5.682×10^{-4}
Tons (short)/sq. ft.	Kgs./sq.meter	9,765.0	Yards	Millimeters	914.4
Tons (short)/sq. ft.	Pounds/sq. in.	2,000.0			
Tons of water/24 hrs.	Pounds of water/hr.	83.333			
Tons of water/24 hrs.	Gallons/min.	0.16643			
Tons of water/24 hrs.	Cu. ft./hr.	1.3349			
Volt/inch	Volt/cm.	0.39370			
Watts	Btu/hr.	3.4129			
Watts	Btu/min.	0.05688			
Watts	Foot–lbs./min.	44.27			
Watts	Foot–lbs./sec.	0.7378			
Watts	Horsepower	1.341×10^{-3}			
Watts	Horsepower (metric)	1.360×10^{-3}			
Watts	Kg.-calories/min.	0.01433			
Watts	Kilowatts	0.001			
Watts(Abs.)	Btu (mean)/min.	0.056884			

Pressure Conversion Factors

To Convert From →	ft³	in³	Gal (US)	Quart (US)	Fluid Ounce (US)	Liter	Mililiter (ml)	Meter³ (m³)
				Multiply by				
ft³	———	1728	7.481	29.92	957.5	28.32	28,320	0,0283
in³	0.00058	———	0.00433	0.01732	0.5541	0.0164	16,39	0,000016
Gal (US)	0.1337	231.0	———	4.0	128.0	3.785	3785	0,00379
Quart (US)	0.0334	57.75	0.250	———	32.0	0.9464	946,4	0,00095
Fluid Ounce (US)	0.00104	1.805	0.00781	0.03125	———	0.0296	29,57	0,00003
Liter	0.0353	61.20	0.2642	1.057	33.81	———	1000	0,001
Mililiter (ml)	0.000035	0.0610	0.00026	0.00106	0.0338	0.001	———	0,000001
Meter³ (m³)	35.31	0.00061	264.2	1057	0.00034	1000	1,000,000	———

Volume Conversion Factors

To Convert From →	lb. per in² (psi)	in. mercury	in. water	ft. water	ATM	kg_f per cm²	kg_f per m²	kPa
				Multiply by				
lb. per in² (psi)	———	2.037	27.68	2.307	0.068	0.0703	703,1	6,895
in. mercury (in. Hg @ 0°C)	0.491	———	13.6	1.133	0.0334	0.0345	345,3	3,386
in. water (in. H₂O @ 4°C)	0.361	0.0736	———	0.0833	0.0025	0.0025	25,4	0,2491
ft. water (in. H₂O @ 4°C)	0.4335	0.8826	12	———	0.0295	0.0305	304,8	2,989
ATM Atmospheric pressure	14.70	29.92	406.8	33.90	———	1.033	10,330	101,3
kg_f per cm²	14.22	28.96	393.7	32.81	0.9678	———	10,000	98,07
kg_f per m²	0.0014	0.0029	0.0394	0.0033	0.00009	0.0001	———	0,0098
kPa	0.1450	0.2953	4.015	0.3346	0.0099	0.0102	102	———

Atmospheric Properties

Location	Alt	Press. psig	Press. in.Hg @32°F	Location	Alt	Press. psig	Press. in.Hg @32°F	Location	Alt	Press. psig	Press. in.Hg @32°F
Baton Rouge	0	14.696	29.92	Souix Falls	1400	13.97	28.44	Reno	4500	12.46	25.37
San Francisco	100	14.64	29.81		1500	13.92	28.33	Denver	5000	12.23	24.90
Tacoma	200	14.59	29.71		1600	13.87	28.23		6000	11.78	23.98
Memphis	300	14.54	29.60	Bismark	1700	13.82	28.13	Flagstaff	7000	11.34	23.09
Syracuse	400	14.48	29.49		1800	13.76	28.02		8000	10.92	22.22
Cincinnati	500	14.43	29.38	Grand Island	1900	13.71	27.92		9000	10.50	21.39
Chicago	600	14.38	29.28	Las Vegas	2000	13.66	27.82		10000	10.11	20.58
Indianapolis	700	14.33	29.17		2200	13.56	27.62		12000	9.35	19.03
Des Moines	800	14.28	29.07		2400	13.47	27.42	Mt. Rainier	14000	8.63	17.58
Fargo	900	14.22	28.96		2600	13.37	27.21		16000	7.97	16.21
Atlanta	1000	14.17	28.86		2800	13.27	27.02		18000	7.34	14.94
Phoenix	1100	14.12	28.75	Billings	3000	13.17	26.82	Mt. McKinley	20000	6.74	13.73
Oklahoma City	1200	14.07	28.65	Roswell	3500	12.93	26.33				
Wichita	1300	14.02	28.54	Salt Lake	4000	12.69	25.84				

Oil Viscosity Recommendations

Crankcase Oils		Antiwear Hydraulic Oils	
Hydraulic System Operating Temperature Range[1]	SAE Viscosity Designation	Hydraulic System Operating Temperature Range[1]	ISO Viscosity Grade
−23°C to 54°C (−10°F to 130°F)	5W, 5W-20, 5W-30	−21°C to 60°C (−5°F to 140°F)	22
−18°C to 83°C (0°F to 180°F)	10W	−15°C to 77°C (5°F to 170°F)	32
−18°C to 99°C (0°F to 210°F)	10W-30, 10W-40	−9°C to 88°C (15°F to 190°F)	46
10°C to 99°C (50°F to 210°F)	20-20W	−1°C to 99°C (30°F to 210°F)	68

Oil Flow Capacity Of Tubing

Figures in the chart are USgpm flow capacities of tubing, and were calculated from the formula: GPM = V × A ÷ .3208, in which V = velocity of flow in feet per second, and A is inside square inch area of tube.

Figures in Body of Chart are USgpm Flows

Tube O.D.	Wall Thick.	2 Ft/Sec	4 Ft/Sec	10 Ft/Sec	15 Ft/Sec	20 Ft/Sec	30 Ft/Sec
1/2"	.035	.905	1.81	4.52	6.79	9.05	13.6
	.042	.847	1.63	4.23	6.35	6.47	12.7
	.049	.791	1.58	3.95	5.93	7.91	11.9
	.058	.722	1.44	3.61	5.41	7.22	10.8
	.065	.670	1.34	3.35	5.03	6.70	10.1
	.072	.620	1.24	3.10	4.65	6.20	9.30
	.083	.546	1.09	2.73	4.09	5.46	8.16
5/8"	.035	1.51	3.01	7.54	11.3	15.1	22.6
	.042	1.43	2.85	7.16	10.7	14.3	21.4
	.049	1.36	2.72	6.80	10.2	13.6	20.4
	.058	1.27	2.54	6.34	9.51	12.7	19.0
	.065	1.20	2.40	6.00	9.00	12.0	18.0
	.072	1.13	2.26	5.66	8.49	11.3	17.0
	.083	1.03	2.06	5.16	7.73	10.3	15.5
	.095	.926	1.85	4.63	6.95	9.26	13.9
3/4"	.049	2.08	4.17	10.4	15.6	20.8	31.2
	.058	1.97	3.93	14.8	9.84	19.7	29.6
	.065	1.88	3.76	14.1	9.41	18.8	28.2
	.072	1.75	3.51	13.2	8.77	17.5	26.4
	.083	1.67	3.34	12.5	8.35	16.7	25.0
	.095	1.53	3.07	11.5	7.67	15.3	23.0
	.109	1.39	2.77	10.4	6.93	13.9	20.8
7/8"	.049	2.95	5.91	14.8	22.2	29.5	44.3
	.058	2.82	5.64	14.1	21.1	28.2	42.3
	.065	2.72	5.43	13.6	20.4	27.2	40.7
	.072	2.62	5.23	13.1	19.6	26.2	39.2
	.083	2.46	4.92	12.3	18.5	24.6	36.9
	.095	2.30	4.60	11.5	17.2	23.0	34.4
	.109	2.11	4.22	10.6	15.8	21.1	31.7

Oil Flow Capacity Of Tubing (Cont'd)

Figures in Body of Chart are USgpm Flows

Tube O.D.	Wall Thick.	2 Ft/Sec	4 Ft/Sec	10 Ft/Sec	15 Ft/Sec	20 Ft/Sec	30 Ft/Sec
1"	.049	3.98	7.96	19.9	29.9	39.8	59.7
	.058	3.82	7.65	19.1	28.7	38.2	57.4
	.065	3.70	7.41	18.5	27.8	37.0	55.6
	.072	3.59	7.17	17.9	26.9	35.9	53.8
	.083	3.40	6.81	17.0	25.5	34.0	51.1
	.095	3.21	6.42	16.1	24.1	32.1	48.2
	.109	3.00	6.00	15.0	22.4	29.9	44.9
	.120	2.83	5.65	14.1	21.2	28.3	42.4
1-1/4"	.049	6.50	13.0	32.5	48.7	64.9	97.4
	.058	6.29	12.6	31.5	47.2	62.9	94.4
	.065	6.14	12.3	30.7	46.0	61.4	92.1
	.072	6.00	12.0	30.0	44.9	59.9	89.8
	.083	5.75	11.5	28.8	43.1	57.5	86.3
	.095	5.50	11.0	27.5	41.2	55.0	82.5
	.109	5.21	10.4	26.1	39.1	52.1	78.2
	.120	5.00	10.0	25.0	37.4	50.0	74.9
1-1/2"	.065	9.19	18.4	45.9	68.9	91.9	138
	.072	9.00	18.0	45.0	67.5	90.0	135
	.083	8.71	17.4	43.5	65.3	87.1	131
	.095	8.40	16.8	42.0	63.0	84.0	126
	.109	8.04	16.1	40.2	60.3	80.4	121
	.120	7.77	15.5	38.8	58.3	77.7	117
1-3/4"	.065	12.8	25.7	64.2	96.3	128	193
	.072	12.6	25.2	63.1	94.7	126	189
	.083	12.3	24.6	61.4	92.1	123	184
	.095	11.9	23.8	59.6	89.3	119	179
	.109	11.5	23.0	57.4	86.1	115	172
	.120	11.2	22.3	55.8	83.7	112	167
	.134	10.7	21.5	53.7	80.6	107	161
2"	.065	17.1	34.2	85.6	128	171	257
	.072	16.9	33.7	84.3	126	169	253
	.083	16.5	32.9	82.3	123	165	247
	.095	16.0	32.1	80.2	120	160	240
	.109	15.5	31.1	77.7	117	155	233
	.120	15.2	30.3	75.8	114	152	227
	.134	14.7	29.4	73.4	110	147	220

ISO/ANSI Basic Symbols For Fluid Power Equipment And Systems

Lines

Line, Working (Main)	
Line, Pilot (For Control)	
Line, Liquid Drain	
Hydraulic Flow, Direction of Pneumatic	
Lines Crossing	or
Lines Joining	
Line With Fixed Restriction	
Line, Flexible	
Station, Testing, Measurement or Power Take-Off	
Variable Component (run arrow through symbol at 45°)	
Pressure Compensated Units (arrow parallel to short side of symbol)	
Temperature Cause or Effect	
Vented Reservoir Pressurized	
Line, To Reservoir Above Fluid Level	
Below Fluid Level	
Vented Manifold	

Pumps

Hydraulic Pump Fixed Displacement	
Variable Displacement	

Motors and Cylinders

Hydraulic Fixed Displacement	
Variable Displacement	
Cylinder, Single Acting	
Cylinder, Double Acting	
Single End Rod	
Double End Rod	
Adjustable Cushion Advance Only	
Differential Piston	

Miscellaneous Units

Electric Motor	M
Accumulator, Spring Loaded	
Accumulator, Gas Charged	
Heater	
Cooler	
Temperature Controller	
Filter, Strainer	
Pressure Switch	
Pressure Indicator	
Temperature Indicator	
Component Enclosure	
Direction of Shaft Rotation (assume arrow on near side of shaft	

Methods of Operation

Spring	
Manual	
Push Button	
Push-Pull Lever	
Pedal or Treadle	
Mechanical	
Detent	
Pressure Compensated	
Solenoid, Single Winding	
Servo Control	
Pilot Pressure Remote Supply	
Internal Supply	

Note

Additional symbols are shown in Vickers Circuitool booklet available for a nominal charge. Ask for circuitool template kit 352.

Valves

Check	
On–Off (manual shut-off)	
Pressure Relief	
Pressure Reducing	
Flow Control, Adjustable–Non-Compensated	
Flow Control, Adjustable (temperature and pressure compensated	
Two Position Two Connection	
Two Position Three Connection	
Two Position Four Connection	
Three Position Four Connection	
Two Position In Transition	
Valves Capable Of Infinite Positioning (horizontal bars indicate infinite positioning ability)	

Color Code For Fluid Power Schematic Drawings

Function	Color
Intensified Pressure	Black
Supply	Red
Charging Pressure	Intermittent Red
Reduced Pressure	Intermittent Red
Pilot Pressure	Intermittent Red
Metered Flow	Yellow
Exhaust	Blue
Intake	Green
Drain	Green
Inactive	Blank

Definition Of Functions

Function	Definition
Intensified Pressure	Pressure in excess of supply pressure which is induced by a booster or intensifier.
Supply Pressure	Power-actuating fluid.
Charging Pressure	Pump-inlet pressure that is higher than atmospheric pressure.
Reduced Pressure pressure	Auxiliary which is lower than supply pressure.
Pilot Pressure	Control-actuating pressure.
Metered Flow controlled	Fluid at flow rate, other than pump delivery.
Exhaust	Return of power and control fluid to reservoir.
Intake	Sub-atmospheric pressure, usually on intake side of pump.
Drain leakage	Return of fluid to reservoir.
Inactive	Fluid which is within the circuit, but which does not serve a functional purpose during the phase being represented.

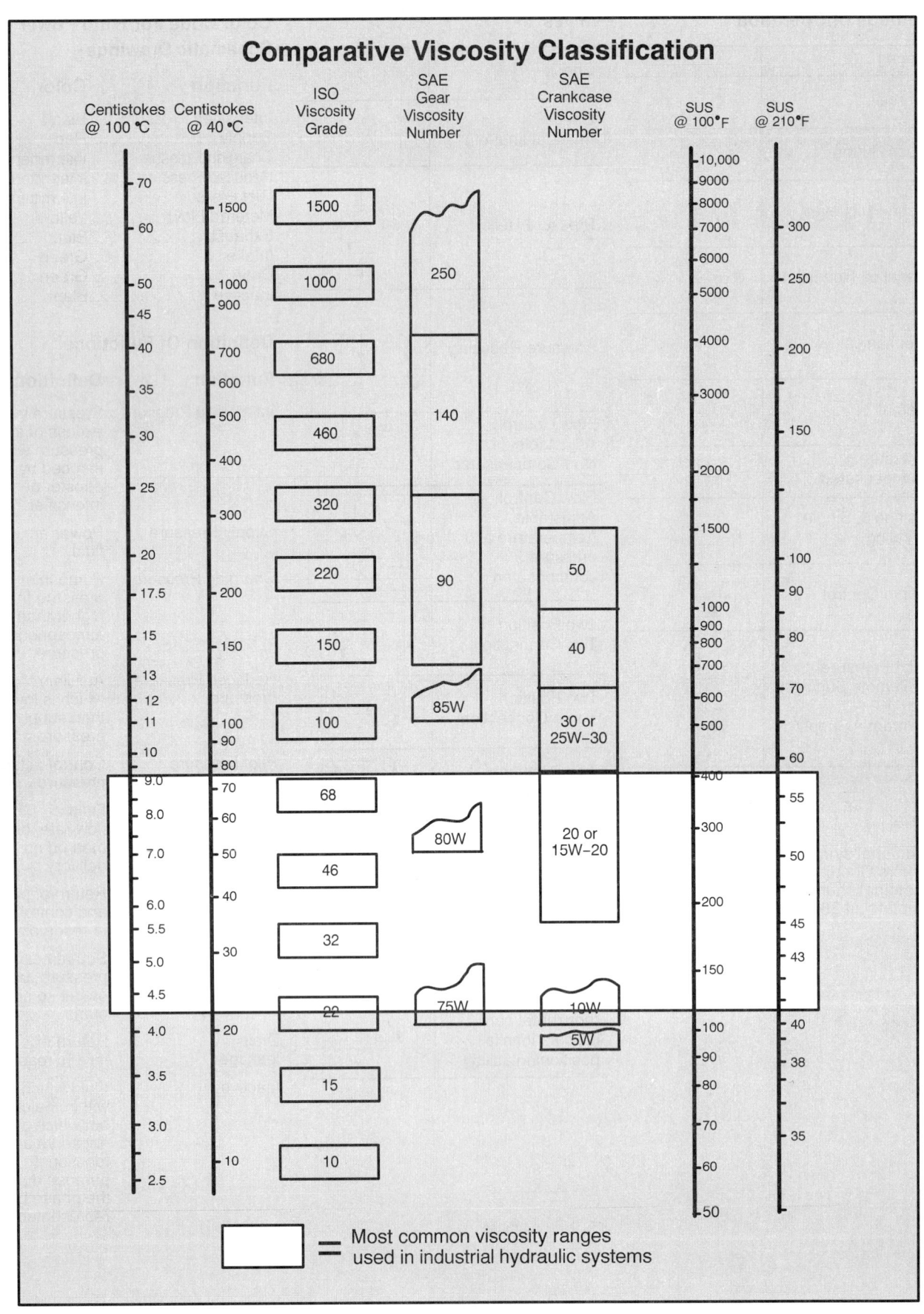

Flow Capacities of Piping

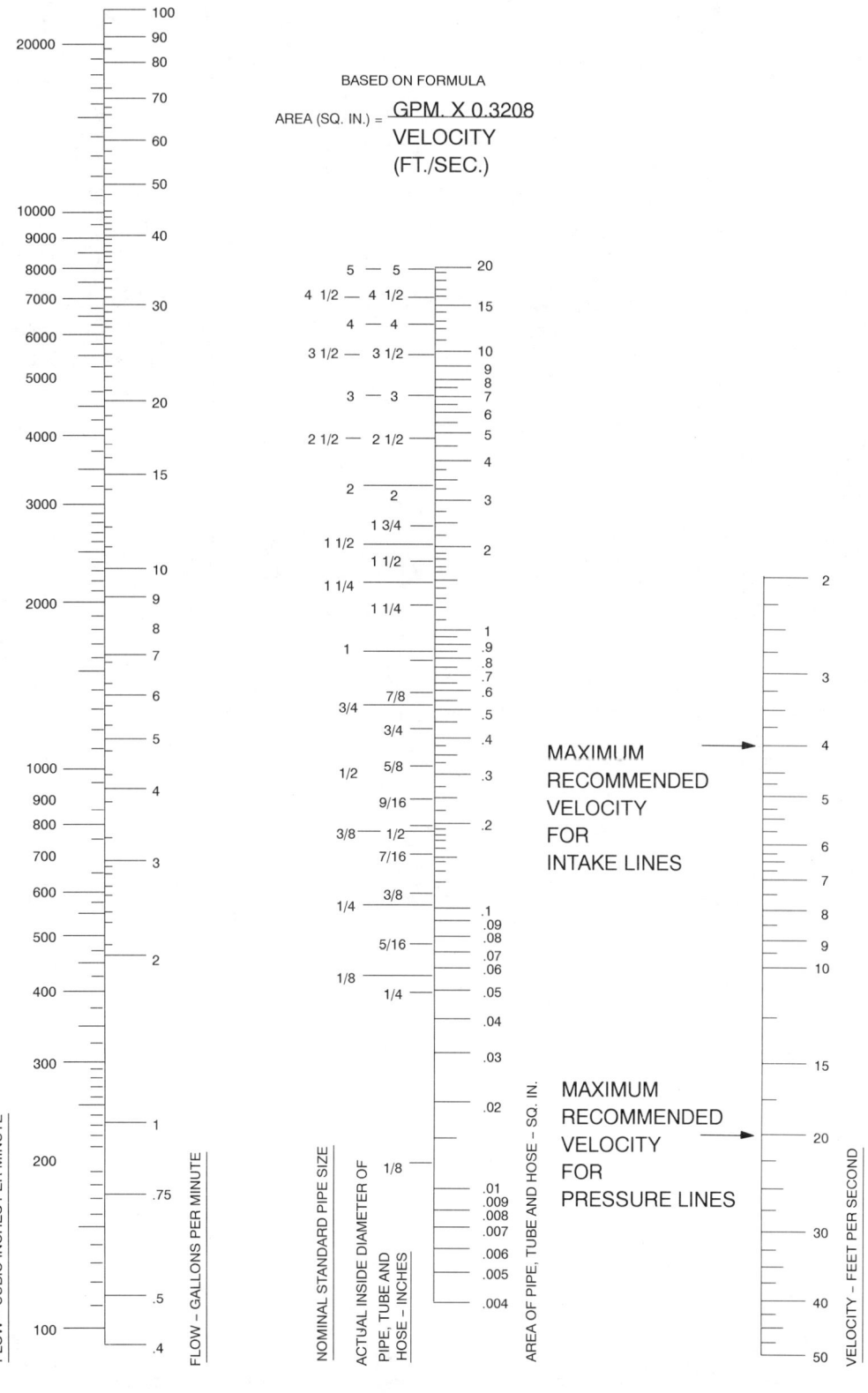

Index

A

absolute rating, 396
absolute zero, 9
AC Fine Test Dust, 378
AC generator, 63
AC solenoids, 64
accumulators, 241
additives, 361
aeration, 15, 270
air removal, 399
alternating current, 53
ampacity, 78
ampere, 48
amplifiers, 71
Andre' Ampe're, 48
angular displacement transducer, 82
angular velocity transducer, 81
anti foaming agents, 362
anti wear additives, 362
anticavitation valves, 152
antioxidant, 362
area, 29
ATF, 365
atmospheric pressure, 8
automatic transmission fluid, 365
auxiliary relief, 149
auxiliary valves, 215
axial piston pump, 264

B

backup rings, 87
baffles, 348
balanced piston relief valve, 220
balanced vane pump, 257
ball joints, 265
barometer, 10
barometric pressure, 10
bending failures, 130
bent axis piston motors, 115
bent-axis piston pump, 265
Bernoulli, Daniel, 36
Bernoulli's Law, 37
Beta rating, 397
bi directional, 107
bias piston, 281
bias spring, 281
bladder accumulator, 245
blocked bypass, 154
boundary lubrication, 353, 384

Boyle, Robert, 244
Brahma, Joseph, 1
breather, 347
breather filter, 347
British Thermal Units, 28
BTU, 28
built in contamination, 375
burst pressure, 334
bus, 176
bypass, 140
bypass valve, 313

C

cables, 76
calorie, 28
carcass, 333
cartridge valve insert, 203
cartridge valve ratio, 203
cartridge valve systems, 179
cartridge valves, 179
catalysts, 360
catastrophic failures, 383
cavitation, 16, 270
centering springs, 156
charge pump, 272, 309
charge relief valve, 310
Charles A. de Coulomb, 47
Charles, Jacques, 244
check valves, 132
chemical contamination, 374
classification of hydrostatic transmissions, 307
clean out plate, 341
cleanliness sample chart, 389
closed center, 140
closed center spools, 145
closed circuit, 43, 309
closed loop, 80
closed loop control, 175
closed loop transmission, 321
CMX series valves, 160
coefficient of friction, 325
combining reservoirs, 343
compensator spring setting, 283
compensators, 281
component cleanliness requirements, 390
compressibility, 28
conductors, 75
confined fluids, 2
connectors, 329
contaminant removal, 355
contamination, 129, 369
control piston, 281

Index

cooling, 355
corner horsepower, 275
corrosion, 360
counterbalance spools, 154
counterbalance valves, 194, 230
cover, 333
cover configuration, 206
crankcase oils, 365
crescent, 261
crossover relief valve, 312
current, 48
cushions, 92
cylinder, 83
cylinder port relief valves, 152

D

DC generator, 63
DC motor, 60
DC solenoid, 54
deadband, 169
deadband compensation, 170
delivery, 250
demulsibility, 361
demulsifiers, 362
diaphragm accumulator, 245
differential, 96, 295
differential operation, 96
differential pressure sensing valves, 199
dip stick, 348
direct acting relief valve, 216
direct current, 53
directional control valves, 131
displacement, 108, 250
distributive control, 176
dither, 171
double acting cylinder, 90
double counterbalance valve, 232
double rod cylinder, 90
drawbar pull, 326
duty cycle, 345
dynamic friction, 109

E

EDC compensator, 290
efficiency, 42, 109
electric analogies, 46
electric fluid level gauge, 348
electric potential (EMF), 48
electric remote control, 71, 157
electrohydraulic compensator, 290
electrohydraulics, 45
electromagnetics, 54

electronic controls, 167
emulsified water, 374
enblock valves, 138
energy, 20, 36
energy storage device, 241
entrained air, 373
ERC, 159
erosion, 385
error signal, 175
error voltage, 321
external contamination, 376
external gear motors, 110
external gear pump, 260

F

face seal, 336
feedback, 165, 321
fieldbus, 176
filler breather, 346
filter, 394
filter media, 394
filter rating, 396
final drive, 300
fire resistance, 361
fittings, 333
flange, 4 bolt, 336
flash point, 361
flat faced coupling, 338
flexible conductor, 330
float spools, 146
flow, 33
flow control valves, 197
flow controls, 234
flow dividers, 234
flow forces, 147
flow transducer, 82
fluid conditioning, 369
fluid conductors, 329
fluid level, 346
fluid power advantages, 5
fluid properties, 356
fluid types, 364
fluid velocity, 330, 331
force, 3, 17
force motor, 57
force motors, 168
force transducer, 81
free air, 373
free water, 374
friction, 109
full film lubrication, 353, 384
full wave rectifier, 68

Index

G

gas charged accumulators, 243
gas contamination, 372
gear motors, 110
gear pump, 260
gear type flow dividers, 234
gerotor pump, 262
gland drain, 86
gross vehicle weight, 326

H

H_2OGate, 347
half wave rectifier, 67
heat generation, 222, 273
high performance pump, 257
high speed hydraulic motors, 110
high torque, low speed motors, 116
high water based fluids, 366
holding current, 65
holding valve, 232
horsepower, 22
hose, 333
hose connectors, 333
hot oil replenishing valve, 312
hot oil valves, 191
HRC, 158
hydraulic analogies, 46
hydraulic cylinder, 83
hydraulic fluid comparisons, 368
hydraulic fluids, 27, 351
hydraulic horsepower, 266
hydraulic leverage, 40
hydraulic logic, 228
hydraulic motor, 108
hydraulic remote control, 157
hydraulic reservoirs, 339
hydrocarbon based fluids, 364
hydrodynamic lubrication, 384
hydrostat, 200, 218
hydrostatic transmission, 293
hysteresis, 171

I

Ideal Gas Law, 244
improper installation, 269
in line check valves, 133
in line configuration, 293
in line piston motors, 114
incompressible, 352
ingression, 377
inlet filters, 348
inlet screens, 348
inline piston pump, 264
inner tube, 333
input horsepower, 127
input power, 266
inrush current, 65
integral valving, 92
intermittent failures, 383
internal feedback, 173
internal gear motor, 111
internal gear pump, 261
internally generated contamination, 379
intra vanes, 257
ISO code, 387
ISO standard, 388
ISO viscosity grades, 359

J

joule, 52

K

kilowatt, 53
kinetic energy, 37

L

laminar flow, 34, 330
law of electric charges, 47
linear actuators, 83
linear displacement transducer, 82
linear variable differential transformer, 173
lip seal, 86
load control valves, 198
load drop check valve, 143
load pressure, 285
load sensing compensator, 283
locating the reservoir, 341
logic valves, 199
long term degradation, 382
low inertia, 296
lubricity, 353
LVDT, 173

M

magnetic drain plugs, 349
main system relief, 148
maintenance generated contamination, 380
maneuverability, 296
manifold block systems, 181
mechanical efficiency, 128, 267

Index

mechanical losses, 267
mechanical power, 266
metallic debris, 268
meter in, 98
meter out, 99
metering, 274
metering point, 275
microbial growth, 374
micrometers, 369
microprocessor, 71, 176
mixed film lubrication, 385
monoblock valves, 139
motor failure, 129
motor speed, 300
motor spool, 145
multipass filter performance test, 396
multiple spool construction, 138
multiple viscosity oils, 357
multipurpose valves, 313

N

needle valves, 236
nominal rating, 396
non compensated flow control, 235
non metallic debris, 269
non positive displacement pumps, 12, 252
non reusable, 333
normally closed, 189
normally open, 189

O

o ring, 86
o ring seal, 336
ohm, 49
Ohm, Georg Simon, 49
Ohm's Law, 49
open center, 140
open circuit, 43
open loop, 79
open loop control, 173
operational amplifier, 73
orbital internal gear motors, 116
orifice, 235
outlet strainers, 348
output flow, 265
output horsepower, 126
over running load, 100
overall efficiency, 129
overall pump efficiency, 266
overrunning loads, 230
oxidation, 360
oxidation resistance, 360

P

parallel circuit, 39
particle contamination, 369
Pascal, Blaise, 1
Pascal's Law, 1
petroleum based fluids, 364
phosphate esters, 366
pilot operated check valves, 135
pilot operated relief valve, 218
pilot operated valves, 169
pilot signal, 186
pilot to close check valve, 136
pilot to open check valves, 135
pintles, 280
pipe, 329
piston accumulator, 244
piston pumps, 262
piston seals, 87
planetary wheel drive, 300
pneumatic remote control, 157
pole piece, 66
positive displacement pumps, 14, 253
potential energy, 37
potentiometer, 71
pour point, 360
power, 21
power beyond, 152
power steering motor, 120
power transmission, 352
power wheel, 300
preassembled cartridge, 258
precharge pressure, 243
pressure, 4, 30
pressure compensated flow control, 197, 237
pressure control valves, 194, 215
pressure differential, 97
pressure drop, 35
pressure intensification, 105
pressure limiting compensator, 282
pressure override, 218
pressure rating, 251
pressure reducer cover, 207
pressure reducing and relieving valves, 195
pressure reducing valves, 194, 225
pressure relief cover, 207
pressure transducer, 71, 81
pressurized reservoirs, 12, 349
preventive maintenance, 386
printed circuitry, 79
priority valve, 235
proactive maintenance, 387
programmable logic control, 176
proportional solenoid control, 172

proportional solenoids, 57
proportional valves, 164
provide dynamic braking, 297
psia, 9
psig, 8
pulse width modulation, 74, 172
pump efficiency, 266
pump failure, 268
pump losses, 267
pump performance curves, 315
purpose of hydraulic fluids, 352
PWM, 74, 172

Q

quick disconnect couplings, 337

R

radial piston motor, 118
radial piston pump, 263
radiation contamination, 374
ram, 89
range code, 387
rapeseed, 366
reactive maintenance, 386
rectifiers, 67
regeneration, 100
regenerative spools, 155
relief valve, 216
remote control valves, 157
reservoir requirements, 340
reservoir size, 345
reservoirs, 339
resistance, 49
resistance in parallel, 51
resistance in series, 50
response time, 322
restrictors, 235
reusable, 333
right angle check valves, 134
right hand rule, 54
rigid conductor, 330
road speed, 302
rod end area, 96
rod seals, 85
roller vanes, 117
rolling radius, 325
rotary actuators, 107
rotary valves, 192
running torque, 109
rust, 360

S

S shape configuration, 294
saddle type yoke, 280
SAE viscosity numbers, 358
sampling frequency, 393
sampling techniques, 392
screw in cartridge valves, 185
screw in valves, 180
seal failure, 103
seal material, 88
sealed filler pipe, 341
sealing, 355
sectional valves, 138
sequence valves, 196, 228
series circuit, 40
series parallel, 142
servo controlled pump, 317
servo pump, 319
servo valves, 172
servo vavles, 165
shading ring, 64
shaft rotation, 258
shuttle check valves, 135
shuttle cover, 206
shuttle valve, 135, 311
sight gauges, 348
single acting cylinders, 89
single acting spools, 146
skid steering, 304
slip in cartridge valves, 203
slip in valves, 179
solenoid control, 172
solenoids, 168
soy, 366
specific gravity, 7, 28
specific heat, 28
speed calculations, 125
split configuration, 294
spool end options, 153
spool type flow dividers, 235
spring loaded accumulators, 243
stalling, 297
stand by pressure, 284
standard atmosphere, 8
standard cavities, 187
Starting torque, 109
static friction, 109
steel tubing, 334
steering, 303
strainer, 394
stroke adjuster cover, 207

Index

stroke limiting, 93
swash plate, 264
synthetics, 366
systemic contamination control, 391

T

tach generator, 71
tandem spools, 144
target cleanliness, 391
telescopic cylinders, 91
temperature compensated flow controls, 239
temperature sensor, 71
thermal contamination, 372
thermal relief valves, 93
three body wear, 385
tie bolt cylinders, 85
torque, 108
torque calculations, 124
torque limiting compensator, 286
torque motor, 59
torque motors, 168
torque/speed ratios, 296
torsional failures, 130
traction, 325
transducers, 81
transformers, 70
transistor, 71
tubing connectors, 335
turbulent flow, 34, 330
two body wear, 383
two speed pump mount, 299

U

U cup, 86
U shape configuration, 293
unbalanced loading, 262
unbalanced vane pump, 255
uni directional, 107
unloading valves, 196, 224

V

V cup, 86
valve cavities, 186
valve spools, 143
Valvistor slip in cartridge, 208
Valvistor technology, 201
vane motors, 112
vane pump, 255
vane pump displacement, 260
variable displacement bent axis piston pump, 278
variable displacement inline piston pump, 279
variable displacement motor, 121
variable displacement pumps, 273
variable displacement radial piston pumps, 277
variable vane pump, 276
vegetable oils, 366
velocity, 33
velocity differential, 97
vented counterbalance valves, 199
VI improvers, 363
viscosity, 356
viscosity index, 357
volt, 49
Volta, Count Alessandro, 49
volume, 31
volume capacity, 340
volume differential, 97
volumetric efficiency, 128, 267, 310
volumetric losses, 267

W

water contamination, 373
water in solution, 374
water removal, 398
water removal elements, 398
watt, 52
Watt, James, 22
weight loaded accumulator, 242
welded cylinders, 85
wheel motors, 298
wheel slip, 297
wheel torque, 325
wire gauge, 76
wires, 76
work, 19

Y

yoke, 280
yoke feedback link, 318
yoke position sensing spool, 287

NOTES

NOTES

NOTES

NOTES